F. WILLIAM CAMBRAY
Dept. of Geological Sciences
Michigan State University

PRINCIPLES
OF STRUCTURAL
GEOLOGY

PRINCIPLES OF STRUCTURAL GEOLOGY

JOHN SUPPE

**Department of Geological and Geophysical Sciences
Princeton University**

Prentice-Hall, Inc. Englewood Cliffs, New Jersey 07632

Library of Congress Cataloging in Publication Data

SUPPE, JOHN.
    Principles of structural geology.

    Bibliography: p.
    Includes index.
    1. Geology, Structural.   I. Title.
QE601.S94 1985   551.8   84-11561
ISBN 0-13-710500-2

Editorial/production supervision
    and interior design: *Kathryn Gollin Marshak*
Cover design: *Diane Saxe*
Manufacturing buyer: *John Hall*

Printed in the United States of America

10 9 8 7 6 5 4 3 2 1

ISBN 0-13-710500-2

Prentice-Hall International, Inc., *London*
Prentice-Hall of Australia Pty. Limited, *Sydney*
Editora Prentice-Hall do Brasil, Ltda., *Rio de Janeiro*
Prentice-Hall Canada Inc., *Toronto*
Prentice-Hall of India Private Limited, *New Delhi*
Prentice-Hall of Japan, Inc., *Tokyo*
Prentice-Hall of Southeast Asia Pte. Ltd., *Singapore*
Whitehall Books Limited, *Wellington, New Zealand*

In His hand are the depths of the earth,
 and the mountain peaks belong to Him.
The sea is His, for He made it,
 and His hands formed the dry land.

Psalm 95.4–5
(New International Version)

CONTENTS

PART II   PRINCIPLES OF DEFORMATION

PART III   CLASSES OF STRUCTURES

PREFACE

Principles of Structural Geology is designed to be a concise introduction to the deformation of the earth's crust, encompassing the wide-ranging subject matter of introductory courses in structural geology. Three principal aspects of structural geology are emphasized:

1. The basic principles of natural rock deformation are presented in Part II, *Principles of Deformation* (Chapters 3, 4, and 5), including stress, strain, and the physical processes of elastic, plastic, and brittle deformation of rock. These chapters present the basic principles of solid-state deformation to the level necessary to comprehend the main phenomena of structural geology. These principles are then applied over and over again in subsequent chapters. A few of the concepts are intrinsically somewhat complex, particularly stress and strain, and cannot be oversimplified without losing the power that they offer for understanding deformation. Special care is taken in these sections and throughout the book to develop a clear physical intuition of the important concepts. Equations are normally used only as a supplement to the text and are given in a form designed to help clarify the underlying physics.

2. The description and origin of the main classes of deformational structures is presented in Part III, *Classes of Structures* (Chapters 6 through 11), including joints, instrusive and extrusive structures, faults, folds, fabrics, and impact structures. Much of this material is descriptive and is important for gaining a realistic comprehension of natural deformation in the earth. A special effort has been made to present real structures, using photographs and well-documented maps and cross sections rather than schematic idealized drawings. Following the basic description, each chapter contains sections on the physical origin of the structures, applying the principles already introduced. For example, the chapter on joints immediately applies material from the preceding chapter on fractures.

Additional material on the geometric methods for the study of structures and some basic structural terminology are introduced earlier in Chapter 2, on map-scale structures, so that the laboratory exercises that are generally part of structural geology instruction may begin early in the course. These geometric methods are presented in a unified fashion, focusing on the common problems of deciphering map-scale structures and inferring subsurface geology.

3. The large-scale deformation within the earth, including its historical and paleogeographic aspects, is presented in the final chapters (12 and 13) as Part IV, *Regional Structural Geology,* after all the basic principles of smaller-scale deformation have been presented. These chapters use the Appalachian and Cordilleran mountain belts in the United States and Canada as examples. The geology of these major deformed regions is introduced in a selective way for the purpose of giving a clear perspective on the scope and methods of regional structural geology, without being encyclopedic. Most of the space is devoted to major throughgoing features and the best-understood aspects of these orogenic belts—for example, the Taconic orogeny of the Appalachians. Topics of current speculation are generally avoided. Some space is devoted to the history of study to give a perspective on how regional structural problems are in fact solved. Additional material on regional structure and tectonics is presented in Chapter 1, "Introduction to Deformation of the Lithosphere." This chapter introduces the major tectonic settings of deformation in the earth, the structure and morphology of continents and ocean basins, and the processes of vertical and horizontal displacement of the lithosphere. The two introductory chapters (1 and 2; Part I) are designed to review and build on the standard background of undergraduate physical geology to provide an overall perspective on deformation of the earth before entering into Part II on physical principles.

A few exercises are placed at the end of each chapter. These vary greatly according to the subject matter. Some are straightforward numeric problems. Others are broad and open-ended essay questions, simply designed to be thought-provoking. Still others provide opportunities for summarizing and reviewing the material already presented.

In keeping with most textbooks, references are kept to a minimum and are largely cited in figure captions. A limited list of selected literature is given at the end of each chapter, giving certain key books, monographs, review articles, and important papers. These selected references should provide a more directed introduction to the fragmented literature of structural geology.

ACKNOWLEDGEMENTS

I am so grateful for all the help and encouragement that I have received in this project. I would also be grateful for suggestions and corrections that might make subsequent editions more useful. A number of people read all or part of the manuscript and offered corrections and insightful suggestions. These include J. Burnsnall, B. Clark, D. Cowan, D. Davis, B. Evans, M. Hamil, J. Rodgers, R. Schweickert, J. Spang, W. Travers, and J. Wickham. Many individuals and organizations provided figures and permission to reproduce them, as acknowledged in the captions. I thank J. Bialkowski, who helped with the many details of completing the manuscript. B. Suppe, my son, wrote a program to help make the index. The patience and encouragement of Logan Campbell, Doug Humphrey, Dave Gordon, and Kathryn Marshak at Prentice-Hall have been great.

John Suppe

PRINCIPLES
OF STRUCTURAL
GEOLOGY

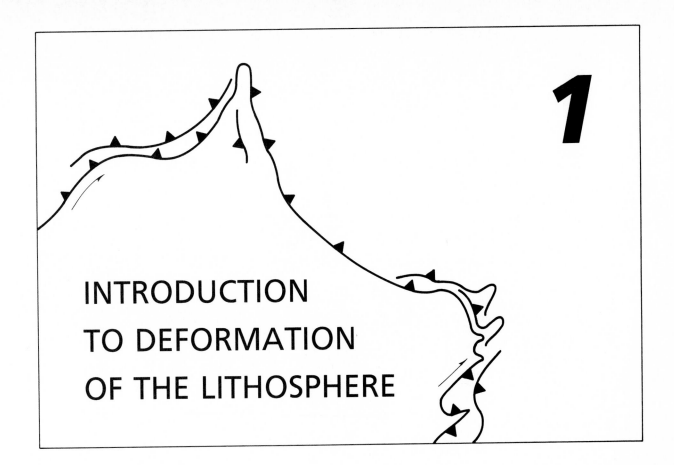

INTRODUCTION TO DEFORMATION OF THE LITHOSPHERE

INTRODUCTION

We begin our study of deformation in the earth from a global, even planetary, perspective by describing the major structural features of the earth's surface and by identifying the large-scale situations and processes by which the upper layers of the earth undergo vertical and horizontal displacement. Beginning with Chapter 2 our emphasis changes from the major settings of global deformation to individual structures, classes of structures, and their mechanisms of formation. We return to large-scale deformation in Chapters 12 and 13 on the Appalachian and Cordilleran mountain belts, which emphasize the historical aspects of the regional deformation of our planet.

Structural geology is the study of the deformed rocks that make up the upper layers of the earth. The subject can be extended to other bodies in the Solar System. Each planet or moon exhibits its own distinctive history and style of deformation, which reflects the degree of mobility of the planetary interior and—to a large degree—its ratio of volume, V, to surface area, S, which is proportional to radius, R.

$$\frac{V}{S} = \frac{\frac{4}{3}\pi R^3}{4\pi R^2} = \frac{R}{3} \qquad (1\text{-}1)$$

As radius increases, the rate of cooling of a planet decreases because there is less surface area to cool across, relative to the total heat content and volume. Therefore, larger planetary bodies are hotter. They are also much more mobile because of the dependence of rock strength on temperature. At high temperatures rock strength decreases exponentially with increasing temperature (Chapter 4), causing the interiors of planets to be weak and able to flow plastically.

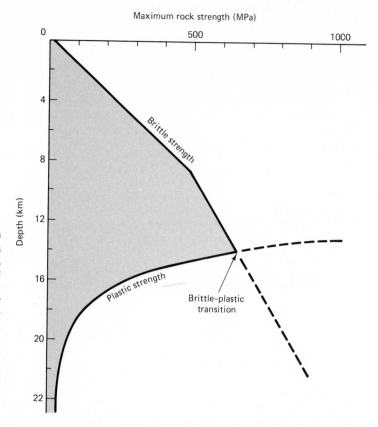

FIGURE 1–1 Maximum rock strength as a function of depth in the earth for a quartz-rich sedimentary rock under a normal 20°C/km geothermal gradient and hydrostatic pore-fluid pressures (Chapters 4 and 5). Higher thermal gradients or pore-fluid pressures would substantially reduce the strength, shifting the curves into the shaded region. Other rock materials show qualitatively similar behavior with a maximum strength at the depth of transition between pressure-dependent brittle behavior and temperature-dependent plastic behavior.

In contrast, rock at the surface of a planet is brittle and relatively weak; however, with increasing depth rock strength increases, which is an effect of pressure (Chapter 5). At some depth the effect of increasing temperature begins to dominate over the effect of increasing pressure, and the strength begins to drop exponentially (Fig. 1-1). Therefore, a layer of high rock strength, called the *lithosphere,* exists near the surface of planets. Hotter planets can be expected to have thinner lithosphere and display plate-tectonic behavior, whereas cooler planets can be expected to have such thick lithosphere that extensive mobility is impossible. Therefore, it makes sense that the moon (R = 1738 km), Mercury (2439 km), and Mars (3393 km) are cold and tectonically inactive, whereas Venus (6055 km) and Earth (6378 km) are hot, internally mobile, and actively deforming.

Beyond properties like radius, mass, and moment of inertia, which are astronomically determined, the best known property of planets that bears on their structural geology is surface morphology. This fact is also true for the earth; its topography and bathymetry are much better known than the deformed rocks at or below the surface. Therefore, let us begin our study of structural geology by considering what the morphology of the earth tells us about its deformation.

From a planetary perspective, the most immediately obvious morphologic feature of the earth is the distinction between continents and ocean basins, which reflects the most fundamental structural property of the earth. The histogram of elevations on the earth is strongly bimodal (Fig. 1-2), with the mean elevation of continents a few hundred meters above sea level, in contrast with the mean elevation of ocean basins at about 4000 m below sea level. The boundary between continents and oceans is commonly taken at 2000 m below sea level, although the lithologic boundary may be higher or lower from place to place. The continents are composed largely of old (100–3000 m.y.) and repeatedly deformed rocks rich in quartz and feldspar, whereas the ocean basins are underlain by young (10–100

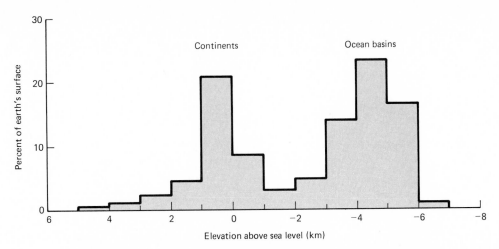

FIGURE 1–2 Histogram of elevations and water depths showing the division of the earth's surface into continents and ocean basins. (From Sverdrup, Johnson, and Fleming, *The Oceans: Their Physics, Chemistry, and General Biology,* © 1942, Renewed 1970, p. 19. Reprinted by permission of Prentice-Hall, Inc., Englewood Cliffs, N.J.)

m.y.), little deformed rocks rich in feldspar and pyroxene that are the upper surface of the convecting interior of the earth.

The morphology of the continents in plan view gives important insight into deformation processes in the earth. Alfred Wegener (1880–1930) devised the theory of continental drift based on the complementary morphology of the edges of South America and Africa across the South Atlantic. Continental drift implies that the continental lithosphere is able to separate into pieces, but that the pieces are able to maintain their shapes with little distortion for hundreds of millions of years, as shown by the fact that South America and Africa are able to fit back together rather precisely. It is also observed that the continent-ocean boundaries on both sides of the South Atlantic have essentially the same shape as the current plate boundary at the Mid-Atlantic Ridge. This observation shows that oceanic lithosphere also maintains its shape without distortion for long periods.

In the following two sections we describe the main morphologic and structural features of the continents and ocean basins. This morphology is strongly affected by erosion and deposition, in contrast with other planetary bodies. Because erosion on the earth is relatively rapid, the mountainous topography that is produced by deformation is being continuously destroyed.

Available data from a variety of geologic settings show that rate of denudation is proportional to relief (Fig. 1-3(a)). This implies that in the absence of uplift, relief, h, decays as a negative exponential function of time, t:

$$h = h_0 e^{-at} \tag{1-2}$$

where h_0 is the initial relief and a is the proportionality between denudation rate and relief. The data suggest that significant topographic relief is destroyed by erosion in tens of millions of years (Fig. 1-3(b)) and that important mountainous regions of the earth record areas of Cenozoic deformation.

MORPHOLOGY AND STRUCTURE OF CONTINENTS

If erosion were the only process actively shaping the surface of the earth, the continents would soon be eroded down nearly to sea level. Therefore, if we look at a regional topographic map of the continents, we apparently see in the

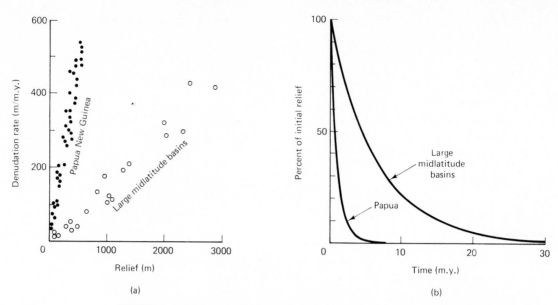

(a)   (b)

FIGURE 1–3 (a) Denudation rate as a function of relief in the tropical and tectonically active Hydrographers Range of Papua, New Guinea, and in a number of large midlatitude drainage basins. (b) Exponential decay of relief in the absence of uplift (Eq. 1-2). (Data from Ahnert, 1970; Ruxton and McDougall, 1967.)

mountainous topography a qualitative picture of places at which relatively recent, large-scale deformation has taken place. This conclusion that mountainous topography is young is supported by the well-known correlation between earthquakes and mountain chains.

What does the distribution of mountainous topography tell us about deformation in the earth? To the first approximation, continents display two morphologic and structural provinces: orogenic belts and cratons. *Orogenic, or mountain, belts* are largely linear zones of intense deformation, including folding, faulting, seismicity, and rugged topography—for example, the Andean orogenic belt (Fig. 1-4). Active orogenic belts are localized along plate boundaries and commonly

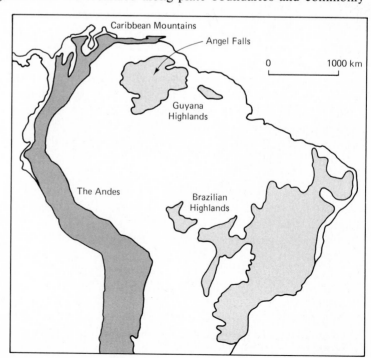

FIGURE 1–4 Topography of northern South America, showing the regions above 500 m in a shaded pattern. The Guyana and Brazilian highlands are epeirogenic uplifts, whereas the Andes and Caribbean Mountains are orogenic uplifts.

along continental margins, apparently reflecting a fundamental instability of continental margins relative to oceanic, and especially continental, interiors. Orogenic belts typically continue to be the locus of deformation for tens to hundreds of millions of years. For example, the active Caribbean orogenic belt along the northern edge of South America (Fig. 1-4) has been deforming since latest Cretaceous (70 m.y.).

In contrast, *cratons* are the stable parts of continents that have not undergone significant orogenic deformation for hundreds of millions of years. Some cratons, such as the Russian platform, are exceedingly flat and lie very close to sea level, as we would expect if only plate-margin deformation and erosion controlled the topography of the continents. In contrast with the Russian platform, most cratons display broad swells in topography unrelated to plate-boundary deformation. For example, the craton of northern South America contains several large domal plateau uplifts, the Brazilian and Guyana highlands (Fig. 1-4). These uplifts are far from any plate boundaries and are aseismic. They appear to be Cenozoic in age based on their topographic relief (Fig. 1-3(b)). For example, the highest waterfall in the world, Angel Falls, is in the Guyana Highlands (Fig. 1-5).

The morphology of the continents indicates two important classes of deformation in the lithosphere: orogeny and epeirogeny. *Orogeny* is the strong, mountain-building deformation that involves folding, faulting, seismicity, and linear mountain chains. *Epeirogeny*, in contrast, is the broad gentle warping of the lithosphere that does not involve local intense folding, faulting, or seismicity—for example, the uplift of the Guyana Highlands. Epeirogenic deformation is best displayed on the cratons because of their otherwise extreme flatness; nevertheless, important epeirogenic deformation is also observed in ocean basins and orogenic belts.

FIGURE 1–5 One-kilometer-high Angel Falls in the Guyana Highlands of the southern Venezuelan craton. The high relief suggests that this is a region of active epeirogenic uplift. (Photograph by R. Hargraves.)

The stable cratons of the world are commonly divided into areas of platforms and shields based on the rocks exposed at the surface. *Platforms* are those areas of the craton in which flat-lying sedimentary rocks are exposed at the surface. The North American craton in the United States is largely a platformal area, shown as the shaded region in Figure 1-6, in which the flat-lying sediments are typically 1 to 2 km thick. *Shields* are those areas of the craton in which peneplained basement rocks are exposed at the surface. For example, much of the North American craton in Canada is an area of shield exposing the eroded roots of several Precambrian orogenic belts, each of which was active for hundreds of millions of years. These ancient orogenic belts can be traced through the entire craton, under the thin platformal sediments (Fig. 1-6).

It is important to realize that younger orogenic belts crosscut, truncate, and redeform rocks of older orogenic belts, as well as their overlying sedimentary cover. This fact has been particularly well demonstrated for the 1-billion-year-old Grenville orogenic belt exposed at the surface in southeastern Canada and buried under much of the eastern and southern United States (Fig. 1-6). The Grenville belt truncates the older north-south Nain orogenic belt at a high angle. The western edge of the Nain belt is the Labrador trough, which is a belt of folding and thrusting from the east toward the 2.6-to-2.4-billion-year-old craton of the Superior province to the west. The geology of this marginal zone of the Nain belt is well known because of extensive deposits of economically important sedimentary iron formation. As shown in Figure 1-7, the iron formation has been traced southward into the Grenville belt, where it has been redeformed and remetamorphosed during the Grenville orogeny. Farther to the southwest, the Grenville belt

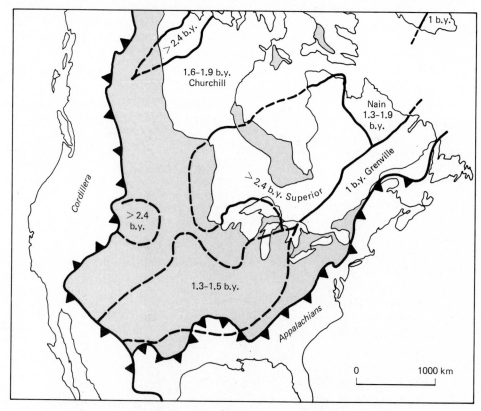

FIGURE 1–6 Precambrian orogenic belts of the North American craton. Shaded regions are areas covered by platformal sedimentary rocks. The edges of the younger Cordilleran and Appalachian orogenic belts are shown with the barbed thrust-fault symbol. (Simplified from King, 1969.)

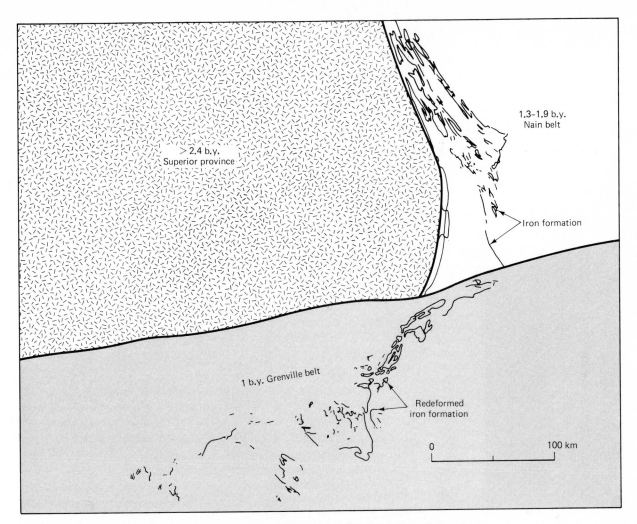

FIGURE 1–7 Junction between Superior, Nain, and Grenville orogenic belts in eastern Canada (see Fig. 1-6). The deformed iron formations in the Labrador trough at the edge of the Nain belt are seen to continue into the younger crosscutting Grenville belt, where they are redeformed and remetamorphosed. Rocks of the older Superior province are also incorporated into the Grenville belt northwest of the iron formation. (Data on iron formations from Gross, 1967.)

includes extensive redeformed and remetamorphosed rocks of the Superior province. The generalization that younger orogenic belts incorporate and redeform rocks of the adjacent older orogenic belts is in a sense made complete by the observation that fragments of the Precambrian Grenville belt are found redeformed and remetamorphosed in the Paleozoic Appalachian orogenic belt to the southeast (Chapter 12).

Thus, the bulk of continental crust is composed of deformed rocks of ancient orogenic belts. With time, typically a few hundred million years, the orogenic belts become stabilized, the locus of deformation moves elsewhere on the earth, and the inactive orogenic belt erodes down to near sea level to become part of the craton.

The thicknesses of the flat-lying platformal sedimentary rocks and the elevation of the unconformity with the underlying peneplained basement rocks provide an important record of epeirogenic deformation of the continents. Figure 1-8 is a *structure contour map*, which shows the present elevation of the eroded top of the Precambrian basement. This eroded surface was generally at or close to sea level when it was first covered, so that—although strata of different ages cover

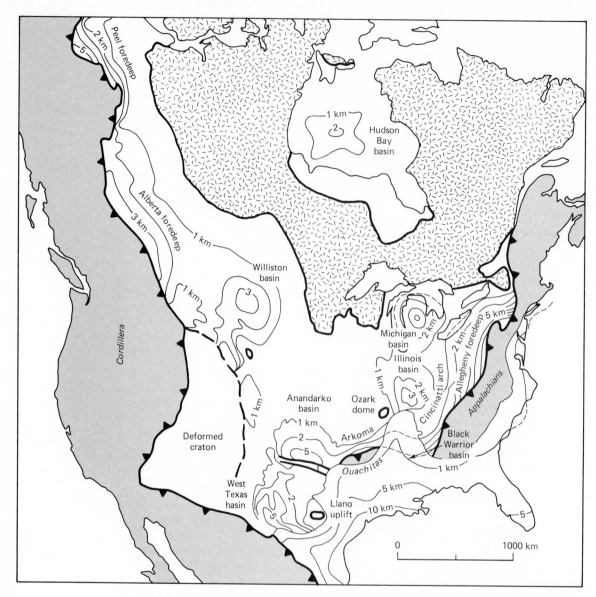

FIGURE 1–8 Structure contour map showing the depth below sea level of the unconformity between Precambrian basement rocks and sedimentary cover of the North American craton. The contours in the Gulf Coast and Atlantic continental margin show the depth to Paleozoic basement rocks. Exposures of Precambrian basement in the Canadian shield are shown with a pattern. (Data from King and Edmonston, 1972.)

the basement from place to place—the structure contours provide a picture of the net epeirogenic deformation since the covering of the basement unconformity. The most characteristic features of epeirogenic deformation visible on structure contour maps are broad, roughly circular basins and domes, including the Michigan, Illinois, Williston, and Hudson Bay basins, each about 500 km in diameter and about 2 to 4 km deep. Domal uplifts include the Ozark dome and Llano uplift of the United States (Fig. 1-8), which expose small areas of Precambrian basement rock. Other uplifts are less pronounced, such as the Cincinnati arch, which displays slower rates of deposition than the surrounding areas, more frequent disconformities, and thinner stratigraphic sequences.

As an example of a cratonic epeirogenic basin, consider the roughly circular Williston basin in northwestern South Dakota and adjacent Montana and Canada

(Fig. 1-8). It is about 700 km in diameter and is approximately 4 km thick at the center, in contrast to a thickness of about 1 to 2 km in the surrounding regions. Studies of the sediments show that this excess subsidence was extremely slow, starting in early Paleozoic and continuing into the Mesozoic, a period of about 250 m.y. The average rate of subsidence was about 10 m/m.y., with the rate of subsidence decreasing with time. This example of the Williston basin shows that epeirogenic subsidence can be extremely slow and very long-lived. In a later section we consider what might be the underlying physical process that produces this subsidence. Similar cratonic basins exist on other continents—for example, the Murzuk and Kufra basins of North Africa, the Amazon basin of South America, the Great Artesian basin of Australia, and the Paris basin of Europe.

Some domal uplifts exist today as topographic features on the cratons, such as the Guyana and Brazilian highlands of South America (Fig. 1-4), the Tibesti and Hoggar uplifts of North Africa, and the Kazakhstan and Putoriana uplifts of central Asia. At least some of these are quite recent in origin, such as the Tibesti and Hoggar uplifts, and must involve rates of uplift on the order of hundreds of meters per million years, substantially faster than the rates of epeirogenic subsidence. This fact may be important in understanding their origin, as discussed later. The North African domal uplifts have associated nonorogenic, midplate volcanism.

In addition to the cratonic basins and domes of epeirogenic origin discussed above, there exist important sedimentary basins at the margins of cratons. Two important types exist, continental-margin basins and foredeep basins. Many cratonic platforms pass gradationally into *continental-margin basins,* which typically contain 10 to 15 km of sediment deposited on the edge of the continent in about 100 m.y. An example of a continental-margin basin is the Gulf Coast basin in the southern United States and Mexico (Fig. 1-8). The total thickness of the continental-margin basins and the rate of subsidence is substantially greater than the cratonic basins, suggesting a somewhat different underlying process, as discussed later.

Foredeep basins lie on the craton adjacent to orogenic belts and are an effect of the orogenic process. The sedimentary fill of foredeeps is contemporaneous with uplift in the adjacent mountains, from which the sediment is derived. Foredeeps are typically 2 to 4 km deep and are deposited in about 10 m.y. North American examples include the Allegheny, Black Warrior, Arkoma, Alberta, and Peel basins (Fig. 1-8), adjacent to the Appalachian, Ouachita, and Cordilleran orogenic belts. Sediments deposited in foredeeps commonly are deformed within a few million years of deposition as they are incorporated into the growing mountain belt. An example of a present-day foredeep is the Persian Gulf and Tigris-Euphrates Valley along the boundary of the Arabian craton and the active Zagros fold belt (Fig. 1-28). The Persian Gulf is the part of the foredeep in which subsidence is more rapid than deposition, apparently due to the dry climate.

To summarize, the structure of continents consists of stable cratons and active orogenic belts. Cratons are underlain by a collage of eroded roots of dead orogenic belts that are overlain in some regions by a thin (< 2 km), nearly flat-lying sequence of platformal sediments. The sediments of the craton are generally deformed into broad epeirogenic domes, arches, and basins, typically about 500 km across and a few kilometers in amplitude.

MORPHOLOGY AND STRUCTURE OF OCEAN BASINS

Like the continents, the bulk of oceanic lithosphere lies far from active plate boundaries and is undergoing only broad epeirogenic deformation. After the

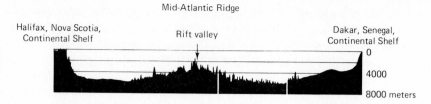

FIGURE 1–9 Bathymetric profile of the North Atlantic Basin between North America and Africa. (After Holcombe, 1977.)

formation of new oceanic lithosphere at an oceanic spreading center, the lithosphere sinks slowly with age, producing the characteristic slopes of ocean basins toward regions of increasing age (Fig. 1-9). The slopes are typically 1:300 to 1:3000, with steeper slopes on the more slowly spreading ridges. The epeirogenic subsidence is proportional to the square root of time (Fig. 1-10) and is largely an effect of cooling of the lithosphere, which is discussed in the next section. The relative smoothness of the ocean floor is mainly interrupted by seamount volcanoes and by fracture zones running parallel to the regional slope, across

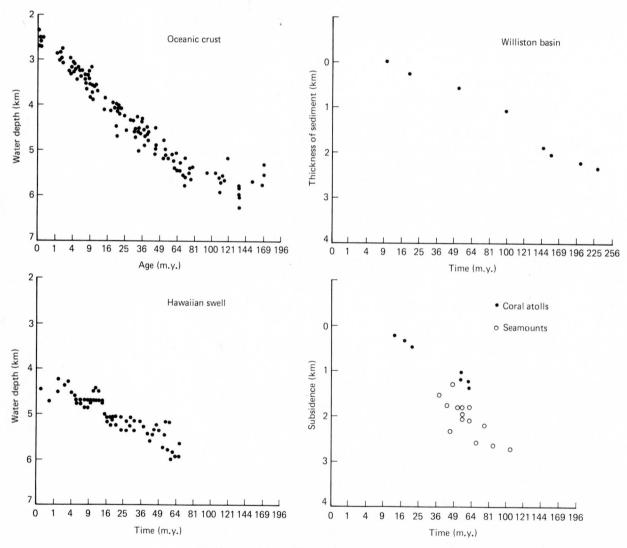

FIGURE 1–10 Epeirogenic subsidence as a function of square root of time. Time for the Hawaiian swell is obtained from the distance along the Hawaiian chain (Figs. 1-11 and 1-12). (Data from Morgan, 1975; Detrick, Sclater, and Thiede, 1977; Detrick and Crough, 1978; and Rocky Mountain Association of Geologists, 1972.)

which the lithosphere shifts in age and water depth. If the volcanoes reach above sea level they erode, producing flat-topped seamounts called *guyots,* which together with associated coral reefs in tropical regions produce excellent sea-level data. Coral atolls subside epeirogenically in the same manner as oceanic lithosphere, with the subsidence proportional to the square root of time (Fig. 1-10).

Ocean basins display epeirogenic updomes that are apparently quite similar to the plateau uplifts of the continents, such as the Guyana Highlands. The largest and best known is the 1000-km-wide updome around Hawaii and extending along the Hawaiian seamount chain (Fig. 1-11). The crest of this broad swell stands 1 to 2 km higher than the normal elevation of oceanic lithosphere of equivalent age. The updoming process takes a few million years at rates of hundreds of meters per million years, whereas the dome subsides at tens of meters per million years, following the characteristic proportionality to square root of time (Fig. 1-10).

The normally smooth regional slope of the ocean floor (Fig. 1-9) is interrupted in some regions by large positive bathymetric features far from present plate boundaries. These features are broadly grouped as *aseismic ridges.* Some are quite linear, such as the Hawaiian-Emperor seamount chain, and some are more equant. Aseismic ridges include at least three distinct phenomena: microcontinental fragments, basaltic plateaus, and linear seamount chains.

It is often said that there are six or seven continents, depending on how you count them, but in fact there are several dozen continental masses, most of which are quite small and largely or entirely submarine. These smaller fragments of continental lithosphere are called *microcontinents.* The microcontinents include New Zealand and its extension to the north as the Lord Howe Rise, Madagascar, Japan, several parts of the Philippines, Corsica and Sardinia, Agulhas Plateau south of Africa, Rockall Bank west of the British Isles, and Lomonosov Ridge in the Arctic Ocean. Most of these fragments are produced by rifting from larger continental masses. Microcontinents are very important in the structural history

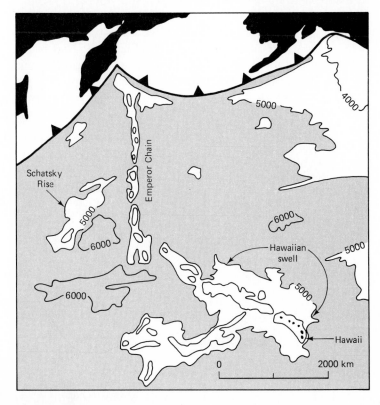

FIGURE 1–11 Map showing broad epeirogenic swell in bathymetry around the Hawaiian Chain. The swell is a much broader feature than the volcanic chain. The normal depth of the old lithosphere of the western Pacific is greater than 5000 m, shown by the shaded pattern.

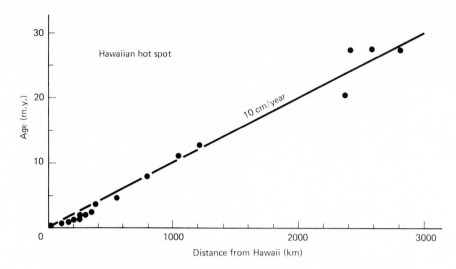

FIGURE 1–12 Steady propagation of volcanism along the Hawaiian Chain at about 10 cm/y. (Data from McDougall and Duncan, 1980.)

of some orogenic belts because they eventually encounter compressive plate boundaries and are accreted.

A number of roughly equant plateaus that are apparently largely basaltic exist in the ocean basins—for example, Schatsky Rise in the North Pacific (Fig. 1-11). These *basaltic plateaus* are apparently great outpourings of basalts but are not well understood. Better understood are the linear seamount chains, which extend for thousands of kilometers. Some, such as the Line Islands, appear to have been active simultaneously throughout, whereas others show a regular progression in age along their length. The latter are designated as *hot-spot chains* because the volcanoes appear to be derived from a fixed spot source of magma lying below the lithosphere. The best known hot-spot chain is the Hawaiian-Emperor seamount chain (Fig. 1-11) which has been propagating across the Pacific for the last 70 m.y., most recently at about 10 cm/y (Fig. 1-12). Apparently the underlying cause of the hot-spot volcanism is also the cause of the epeirogenic updoming (Figs. 1-10 and 1-11). Another important aseismic ridge of hot-spot origin is the Ninetyeast Ridge in the eastern Indian Ocean (Fig. 1-13). The present position of the hot-spot source is apparently near Kerguelen Island in the south

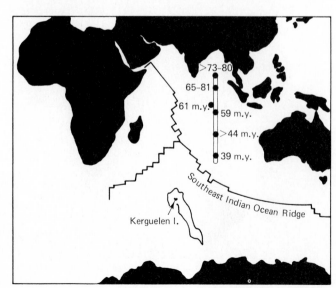

FIGURE 1–13 Southward propagation of the Ninetyeast Ridge hot spot in the Indian Ocean. The present position of the hot spot is apparently near Kerguelen Island, separated from Ninetyeast Ridge by later spreading on the Southeast Indian Ocean Ridge. (Data from Curray and others, 1981.)

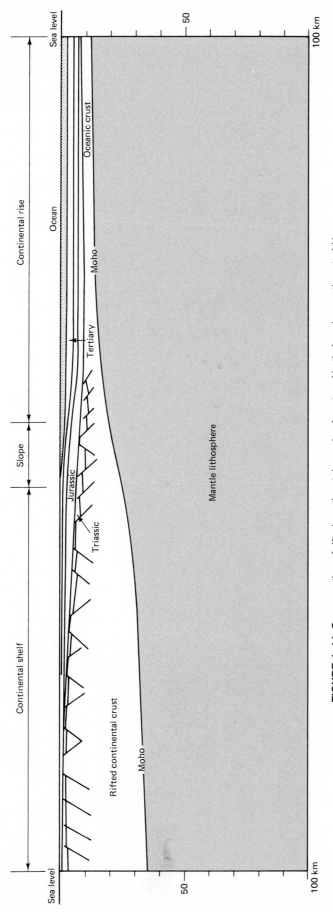

FIGURE 1–14 Cross section of rifted continental margin of eastern North America southeast of New England. (After interpretation of CDP seismic line in Grow, Mattick, and Schlee, Amer. Assoc. Petrol. Geol. Memoir 29, p. 77, 1979.)

Indian Ocean, separated from the Ninetyeast Ridge by new oceanic lithosphere formed along the Southeast Indian Ocean Ridge.

The oceanic lithosphere normally slopes toward the continents (Fig. 1-9); however, near the abundant sediment sources of the continents, it is covered by great submarine-fan complexes of the *continental rise,* which slope away from the continent at about 1:300 to 1:6000. Farther from the continent the rise merges with the extremely flat *abyssal plain* (1:10,000 or less). The steepest topography is at the edge of the continent, which is the *continental slope* (Fig. 1-14), commonly having a slope of about 1:300; in some areas, it may be even clifflike. Because of the relatively steep slopes and thick sediments, the continental slope and—to a lesser extent—the continental rise are important sites of submarine land sliding. In some areas—for example, in the Mississippi and Niger deltas—the sliding is so extensive that major structures are produced (Chapter 8).

The final major morphologic and structural features of ocean basins that we discuss are orogenic features that lie along convergent plate boundaries; a cross section is shown in Figure 1-15. As the oceanic lithosphere approaches the convergent boundary, it flexes gently upward at the *outer rise* (Fig. 1-22) and then downward into the *trench,* which is the oceanic equivalent of the foredeep. The region between the trench and the volcanic arc is called the *forearc,* which generally grows in width with time (Fig. 1-16) as a result of accretion of deformed oceanic and trench sediments into the *accretionary wedge* and deposition of

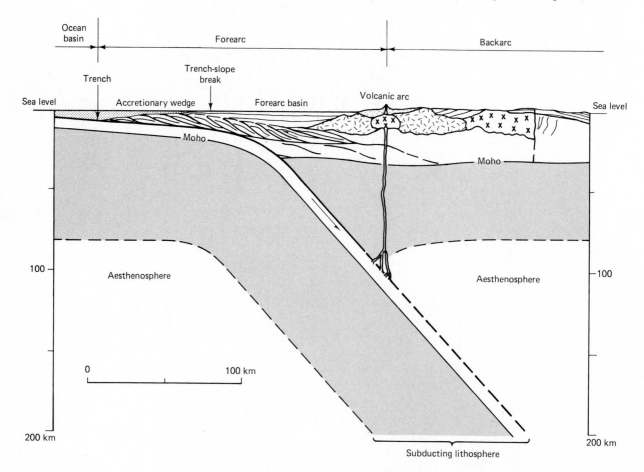

FIGURE 1–15 Cross section of convergent continental margin showing the main morphologic and structural features. (Based on data on Guatemalan continental margin, from Seely, Vail and Watson, 1974, Trench slope model: in Burk and Drake eds., *The Geology of Continental Margins,* Springer-Verlag, New York, p. 249–260.)

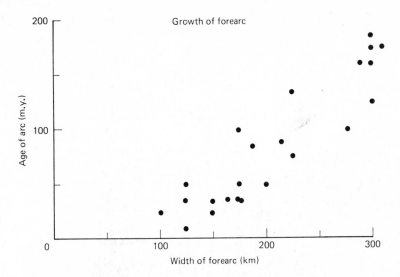

FIGURE 1–16 Expansion of forearc with time; width of forearc measured from trench to volcanic axis. (From Dickinson, J. Geophysical Research, v. 78, p. 3376–3389, 1973, copyrighted by the American Geophysical Union.)

sediments in the *forearc basin*. The *trench-slope break, or outer high,* the boundary between the forearc basin and the accretionary wedge, generally propagates seaward with time as a result of the accretionary processes. Some forearcs undergo little accretion because much of the sediment subducts, and the subduction zone may actually erode some forearcs.

The region beyond the volcanic arc is called the *backarc*. In some continental areas, such as the Andes, the backarc region is a compressive mountain belt. In other continental and oceanic areas, the backarc is an area of horizontal extension in which rift valleys and new ocean basins, called *backarc basins,* form. Examples of backarc basins include the Japan Sea between Japan and Korea, Okinawa Trough, and the basins just west of the Mariana arc.

In the following sections we consider the most important physical processes that control the morphology of the earth by producing vertical and horizontal displacements of the lithosphere.

VERTICAL DISPLACEMENTS OF THE LITHOSPHERE

One of the most remarkable insights into the origin of the topography and bathymetry of the earth and its underlying structure came in 1854–1855 when John Henry Pratt, Archdeacon of Calcutta, attempted to reconcile a large discrepancy in some precise geodetic measurements made in India under the direction of Col. Everest, for whom the mountain is named. The angular length of a 600-km-long North-South survey line was measured by two methods, precise surveying along the line and angular difference between astronomical sightings at both ends. By angular length we mean the angle between the two end points and the center of the earth ($\beta' - \beta$ in Fig. 1-17). Pratt considered that the mass of the Himalayan Mountains and the Tibetan Plateau to the north would deflect the plumb line used to determine the astronomical vertical in the proper sense to account for the discrepancy. Nevertheless, after carefully integrating the gravitational attraction of the mountain mass, Pratt found that the discrepancy should be 15″ of arc rather than the 5″ of arc that was observed. Pratt had no explanation.

It was left to G. B. Airy, the Royal Astronomer in Greenwich, to resolve the difficulty with what is in retrospect a remarkable bit of intuition into the mechanical behavior of the earth. Airy recognized that the interior of the earth is weak and able to flow viscously, whereas rocks near the surface have some finite strength. He considered that the earth might be approximated by the model shown

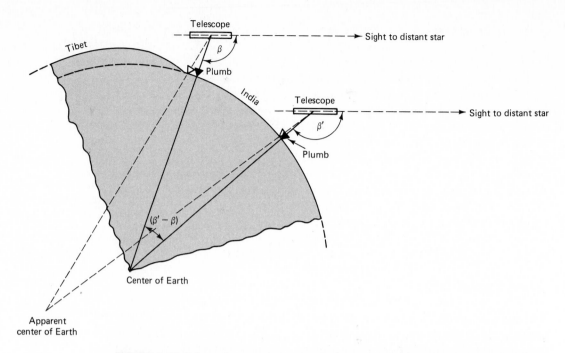

FIGURE 1–17 Schematic cross section of the earth showing Pratt's hypothesis for the incorrect measurement of arc length ($\beta' - \beta$) caused by the gravitational deflection of the plumb line by the mass of the Himalayan Mountains and Tibetan Plateau. The plumb line is used to locate the center of the earth in the astronomical sightings.

in Figure 1-18, after his original paper. The inside was considered to be a viscous fluid, whereas the outermost part was taken to be a less-dense shell of finite strength containing one mountain mass. The situation on the right is in effect the case considered by Pratt. Airy argued that this is an impossible state of affairs; he showed that the rigid shell could not support a surface load of the magnitude of the Tibetan Plateau and would fracture and settle down into the more-dense, viscous interior. Airy reasoned that a mountain mass must be accompanied by a root, such as the one shown on the left; mountains are buoyantly supported like icebergs. The downward projecting root of rock less dense than the viscous interior would reduce the net gravitational attraction of the mountain mass and explain the astronomical measurements (Fig. 1-17). What began with Pratt as an investigation of discrepancies in some geodetic observations became with Airy a real insight into the mechanical behavior of the earth.

Airy's view of the earth has been shown to be essentially correct. During the Ice Ages surface loads in the form of ice caps and large lakes rested on the surface

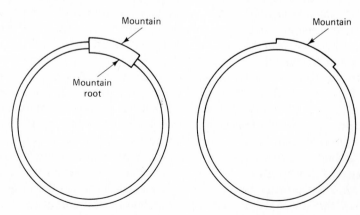

FIGURE 1–18 Airy's cross sections of the earth, showing one schematic mountain mass. (After Airy, 1855.)

of the continents and later returned to the ocean basins. Nonelastic deformation resulted from loads as small as 100 m thick and a few hundred kilometers in diameter. Deformation still going on as a result of this loading cycle is a major source of information on the viscosity of the interior of the earth.

The inner viscous region is now called the *aesthenosphere,* and the outer shell of finite strength is called the *lithosphere.* The boundary between them is, of course, a boundary in mechanical properties, which may not be sharp (Fig. 1-1). It is mainly temperature-dependent and may be approximated by a surface something like 0.9 T_{melting} (Kelvin). Our present view of the lithosphere is more complicated than Airy's; the earth is now known to contain several important compositional layers: the iron-nickel core; the mantle, composed of dense silicates and oxides; and the outer crust, composed of lighter silicates. The boundary between the aesthenosphere and lithosphere normally lies within the upper mantle; however, it can move up and down with time as a result of temperature changes and deformation of the lithosphere.

Differential Isostasy

The most important discovery of Airy and Pratt was that regional topography and bathymetry of the earth and other planetary bodies is to the first approximation supported by buoyantly stable density distributions rather than by the strength of the lithosphere. This is the law of *isostasy,* which states that horizontal surfaces in the aesthenosphere will be surfaces of constant pressure, which implies that the mass of the overlying column of rock is everywhere the same. Regional differences in elevation reflect regional differences in mean density of the underlying rock. In mathematical terms the sum of the masses of each layer in any column of rock above a datum in the aesthenosphere is constant:

$$M_w + M_s + M_c + M_m + M_a = \text{constant} \qquad (1\text{-}3)$$

where the layers are sea water, w, sediment, s, igneous and metamorphic crust, c, mantle lithosphere, m, and mantle aesthenosphere, a (Fig. 1-19). Removing the area of the column of rock, we can write

$$\rho_w h_w + \rho_s h_s + \rho_c h_c + \rho_m h_m + \rho_a h_a = \text{constant} \qquad (1\text{-}4)$$

where h is the thickness of each layer and ρ is its density. The processes that control the regional morphology of the earth are the processes that modify the mean density and thickness of one or more parts of the lithosphere.

There are many mass distributions that can produce the same topography. Airy's iceberg model of isostasy was that the lithosphere is everywhere the same density and that topography reflects differences in thickness of the lithosphere (Fig. 1-19). In contrast, Pratt in later work considered that regional differences in elevation mainly reflected differences in density of the lithosphere, with a horizontal base to the lithosphere (Fig. 1-19). Our present view of isostasy is more elaborate, reflecting our more detailed knowledge of the lithosphere and the processes that shape it. In the following discussion we make use of a four-layer model of the lithosphere (Fig. 1-19), composed of sea water, sediments, igneous and metamorphic crust, and mantle lithosphere. Each of these layers may vary their thicknesses and densities as a result of various geologic processes, which through isostasy control the topography and bathymetry of the earth. Some of the important processes are (1) deposition of sediment and displacement of sea water, (2) evaporation of sea water, (3) erosion of mountains, (4) addition to crust or mantle through igneous intrusion, (5) thermal expansion or contraction associated with temperature changes, (6) thickening or thinning of crust or mantle by tectonic compression or extension, and (7) chemical phase changes. Because the base of

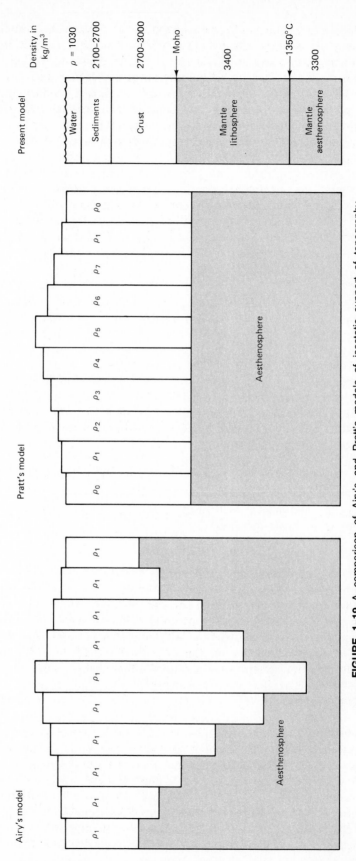

FIGURE 1–19 A comparison of Airy's and Pratt's models of isostatic support of topography. Topographic relief is produced by varying the thickness of the lithosphere in Airy's model, but by varying density in Pratt's model. In contrast, in present models, each segment of lithosphere is potentially variable in thickness or density of sea water, sediments, igneous and metamorphic crust, and mantle lithosphere.

the lithosphere is essentially a temperature boundary ($T \approx 0.9T_{\text{melting}}$ Kelvin), mantle lithosphere is more dense than mantle aesthenosphere (Fig. 1-19). Therefore, thickening of the crust causes uplift of the surface in the manner of Airy's model, whereas thickening of the mantle lithosphere causes sinking of the surface.

In general we have incomplete knowledge of the thickness and density distribution of the lithosphere. Fortunately, for many of the processes mentioned above, we need to consider only changes in thickness, elevation, and density, and not their absolute values. We might call this subject *differential isostasy,* which is based on two relationships:

1. The sum of the changes in mass in a column above any datum in the aesthenosphere is zero, based on Equations 1-3 and 1-4:

$$\Sigma \Delta M = \Delta M_w + \Delta M_s + \Delta M_c + \Delta M_m + \Delta M_a = 0 \qquad (1\text{-}5)$$

or

$$\Delta(\rho_w h_w) + \Delta(\rho_s h_s) + \Delta(\rho_c h_c) + \Delta(\rho_m h_m) + \Delta(\rho_a h_a) = 0 \qquad (1\text{-}6)$$

2. The change in elevation of the surface, ΔE, is equal to the sum of the changes in thickness of the crust and mantle layers:

$$\Delta E = \Delta h_w + \Delta h_s + \Delta h_c + \Delta h_m + \Delta h_a \qquad (1\text{-}7)$$

These are the fundamental equations of differential isostasy.

As an example let us compute the excess thickness of the mountain root under the Tibetan Plateau that would be predicted by Airy (Fig. 1-19), assuming the root is entirely thickened crust of normal density. The Tibetan Plateau is 5 km high. Five kilometers is also the change in elevation, ΔE, using Equation 1-7,

$$\Delta E = 5 \text{ km} = \Delta h_c + \Delta h_a$$

If 2800 kg/m^3 is the density of the crust and 3300 kg/m^3 is the density of the aesthenosphere, then by Equation 1-6,

$$\Delta(\rho_c h_c) + \Delta(\rho_a h_a) = (2800 \text{ kg/m}^3)\Delta h_c + (3300 \text{ kg/m}^3)\Delta h_a = 0$$

and

$$\Delta h_c = 33 \text{ km}$$

According to this estimate, Tibet should have a mountain root about 30–35 km thick. Normal cratonic crust whose surface is at sea level is 35–40 km thick, based on seismology. Therefore, a double crustal thickness of approximately 70 to 80 km is predicted for Tibet, which is in agreement with crustal thicknesses estimated by geophysical methods. Other major active orogenic belts with high elevation, such as the Alps and Andes, have crustal thicknesses of about 70 km.

Thermal Topography

The next major discoveries of processes that control regional topography in the earth came over a hundred years after the discovery of isostasy. Soon after the discovery of plate tectonics and sea-floor spreading, it was found that the ocean floor subsides with age in a very regular fashion; the water depth is proportional to the square root of the age of the lithosphere (Fig. 1-10). This characteristic pattern of subsidence is due to cooling and thickening of the mantle lithosphere with time. If we assume that aesthenosphere extends nearly to the sea bottom under midocean ridges, then it initially has the temperature of the aesthenosphere, about

1350°C, nearly to the surface; it will gradually cool and thermally contract with age. The change in elevation is

$$\Delta E = \frac{\rho_a}{\rho_a - \rho_w} \left\{ 2\alpha(T_w - T_a) \sqrt{\frac{kt}{\pi}} \right\}$$ (1-8)

where the terms in the braces describe the thermal contraction as a function of time, t, and the ratio of densities of aesthenosphere and sea water describe the isostatic amplification of the subsidence that is caused by the weight of the water filling the ocean basin. Here T_a is the initial temperature of the rock at the ridge crest, which is the temperature of the aesthenosphere, 1350°C, T_w is the temperature of sea water, α is the volumetric coefficient of thermal expansion, $3.2 \times 10^{-5}/°C$, and k is the thermal diffusivity of the lithosphere, 8×10^{-7} m²/s. The fact that water depths in the oceans closely obey Equation 1-8 indicates that heating and cooling is a major cause of vertical displacement in the earth.

Thermal expansion and contraction is not limited to cooling of normal oceanic lithosphere; it is now recognized as an important process in the subsidence of sedimentary basins on continental margins and of epeirogenic updoming and subsidence on the cratons and in ocean basins because they subside with the characteristic dependence on square root of time (Fig. 1-10).

Subsidence of Continental Margins

Many important sedimentary basins, especially those on stable continental margins, display important normal faulting at the initial stages of formation of the basin (Fig. 1-14). There is an initial rapid subsidence, followed by a diminishing rate of subsidence during the next 100 to 200 m.y. The total subsidence is generally substantial; 10 to 15 km of sediment are deposited. The rifted continental margin of the eastern United States and Canada is typical of these basins (Fig. 1-14). Three physical processes in combination are important in producing the subsidence of rifted continental margins: (1) initial subsidence in response to horizontal extension of the lithosphere, (2) thermal subsidence in response to the disturbed geothermal gradient, and (3) isostatic subsidence in response to the weight of the sediment deposited in the basin.

Initial Subsidence. Consider a segment of the continental lithosphere of thickness a and width a prior to horizontal extension (Fig. 1-20). After extension during rifting, there is no loss of mass or area, so the new width is increased to $a\beta$ and the thickness of the lithosphere is reduced to a/β, where β describes the amount of extension. What is the subsidence of the top of the lithosphere in response to this stretching? Applying the two equations of differential isostasy (Eqs. 1-6 and 1-7), we find that the initial subsidence, Z_i, is

$$Z_i = \frac{(\rho_c - \rho_a)h_c + (\rho_m - \rho_a)h_m}{\rho_w - \rho_a} \left(1 - \frac{1}{\beta} \right)$$ (1-9)

where the densities of sea water, crust, mantle lithosphere, and aesthenosphere are $\rho_w = 1030$ kg/m³, $\rho_c = 2800$ kg/m³, $\rho_m = 3400$ kg/m³, and $\rho_a = 3300$ kg/m³. The important thing to note is that there is a linear relationship between the subsidence and the thinning function $(1 - 1/\beta)$.

Thermal Subsidence. A sudden horizontal tectonic extension of the lithosphere produces an increase in the geothermal gradient beyond the equilibrium gradient under cratons, as shown in Figure 1-20. The lithosphere acts as a thermal boundary layer, across which heat is conducted, whereas the aesthenosphere has close to an adiabatic thermal gradient because of convection. The base of the

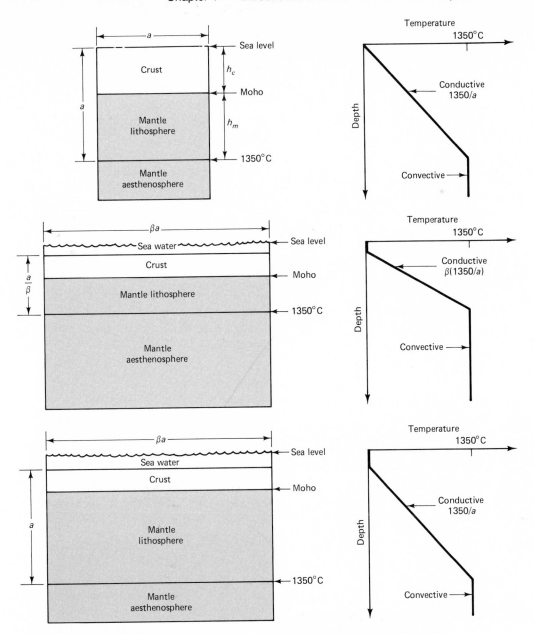

FIGURE 1–20 Model of subsidence and thermal gradient produced by a sudden stretching of the lithosphere. The top figure shows a square segment of lithosphere of thickness a. The middle figure shows the same segment of lithosphere after rapidly stretching by an amount β. The bottom figure shows the lithosphere sometime later, after the thermal gradient has returned to its initial value. (Based on McKenzie, Some remarks on the development of sedimentary basins: Earth Planet Sci. Lett., v. 40, p. 25–32, 1978.)

lithosphere is defined as the temperature at which the mantle is able to convect (about 1350°C). If lithosphere of thickness a is suddenly (less than 20 m.y.) thinned to a/β by horizontal extension, the average geothermal gradient is changed in proportion to the extension β, from (1350°C/a) to β(1350°C/a). The decay of this excess thermal gradient causes a subsidence of the lithosphere that is similar to the thermal subsidence of oceanic crust, exhibiting the same characteristic proportionality between subsidence and square root of time (Eq. 1-8).

Sediment Loading. There is additional subsidence in response to the load of any sediments deposited. If there is an abundant supply of sediment, then the amount deposited is isostatically controlled in a remarkable way by the water depth. Let us assume that no sediment will be deposited above sea level. Given an initial water depth h_w—for example, 2 km—what thickness of sediment h_s must be deposited to fill the basin up to sea level? There is no change in the surface elevation; $\Delta E = 0$ in Equation 1-7 because water is displaced by sediment up to the original water surface. From Equations 1-6 and 1-7 of differential isostasy, we have

$$\rho_s h_s - \rho_w h_w + \rho_a \Delta h_a = 0 \qquad (1\text{-}10)$$

and

$$\Delta E = h_s - h_w + \Delta h_a = 0 \qquad (1\text{-}11)$$

Equating 1-10 and 1-11, we obtain the isostatic relationship between initial water depth, h_w, and maximum sediment thickness, h_s:

$$h_s = \frac{(\rho_w - \rho_a)}{(\rho_s - \rho_a)}\, h_w \qquad (1\text{-}12)$$

Thus if the initial water depth, h_w, is 2 km and the density of the sediment is 2500 kg/m³, then 5.75 km of sediment are needed to fill the basin up to sea level. Equation 1-12 explains why continental margins have the greatest sedimentary thicknesses; they have both great initial water depths and abundant supplies of sediment.

Flexure of the Lithosphere

So far in our discussion of isostasy and vertical displacement of the lithosphere, we have in fact assumed that the lithosphere is only strong enough to resist convection. We have assumed with Airy that each crustal column is completely independent and not strong enough to support any adjacent loads (Figs. 1-18 and 1-19), which is the assumption of *local isostasy* (Fig. 1-21). The lithosphere in fact

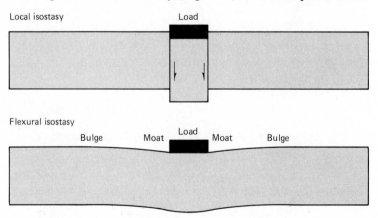

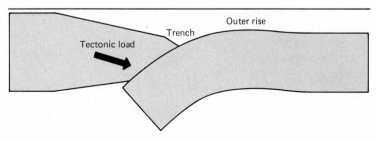

FIGURE 1–21 Schematic diagram of the support of a load, such as a volcanic mountain mass or ice cap, by local Airy isostasy and by flexural isostasy. Bottom diagram shows plate flexure in response to a tectonic load.

has some substantial strength (Fig. 1-1); therefore, a surface load will not be supported solely by the buoyant forces of local isostasy, but will be partially supported by the strength of the lithosphere adjacent to the load, as shown in Figure 1-21. This flexure of the lithosphere is observed adjacent to major loads on the earth's surface. The most straightforward case of plate flexure is the loading of the oceanic lithosphere by seamount volcanoes. There is a well-defined moat around the volcano caused by partial support of the volcano by the strength of the adjacent lithosphere. For example, this moat can be seen in the bathymetry adjacent to the Island of Oahu (Fig. 1-22). In other cases the moat is filled with sediment and volcanic rocks.

The mechanics of the flexure are closely related to the mechanics of bending of beams studied in engineering and to the mechanics of folding, discussed in Chapter 9. If we model the lithosphere as an elastic plate lying on a more dense fluid, then there are two major contributions to the vertical deflection Z: (1) an exponential decrease in deflection away from the load because the load is supported most by the nearest lithosphere, and (2) a sinusoidal deflection caused by the buoyant resistance of the aesthenosphere to the bending. The two effects combine by multiplication to produce a decaying sinusoidal deflection

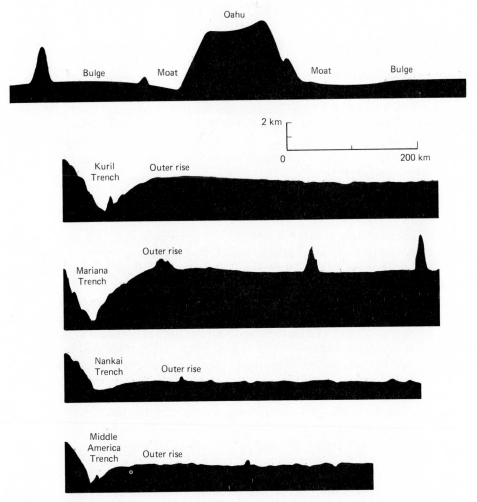

FIGURE 1–22 Bathymetric profiles exhibiting plate flexure. (From Bodine, Steckler, and Watts, J. Geophysical Research, v. 86, p. 3695–3707, 1981; Caldwell and Turcotte, J. Geophysical Research, v. 84, p. 7572–7576, 1979; copyrighted by The American Geophysical Union.)

$$Z = Z_0 e^{-x/\alpha}\left(\cos \frac{x}{\alpha} + \sin \frac{x}{\alpha}\right) \qquad (1\text{-}13)$$

where x is the horizontal distance from the center of the load, $\alpha = [4D/(\rho_a - \rho_w)g]^{1/4}$, D is the flexural rigidity, $D = [Eh_e^3/12(1 - \nu^2)]$, E and ν are elastic constants, h_e is the elastic thickness of the lithosphere, and Z_0 is a constant related to the weight of the load. Normally only the first sinusoidal cycle is observed in bathymetry because of the rapid exponential decay.

Plate flexure is also well known to produce foredeep basins and deep-sea trenches adjacent to active mountain belts. The mechanics of plate flexure adjacent to active mountain belts differs in detail from the flexure caused by volcanic islands because the load on the plate is not just the vertical weight of the mountain mass, but also a horizontal force of plate compression (Fig. 1-21). Furthermore, a force is applied by the pull of the subducting lithosphere. Nevertheless, the form of the deflection is qualitatively similar to that caused by an island load. Figure 1-22 shows the shapes of plate flexures adjacent to deep-sea trenches. It should be noticed that the flexure is much broader where old, thick lithosphere of early Cretaceous age is being flexed, such as the Kuril and Mariana trenches, relative to young mid-Cenozoic lithosphere, such as the Middle America and Nankai trenches. This fact is due to the thickening of lithosphere with age.

HORIZONTAL DISPLACEMENTS OF THE LITHOSPHERE

We have considered the main mechanisms by which vertical displacements of the lithosphere of regional dimensions are produced, together with the principal associated structures of cratons, continental margins, and ocean basins. We now turn to the horizontal displacements of the lithosphere—that is to continental drift, plate tectonics, and regional aspects of plate-boundary deformation. These topics will complete this first chapter introducing the major settings of deformation in the lithosphere.

Kinematic Model of Plate Tectonics

It has long been recognized that present-day orogenic deformation is concentrated in a narrow belt of rugged topography and intense seismicity, with little deformation away from these zones. With the discovery of sea-floor spreading and the recognition of the significance of marine magnetic anomalies, caused by the combination of sea-floor spreading and geomagnetic reversals, it became possible to evaluate quantitatively the motions over the surface of the earth because this surface constitutes a closed kinematic system; the surface of the earth maintains a constant surface area. Therefore, the rate of creation of surface area by sea-floor spreading and backarc extension is equal to the rate of destruction of surface area by compression and subduction. Furthermore, the facts that the rifted continents fit back together again and that magnetic anomalies on opposite sides of a spreading ridge have complementary shapes show that the lithosphere acts as a set of nearly rigid plates.

The fact of rigid plates moving on the surface of a sphere of fixed surface area gives rise to the important kinematic properties of plate tectonics. At any instant the relative motion of two plates, A and B, on the surface of a sphere can be described as a rotation about an axis passing through the center of the sphere. The rate of rotation can be described as a vector $_A\Omega_B$

$$_A\Omega_B = \omega \mathbf{k} \qquad (1\text{-}14)$$

where k is a unit vector along the rotation axis and ω is the magnitude of the angular rotation rate (for example, in degrees per million years). If we wish to know the rate of relative motion at a point P, then

$$\nu = {}_A\Omega_B \times r_i \qquad (1\text{-}15)$$

where r_i is the radius vector extending from the center of the earth to the point P (Fig. 1-23). The linear magnitude of the relative motion at point P is

$$v = \omega R \sin \theta \qquad (1\text{-}16)$$

where R is the radius of the earth and θ is the angle between the axis of rotation and the radius vector, r_i. Note that rigid rotation implies that the magnitude of relative motion should vary systematically with angular distance θ from the pole of rotation according to $\sin \theta$. Therefore, the relative motion velocity is zero at the pole and a maximum ωR at 90° from the pole. This prediction is compared with data from the Atlantic Ocean in Figure 1-23.

Rigid plates and fixed surface area of the earth require that if we make a vector sum of rotation rates of n plates starting and ending with plate 1, then the vector sum is zero:

$$_1\Omega_2 t + {}_2\Omega_3 t + \cdots + {}_n\Omega_1 t = 0 \qquad (1\text{-}17)$$

where t is time. This fact allows us to predict relative motions along plate boundaries that do not have seafloor spreading anomalies, either for lack of data or nonspreading boundaries. The relative motions of plates surrounded by trench and transform boundaries are least well known—for example, the Philippine Sea plate. A best fit solution for the present-day relative rotation vectors is given in Table 1-1.

Equation 1-17 is valid only for small or *infinitesimal rotations*. As rotations become large with time, the sequence in which the rotations are performed or the sequence traveled in the circuit of plates affects the final result; these are called *finite rotations*. For example, in Figure 1-24 a plate undergoes large finite rotation about each of two orthogonal axes; the order in which the two rotations are performed controls the final position and orientation of the plate. The actual path of plate rotation may be quite complex; it is the sequential sum of the entire

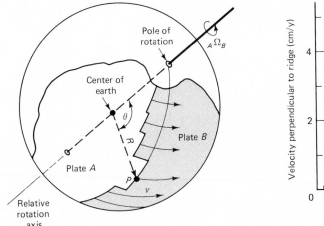
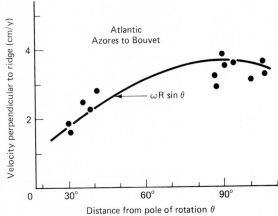

FIGURE 1–23 Geometry of instantaneous relative rotation vector ${}_A\Omega_B$ between two plates, *A* and *B*. Comparison of predicted and measured dependence of velocity, *v*, on angular distance from pole of rotation, θ, for the Mid-Atlantic Ridge. (From Morgan, W. J., J. Geophysical Research, v. 73, p. 1959–1982, 1968, copyrighted by the American Geophysical Union.)

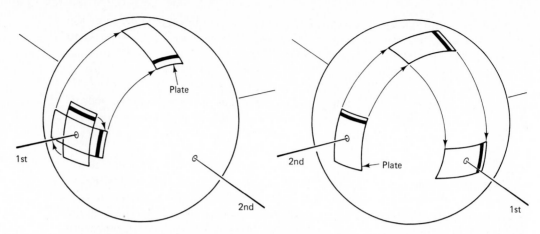

FIGURE 1–24 Effect of the sequence of finite rotations on the final position and orientation of a plate.

history of instantaneous rotations about poles of changing orientations. Nevertheless, no matter what the actual sequence of relative motions between two plates during a time interval, there exists a single equivalent finite rotation about a fixed axis that can move the two plates from their initial to final positions (LePichon and others, 1973). These equivalent finite rotations are important for making plate tectonic reconstructions.

Normally, plate kinematics are studied in a relative reference frame; one plate is considered fixed and the motions of other plates relative to it are determined. Occasionally other frames of reference are utilized. For example, paleomagnetic pole positions provide an estimate of the earth's rotational axis and may be used as a partial frame of reference. It is incomplete because the earth's rotational or magnetic axis does not uniquely define a frame of reference because the paleolongitude is undefined.

TABLE 1–1   Instantaneous Relative Rotation Vectors*

Plate Pair	Latitude °N	Longitude °E	ω (deg/m.y.)
North America–Pacific	48.77	−73.91	0.852
Cocos-Pacific	38.72	−107.39	2.208
Nazca-Pacific	56.64	−87.88	1.539
Eurasia-Pacific	60.64	−78.92	0.977
India-Pacific	60.71	−5.79	1.246
Cocos–North America	29.80	−121.28	1.489
Africa–North America	80.43	57.36	0.258
Eurasia–North America	65.85	132.44	0.231
North America–Caribbean	−33.83	−70.48	0.219
Cocos-Nazca	5.63	−124.40	0.972
Caribbean–South America	73.51	60.84	0.202
Nazca–South America	59.08	−94.75	0.835
Africa–South America	66.56	−37.29	0.356
Antarctica–South America	87.69	75.20	0.302
India-Africa	17.27	46.02	0.644
Arabia-Africa	30.82	6.43	0.260
Africa-Eurasia	25.23	−21.19	0.104
India-Eurasia	19.71	38.46	0.698
Arabia-Eurasia	29.82	−1.64	0.357
India-Arabia	7.08	63.86	0.469
Nazca-Antarctica	43.21	−95.02	0.605
India-Antarctica	18.67	32.74	0.673

*After Minster and Jordan, 1978

Long-lived hot-spot volcanic sources such as Hawaii (Figs. 1-11 and 1-12) and Kerguelan (Fig. 1-13) would provide a frame of reference if the sources were deep within the mantle and fixed relative to one another. On the other hand, if the sources were moving with the convecting mantle at a rate that was fast relative to plate motions, then a hot-spot frame of reference would be impossible. Available information suggests that hot spots are approximately fixed relative to one another over long periods—for example, see Figure 1-25. Hot spots apparently are the effects of some deep mantle convection, although their specific origin is not well known.

Orientation of Plate Boundaries

The orientation of plate boundaries relative to the local relative-motion vector has a fundamental effect on the nature of the plate boundary and the kind of deformation structures produced in the rock. The three end-member classes of plate boundaries are *ridge, transform fault,* and *trench,* in which transform faults are parallel to the relative motion vector and end-member ridges and trenches are perpendicular (Fig. 1-26). Plate boundaries in oceanic regions tend to be closer to the end-member geometry than in continental areas, where deformation is more

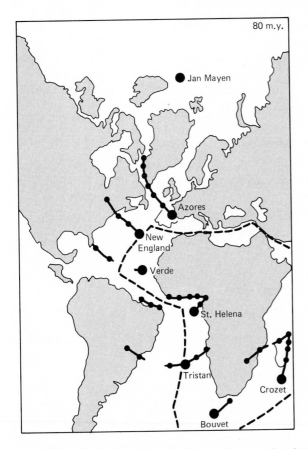
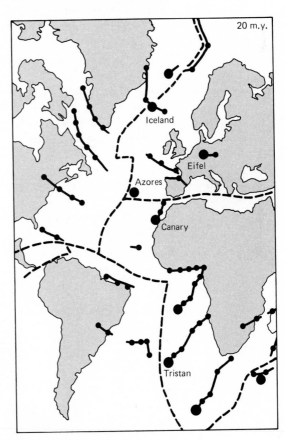

FIGURE 1–25 Motion of long-lived hot spots, assuming they are fixed with respect to each other, which is in good agreement with actual tracks. Reconstructions for 80 and 20 m. y. ago with generalized plate boundaries shown as dashed lines. Large dots show the position of the hot spot at the time of the reconstruction. Small dots show the predicted positions each 10-m.y. period earlier. (After "Hotspot tracks and the opening of the Atlantic and Indian Oceans", Morgan, W. J., p. 443–487 in The Sea, vol. 7, C. Emiliani, ed., copyright © 1981. Reprinted by permission of John Wiley & Sons, Inc.)

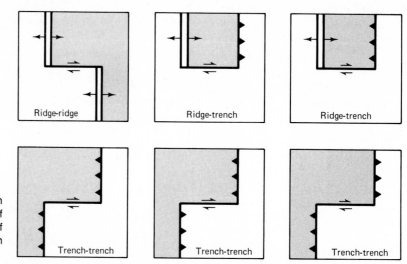

FIGURE 1–26 Six classes of transform faults. In addition, the mirror images of these produce a total of 12 classes of transform faults. The teeth on the trench boundaries lie on the overriding plate.

complex. Transform faults tend to be close to the ideal orientation, parallel to the relative motion vector. Ridge crests also tend to be perpendicular to relative motion vectors and to adjacent transform faults. Trenches are commonly not perpendicular to the relative motion.

Plate tectonics requires that plate boundaries be a closed and interconnected system of ridges, trenches, and transform faults. Six fundamental classes of transform faults, plus their mirror images, are possible, including one ridge-ridge, two ridge-trench, and three trench-trench (Fig. 1-26). Similarly, there are sixteen conceivable triple junctions between three plate boundaries, although only a few of them can exist stably (McKenzie and Morgan, 1969).

Forces Acting on Plates

The motion of plates of lithosphere over the surface of the earth is the net effect of several driving and resisting forces acting on the plates. The most important forces are (1) the gravitational push down the flanks of oceanic ridges, (2) the resistance of trenches and compressive mountain belts, and (3) the gravitational pull of the subducting lithosphere caused by its greater density in comparison to the aesthenosphere.

SETTING OF OROGENIC DEFORMATION ALONG PLATE BOUNDARIES

Most of the natural deformation of rocks that is of concern in structural geology, except for epeirogenic deformation, apparently took place at or near plate boundaries, although the actual style of tectonics is potentially different in the early Precambrian, when the lithosphere was possibly thinner because of a hotter earth. Nevertheless, deformation of the lithosphere in the last one to two billion years appears to be analogous to present-day deformation. For this reason, we end this first chapter introducing the major aspects of deformation of our planet with a discussion of the setting of orogenic deformation along plate boundaries. This will serve to forge a link between our discussion of the kinematics of plate tectonics in the preceding section and the detailed exploration of the structures that form as plates scrape past one another, which follows in Chapters 2 and 6 through 10.

There is a correspondence between the three end-member types of plate boundaries: (1) extensional (ridge), (2) compressional (trench), and (3) transform

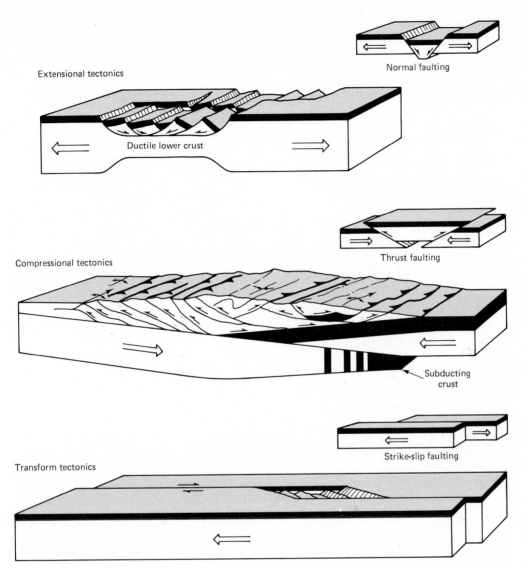

FIGURE 1–27 Block diagrams illustrating the three classes of plate-boundary tectonics and the analogous three classes of faulting. Extensional tectonics make the lithosphere thinner and localize deformation, whereas compressional tectonics make the lithosphere thicker and widen the deformed zone.

and the three fundamental types of faults: (a) normal, (b) thrust, and (c) strike slip (Fig. 1-27). Normal faults along plate boundaries are produced in horizontal extension, thrust faults in horizontal compression, and strike-slip faults in no horizontal extension or compression normal to the fault. Nevertheless, plate boundaries are not just a single fault, but are complex zones of deformation containing many faults, folds, and other deformational structures. The widths of extensional and transform plate boundaries range from less than 25 km to more than 100 km. In some cases, compressional plate boundaries can have active deformation over a zone of more than 1000 km wide. For example, in the Middle East we see that the extensional plate boundaries of the Red Sea and Gulf of Aden are narrow in comparison with the broad zone marking the Alpine-Himalayan orogenic belt to the north (Fig. 1-28).

The relative widths of different classes of plate boundaries have a simple mechanical explanation related to the change in strength of the plate during continued motion. The strength of a lithospheric plate, to the first approximation,

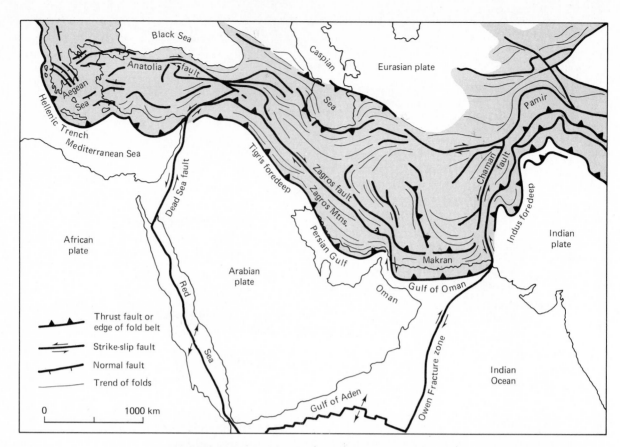

FIGURE 1–28 Tectonic map of the Alpine-Himalayan orogenic belt between Greece and India. This is a broad zone of convergence between the Eurasian plate on the north and the fragments of the former continent Gondwanaland on the south, the African, Arabian and Indian plates. (Compiled from Berberian, 1981, and McKenzie, 1978b.)

is dominated by its thickness, just as a thicker sheet of aluminum is stronger than a thinner sheet. If a segment of continental lithosphere is extended horizontally as in Figure 1-27, it will first deform at the weakest point, causing the lithosphere to be thinned by localized extension. The thinning makes the weakest point still weaker; therefore, continued deformation tends to localize extensional plate boundaries within the lithosphere. In contrast, deformation along compressive plate boundaries generally thickens and therefore strengthens the lithosphere, causing the zone of deformation to grow in width. Transform plate boundaries (Fig. 1-27) have little thickening or thinning of the lithosphere; their width is no greater than the thickness of the lithosphere. Plate boundaries in continental lithosphere are generally wider than in oceanic lithosphere because the continents are much more heterogeneous mechanically (for example, Fig. 1-7).

The dominant faulting in extensional plate boundaries is normal faulting (Fig. 1-27), but strike-slip faults that form a connection between different normal faults are also common. The dominant faulting in transform plate boundaries is strike-slip faulting (Fig. 1-27), but thrust faults and normal faults commonly form connections between different strike-slip faults. For example, the transform boundary between the Arabian and African plates (Fig. 1-28) that extends north from the Red Sea through the Dead Sea is a complex zone of deformation about 100 km wide, marked by abundant compressive folds, thrust faults, and normal faults, in addition to strike-slip faults (Fig. 1-29). The dominant faulting along compressive plate boundaries is thrust faulting (Fig. 1-27), with strike-slip faults forming connections between different thrust faults. However, in addition some

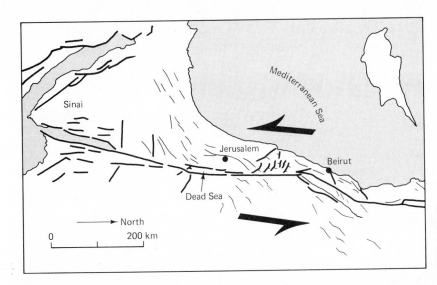

FIGURE 1–29 Structures around the Dead Sea transform zone between the African and Arabian plates. Faults are shown as heavier lines, folds as lighter lines. (Simplified from Vroman, 1967.)

compressive plate boundaries have major zones of strike-slip faulting and normal faulting that are altogether unexpected from the simple analogy between classes of plate boundaries and classes of faults. In a nutshell, this more-complex behavior is a result of the mechanical heterogeneity of orogenic belts and is illustrated in the following pages with an introduction to a segment of the active Alpine-Himalayan orogenic belt.

Active Plate Compression: Greece to India

The zone of active plate compression extending from Greece to India in Figure 1-28 is just one segment of the great Alpine-Himalayan orogenic belt, which stretches from the western Mediterranean to Indonesia with extensions into the western Pacific and Central Asia. Within our segment, it is a zone of plate convergence between two formerly separated continental masses, Eurasia to the north and several fragments of the Mesozoic continent Gondwanaland to the south, including Africa, Arabia, and India (Fig. 1-28).

The Mesozoic ocean that intervened between Eurasia and Gondwanaland is commonly called *Tethys*. The floor of the eastern Mediterranean is apparently a last vestige of Tethys, where Africa has not yet collided with Greece and western Turkey. This piece of ocean floor is subducting under the Aegean island arc along the Hellenic Trench. Similarly, there may be a vestige of Tethyan oceanic crust in the Gulf of Oman, subducting under the Makran Coast of Iran and Pakistan (Figure 1-28). The Indian Ocean proper, the Gulf of Aden, and the Red Sea are new Cenozoic ocean basins that continue to grow as the rifted fragments of Gondwanaland disperse.

The Tethyan ocean basin is now completely destroyed in the Arabian and Indian sectors of the orogenic belt as a result of plate convergence. Deformed fragments of the oceanic crust and mantle, called *ophiolites,* are exposed in a linear belt running through southeastern Turkey and the Zagros Mountains of Iran in the Arabian sector and through Baluchistan and northern Himalayan Mountains in the Indian sector (Fig. 1-30). These ophiolite belts are called *suture zones* because they mark the site at which the two sides of an ocean are closed up between continental masses or between quasi-continental masses such as island arcs or small fragments of continental crust.

Deformed material south of the suture zone is sedimentary rock and underlying basement of the former Tethyan margins of the Arabian and Indian

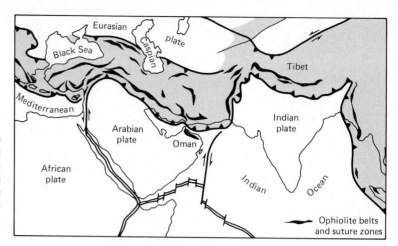

FIGURE 1–30 Ophiolite belts and suture zones of the Alpine-Himalayan mountain belt between Yugoslavia and Burma, which mark the traces of closed ocean basins that formerly existed between the Eurasian craton to the north and the continent of Gondwanaland to the south. (Simplified after Coleman, 1981.)

continents. Figure 1-31 is a satellite image of part of the Zagros fold-and-thrust-belt, which contains deformed sediments of the Tethyan margin of Arabia. To the south is the Persian Gulf, which—together with the Tigris-Euphrates valley—marks the foredeep at the edge of the Arabian craton. Another foredeep exists at the edge of the Indian craton adjacent to the Himalayan and Baluchistan fold-and-thrust belt (Fig. 1-28).

Other ophiolite belts exist to the north of the main Tethyan ophiolite belt (Fig. 1-30), which are suture zones marking the sites of other closed ocean basins. In a broad sense these are also parts of the Tethyan ocean basin, but with intervening fragments of continental lithosphere. The floors of the Black Sea and southern Caspian Sea are apparently also oceanic crust, enclaves of not-yet-destroyed oceanic lithosphere.

In summary, we have built up an image of the Alpine-Himalayan orogenic belt as a 500–1500 km-wide compressive boundary zone between the Eurasian plate on the north and the African, Arabian, and Indian plates to the south. This boundary zone is composed of the deformed margins of the megacontinental masses, suture zones of various formerly intervening ocean basins, and some microcontinental fragments. It is in this heterogeneous, broadly compressive plate-boundary zone between cratonic masses that we observe a complicated pattern of internal deformation, not just the thrust faults and associated folds that mark the margins of the zone.

One of the most unexpected features of the orogenic belt is major zones of strike-slip faulting running roughly parallel to the margins of the orogenic belt and to the thrust faults and folds. This pattern of faulting indicates that large slivers of rock within the orogenic belt are moving laterally away from the narrowest, most constricted parts of the mountain belt in eastern Turkey and in the Pamirs of northern Pakistan and northwest India (Fig. 1-28). Apparently the Indian and Arabian cratons are substantially stronger than the material making up the orogenic belt and are acting as relatively rigid indenters. This causes slices of the crust to move laterally from the points of indentation to regions in which the lithosphere is less constrained, in the oceanic margins of the fold belt in the eastern Mediterranean and along the Makran Coast of Iran and Pakistan. The orogenic belt also appears to be relatively unconstrained along the Zagros fold belt because of low friction caused by the large amount of mechanically weak salt within the stratigraphic section. Some of the salt flows to the surface as domal intrusions (Fig. 1-31).

The concept that cratons act as rigid indentors that cause lateral extrusion of crustal slices is illustrated in the laboratory experiment shown in plan view in Figure 1-32. The slices of plasticine in the model have extruded toward the

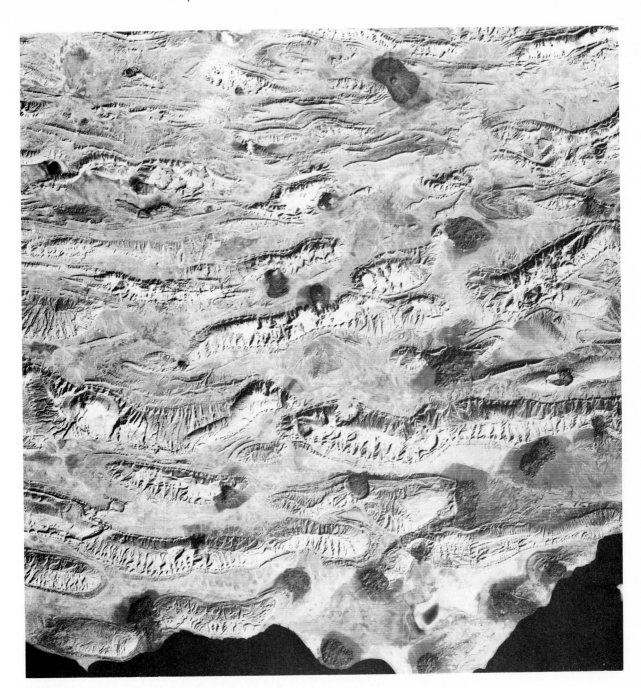

FIGURE 1–31 Satellite image of the Zagros fold-and-thrust belt in southeastern Iran. The black area at the bottom is water of the Persian Gulf. The black spots on land are salt domes that have pierced to the surface. The width of the image is about 175 km.

unconstrained free side of the apparatus, with slip between slices taken up along strike-slip faults. This concept of rigid indentation also helps understand the broad expanse of Cenozoic and present-day continental deformation in eastern Asia north and east of the Himalayas, in addition to the areas already discussed. Large crustal blocks bounded by strike-slip faults in China and Indochina appear to have been extruded toward the east and southeast in response to indentation by the Indian craton (Molnar and Tapponnier, 1975; Tapponnier and others, 1982).

In the Aegean area the Alpine orogenic belt lies below sea level as a result of extensive normal faulting. The Aegean Sea is a backarc basin of the Hellenic

FIGURE 1–32 Plan view of laboratory deformation experiment involving indentation of plasticine by a rigid rectangular indenter (5 cm wide). The plasticine, composed of light and dark layers, has extruded toward the unconstrained side of the apparatus with slip along strike-slip faults. (Experimental details and applications to tectonics are given by Peltzer and others, 1982, and Tapponnier and others, 1982. Photograph courtesy of G. Peltzer.)

Trench. Extension on this scale within the mountain belt implies that the belt is widening faster than the rate of plate convergence.

To proceed much farther with an understanding of the deformation within complex orogenic belts, we need to know much more about the mechanical properties of rocks and the deformational structures that exist. Therefore, in the following chapters we proceed first to the geometric properties of deformational structures and then to the basic material processes by which rocks deform and to the origin of the specific classes of common structures. Finally, in the last two chapters, we return to deformation on the scale of orogenic belts.

EXERCISES

1–1 The Mediterranean Sea, now approximately 3 km deep, dried up in the Pliocene. In the process of drying up, about 1 km of salt was deposited (ρ = 2200 kg/m^3). Estimate the depth of the sea bottom below worldwide sea level when it was dry. Also estimate the depth prior to drying up.

1–2 An ocean basin 6 km deep lies adjacent to a continent and collects a lot of sediment (ρ = 2600 kg/m^3). What thickness of sediment is needed to fill the basin to sea level? Note that 6 km is close to the maximum depth of an ocean basin in a nonorogenic environment.

1–3 Carbonate sediments are substantially more dense than silicate sediments (limestone, ρ = 2700 kg/m^3). What is the effect of this difference in density on the thickness of sediment that can be deposited?

1–4 The only other terrestrial planet that might be undergoing plate tectonics is Venus. If this turns out to be true, what would the topography due to thermal subsidence look like compared with Earth? There are no oceans on Venus and the surface temperature is 740K, as opposed to 275K on Earth.

1–5 Derive Equation 1-9 for the initial subsidence in response to horizontal stretching of the lithosphere based on Figure 1-20 and the equations of differential isostasy. If the crust is 35 km thick and the lithosphere is 125 km thick, what is the initial subsidence?

1–6 How does isostasy affect the rate at which mountainous topography decays? Modify the relationship between denudation rate and relief (Eq. 1-2) to include the effect of local isostasy. Under these conditions, how much time is required to denude a mountain belt from 1 km relief to 100 m relief using the two sets of denudation rates given in Figure 1-3?

1–7 Which classes of transform faults (Fig. 1-26) maintain a constant length through time? Which lengthen and which contract? What is the rate of change of length in relation to the rate of slip?

SELECTED LITERATURE

KENNETT, J., 1982, *Marine Geology,* Prentice-Hall, Englewood Cliffs, NJ, 813 p.

KING, P.B., 1977, *The Evolution of North America,* rev. ed., Princeton University Press, Princeton, NJ, 197 p.

LePICHON, X., FRANCHETEAU, J. AND BONNIN, J., 1973, *Plate Tectonics,* Elsevier, New York, 300 p.

TURCOTTE, D. L. AND SCHUBERT, G., 1982, *Geodynamics: Applications of Continuum Physics to Geological Problems,* Wiley, New York, 450 p.

WINDLEY, B.F., 1977, *The Evolving Continents,* Wiley, New York, 385 p.

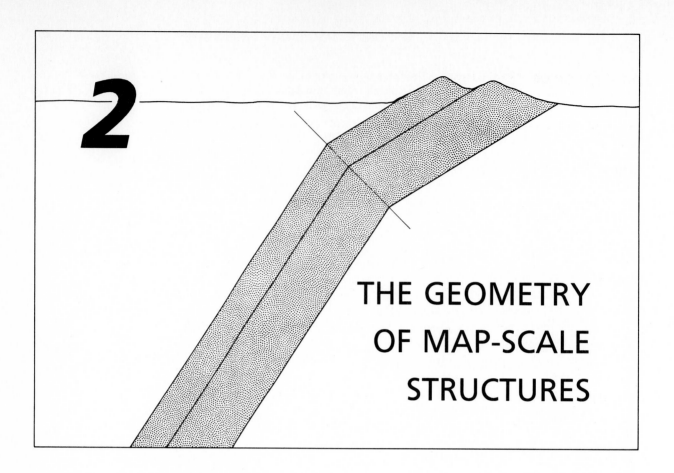

THE GEOMETRY
OF MAP-SCALE
STRUCTURES

INTRODUCTION

It is not a trivial task to observe the large-scale structure of the earth's crust. The fundamental problem is that rocks are opaque. For example, we can directly observe the fold in Figure 2-1 only at the eroded ground surface; the part of the fold that is underground must be inferred by extrapolation and interpolation of surface observations, such as the orientations of beds and the map distribution of rock units. Our knowledge of map-scale structure in three dimensions is substantially limited by the difficulty of making accurate predictions of subsurface structure based on surface geologic data.

Occasionally, if there is enough economic or scientific incentive, the surface data may be augmented by drilling, excavation, and by geophysical measurements such as reflection seismology, gravity, magnetics, and electrical resistivity. Each of these methods is expensive and time-consuming, and each has its own difficulties in interpretation (for example, Tucker and Yorston, 1973). Reflection seismic profiling in little-deformed regions has been the most successful. The fact remains however, that most knowledge of large-scale structures comes from rock exposures at the ground surface.

It is our purpose in this chapter to introduce the more straightforward and powerful methods of inferring the three-dimensional geometry of structures from earth-surface geologic data. These data and methods form the basis for much of what we know about the structure of continental crust. In addition to discussing these methods, it will be convenient to introduce the principal descriptive terminology of folds and faults, which are the principal map-scale structures.

ORIENTATION OF LINES AND PLANES

Any attempt to infer the three-dimensional geometry of a structure requires knowledge of the orientation of the geologic planes or surfaces that compose the

FIGURE 2–1 Aerial photograph of map-scale anticline in south central Wyoming.

structure—for example, stratigraphic contacts, beds, faults, schistosity, veins, and joints. Once we know the orientation of layers, we may be able to determine their position at depth by extrapolation or interpolation. The orientation of these planes may be specified in several ways. Three ways that are particularly useful to structural geology are (1) strike and dip, (2) dip direction, and (3) pole to plane.

The *strike* of a plane is the orientation of its line of intersection with any horizontal plane (see Fig. 2-2). The strike direction can be measured in the field with a geologic compass because the level bubble defines a horizontal plane. If the axis of a level compass is placed parallel to a strike line, then the angle between the strike line and north can be measured; this angle is the *azimuth* of the strike line (see Fig. 2-2). Azimuths are measured clockwise from north and range between 0° and 359°.

The *dip* of a plane is the acute angle between the plane and any horizontal plane (Fig. 2-2). Dips range between 0° and 90°. By definition, the angle between two planes is measured in a plane normal to the line of intersection of the planes; therefore, the dip angle is measured in a plane normal, or perpendicular, to the strike line. The dip can be measured in outcrops with an inclinometer on a geologic compass. Note that for a given strike, there are two possible planes with the same angle of dip; for example, a plane with a strike of 128° could have a dip of 48° to the southwest (48°SW) or 48° to the northeast (48°NE). The particular case must be specified. A typical strike-and-dip measurement would be specified, for example, as 128°, 48°SW. Alternatively, the measurement can be written N 52°W, 48°SW; that is, 52° west of north is the same as an azimuth of 128°.

The measurement of strikes and dips of surfaces need not be confined to outcrops. If the orientation of a bed or other geologic surface is constant on a map scale, as is the case in parts of Figure 2-1, the strike and dip may be calculated by standard trigonometric means or determined graphically if we know the elevation of the bed at three points that do not lie in a line, which is the well-known *three-point problem* outlined in Box 2-1. The elevation of a layer may be read from a geologic map at points where the contact intersects the contour lines of the topo-graphic base. If the layer is visible from the air, its elevation at several points may be measured using a pair of vertical aerial photographs viewed stereoscopically.

As an alternative to strike and dip, the orientation of a surface can be defined by its *dip direction,* which is the orientation of a downward-directed line lying in the surface and perpendicular to the strike (Fig. 2-2). Dip direction may be specified with two numbers: the azimuth of its horizontal projection and the angle of dip. A typical dip direction is written 218°, 48°; thus it is somewhat more economically stated than the equivalent strike and dip (128°, 48°SW) because there is no ambiguity in direction of dip. Dip direction is inconvenient to measure in the field using a Brunton compass, which is the standard geologic compass in North America; however, dip direction can be measured directly in a single operation using a Clar compass (see Fig. 2-2).

A third way in which the orientation of a plane is commonly specified in structural geology is by its pole. A *pole* is a line normal to a plane and is perpendicular to both the strike and the dip direction. The downward-directed pole is normally used in structural geology. The pole to a plane is rarely measured in the field, but is useful in manipulating data, as we shall see later in this chapter.

In addition to dip directions and poles to planes, there are a variety of linear entities whose orientations need to be specified in structural geology—for example, gouge marks and scratches on fault surfaces, mineral lineations, fold hinges, paleomagnetic vectors, drill holes, and sedimentary structures that record depositional-current directions. The orientations of lines in space are commonly specified in one of two ways in structural geology: (1) bearing and plunge or (2) rake. Bearing and plunge are analogous to the two numbers used to specify dip

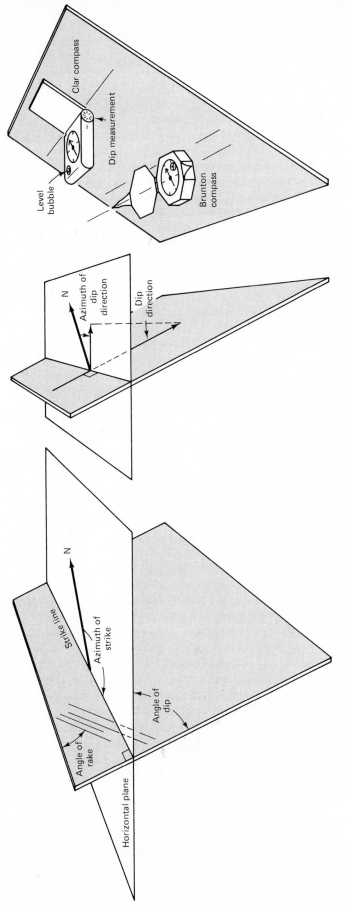

FIGURE 2–2 Definitions of strike, dip, and dip direction of an inclined plane and the rake of lines lying within the plane. Azimuths of lines are measured clockwise from north. The azimuth of the dip direction and the angle of dip are measured simultaneously with the Clar compass. The strike is first measured with the Brunton compass, as shown, then the compass is rotated to the plane perpendicular to the strike and the dip is measured, using an inclinometer.

BOX 2–1   The Three-Point Problem

If the elevation of a planar surface, such as a bedding surface, is known at three points that do not lie along a straight line, then its strike and dip may be calculated or determined graphically. In the figure of this box, the elevation of the top of a formation is known in wells at *A, B,* and *C.* On the top of the formation, there is a point *B'* at the same elevation as in well *B* and lying along the line *AC.* The location of point *B'* can be found by interpolation of elevations between *A* and *C* because we are assuming that the bed is a planar, uninterrupted surface of constant strike and dip. The line *BB'* connects two points on the bed of the same elevation and is therefore a strike line. The dip direction is perpendicular to the strike. The dip angle can be measured by constructing a cross section perpendicular to *BB'*—for example, through point *A.* Alternatively, the dip may be computed by trigonometry.

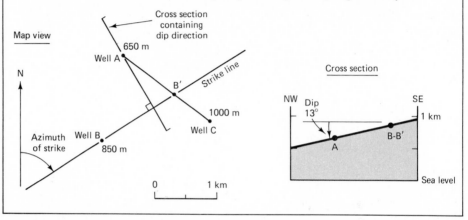

---

direction (see Fig. 2-2). *Bearing* is the azimuth of the horizontal projection of the downward plunging line. *Plunge* is the acute angle between the line and any horizontal plane, by definition measured in the vertical plane containing the line and its horizontal projection (Fig. 2-3). Bearing and plunge can be measured directly and simultaneously using the Clar compass, for which it is specifically designed. Bearing and plunge also may be measured with a Brunton compass, although it is less convenient and in some cases subject to larger errors. The axis of the level Brunton compass is placed so that it appears parallel to the horizontal projection of the line. The plunge is measured with the inclinometer.

If a line lies within a plane—for example, a scratch mark on a fault surface—its orientation may be specified by its rake, as an alternative to its bearing and plunge. The *rake* of a line is the acute angle between the downward-directed line and the strike of the plane containing it (see Fig. 2-2)*. Note that for a given plane there are two possible lines with the same angle of rake; for example, a plane oriented 128°, 48°SW may contain lines with rakes of 36° to both the southeast (36°SE) and the northwest (36°NW). The particular case must be specified.

A variety of symbols are used on geologic maps to indicate the orientation of geologic planes or lines measured in outcrop. Some of the more commonly used symbols are shown in Figure 2-4. The strike of a plane is given by the orientation of the long bar in the symbol and the direction of dip is shown by the short mark to

---

*The term *pitch* is synonymous with *plunge* in some contexts and with *rake* in other contexts; the term is best avoided.

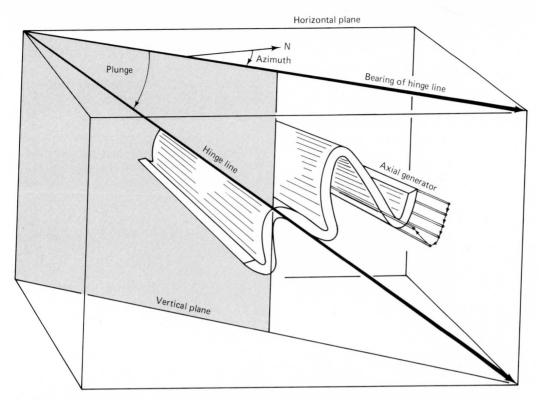

FIGURE 2–3 Definitions of bearing and plunge of a line for example, a hinge line of a fold. The plunge is the angle between the line and its horizontal projection, measured in a vertical plane. The bearing of the line is the orientation of its horizontal projection. Also illustrated is the fact that cylindrical folds have a shape that can be generated geometrically by sweeping a line, the axial generator, through space parallel to itself.

the side of the bar. The bearing of a line and its sense of plunge are shown with an arrow. The angle of dip or plunge is written next to the symbol.

We are now in a position to consider the geometry of folds.

THE CYLINDRICAL NATURE OF FOLDS

Naturally folded rock layers take on a wide variety of shapes, as is illustrated in Chapter 9 on folding. Nevertheless, certain shapes are far more common than others, and most shapes that we might envisage do not exist at all. For example, folds of the sort you make by crumpling a sheet of paper in your fist do not exist in the earth because there is no physical process analogous to the crumpling-in-the-fist operative in the earth. Rocks were not cast in their present folded shape, but were deformed from an initial state, commonly relatively planar horizontal layers. The shapes that folds assume reflect the mechanics by which they were formed and must be explained by any successful theory of folding (Chapter 9).

Naturally folded rock layers normally have a pronounced cylindrical symmetry; the layers have been bent about a single axis or direction, called the *fold axis*. To the first approximation, the shape of a single folded surface can be generated by sweeping a straight line, called an *axial generator*, through space parallel to itself (see Fig. 2-3); each segment of the folded surface has this one line or orientation in common. Such folds are called *cylindrical* folds. Few folds are

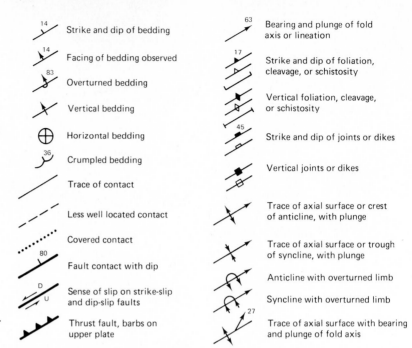

FIGURE 2–4 Commonly used geologic-map symbols.

14 Strike and dip of bedding

14 Facing of bedding observed

83 Overturned bedding

Vertical bedding

Horizontal bedding

36 Crumpled bedding

Trace of contact

Less well located contact

Covered contact

80 Fault contact with dip

D
U Sense of slip on strike-slip and dip-slip faults

Thrust fault, barbs on upper plate

63 Bearing and plunge of fold axis or lineation

17 Strike and dip of foliation, cleavage, or schistosity

Vertical foliation, cleavage, or schistosity

45 Strike and dip of joints or dikes

Vertical joints or dikes

Trace of axial surface or crest of anticline, with plunge

Trace of axial surface or trough of syncline, with plunge

Anticline with overturned limb

Syncline with overturned limb

27 Trace of axial surface with bearing and plunge of fold axis

precisely cylindrical; nevertheless, most folds are to the first approximation cylindrical over most of their surface (Chapter 9).

The orientation of the approximate axial generator, or fold axis, can be determined statistically from measurements of bedding at different points around a fold, using methods introduced later in this chapter. Once determined, the cylindrical axis is of great value in determining subsurface structure, especially if the fold is plunging, because it allows us geometrically to project a folded layer down into the earth from its exposure at the surface. The degree to which such subsurface predictions are correct depends only on the degree to which the structures are cylindrical. For a truly cylindrical fold, all cross sections, or slices, cut perpendicular to the fold axis are identical; in real cylindrical folds, parallel slices are only approximately the same.

The most common way of representing the three-dimensional geometry of structures is to construct geologic cross sections. These sections are commonly constructed on vertical planes; however, if the structures are strongly plunging, a vertical cross section will present a distorted view of the structure with apparent changes in stratigraphic thickness. In these cases it is more appropriate to construct a section perpendicular to the cylindrical axis, which is called a *profile section*. There is little difference between a vertical section and a profile section if the plunge is less than 30° to 35°. Either kind of section is constructed by projecting each point on a structure exposed at the surface of the earth parallel to the fold axis, into the plane of the section (Fig. 2-5). The construction of cross sections is discussed more completely later in this chapter.

The fold axis is also very useful in viewing folds qualitatively. If a fold is plunging into the earth, you may look at a map or an outcrop of the fold in the direction of the plunging axis and see a profile section directly. This method of viewing geologic maps along the cylindrical axis is called the *down-plunge*, or *down-structure*, method and allows you quickly to visualize the three-dimensional geometry of a structure. For example, the schists and phyllites in the Kvarnbergs-vattnet area of northern Sweden shown in Figure 2-6 exhibit a regional north-northeasterly plunge of about 20°. By looking at the map toward the northeast at

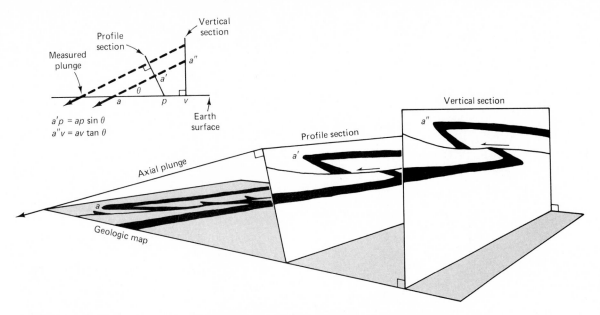

FIGURE 2–5 Two types of cross sections constructed by projection along the cylindrical fold axis; for example, point *a* on the geologic map projects along the axis to point *a'* on the profile section and to point *a"* on the vertical cross section. The profile section is constructed perpendicular to the cylindrical axis, whereas the vertical section is constructed perpendicular to the horizontal projection of the axis.

an angle of 20° to the page, we see the structure approximately as it would appear in profile. We can check our ability at viewing maps down structure by looking at the cross sections (Fig. 2-7). The down-structure method of viewing maps is not restricted to folded rocks, but can be fruitfully applied to a variety of structures (Mackin, 1950).

The layers in many folds are bent about sharp hinges where the layers are most strongly curved. The locus of maximum curvature along a folded surface is called the *hinge line* (Fig. 2-3). Local maxima may exist and they may even bifurcate. Therefore, many hinge lines generally exist along a single surface (for example, see Fig. 9-13). The hinge line of a fold is generally parallel or approximately parallel to the fold axis; indeed, the term *fold axis* is sometimes used loosely as a synonym of hinge line. Thus the axis of a small fold may be determined by directly measuring the orientation of its hinge line as an alternative to calculating the fold axis from bedding orientations.

There are two other important methods of determining the cylindrical axis of a fold: (1) the study of associated smaller-scale folds; (2) the study of associated cleavage, or schistosity. These two methods make use of phenomena that are also important to understanding the mechanics of folding and will be explored more fully in Chapters 9 and 10; we merely introduce them below.

As a region deforms, folds often develop simultaneously on a variety of scales. The wavelength of each fold is proportional to the thickness of the layer being buckled; microscopic layers are folded on a microscopic scale, layers on the scale of beds are folded on an outcrop-scale, and mechanical layering on the scale of whole formations gives rise to map-scale folds (for example, see Fig. 9-22). A hierarchy of fold sizes is associated with a hierarchy in scales of interlayering of stiffer and softer materials. What is important to our study of the geometry of folds in the present chapter is that all scales of folds that are produced in a single deformation of initially parallel layers commonly have the same axis; thus the axis

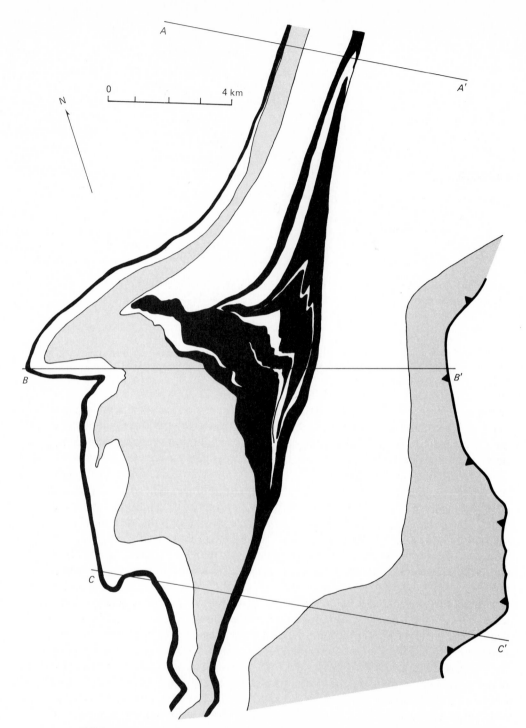

FIGURE 2–6 Geologic map of schists and phyllites of the Kvarnbergsvattnet area in the Caledonian mountain belt of northern Sweden. The fold axes and lineations exhibit a regional north-northeasterly plunge of about 20°; therefore, by viewing the map down structure in a northeasterly direction at about 20° to the page, we can see the structure approximately as it would appear in profile section. The cross sections are shown in Figure 2-7. (Simplified from Sjöstrand, 1978.)

of a map-scale fold may be determined by measuring the orientations of hinge lines of outcrop-scale or even microscopic folds.

In addition to the buckling of rock layers during compression in the earth, layers commonly flow as extremely viscous or plastic substances. This flow may

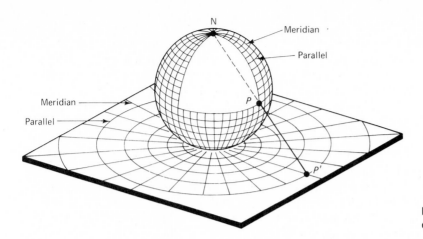

FIGURE 2–11 Projection of a spherical coordinate net onto the image plane.

projection. *Polar nets* are centered on the pole of the coordinate net and *meridional nets* are centered on a point on the equator. Figures 2-12 and 2-13 show the polar and meridional stereographic and equal-area nets. Only half the sphere is projected because that is all that is needed to plot the orientation of any line or plane.

The spherical coordinate net is used in practical problems much as you would use a ruler or a piece of graph paper to locate positions in geographic space; in this case we are locating positions in angular space. Normally, a piece of tracing paper is laid over the net and the actual work is done on the tracing paper. If we take the circumference of the net to be the trace of a horizontal plane, as is usual, and mark some point along it as geographic north, then the orientations of all horizontal lines plot as points along the circumference. Thus, the azimuth of any strike line or the bearing of any plunging line can be measured off along the circumference. For example, a bearing of 72° is shown as *B* in Figure 2-14. Note that the other end of the bearing, or strike, line *B'* hits the circumference at the opposite side of the net (72° + 180° = 252°).

The plunge of a line or the dip of a dip direction or a pole may be measured directly from the net if we use a polar projection. This is because all great circles shown on a polar net, except for the circumference, are the traces of vertical planes passing through the center of the sphere. Therefore, plunges and dips may be measured directly down from the circumference along the appropriate great circle. For example, in Figure 2-14 a line bearing 72° and plunging 45° is shown as point *C*. As a second example, a bed with a dip direction of 227°, 43° is plotted at point *Q*. The pole to bedding of this orientation is a line perpendicular to both the dip direction and the strike and therefore plots as the point 90° to each of them (point *S*). Thus we have used two different—but equivalent—methods of representing the orientation of a plane, dip direction, and pole to bedding.

A third method of representing the orientation of a plane is to plot its intersection with the sphere—that is, as a great circle. For this we may no longer conveniently use a polar net but should use a meridional net. The usual operation is to rotate the tracing paper until the strike of the plane to be plotted is parallel to the axis of the underlying net; in this position the family of all planes with the strike in question are displayed. The great circle corresponding to the desired plane is then located by measuring the dip in a plane perpendicular to the strike or along any of the parallels of the net. For example, a plane oriented 65°, 20° SE is shown in Figure 2-15. The pole to the plane, *P,* and the dip direction, *Q,* are also shown.

We are now able to represent the orientation of any line or any plane on a spherical projection and are in a position to measure the angles between them and

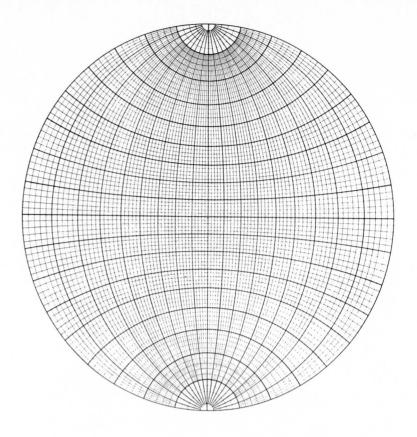

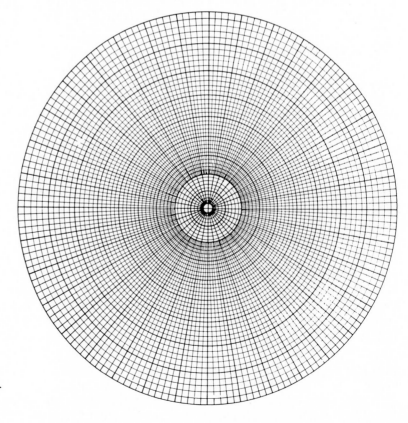

FIGURE 2–12 Meridianal and polar ste-reographic (equal angle) nets.

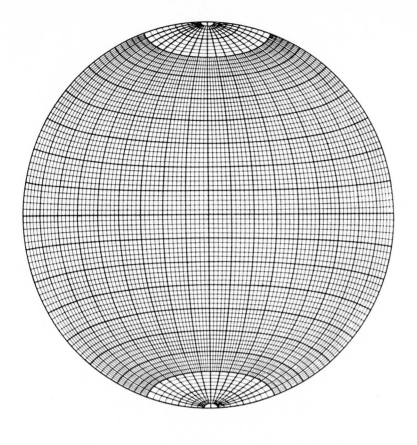

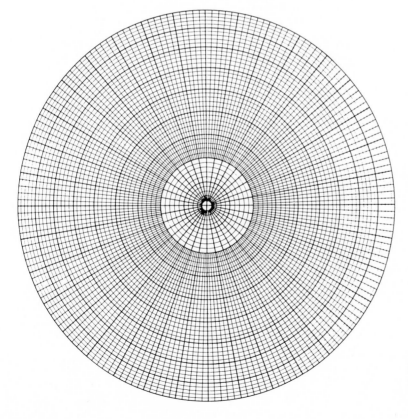

FIGURE 2–13 Meridianal and polar equal-area nets.

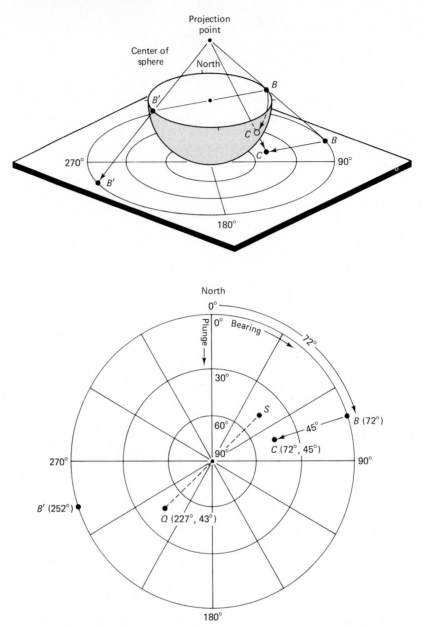

FIGURE 2–14 Representation of orientation of lines on a polar spherical projection. Azimuths are measured clockwise from north and plunges from the outside in. For example, the center of the projection represents a vertically plunging line. Points Q and S lie on the same vertical great circle and 90° apart.

to manipulate such data in a variety of useful ways. If we wish to find the orientation of the line of intersection of two planes—for example, the intersection of cleavage and bedding, to determine the cylindrical axis of a fold—we simply plot the two planes as great circles, and their point of intersection represents their line of intersection, the cylindrical axis. For example, in Figure 2-16 a bedding plane oriented 162°, 46°SW and a cleavage plane oriented 102°, 80°SW have a line of intersection P, oriented 272°, 45°.

The angle between the two planes may be measured in two ways. The most general way is to plot the poles to the two planes and then rotate the meridional net until both poles lie along the same great circle. The great-circle distance between the two poles is the angle between the two planes; in the case above the

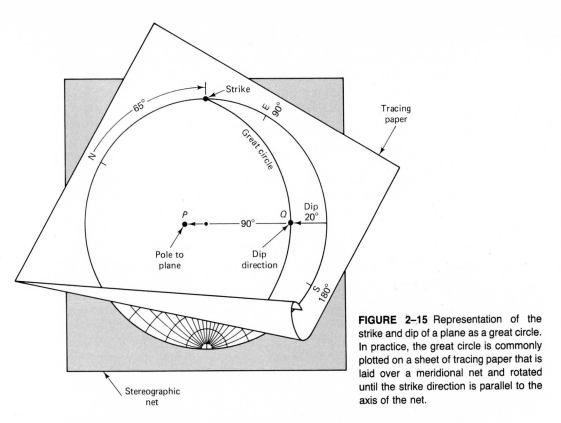

FIGURE 2–15 Representation of the strike and dip of a plane as a great circle. In practice, the great circle is commonly plotted on a sheet of tracing paper that is laid over a meridional net and rotated until the strike direction is parallel to the axis of the net.

angle is 62° (Fig. 2-16). The second method may be used only with an equal-angle stereographic projection; on this projection the angle between the two planes may be measured directly with a protractor at their point of intersection (point *P* in Fig. 2-16).

In analogous fashion, a number of other geometrically distinct classes of problems of practical importance to structural geology may be solved using spherical projections; these classes are illustrated by the exercises at the end of this chapter and contain brief suggestions for their solutions. Almost all problems can be solved equally well using either the stereographic or equal-area projections. The stereographic projection can be more useful in problems requiring

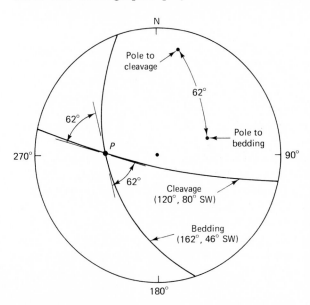

FIGURE 2–16 The orientation of the line of intersection *P* between two planes is easily determined by plotting the two planes as great circles. The angle between the two planes is the great-circle distance between the poles to the planes. It is also the angle between the tangents to the two planes at their point of intersection *P*, but only on an equal-angle stereographic projection.

rotation of lines about an axis—for example, in plate tectonics, because they involve small circles, which are not distorted in stereographic projections (see Exercises 2-9 and 2-10). Lambert's projection, because of its equal-area property, takes on major importance in the statistical determination of the axis of symmetry of folds.

Statistical Measurement of Fold Axes

One of the most widespread uses of spherical projections in structural geology is in determining the mean orientation of fold axes in a region and in presenting a visual representation of the dispersion of the axes about that mean. In a previous section we saw that four types of data may be used to determine the orientation of fold axes: (1) bedding orientations, (2) cleavage orientations, (3) hinge lines of small folds, and (4) cleavage-bedding intersections. With our knowledge of the spherical projections, we may now represent these data as poles or as great circles.

The axis of a cylindrical fold is the line in common to all bedding orientations around the fold. Therefore, if we plot two measurements of bedding as great circles on a spherical projection, their point of intersection will be the fold axis (sometimes labeled β). This estimate of the fold axis may not be very accurate because of human errors in measurement, natural irregularity of the bedding surface, undetected slumping of outcrops, and noncylindrical aspects of the folding. A better estimate of the mean fold axis can be made with a larger number of bedding measurements. If each measurement is plotted on a lower-hemisphere spherical projection as a great circle, then each circle will have one β intersection with every other circle (Fig. 2-17). A plot of bedding or cleavage as great circles is therefore called a β *diagram*. The number of intersections increases greatly with the number of bedding measurements, *n:*

$$\text{Number of intersections} = \frac{n(n - 1)}{2} \tag{2-1}$$

Each of these intersections is an estimate of the orientation of the fold axis, although each estimate is not independent of the others. If the data were perfect and the folds were precisely cylindrical, all the great circles would intersect at a single point, the fold axis (Fig. 2-17). With actual data the fold axis is chosen to be in the center of the greatest concentration of intersections (Fig. 2-17), ignoring spurious intersections.

An alternative method of determining the fold axis from a number of bedding measurements is to represent the orientations as poles to bedding. Such a representation of bedding or cleavage orientations as poles is called a π *diagram,* the Greek letter π signifying pole. If the data were without errors and the folds were perfectly cylindrical, all poles would plot along a single great circle with the fold axis being the pole to the great circle (Fig. 2-17). With actual data, the best estimate of the fold axis is chosen to be the pole to the best fitting great circle. Figure 2-17 shows the equivalent π and β diagrams for a single set of data.

Just as the axis of dispersion of bedding orientations is the fold axis, so is the axis of dispersion of cleavage, if the cleavage formed during the folding. The fold axis may be determined by plotting the cleavage measurements as a π or β diagram; however, the accuracy of the fold-axis determination is significantly less than the accuracy of the determination based on bedding measurements because the variation of bedding orientations about a fold is generally significantly greater than the variation of cleavage orientations. A generally more-satisfactory method

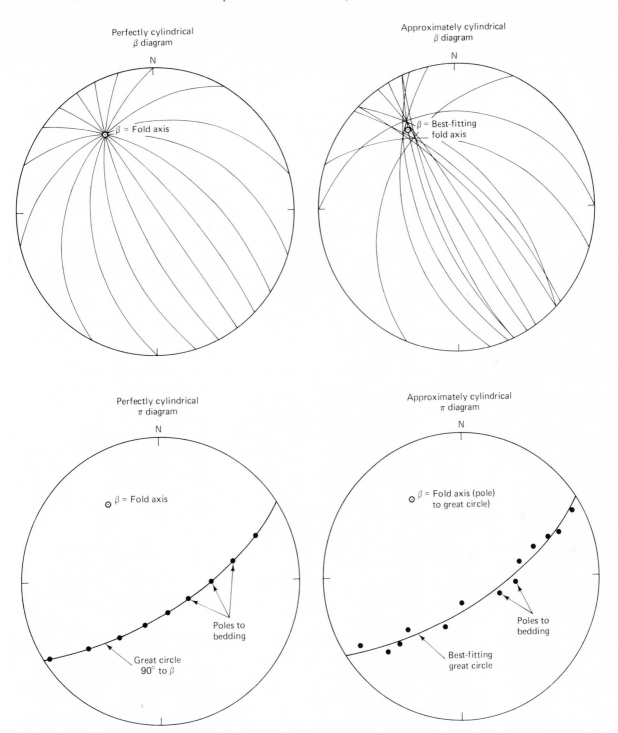

FIGURE 2–17 β and π diagrams for determining the fold axes, shown for a perfectly cylindrical
fold and for an approximately cylindrical fold.

of using cleavage to determine the fold axis is to determine the cleavage-bedding
intersection either from direct measurements of the intersection in outcrop or
from measurements of cleavage-bedding pairs.

Linear data on fold-axis orientation, principally cleavage-bedding intersections and hinge lines of outcrop-scale folds, may also be plotted on spherical
projections. The best estimate of the fold axis is then taken as the center of the

greatest concentration of data. All the data from a single area, including the fold axis determined from bedding measurements, cleavage-bedding intersections, and hinge lines, may be summarized on a single diagram, called a *synoptic (view together) diagram.*

The choice of the best fold axis to use for the purpose of projecting geology to depth (for example, Fig. 2-5) is usually made by inspection of the data plotted on a spherical projection—for example, in a π, β, or synoptic diagram (Fig. 2-18). More precise statistical methods are available for estimating the mean orientation of linear-orientation data and are used especially in the study of paleomagnetism. Formal statistical methods of locating the mean orientation of a set of data assume isotropic dispersion about the mean. This assumption is adequate for regions of cylindrical folding, but for more complex regions of refolded or noncylindrical folding, the statistical methods are inappropriate. In these more-complex cases it is usual to perform most analysis by inspection of the data in a contoured form.

It is common to present orientation data on spherical projections in a contoured form for ease in locating the concentrations of data, particularly in polydeformed regions, and also as a convenient graphical display. The contours signify the number of times the data are more concentrated than a random orientation of the data (Fig. 2-18). Suppose, as an unlikely example, that we have 100 measurements of randomly oriented bedding, then one pole to bedding should on the average appear in each 1 percent of the area of an equal-area spherical projection. If the bedding were not randomly oriented, then significant regions would exist with less than one or more than one pole to bedding. A region containing two or more poles per 1-percent area would be enclosed within a so-called two-times random contour.

Several methods of contouring by hand exist, and others are designed for computer use (for example, Robinson and others, 1963). The hand methods generally make use of a circular counter with an area that is 1 percent of the area of the equal-area net (see Turner and Weiss, 1963). As the counter is moved, the number of poles within any 1-percent area of the net can be counted. The counter may be moved over a grid pattern with counts made at each point on the grid. The counts may then be contoured as numbers of times random concentration.

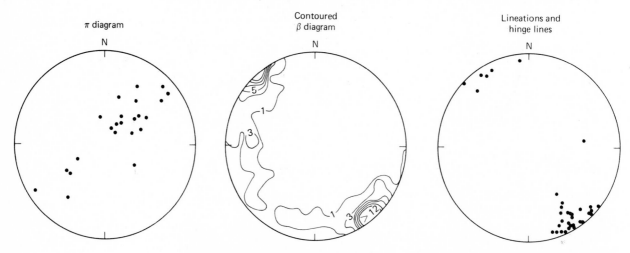

FIGURE 2–18 Coincidence of cylindrical axis determined by several methods in the Otago Schists of New Zealand: π diagram of poles to bedding and schistosity, β diagram with contours in percent, and lineation and hinge lines of minor folds. (Simplified from Robinson, Robinson, and Garland, 1963.)

CONSTRUCTION OF RETRODEFORMABLE CROSS SECTIONS

Most attempts to infer the complete shape of a map-scale structure involve the construction of cross sections—for example, Figure 2-7. These sections are generally vertical and oriented perpendicular to the regional strike, and are called *transverse sections*. A reasonably complete, three-dimensional understanding of the shape of a structure can be obtained by drawing a series of closely spaced transverse sections and perhaps additional sections parallel to the regional strike, which are called *longitudinal, or strike, sections*. Cross sections can sometimes be useful in other orientations, as well.

There are three principal facts that we may use in the construction of cross sections of layered rocks: (1) the orientation of bedding, cleavage, and fold axes at specific places, (2) the distribution and thickness of stratigraphic units, and (3) the originally undeformed nature of the rocks. The fact that the rocks were originally undeformed may seem trivial to mention, but it is actually an important key to the solution of many structural problems. It must be geometrically possible to undeform any valid cross section to an earlier less deformed or undeformed state; the cross section must be *retrodeformable*. Retrodeformable cross sections are commonly called *balanced sections* in the petroleum industry (Dahlstrom, 1969); an example is given in Figure 2-19. Interpretation A is a cross section of a faulted anticline. If we attempt to slip the fault back to an original unslipped state with the beds matching across the fault, we find that it is geometrically impossible. Therefore, the cross section is not retrodeformable and is apparently an impossible solution to the available geologic data. Interpretation A may be easily modified to a retrodeformable solution that satisfies the basic data (for example, interpretation B), which is therefore a possible solution. We have seen from this example that the fact of originally undeformed rocks can be important in removing some structural solutions from further consideration. Many cross sections are found on close examination to be not retrodeformable.

There are four principal aspects to the construction of cross sections: (1) assembly of the basic data, (2) extrapolation and interpolation, (3) termination of the structure, and (4) check for retrodeformability. For example, the top cross

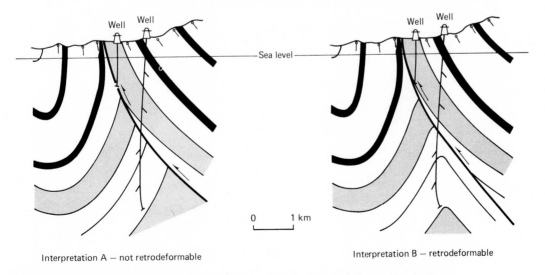

Interpretation A — not retrodeformable

Interpretation B — retrodeformable

FIGURE 2–19 Examples of retrodeformable and nonretrodeformable solutions to the structure of a faulted anticline, western Taiwan.

section of Figure 2-20 shows the basic structural and stratigraphic data assembled onto a topographic profile. This information is extended into unknown regions in interpretation A of Figure 2-20 by extrapolation and interpolation. Structures eventually terminate or are substantially modified from their near-surface shape in a way that may not be determinable by direct extrapolation; therefore, we must decide where to stop extrapolating. Most cross sections are terminated at the limits of the region of immediate interest or at some arbitrary depth, such as sea level. The cross sections in Figure 2-19, for example, are terminated on the flanks of the anticline that is being explored for petroleum and near the maximum depth accessible to drilling. These are examples of *incomplete cross sections,* which terminate before the structure terminates (see also interpretation A, Fig. 2-20). An incomplete cross section is, in general, less enlightening than a *complete cross section,* which includes the entire structure, because a consideration of how a structure may terminate can lead to cross-sectional shapes that might not be

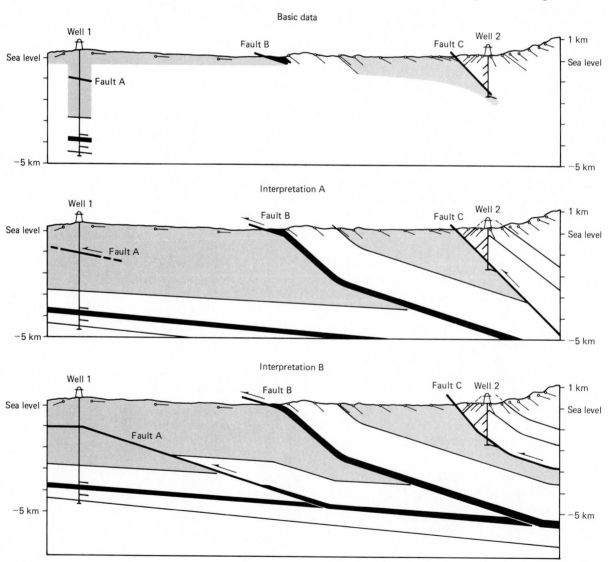

FIGURE 2–20 Construction of a cross section across part of the western Taiwan fold-and-thrust belt. The top section gives the basic data, the middle section gives a simple extrapolation of the data to 5 km below sea level, and the bottom section gives a more-complete interpretation, which is discussed in the text.

expected from extrapolation and interpolation alone. A complete cross section can give insight into the fundamental nature and origin of the structure.

Interpretation A of Figure 2-20 is an extrapolation of the basic data to fill in the area of the cross section to a depth of 5 km below sea level, just below the bottom of well 1. If we extend the depth of the cross section to include more of the structure and if we think more about the origin of the structure, we can make a more enlightening interpretation—for example, interpretation B.

How was this second interpretation constructed? First, fault A was considered to be a minor fault, not worth extrapolating in interpretation A because of its minor apparent displacement. However, the anticline on which well 1 sits does not exist below fault A, so the fault might somehow be important to the existence of the anticline. In interpretation B the fold is produced as the result of a bend in fault A, just below the crest of the anticline (a fault-bend fold; see Chapter 9). Second, fault B runs parallel to bedding in its upper plate, but crosscuts bedding in its lower plate. Therefore, where the fault intersects the base of the black layer in the lower plate, it must flatten and run parallel to bedding, as is shown in interpretation B. A fault that runs along a weak stratigraphic horizon is sometimes called a *décollement*, from the French *unglue*, which has the same root as collage. It is known that many thrust faults in a region will step up toward the surface from a single décollement; therefore, we try the idea that fault A may also run along the black décollement, as is shown in interpretation B. This idea is successful because it predicts the observed near-surface steepening of fault B caused by bending of the upper sheet of fault A. In this example of Figure 2-20, we see that some consideration of the possible origin of structures and some expansion of the scope of the cross section leads to a more complete and enlightening solution. However, to do this in interpretation B required more geological experience and knowledge of how structures form than the simple extrapolation of interpretation A. More experienced geologists can make better cross sections.

We can test interpretation B of Figure 2-20 further by graphically retrodeforming it, as is done in Figure 2-21. Here we see two features that may cause us to reevaluate the details of interpretation B. First, fault A is seen to have a fairly large slip, about 3 km. Therefore, we can check interpretation B further by extending the cross section to the left to see if there is any structure consistent with the predicted 3 km of slip. Second, fault B is still observed to steepen near the surface, even after retrodeformation; therefore, the steepening is not solely the result of stepping up of fault A from the décollement, as was originally suggested. We will not further explore the details of this cross section; these examples are sufficient to show that quantitative graphical retrodeformation of a cross section allows both further verification of the internal consistency of the structural interpretation and insight into the less-obvious structural details. In the case of Figure 2-21, it also gives a quantitative estimate of the minimum slip on the faults and the amount of material eroded.

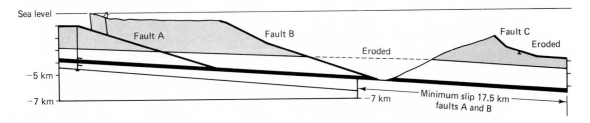

FIGURE 2–21 A retrodeformed reconstruction of interpretation B of Figure 2-20. The upper plate of fault C is not shown because it comes from far to the right.

Both interpretations A and B of Figure 2-20 appear to be consistent with the data. Any complete, retrodeformable cross section may be considered a possible cross section. In general, we can draw more than one possible cross section that will satisfy the available information. However, the fact that multiple possible cross sections can be constructed should not be viewed with alarm or pessimism because with reasonable amounts of data there are generally only a limited number of fundamentally different solutions that are complete and retrodeformable. Furthermore, multiple cross sections are sometimes drawn to help in determining the need for additional drilling and geophysical exploration and for evaluating the economic risk of further investment. Multiple cross sections also can help us identify the basic scientific issues to be investigated.

Assembly of Data on a Cross Section

The construction of a geologic cross section is begun with the selection of a line of section through the region of most complete data and least minor complications. Sections are generally constructed approximately perpendicular to the regional fold axis so that folds will be seen in symmetrical profile section and the layers will appear in their true thicknesses. For convenience of construction, sections are usually vertical. However, if the fold axis is strongly plunging, an inclined profile section perpendicular to the fold axis may be required. Sections drawn at an irrational, nonsymmetric direction to the cylindrical axis generally give little help in understanding the three-dimensional shape of a structure.

An accurate topographic profile is constructed showing the elevation at each point along the line of the section (Fig. 2-20). Normally, the vertical and horizontal scales are equal because unequal scales produce distortions in the apparent shape of the structure, which usually appear as vertical stretching. Vertically exaggerated scales are occasionally effective in areas of very gentle structure, with dips of a few to 10 degrees, enabling slight undulations of the layers to be illustrated. With steep dips or large vertical exaggeration, unsatisfactory distortion is produced; beds do not appear at their true dips, apparent thicknesses become a function of dip, and moderately dipping beds and faults appear to be nearly vertical. Vertically exaggerated cross sections of complex structures are difficult to construct accurately and they give a distorted, generally confusing picture of the structure.

Once a topographic profile is constructed (Fig. 2-20), the basic geologic data may be plotted along it: principally, measurements of strike and dip and intersections of stratigraphic contacts and faults with the line of section. Nearby measurements of strike and dip, off the line of section, may be projected into the section along the strike line at the elevation of the measurement. In mountainous regions some measurements may appear up in the air or down in ground on the cross section because of projection from adjacent peaks or valleys. If the strike is not perpendicular to the section, the apparent dip plotted on the section will be less than the true dip. The apparent dip may be determined using a spherical projection (Exercise 2-3) or a graph (Fig. 2-22). The apparent-dip correction is small if the angle θ between the section and the dip direction is less than about 30° (Fig. 2-22).

If the section is approximately parallel to the strike, it may be better to project data up or down the dip direction rather than along the strike. Strike-and-dip measurements are often plotted as a short line to indicate the apparent dip, with a small circle or dot where the measurement projects into the section (for example, Fig. 2-20). The plotting of data directly on the section helps preserve a distinction between observation and interpretation in the completed cross section.

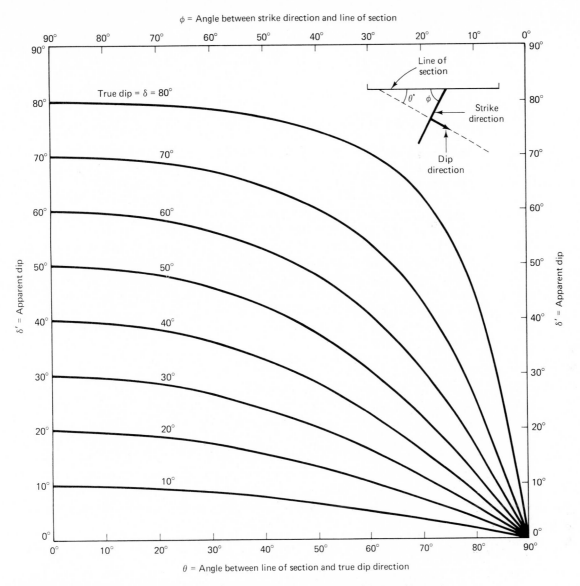

FIGURE 2–22 Change in apparent dip, δ', in a vertical cross section as a function of its angle with the strike and dip directions (ϕ and θ); $\delta' = \tan^{-1} (\tan \delta \sin \phi) = \tan^{-1} (\tan \delta \cos \theta)$.

Extrapolation and Interpolation: Busk and Kink Methods

We may now extend our cross section into unknown regions by extrapolation and interpolation. Note that there is little distinction between extending a cross section up into the air or down into the ground; the buried rocks are as inaccessible as the eroded ones unless there is enough economic incentive to explore the subsurface regions. The part of a cross section extended up into the air is no more speculative than the part underground.

It is best to concentrate at first on extrapolation and interpolation of beds and stratigraphic contacts rather than faults. Much less is generally known about the orientations of faults; indeed, many faults are not observed in the field at all because fault zones are easily eroded. The existence of unexposed faults is inferred from discordences in the structure and stratigraphy.

Extrapolation and interpolation is most successful when each successive layer takes on a shape that is closely conformable to the adjacent layers; that is, the structure is *harmonic* (for example, Fig. 9-23). If the nearby layers assume quite different shapes—that is, the structure is *disharmonic* (for example, Fig. 9-18)—only the more general aspects of the structure may be extrapolated into unknown regions. Disharmonic structure commonly reflects some strong contrast in mechanical properties between adjacent layers.

Any method of extrapolation or interpolation requires that we make assumptions. The variety of geologic situations is diverse; therefore, no universally applicable methods exist. The Busk and kink methods are two methods that may be used in layered rocks that have not flowed extensively—for example, many unmetamorphosed sediments. Both methods assume that the layers are of constant thickness. With some modification of the methods, certain cases of nonconstant layer thickness may also be treated.

The well-known *Busk method* (Busk, 1929) assumes that folds are parallel and concentric. *Parallel folds* have constant layer thickness measured perpendicular to bedding. *Concentric folds* are composed of pie-shaped segments of circular arcs or shells (Fig. 2-23). Each arc segment has a unique center of curvature. At the boundary between two segments, the beds undergo an abrupt inflection along an *inflection line,* which contains the centers of curvature of the two abutting segments (Fig. 2-23).

The Busk method is based on the important fact that bedding measurements in concentric folds are tangents to the circular arcs; therefore, if we draw lines perpendicular to the bedding measurements, we may locate the centers of curvature (see Fig. 2-23). Once the centers of curvature are located, stratigraphic

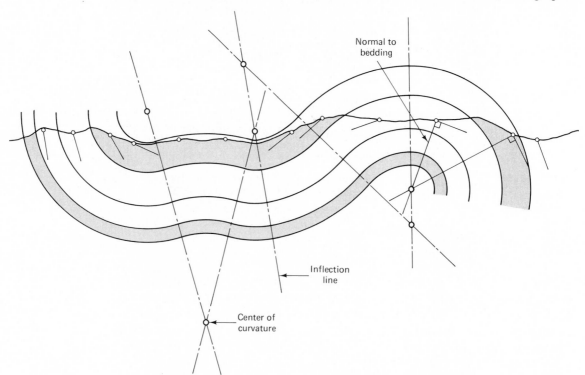

FIGURE 2–23 Extrapolation and interpolation of folded structure using the Busk method of circular arcs. The Busk method commonly makes erroneous predictions because most folds are not composed of segments of circular arcs; that is, they are not concentric.

contacts may be easily extrapolated and interpolated using a drawing compass. The Busk method gives excellent results in structures or parts of structures that closely approximate a concentric geometry. Unfortunately, most folds are not concentric, and the Busk method makes erroneous predictions in many cases, as shown by such expensive methods as drilling.

The *kink method,* less well known than the Busk method, also assumes that the folds are parallel, but instead assumes they have straight limbs and sharp angular hinges (kink, or chevron, folds). The kink method is motivated by the fact that detailed measurements show that many large folds in sedimentary rocks are composed of a series of sharp bends. For example, in Figure 2-24 we see a series of closely spaced bedding measurements along a line of section and down a bore hole, made with a well-logging device called a *dipmeter*. Both the surface measurements and the dipmeter survey show a relatively constant dip over a considerable distance and then a relatively sudden change at a new constant dip. We may easily extrapolate such angular folds using the kink method.

If bed thickness is constant (parallel fold), the axial surface bisects the angle between the fold limbs—that is, $\gamma_1 = \gamma_2$ (Fig. 2-24). The angle γ between the limb and the axial surface is called the *axial angle*. If we have enough data to define clearly the limb dips, we can determine the orientation of the axial surface with considerable precision; the trace of the axial surface may then be easily located with a protractor and all the layers extrapolated. Where two axial surfaces intersect, a new axial surface is formed, also satisfying the equal-angle rule $\gamma_1' = \gamma_2'$ (Fig. 2-24).

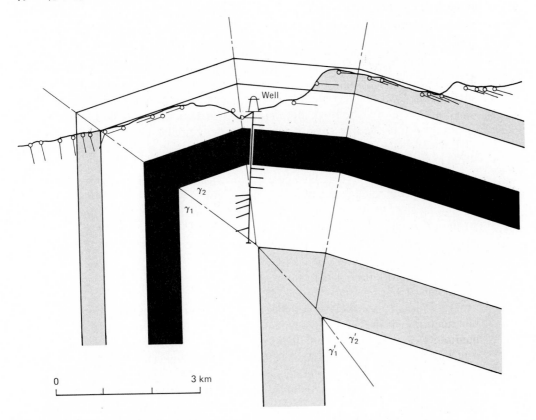

FIGURE 2–24 Extrapolation and interpolation of folded structure by the kink method. Note that the axial surface must bisect the angle between the two limbs of the fold such that $\gamma_1 = \gamma_2$ and γ_1' = γ_2' in order for layer thickness to be equal on both limbs.

The kink method has made good predictions in many cases tested by drilling. It can even work well in folds that are composed of smooth curves rather than straight segments, given moderately complete data, because any smooth curve can be approximated by a series of straight-line segments (Fig. 2-25). After

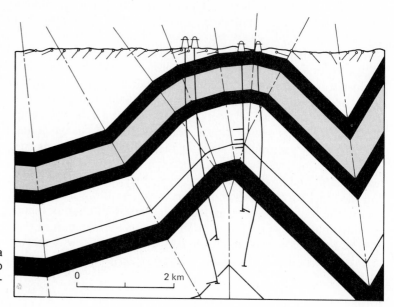

FIGURE 2–25 Approximation of a rounded fold crest by a series of sharp kinks; Chinshui anticline, western Taiwan. (Simplified from Namson, 1981.)

constructing the approximate shape by the kink method, the bed shape may be smoothed by hand if desired. The kink method predicts the same shape as the Busk method for truly concentric folds because the axial surfaces of the kink method intersect at the center of curvature, just as the bed normals do in the Busk method. Therefore, the Busk method has no fundamental advantage over the kink method even in the one case to which it is truly applicable. The kink method is also useful because it is relatively easy to retrodeform a structure composed of straight-bed segments.

The kink method may also be applied to folds exhibiting a sudden, but fixed, change in layer thickness across the hinge (for example, Fig. 9-10). Some folds exhibiting slaty cleavage display this property. In this case either the orientation of the axial surface or the ratio of bed thicknesses must be known in addition to the limb dips. The two axial angles, γ_1 and γ_2, are given by

$$\frac{\sin \gamma_1}{\sin \gamma_2} = \frac{T_1}{T_2} \tag{2-2}$$

where T_1 and T_2 are the thicknesses of single layers on the two limbs.

Structures that involve substantial heterogeneous flow—for example, many folds in metamorphic rocks (Chapter 9)—cannot generally be extrapolated with such simple methods as the Busk or kink methods. Nevertheless, these structures generally have a well-developed cylindrical axis, which is a basis for satisfactory extrapolation, especially if the fold axes plunge. Any observed point along a distorted layer may be extrapolated into a profile section by projection parallel to the fold axis, as already discussed under the heading "Cylindrical Nature of Folds" (see Fig. 2-5). A great variety of special methods of extrapolation and interpolation may be adapted, as appropriate, to specific structural situations.

Methods of Retrodeformation

If a structure is the result of deformation, it must be geometrically possible to retrodeform it to an earlier shape. Retrodeformation is, of course, possible only geometrically; rocks generally are incapable of mechanical retrodeformation. Therefore, if a cross section is not retrodeformable, either the rocks were originally cast in their present shape or the section is an impossible solution to the available data and is incorrect.

Direct graphical retrodeformation of cross sections, involving unfolding folds, unslipping faults, removing intrusions, and undistorting the rock, is not commonly attempted. Complete retrodeformation is particularly difficult in complexly distorted rocks, especially many metamorphic rocks, for which the distortion varies from point to point in the rock and must be measured in some way. This distortion is a special topic to be discussed in Chapter 3.

Structures that formed with only relatively simple or little internal distortion can be directly retrodeformed much more easily. For example, cross sections of parallel folds, constructed with the kink method, can be easily retrodeformed because the fold is composed of straight-line segments rather then curves (for example, Fig. 2-24). We must stress, however, that any retrodeformation requires either knowledge or assumptions about the initial undeformed shape of the rocks (see also Chapter 3).

Perhaps the most fundamental assumption we might make about retrodeformation is *conservation of mass;* the mass of the rock is conserved during the deformation. Even this assumption is invalid in some cases, through volume change caused by loss of pore water and dissolution (Chapter 10); however, in many cases mass as well as volume *(conservation of volume)* are approximately conserved during tectonic deformation because most compaction takes place during the first kilometer of initial burial. Volume is three-dimensional, which makes application of conservation of volume awkward in evaluating structural solutions, but not impossible. Fortunately, many problems need be considered only in cross section. For example, in the common case of cylindrical folds, the assumption of conservation of volume may be reduced to *conservation of area,* if there is little stretching or compression parallel to the fold axis. If the folds are parallel (constant bed thickness measured normal to bedding), the assumption is further reduced to *conservation of bed length.* With such assumptions, where appropriate, we are now in a position to make some practical test of retrodeformation.

To make use of conservation of bed length in checking structural interpretations, we must establish reference lines on either side of the cross section; these were presumably parallel before deformation. The reference lines are shown as nails in the schematic drawings at the top of Figure 2-26. After deformation, the bed lengths between the reference lines should all be equal. If the bed lengths are not equal, then the cross section is wrong, or at least one reference line has been deformed. In the bottom cross section of Figure 2-26, there is a bedding-plane fault (décollement) in the gypsum layer (Muschelkalk); therefore, the bed lengths change abruptly across this décollement. Reference lines do not generally exist naturally in the rock. They must be chosen, preferably in areas of no interbed slip—for example, horizontal beds, flat synclines, or possibly flat-topped anticlines (Fig. 2-26).

The constraints of conservation of area and conservation of bed length may be applied together. Consider a structure such as Figure 2-26 that involves folding above a décollement horizon. If bed length is conserved, we may measure the

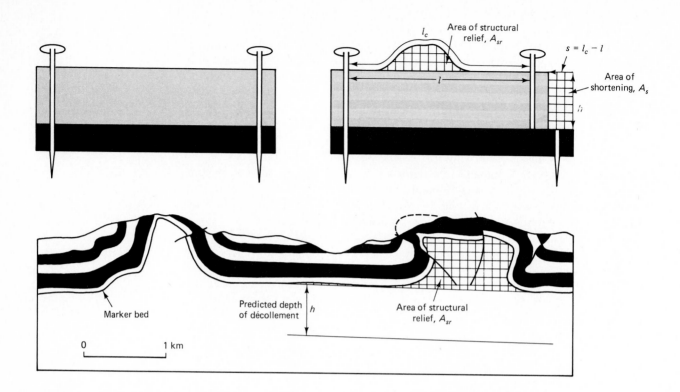

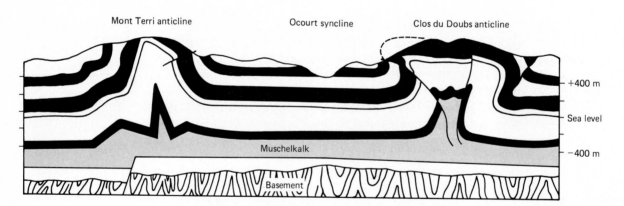

FIGURE 2–26 Relationship between area of structural relief of a marker bed above its undeformed level, A_{sr}; shortening, s; and depth to décollement, h (Eq. 2-3 and 2-4). In the middle cross section, the area of structural relief to the base of the lower Doggar Limestone in the Clos du Doubs anticline of the Jura Mountains is shown. With the additional measurement of the curvimetric shortening, the depth of décollement, h, in the evaporitic Muschelkalk (Triassic) may be calculated. A completed cross section is shown at the bottom. (Modified after Laubscher, 1961.)

shortening to be the difference between the straight-line distance between two points on a bed l and the curved distance measured along the bed l_c (Fig. 2-26); shortening measured in this way is called *curvimetric shortening*: $s_c = l_c - l$. Curvimetric shortening may be determined in cross section with a map-measuring device. In Figure 2-26 the curvimetric shortening is the same for every bed above the décollement.

Every point along a deformed bed in Figure 2-26 stands at or above the original elevation of the bed; the area between the original and present elevation of a bed is called the *area of structural relief, A_{sr},* for that bed. The area of structural relief may be measured with a planimeter or by counting squares on graph paper. If we have conservation of area and bed length, then the curvimetric shortening, s_c, and the area of structural relief, A_{sr}, are related as follows:

$$A_{sr} = s_c \cdot h \qquad (2\text{-}3)$$

where h is the undeformed depth of the décollement below the bed. This equation makes no additional assumptions about the details of the deformation above the décollement except that the shortening is the same in all layers. This assumption is known to be valid in many cases (Laubscher, 1965). This equation has several practical applications. If the area A_{sr} and the shortening s_c are known, the depth of an unknown décollement may be estimated (Fig. 2-26). If the depth h and the area A_{sr} are known, we may estimate a shortening s_a by

$$s_a = \frac{A_{sr}}{h} \qquad (2\text{-}4)$$

which we call the *planimetric shortening, s_a.* If both the curvimetric and planimetric shortening are independently known, they may be compared as a test of retrodeformability, as illustrated with an example from the Jura Mountains of Switzerland.

Figure 2-27(a) is a transverse cross section of a cylindrical fold in the Jura Mountains with at least part of the stratigraphic section displaying parallel folding. The area of the structural relief is 1.55 km^2 and the depth of the décollement is 0.6 km; therefore, the planimetric shortening is 1.55 km^2 ÷ 0.6 km ≅ 2.6 km. In contrast, the curvimetric shortening from bed length measurements is 0.95 km. Therefore, the cross section in 2-27(a) is not retrodeformable; it apparently has too much area of the lower stratigraphic unit. An alternative cross section, Figure 2-27(b), satisfies the same geologic data and has concordant planimetric and curvimetric shortening of about 2.0 km. This second solution, involving the stepping up of a major thrust fault, is therefore retrodeformable (see Laubscher, 1965, for other examples).

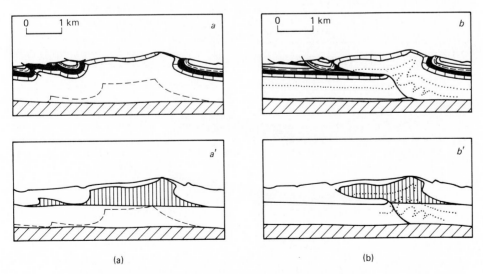

(a)   (b)

FIGURE 2–27 Two structural interpretations of the Bourrignon anticline in the Jura Mountains. Cross section *a* is internally inconsistent, whereas cross section *b* is consistent, as discussed in the text. (After Laubscher, 1965.)

Retrodeformation of Faults

Imagine a fault before it even exists, before it has undergone any displacement. The geologic structures, including all bedding, intrusive contacts, folds, and older faults, pass continuously through the nonexistent fault surface. Every point along the fault is in contact with an immediately adjacent point on the opposite wall of the fault. After the fault has slipped, each pair of points is separated by the amount and direction of the fault slip. The line in the plane of the fault connecting a pair of points is called the *net slip* for that pair of points.

The net slip is not easy to determine. Each measured net slip requires that we locate, in three dimensions, two points in the rock on opposite sides of the fault that were together before the fault existed. In practical cases we must identify a geologic line—for example, the intersection of a specific identifiable bed with the dike in Figure 2-28—to locate two points that were originally connected prior to fault slip. The geologic line pierces the fault prior to slip, thereby defining the two points, which are therefore called *piercing points* (Fig. 2-28). Other geologic lines that may be used to define the net slip include hingelines of preexisting folds on an identifiable bed, angular unconformities overlapping identifiable geologic contacts, and changes in rock type along a layer of specific age (facies change)—for example, a shoreline-beach facies of specific age. In practical problems, the geologic lines should intersect the fault at a relatively high angle in order to measure the net slip accurately. If the geologic line, such as a shoreline facies, weaves back and forth across the fault, it may be uncertain how many piercing points there are and which points actually match.

The net slip lies within the plane of the fault. Therefore, we might expect that net slip generally cannot be observed and measured in cross section because the cross section may not contain any matching piercing points. The offset of a geologic contact along a fault, as viewed in cross section or map view, is only an *apparent offset* and is not in general the net slip. This is because a geologic contact is a surface in three dimensions and its intersection with a fault is a line before slip and two offset lines after slip (Fig. 2-28). Each line is called a *cutoff line,* or simply *the cutoff.* If a fault is not vertical, the lower fault block is called the *footwall,* and

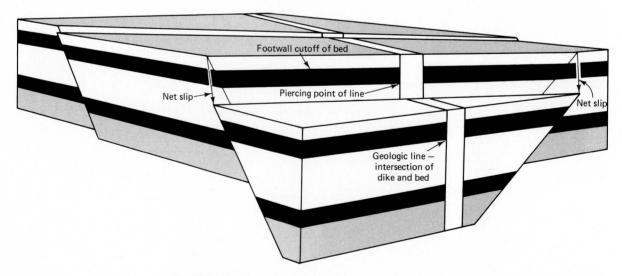

FIGURE 2–28 The net slip on a fault is the line joining two points on opposite walls of the fault that were originally connected. In practice, net slip is determined by determining the points on opposite walls of a fault where a geologically defined line pierces the fault—for example, the intersection of a dike and a bed.

the upper fault block is called the *hanging wall;* in mining, the footwall is the side of the fault you stand on. Therefore, the cutoff line of a formation on the hanging wall is called the *hanging-wall cutoff;* on the footwall it is the *footwall cutoff* (Fig. 2-28). We do not know which points along the hanging wall and footwall cutoffs originally were joined. Only if we independently know or are willing to assume the direction of slip along the fault are we able to determine the net slip from offset cutoff lines of geologic contacts.

It is only in cross sections drawn parallel to the slip vector that the apparent offset of a geologic contact is equal to the net slip. Fortunately, many faults have slip vectors parallel or nearly parallel to the dip direction. These are called *dip-slip* faults and include normal and thrust faults. The hanging wall slips downdip relative to the footwall in *normal faults;* the hanging wall rides updip over the footwall in *thrust faults*. If a cross section is drawn perpendicular to the strike of a normal or thrust fault, then the slip vectors probably lie approximately within the plane of the cross section; therefore, the apparent offset of the hanging wall and footwall cutoffs is a good estimate of the net slip. In making such estimates we should realize that we are in effect assuming slip direction unless we have direct evidence of dip-slip motion.

Steeply dipping faults with horizontal slip are called *strike-slip faults;* they, together with normal and thrust faults, compose the three principal categories of simple faults (Chapter 7). Two fundamental geometric classes of strike-slip faults exist based on the sense of slip: right lateral and left lateral. Slip is *right lateral* if an observer standing on one side of the fault observes the opposite side moving to the right. Conversely, if the opposite side moves to the left, the slip is *left lateral*. Strike-slip faults are the only common class of simple faults that do not display the net-slip vectors as apparent offsets in standard vertical cross sections drawn perpendicular to strike of the fault. Estimates of net slip on strike-slip faults are best made with a pair of vertical cross sections drawn parallel to the fault surface on each side. The details of the geology can then be matched.

The most obvious and common mistake of nonretrodeformable cross sections involves faults that cannot be slipped back to an undeformed state (for example, Fig. 2-19). Checking fault slip for retrodeformability is reasonably straightforward in sections that contain the net-slip vector, but it is commonly difficult in other, arbitrary sections. For this reason it is very helpful to draw cross sections through thrust and normal faults perpendicular to the fault strike.

In many cases, specific piercing points cannot be determined from available geologic data. For example, apparent offsets of contacts in map view may be all that can be determined. Nevertheless, useful and enlightening estimates of fault

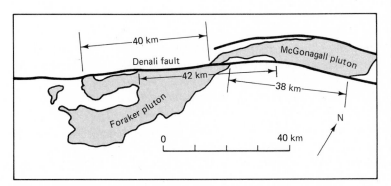

FIGURE 2–29 An example of estimation of net slip without precise identification of piercing points. The McKinley segment of the Denali fault in Alaska is generally hidden in broad, glacier-filled valleys. Nevertheless, the contacts between the granodiorites of the McGonagall and Foraker plutons and their sedimentary and metamorphic country rock allow estimation of the horizontal component of the net slip because equivalent contacts all have apparent offsets of 40 ± 2 km. The two plutons have nearly identical mineralogy, chemistry, and Oligocene radiometric dates. (Modified after Reed and Lanphere, Geol. Soc. Amer. Bull., v. 85, p. 1883–1892, 1974.)

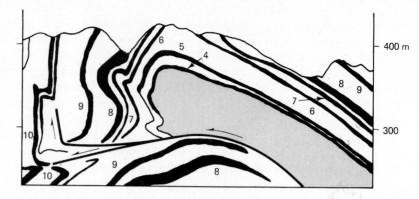

FIGURE 2–30 Example of nonconstant fault slip as a result of displacement being taken up by folding and subsidiary faulting. Coal beds 4 through 9 show over 200 m of slip, whereas bed 10 shows almost no slip. Suffolk fault, anthracite coal basin, Pennsylvania. (Modified from Danilchik, Rothrock, and Wagner, 1955.)

slip based on circumstantial evidence can be made in some cases. For example, Figure 2-29 shows a segment of the Denali fault zone near Mt. McKinley in Alaska, where the fault is generally hidden from direct observation by broad, glacier-filled valleys and low passes covered by surficial deposits. Nevertheless, the contacts between two Oligocene granodiorite plutons and their sedimentary and metamorphic country rock suggest they may be offset fragments of an originally single pluton. The circumstantial evidence is that the two plutons have nearly identical mineralogy, chemistry, and radiometric dates and that their contacts with the country rock have nearly identical apparent offset, about 40 km right lateral. Therefore, the Denali fault, which is generally considered to be a right-lateral strike-slip fault, probably has about 40 km of post-Oligocene, right-lateral slip in this region, even though we have not located any piercing points and do not know the dips of the contacts.

Cross sections are sometimes constructed or checked with the assumption of constant slip along a fault. This assumption may be valid in certain specific cases, but in general faults do not have constant slip. As the net slip changes along a fault surface, the changes in slip are taken up in bulk deformation of the fault blocks or by transfer of slip to other faults or folds. One important cause of nonconstant fault slip is nonplanar faults; as the two blocks of a nonplanar fault slip past one another, there must be distortion of at least one of the blocks because rocks are not strong enough to support major voids along faults. Another important cause of nonconstant slip is folding at the tip of propagating thrust faults, which is called *fault-propagation folding* (Chapter 9). For example, Figure 2-30 is a cross section showing an abrupt change in slip along the Suffolk fault in the Pennsylvania Appalachians because of fault-propagation folding.

Cross sections may be checked across strike, based on the consistency of fault slip from section to section in a series of closely spaced transverse cross sections. (Dahlstrom, 1969, Laubscher, 1965).

STRUCTURE-CONTOUR MAPS AND BLOCK DIAGRAMS

If a structure is especially complex in three dimensions or if precise visualization of the structure is especially important for practical purposes, such as mining or petroleum exploration, cross sections alone may be insufficient to portray the important aspects of the structure fully. In these special cases other visual means, especially block diagrams and structure-contour maps are sometimes employed.

Block diagrams are simply three-dimensional cross sections portraying the structure on the faces of the block (for example, Fig. 2-31). Block diagrams are easily drawn if sections have already been constructed; the cross sections are simply redrawn in perspective or using other projections. A variety of types of

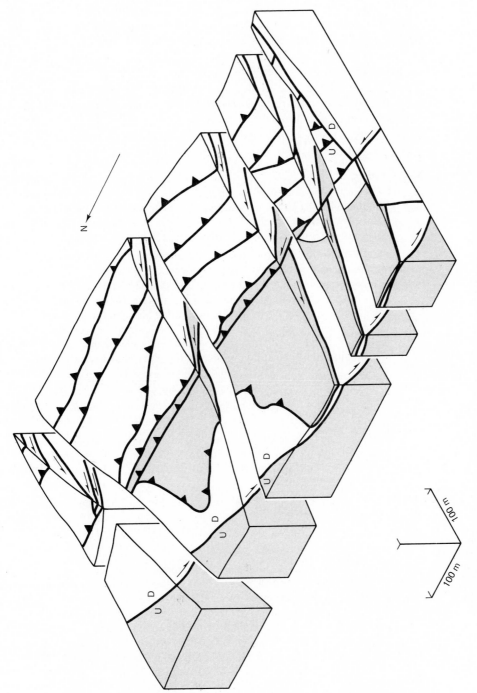

FIGURE 2-31. Example of a block diagram, showing complex fault geometry in The Decaturville, Missouri, meteorite-impact structure. (Simplified after Offield and Pohn, 1979.)

100 m

100 m

N

U
D

U
D

U
D

U
D

block diagrams exist. One particularly useful type portrays the shape of a single, deformed surface in perspective.

Structure-contour maps are contour maps that represent the shape of a structure by showing the elevation of every point on a deformed surface, generally a bed, fault, or intrusive contact. Figure 7-32 is a structure-contour map of a salt dome. Figure 1-8 is a regional structure contour map of the top of the Precambrian basement in the North American craton. If extensive drilling data exist, structure-contour maps can be easily constructed by directly contouring the measured depths of the chosen bed. In most cases, however, closely spaced cross sections are first constructed, the depths of the chosen bed are measured on the cross sections, and the data are transferred to map view and contoured. In petroleum exploration and development, structure-contour maps are commonly made for each reservoir horizon in a structure to estimate the position of the crest of the structure and the possible volume of hydrocarbons.

EXERCISES

2–1   Plot the following strikes or bearings as dots on a spherical projection: 162°, N35°W, S80°E, N2°E, 273°. Plot the following bearings and plunges and dip directions using a polar net: 162°, 30°; N35°W, 65°; S80°E, 4°; N2°E, 85°; 273°, 122° (overturned beds).

2–2   Plot the following dips and strikes as great circles, dip directions, and as poles using a meridional net: N38°W, 43°N (older beds) and 163°, 60°E (unconformably overlying beds). What is the bearing and plunge of the line of intersection of the older beds with the unconformity and the angle between them? Assuming that the angle between the beds did not change during tilting of the unconformity, what was the dip direction and dip of the older beds when the younger beds were first deposited? Solve by rotating the younger beds to horizontal about their strike line and simultaneously rotating the pole of the older beds about the same axis.

2–3   If a sequence of beds is oriented N25°E, 75°SE (strike and dip), what will be their apparent dip in a vertical cross section striking N55°E? Solve this problem using a spherical projection, representing the orientation of the bedding and the section both as great circles.

2–4   Consider a bed whose surface is not exposed, but the trace of bedding is exposed on joint faces. What is the orientation of the bed if its apparent dip is 26°SW on a vertical joint face striking N44°E and its rake is 28°SE on an adjacent joint face oriented N56°W, 30°N?

2–5   Complete the unfinished cross section in Figure 2-32 using the kink method of extrapolation.

2–6   Determine the strike and dip of the thrust fault shown in Figure 2-33 using the methods of the three-point problem.

2–7   Groove casts are linear marks on the bottoms of beds that record the orientation of the current that formed the bottom mark. Groove casts are important in determining current directions; however, if the beds are folded or tilted, the beds—together with their groove casts—must be rotated to horizontal, usually using a spherical projection, before the horizontal current direction may be determined. The strikes and dips of bedding and associated rakes of groove casts are as follows: N38°W, 90°; 70°N; N29°W, 24°E; 70°N; N38°W, 58°E; 86°N; N38°W, 46°E; 88°N; N36°W, 66°E; 88°S; N42°W, 66°W; 88°N; N44°W, 40°W; 85°N; N36°W, 78°E; 62°N; N25°W, 20°W; 70°N; N46°W, 80°E; 90°; N40°W, 82°W; 80°N; N34°W, 80°E; 78°N. Find the fold axis and the current direction for these folded sediments.

2–8   In paleomagnetic studies of folded strata, it must be determined whether the rocks were magnetized before or after folding. This question is commonly answered with a *fold test,* in which the magnetization vectors are rotated with the beds to horizontal; it is observed whether the magnetization vectors are more tightly clustered in the folded or unfolded state. For the following eight sites in a folded terrain, dip

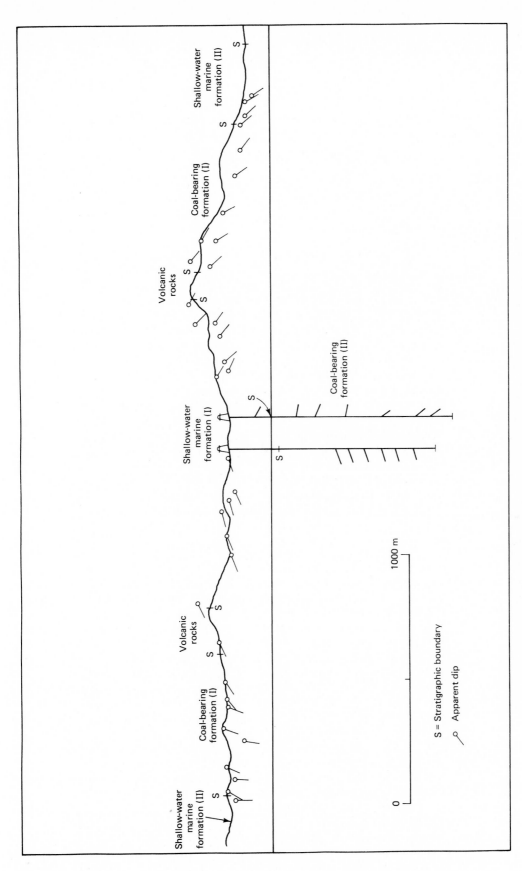

FIGURE 2-32 Unfinished cross section of the Shantzechiao anticline, Taiwan, to be completed as Exercise 2-5. Use the kink method of extrapolation, first choosing four or five domains of roughly constant dip. Bisect the angles between the domains to determine the orientations of the axial surfaces. The positions of the axial surfaces may be adjusted to best fit the data, particularly to get equivalent beds to match across the crest of the anticline. Parallel straight lines may easily be drawn using two triangles, one held fixed and one movable.

73

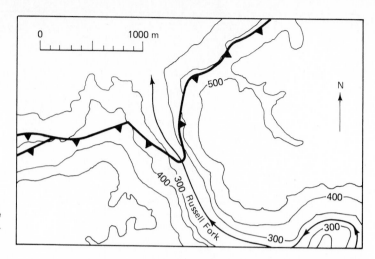

FIGURE 2–33 Map of a thrust fault to be
used in Exercise 2-6. The contour interval is 100 m.

directions and associated magnetization vectors are given: (331°, 05°; 14°, 16°),
(337°, 22°; 12°, 10°), (08°, 48°; 54°, 43°); (89°, 68°; 180°, 53°), (26°, 56°; 85°, 20°), (109°,
48°; 160°, 32°), (136°, 32°; 174°, 20°), and (147°, 07°; 357°, 01°). Is there any reason to
believe these rocks were magnetized before folding? The present latitude of these
rocks is 46°N; what is the apparent change in latitude since magnetization assuming
(1) postfolding magnetization and (2) magnetization when flat-lying? Note that the
relationship between the plunge P of the magnetization vector and the latitude θ of
the rock during magnetization can be approximated by $\tan P = 2 \tan \theta$, which holds
for a uniformly magnetized sphere. If the apparent change in latitude is due to
continental drift, what is the minimum displacement in metric units? Is any other
motion suggested by the data?

2–9   If an unoriented core is recovered from a drill hole, the true orientation of the
layering is ambiguous. The set of all possible orientations of the normal to the
layering, for example, describes a cone about the axis of the drill hole with a half
angle equal to the apparent dip of the layering in the core. This cone of possible
orientations may be represented on an equal-angle projection as a small circle. The
center of the circle on the sphere is the axis of the bore hole (axis of the cone) and
the angular radius of the small circle is the half-angle of the cone. Nevertheless, the
center of the small circle on the equal-angle projection is not the axis of the cone
because of distortion in the projection. A small circle of radius r about a pole p is
constructed on an equal-angle projection as follows. Plot the pole p and measure off
a distance r to both sides of p along a single great circle passing through p. These two
measured points, r_1 and r_2, lie on the circumference of the small circle on opposite
sides of a diameter, on both the sphere and the projection. The center of the circle on
the sphere is p. The center of the circle on the projection is the midpoint of the
straight line connecting the two points r_1 and r_2; once the center is located, the small
circle is constructed with a compass. This small circle represents the locus of all
directions of angular distance r from the pole p. Problem: Given three nonparallel
drill holes encountering the same unfolded layer, what constraints can be placed on
the orientation of the layer if the bearings and plunges of the holes and apparent dips
of the layer on the cores are as follows: (N46W, 45; 20), (N70W, 20; 28), (S36W, 38;
49)?

2–10   The Philippine Sea plate is presently rotating relative to the Asian plate about the
pole 45°N, 150°E at a rate of 1.2°/m.y. in a clockwise direction, if the pole is viewed
from above. Assuming the Asian plate is fixed and the pole and rate of rotation are
constant, what was the position of the island of Luzon (17°N, 122°E) on the
Philippine Sea plate 20 m.y. ago? Use the methods of Exercise 2-9. What would be
the present paleomagnetic pole of a rock on Luzon magnetized 20 m.y. ago (see
Exercise 2-8)? Note that angles between lines on the surface of a sphere are
preserved on an equal-angle projection.

SELECTED LITERATURE

Bishop, M.S., 1960, *Subsurface Mapping,* Wiley, New York, 198 p.

Busk, H.G., 1929, *Earth Flexures,* Cambridge University Press, Cambridge, 106 p.

Dahlstrom, C. O. A., 1969, Balanced cross sections: Canadian J. Earth Sci., v. 6, p. 743–757.

Elliot, D., and Johnson, M. R. W., 1980, Structural evolution of the northern part of the Moine thrust belt, N.W. Scotland: Transactions Royal Society of Edinburgh: Earth Sci., v. 71, p. 69–96.

Gwinn, J. E., 1970, "Kinematic patterns and estimates of lateral shortening, Valley and Ridge and Great Valley provinces, south-central Pennsylvania," in Fischer, G. W., Pettijohn, F. J., Read, J. C., Jr., and Weaver, K. N., ed., *Studies of Appalachian Geology,* Central and Southern (Cloos Vol.), Wiley-Interscience, New York, p. 127–146.

Higgs, D. V., and Tunell, G., 1966, *Angular Relations of Lines and Planes,* W. H. Freeman, San Francisco, 43 p.

Laubscher, H. P., 1965, Ein kinematisches Modell der Jurafaltung: Eclogae Geologicae Helvetiae, v. 58, p. 231–318.

Mackin, J. H., 1950, The down-structure method of viewing geologic maps: J. of Geol., v. 58, p. 55–72.

Ragan, D. M., 1973, *Structural Geology, an introduction to geometrical techniques,* Wiley, New York, 208 p.

Tucker, P. M., and Yorston, H. J., 1973, Pitfalls in seismic interpretation: Society of Exploration Geophysicists Monograph 2, 50 p.

Turner, F. J., and Weiss, L. E., 1963, *Structural Analysis of Metamorphic Tectonites,* McGraw-Hill, New York, 545 p.

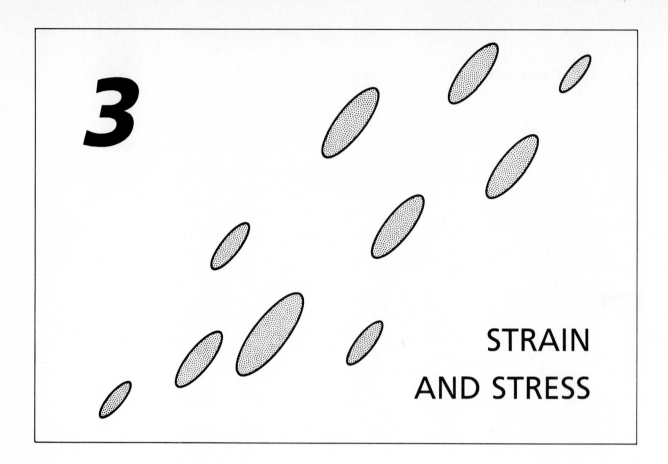

3

STRAIN
AND STRESS

INTRODUCTION

We need to fortify ourselves with a basic understanding of the deformation of solids before we proceed much further with questions of why and how rocks flow, buckle, fracture, and slide past one another in the great deformed belts of the earth. Mechanical principles will also help us understand why cratons and ocean basins experience so little deformation. Let's step back and consider three basic questions of the deformation of solids, which are the subject of Chapters 3, 4, and 5.

First, what do we really mean by deformation? How do we know a rock is deformed and how do we describe this deformation? These are essentially questions in geometry.

Consider an object such as the *Olenellus* trilobite in Figure 3-1, which has been deformed with the rock containing it. We know the original shape of *Olenellus* because it has been found in undeformed rocks. If we abstractly call the original form of an object $\mathscr{X}$ and its final form $\mathscr{X}'$, then we might write $\mathscr{X}' = \mathbf{E}\mathscr{X}$, where $\mathbf{E}$ is something that describes the change in size and shape. If we could multiply the original shape $\mathscr{X}$ by $\mathbf{E}$, we would get the final shape $\mathscr{X}'$ as the answer.

The trilobite has quite a complex shape, but the rock itself may have undergone a rather simple and uniform deformation—it has been shortened in the 1 direction and elongated in the 2 direction in Figure 3-1. But suppose the trilobite had originally been oriented parallel to the 1 direction; in this case, it would now be short and wide. If it originally had been perpendicular to the 1 direction, it would now be long and narrow. In each case the overall deformation of the rock is the same, even though the new shape of the fossil is different. The trilobite is important in this context only because we know its original shape. It allows us to

FIGURE 3–1 A deformed *Olenellus* trilobite from Lancaster Valley, Pennsylvania Appalachians. Principal-stretch directions are shown.

determine **E**, which is what we really want to know. No matter what the object, **E** is the same.

The relationship **E** between the initial and final size and shape of a solid is what is meant by strain, a major concern of this chapter. Strain plays a central role in answering what deformation is and how we measure it.

The second concern of the chapter is stress; it plays a central role in answering our second question: What makes a solid deform?

Something has forced the atoms of our trilobite to slip past one another until they arrived at their present relative positions. Since the deformation is permanent, it required breaking and re-forming many bonds. Whether or not a bond breaks depends upon the intensity and orientation of the forces acting on it. We would like to know these interatomic forces, but it would be impractical to consider individually, for instance, all 10^{22} bonds in a cubic centimeter of rock. Some average measure of the interatomic forces is needed, and for this we shall use stress.

Stress, σ, is the force exerted by all the atoms of one side of any arbitrary plane within a body on the atoms immediately on the other side, divided by the area of the plane. Stress is proportional to the average force on a bond because area is proportional to the number of bonds. We shall see later that stress has a number of special properties that are not immediately apparent here.

If strain describes the change in size and shape of a solid in response to forces and if stress is a measure of the intensity and orientation of these forces, what actually goes on physically—microscopically and submicroscopically—to produce the strain? That is, what are the deformation mechanisms? This is our third question, pursued in Chapter 4 (devoted to mechanisms of distributed deformation) and Chapter 5 (devoted to fracture).

Specific deformation mechanisms include brittle fracture of mineral grains, elastic deformation, slip along atomic planes within crystals or along grain boundaries, rigid rotation of grains, and diffusion of material from regions of high stress to regions of low stress. Several mechanisms generally operate simultaneously, and they may be dependent or independent of one another. For example, the sliding of one grain past another in a rock may depend on the mechanism of deformation at the end of the sliding surface. If two simultaneous mechanisms are independent, the deformation is controlled by the fastest; if the mechanisms are interdependent, the deformation is controlled by the slowest. Changes in factors such as temperature, pressure, chemical environment, or rate of deformation result in changes in the dominant deformation mechanism. We can often establish which mechanisms have operated during the deformation of a rock by examination under an optical or electron microscope.

We would like to establish relationships between the deformation of rocks and the stresses responsible. To do this we need stress-strain relationships, also known as rheological models or constitutive relationships, that are valid for the particular materials. These relationships are equations that define the mechanical behavior of an idealized material and may be empirical or theoretical in origin. They describe the net effect of all the deformation mechanisms. A well-known example is viscous behavior for which stress is proportional to strain rate ($\sigma = \eta\dot{\epsilon}$, where σ is stress; η is the constant of proportionality, that is, the viscosity; and $\dot{\epsilon}$ is the strain rate, the dot signifying rate). Another example is elastic behavior, for which stress is proportional to strain ($\sigma = E\epsilon$, where E is the constant of proportionality, Young's modulus). More-complicated mathematical relationships between stress and strain can of course be written and do approximate the more complex behavior of many rocks.

We are mainly concerned with stress-strain relationships because they allow us to make mathematical or physical models of the deformation of rocks. For example, we can compare the geometry or strain of a model fold with the geometry of real folds and see if the deformation mechanisms of the real fold are consistent with the model stress-strain relations. Our use of stress-strain relations is different from that in rock and soil mechanics, which are engineering disciplines. The engineering problem is to predict the deformation in response to a load that will be applied; often the goal is to prevent failure of a structure. In contrast, our problem is to take deformed rocks post-mortem and to learn how and why they got that way.

It should be noted that stresses act instantaneously; a deformed rock is the result of a whole history of stresses of changing magnitude and orientation. Strain, in contrast, is the sequential summation of all the instantaneous deformations due to the instantaneous stresses. For this reason there is, in principle, not enough information contained within the strain of a deformed rock to reconstruct the entire history of stresses that produced the deformation.

GENERALITIES

Physical properties of materials and fields, such as electric fields, exist apart from any reference frame or coordinate system with which we may wish to consider them. In fact, a coordinate system is just a fancy ruler and direction finder. We shall use a Cartesian coordinate system with axes labeled 1, 2, and 3 (Figure 3-2). For special problems that have intrinsic symmetries, other systems to locate positions could be useful—for example, cylindrical, polar, or curvilinear orthogonal. Regardless of the system or its orientation, three numbers are needed to locate each point in space; for us these are the coordinates X_1, X_2, and X_3. Rather than write out all three coordinates, we shall generally write X_i, or sometimes X_j, to signify the location of a point, implying that the subscripts i or j take on all three

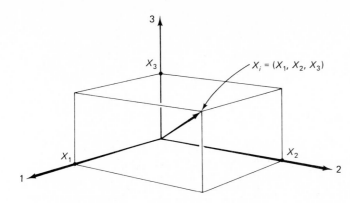

FIGURE 3-2 Location of the position of a point X_i by its coordinates X_1, X_2, and X_3 in a Cartesian coordinate system.

values 1, 2, and 3; that is, X_i means (X_1, X_2, X_3). The position of any point X_i may be thought of as a vector, or arrow, reaching from the origin of the coordinate system ($X_1 = 0$, $X_2 = 0$, $X_3 = 0$) to the point in question (Fig. 3-2).

There is a specific value of any physical property or field at each point in space. Depending on the nature of the phenomenon, one or more numbers are required to define the property at each point. For example, only a single number is needed to define temperature or density at each point in space, whereas three numbers are needed to define heat flow or gravitational field because they have a direction. Six independent numbers are required to define strain, thermal or electrical conductivity, and stress. Some physical properties, such as elasticity, may require as many as 21 independent numbers to be completely defined.

Single-number properties (scalars) might be visualized as a dot of specific darkness (magnitude) at each point in space. A photographic negative is, for example, a scalar field of opacity that represents a scalar field of light intensity. Three-number properties (vectors) can be thought of as arrows of specific lengths and orientations at each point in space; a map of ocean currents is an example of a vector field of velocity. A six-number property (tensor) can be visualized as an ellipsoid of specific size, shape, and orientation at each point in space. A map of a strain field may be represented by ellipsoids (see Fig. 10-9). Twenty-one–number properties also have geometric representations, but they are quite involved, and—in contrast with three- and six-number properties—they will be of no particular use to us here. Because stress and strain are six-number properties, they are intrinsically somewhat complicated and cannot be oversimplified without losing the power they have to offer in understanding the deformation of rocks.

A material property or field is *homogeneous* if it has the same value or set of values at every location in a body. The material property is called *isotropic* if it is independent of direction at each point.

STRAIN AND RIGID MOTION

We shall now introduce the geometry of deformation in a relatively precise manner, for two principal reasons. First, we want to be able to think about it clearly. Second, we may wish to measure the strain in rocks to help understand how they were deformed.

Strain, in the most general sense, is a relationship between the size and shape of a body before and after deformation. This relationship may be either homogeneous or inhomogeneous, as illustrated in Figure 3-3. Any precise treatment of inhomogeneous strain is difficult mathematically, and in practical problems it is usually approximated by a number of small homogeneous domains. We shall deal only with homogeneous strain directly.

FIGURE 3–3 Homogeneous and inhomogeneous strain.

Many measures and definitions of strain have been developed for specific problems, but all can be generalized to the definition given above. Here are a few examples of commonly used measures of strain. Consider, in one dimension, a line of length L that is stretched to length L' (Fig. 3-4). One of the most important measures of the stretching is the *extension* $\epsilon = (L' - L)/L$. Another measure important in structural geology is *stretch,* $S = L'/L = (1 + \epsilon)$. Still another measure useful in some problems is *quadratic elongation,* λ, defined by $\lambda = (L'/L)^2 = (1 + \epsilon)^2$. Yet another measure of the stretching is *natural,* or *logarithmic, strain,* $\bar{\epsilon} = \log_e(L'/L) = \log_e(1 + \epsilon)$. These examples emphasize that several alternative measures of strain are possible; which to use is a question of pragmatism.

Next let's consider, in two dimensions, a line of length L that is bent sideways (see Fig. 3-4) with one end displaced a distance D relative to the other, the final line being L'. The angle between the initial and final line is ψ. A measure of this sideways deflection is the *angular,* or *engineering, shear strain,* $\gamma = D/L = \tan \psi$, where ψ is the *angular shear.*

Finally, consider a body of volume V deformed to a new volume V'. A measure of the change in volume is the *dilation* $\Delta = (V' - V)/V$. Another measure of the change in volume is the *volume ratio* V'/V, which is a volumetric analog of stretch. In all these examples we see that strain provides a link between an initial and a final state and that strain is a dimensionless quantity. Strain is sometimes expressed in percent.

We now look at three-dimensional strain in quite a general way to learn the relationship between strain and displacement. This requires some care and attention to details; it is most easily accomplished using a mathematical notation, introduced later, that allows us to focus on the physical meaning of each equation rather than the many algebraic details. This notation will also help when we consider stress and stress-strain relations. Some readers, less accustomed to the mathematics, may wish to skim the next few pages for a first reading and return later for a more detailed inspection.

We need to distinguish four independent geometric processes that contribute to the total displacement of any part of a material during deformation; they are *rigid-body translation, rigid-body rotation, distortion* (change in shape), and

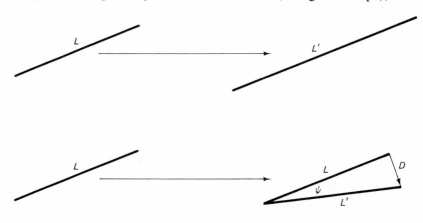

FIGURE 3–4 Deformation of a line from initial length and orientation, *L,* to a final one, *L'.*

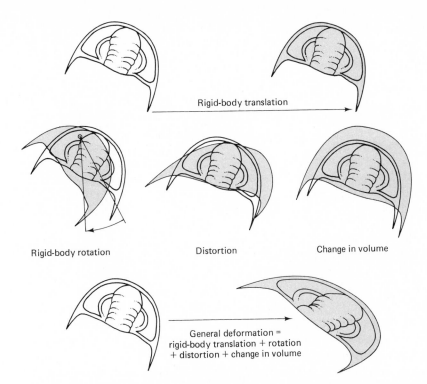

Rigid-body translation

Rigid-body rotation   Distortion   Change in volume

General deformation =
rigid-body translation + rotation
+ distortion + change in volume

FIGURE 3–5 Illustration of the fact that displacement in a general deformation is the sum of four geometric processes: rigid translation, rigid rotation, distortion, and change in volume.

change in volume (see Fig. 3-5). The last two, which together compose the strain, involve relative motions of adjacent atoms. In contrast, the first two involve only *rigid-body motion* with respect to an external reference. Each of these geometric processes is separable; for example, the homogeneous distortion recorded by the fossil in Figure 3-1 tells us nothing about rigid translation or rotation, either during deformation or since removal from the outcrop.

In the following paragraphs we shall separate displacement into its geometric parts and thereby obtain a relationship between displacement and strain. This relationship is important because it allows us to calculate the change in shape of an object due to any strain or allows us to measure the strain recorded by distorted fossils or other objects of known initial shape.

First, we must be able to specify the shape of an object. Let us consider a body atomistically as a continuous array of points. If we establish a coordinate system, we may then represent the position of some arbitrary atom or material point within the body by X_1, X_2, and X_3 (or X_i for short, where i takes on the values 1, 2, and 3); see Figure 3-2. The size and shape of any object, such as a fossil, in the vicinity of this material point X_i may be described by an appropriate set of lines or vectors $\langle dX_i \rangle$ emanating from the point X_i and extending to specific points on the object whose positions we wish to specify (see Fig. 3-6). The brackets, $\langle \; \rangle$, are used to signify the entire *set* of lines or vectors; thus $\langle dX_i \rangle$ signifies all the points that we specify on an object. For the moment we need only consider one member of the set, some point or vector dX_i (with components dX_1, dX_2, dX_3)*, because all other members will have analogous behavior. Thus we have greatly simplified the problem of describing the deformation of an object. We need only learn how to describe the deformation of a line segment or vector dX_i in three dimensions.

The material point X_i is displaced to a new position X_i' (that is, $X_i' = X_1'$, X_2', X_3') as a result of any specific deformation, and the attached material line segment

---

*If $Y_i = (Y_1, Y_2, Y_3)$ is the end point of the vector dX_i, then the components of the vector may be calculated as $(Y_1 - X_1, Y_2 - X_2, Y_3 - X_3)$.

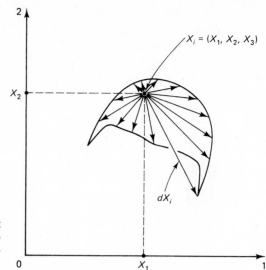

FIGURE 3–6 The shape of an object may be specified as a set of vectors $\langle dX_i \rangle$ radiating from an arbitrarily chosen material reference point X_i.

or vector dX_i is both stretched and rotated to become a new vector, dX_i' (see Fig. 3-7). In analogous fashion, the entire set of vectors, $\langle dX_i \rangle$, will be stretched and rotated to become a new set of vectors, $\langle dX_i' \rangle$, that describes the size, shape and orientation of the original object, $\langle dX_i \rangle$, after it is deformed.

Let us now look at the displacement in more detail. The material point X_i undergoes a displacement $(U_0)_i$ during deformation, which is just the final position minus the initial position:

$$(U_0)_i = X_i' - X_i \tag{3-1}$$

The displacement $(U_0)_i$ of the point X_i is common to all the material lines or vectors in the body $\langle dX_i \rangle$; thus it is called *rigid-body translation*, as is shown in Figure 3-7.

We now consider how the displacement U_i of the material point at the end of any vector dX_i differs from the displacement of the point X_i. The displacement of the endpoint is just the rigid-body translation $(U_0)_i$ plus an additional displacement $E_{ij}dX_j$ that depends upon its position in the material (see Box 3-1 for a discussion of the mathematical notation):

$$U_i = (U_0)_i + E_{ij}dX_j \tag{3-2}$$

The set of nine numbers E_{ij} describes all motion that is not rigid-body translation, that is, all displacement that varies with position. For this reason the terms of E_{ij} are called the *displacement gradients*.

FIGURE 3–7 The problem of describing the deformation of an object can be reduced to the problem of describing the displacement of a material reference point, X_i, and the deformation of any line segment, dX_i, because the entire object can be described as a set of analogous line segments $\langle dX_i \rangle$ (see Fig. 3-6).

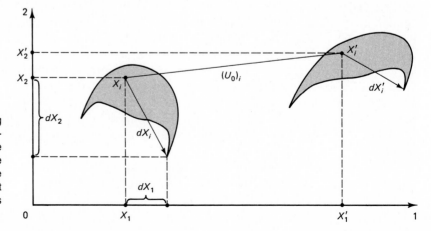

BOX 3–1   Einstein Summation Convention and Matrix Notation

The notation we are using is a shorthand way of writing many equations with many terms. It is helpful because it allows us to focus more on the physical meaning of the basic equation than on all the individual equations and terms. The notation should be memorized because we shall use it from time to time. It is also common in the literature of structural geology. Equation 3-2,

$$U_i = (U_0)_i + E_{ij}dX_j \qquad i, j = 1, 2, 3$$

represents the following three equations for the components of displacement **U** in the 1, 2, and 3 directions:

$$U_1 = (U_0)_1 + E_{11}dX_1 + E_{12}dX_2 + E_{13}dX_3$$
$$U_2 = (U_0)_2 + E_{21}dX_1 + E_{22}dX_2 + E_{23}dX_3$$
$$U_3 = (U_0)_3 + E_{31}dX_1 + E_{32}dX_2 + E_{33}dX_3$$

The letter subscripts in Equation 3-2 take on all possible values (in this case, 1, 2, and 3, referring to the different coordinate directions). When a letter subscript occurs twice in the same term, summation with respect to that subscript is automatically understood. For example, j is repeated twice in the term $E_{ij}dX_j$, which means $E_{i1}dX_1 + E_{i2}dX_2 + E_{i3}dX_3$ for each of the three equations ($i = 1, 2, 3$); j is called a *dummy subscript* and has no meaning beyond indicating summation. This is called the *Einstein summation convention*, named for Albert Einstein, who introduced it.

As an example of the meaning of the individual terms, $E_{21}dX_1$ is that part of the displacement in the 2 direction that depends on or varies with position in the 1 direction. Thus U_2, the displacement of a vector or point dX_i in the 2 direction, is the rigid translation of the entire body in the 2 direction $(U_0)_2$ plus those parts of the displacement in the 2 direction that vary with position in the 1, 2, and 3 directions:

$$U_2 = (U_0)_2 + E_{21}dX_1 + E_{22}dX_2 + E_{23}dX_3$$

It is also sometimes useful to write such equations in matrix notation:

$$\begin{bmatrix} U_1 \\ U_2 \\ U_3 \end{bmatrix} = \begin{bmatrix} (U_0)_1 \\ (U_0)_2 \\ (U_0)_3 \end{bmatrix} + \begin{bmatrix} E_{11} & E_{12} & E_{13} \\ E_{21} & E_{22} & E_{23} \\ E_{31} & E_{32} & E_{33} \end{bmatrix} \begin{bmatrix} dX_1 \\ dX_2 \\ dX_3 \end{bmatrix}$$

is equivalent to $U_i = (U_0)_i + E_{ij}dX_j$.

Once again, our notation is a shorthand way of writing many equations and variables, which allows us to focus on the physical meaning of the equations. Actual calculations with such equations are, of course, cumbersome and quite time-consuming if done by hand; they are more easily done with a calculator capable of matrix multiplication and addition. If calculations are done for many sets of values—for example, a large set of material points or vectors $\langle dX_i \rangle$—they are best done with a computer.

The displacement gradients E_{ij} will give the motion that is not rigid-body translation of any point or vector dX_j if multiplied by the components of the vector (dX_1, dX_2, dX_3) in the proper manner (see Box 3-1). As a two-dimensional example, ignoring rigid translation, suppose

$$E_{ij} = \begin{bmatrix} E_{11} = 2.0 & E_{12} = 0.5 \\ E_{21} = 1.5 & E_{22} = 1.5 \end{bmatrix}$$

and suppose we wish to know the displacement of a point with coordinates $(dX_1 = 7.2\ dX_2 = 4.8)$; then:

$$U_i = E_{ij}dX_j = \begin{bmatrix} U_1 \\ U_2 \end{bmatrix} = \begin{bmatrix} E_{11}dX_1 + E_{12}dX_2 \\ E_{21}dX_1 + E_{22}dX_2 \end{bmatrix}$$

$$= \begin{bmatrix} (2.0)(7.2) + (0.5)(4.8) \\ (1.5)(7.2) + (1.5)(4.8) \end{bmatrix} = \begin{bmatrix} 16.8 \\ 18.0 \end{bmatrix}$$

The new position of the point would be

$$dX_i' = dX_i + E_{ij}dX_j = \begin{bmatrix} dX_1' \\ dX_2' \end{bmatrix} = \begin{bmatrix} 7.2 + 16.8 \\ 4.8 + 18.0 \end{bmatrix} = \begin{bmatrix} 24.0 \\ 22.8 \end{bmatrix}$$

For any homogeneous deformation, each point will have a different displacement, but the values of E_{ij} will be identical everywhere. Thus E_{ij} and $(U_0)_i$ completely describe the deformation.

At this point in the mathematical description of strain and its relation to displacement, we have a great parting of the ways. If the displacements and displacement gradients are sufficiently small, then considerable simplification in the mathematics is possible, leading to *infinitesimal-strain theory*. In infinitesimal strain the final positions are very close to the initial positions, and the final

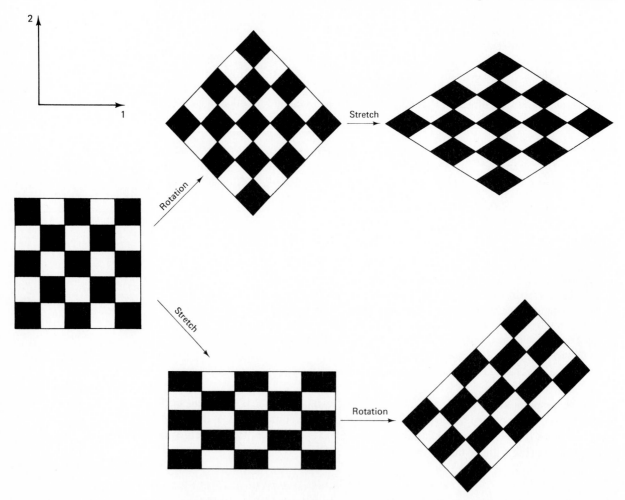

FIGURE 3–8 Illustration of the fact that the sequence in which finite stretches and rotations are performed affects the final result (also see Fig. 1-24).

position is independent of the details of the actual path of displacement. This is the strain theory commonly developed in engineering and physics. The simplifications of infinitesimal strain are valid in the case of the purely elastic deformation of rocks for which strains are generally less than 1 percent; however, they are not valid for the deformations that accompany the large-scale flow of rocks, which must be described by a more-complex *finite-strain theory*.

The reason for this mathematical distinction between very small and larger strains may be simply illustrated. The displacement gradient E_{ij} includes two distinct geometric processes: strain, which is all the relative motion of material particles, and rigid-body rotation, which is rigid motion relative to an external frame of reference. Consider the square in Figure 3-8. If we first rotate it 45° clockwise about its center and then stretch it horizontally, the result is different from first stretching horizontally and then rotating. These deformations are examples of finite deformation, and the sequence in which rotations and strains are applied is obviously important. In contrast, if the rotation and strain are infinitesimally small, the sequence in which they are applied is unimportant.

In infinitesimal-strain theory, the final position, $(U_0)_i + dX_i'$, of a material point or vector is simply the sum of the initial position, dX_i, the rigid translation, $(U_0)_i$, the displacement due to rigid rotation, $\omega_{ij}dX_j$, and the displacement due to strain, $\epsilon_{ij}dx_j$ (the order of addition is unimportant):

$$[(U_0)_i + dX_i'] = [dX_i + (U_0)_i + \omega_{ij}dX_j + \epsilon_{ij}dX_j] \tag{3-3}$$

where

$$\omega_{ij} = \frac{(E_{ij} - E_{ji})}{2} \tag{3-4}$$

is called the *infinitesimal rigid-body rotation vector* and

$$\epsilon_{ij} = \frac{(E_{ij} + E_{ji})}{2} \tag{3-5}$$

is called the *infinitesimal-strain tensor.* That ω_{ij} and ϵ_{ij} indeed describe rigid-body rotation and strain is proved by Nye (1957) and Malvern (1969), for example; it

---

*In matrix form

$$\omega_{ij} = \begin{bmatrix} 0 & \dfrac{E_{12} - E_{21}}{2} & \dfrac{E_{13} - E_{31}}{2} \\ \dfrac{E_{21} - E_{12}}{2} & 0 & \dfrac{E_{23} - E_{32}}{2} \\ \dfrac{E_{31} - E_{13}}{2} & \dfrac{E_{32} - E_{23}}{2} & 0 \end{bmatrix}$$

Note that $\omega_{ij} = -\omega_{ji}$ (for example, $\omega_{32} = -\omega_{32}$), so ω_{ij} is called *antisymmetric* and has only three independent components. It is a vector. In matrix form

$$\epsilon_{ij} = \begin{bmatrix} E_{11} & \dfrac{E_{12} + E_{21}}{2} & \dfrac{E_{13} + E_{31}}{2} \\ \dfrac{E_{21} + E_{12}}{2} & E_{22} & \dfrac{E_{23} + E_{32}}{2} \\ \dfrac{E_{31} + E_{13}}{2} & \dfrac{E_{32} + E_{23}}{2} & E_{33} \end{bmatrix}$$

Note that $\epsilon_{ij} = \epsilon_{ji}$ (for example, $\epsilon_{23} = \epsilon_{32}$), so that ϵ_{ij} is called *symmetric* and has six independent components. It is a tensor. The displacement gradient $E_{ij} = \omega_{ij} + \epsilon_{ij}$ is sometimes called the *rotational strain* because it includes rigid rotation as well as strain; it has nine independent components.

also may be demonstrated numerically by multiplying any ω_{ij} or ϵ_{ij} by some set of vectors $\langle dX_i \rangle$ and see that they rotate or strain.

When the displacement gradients E_{ij} are finite, the definition of the strain tensor ϵ_{ij} given in Equation 3-5 is no longer valid; a more-complex finite-strain definition involving extra terms and careful attention to frames of reference is necessary (see Malvern, 1969). Furthermore, with finite deformations the displacement gradients E_{ij} can no longer be factored simply into pure strain plus pure rigid-body rotation (Eqs. 3-3 to 3-5). Nevertheless, any finite deformation can still be separated into a rigid-body translation, a rigid-body rotation, and a stretch, which is another measure of strain; the operations must be performed in a specific sequence, as indicated below.

Consider once again a set of small material lines or vectors $\langle dX_i \rangle$ that describe the shape of an object (Fig. 3-6). During any finite deformation each material line dX_i is, in general, simultaneously stretched, rotated, and translated to coincide with a new vector, dX_i'. The motion in detail may be complex; nevertheless, the net deformation of the small material line dX_i can be treated as if it were the effect of one of the following two sets of equivalent operations done in sequence:

1. a stretch S_{jk} at point X_i,
2. a rigid-body rotation about point X_i,   (3-6a)
3. a rigid-body translation to X_i',

or, alternatively:

1. a rigid-body translation to X_i',
2. a rigid-body rotation about point X_i',   (3-6b)
3. a stretch S_{ij}'.

Neither of the two sets of operations will in general describe the actual sequence of intermediate steps in the deformation; they are simply two straightforward sets of operations that produce equivalent net deformations. The question of actual intermediate steps is considered later in this chapter in the section, "Deformation Paths."

Ignoring translation, the two sets of operations that describe the net deformation may be written as the following equivalent equations:

$$dX_i' = R_{ij}S_{jk}dX_k \qquad (3\text{-}7a)$$

and

$$dX_i' = S_{ij}'R_{jk}'dX_k \qquad (3\text{-}7b)$$

where R_{ij} is called the *rigid rotation matrix*, S_{ij} is the *right-stretch tensor*, and S_{ij}' is the *left-stretch tensor*. If there is no rigid-body rotation, R_{ij} takes on the value 1 and the right and left stretches are equal. Similarly, if there is no strain, the stretch tensors take on the value 1; for example, if $L = L'$ in Figure 3-4, then the stretch parallel to the line is $S = L'/L = 1$.

We are now able, at least in theory, to separate any homogeneous deformation of a small region into an equivalent rigid-body translation, a rigid-body rotation, and a strain (stretch). To do this for an actual geologic structure with nonhomogeneous deformation, such as even a simple fold, requires more theory (see Malvern, 1969). Deformation involving mainly rigid-body motion was already considered implicitly in our discussion of motion of fault blocks in Chapter 2; the material in one fault block moves approximately in unison by rigid-body translation and rotation, with little strain, relative to an arbitrary reference in another fault block. Similarly, on the scale of lithospheric plates, most motion in the crust of the earth is rigid-body motion relative to some arbitrary reference on one plate or relative to the hot-spot frame of reference (Chapter 1). In these examples the displacement gradients due to strain are concentrated near the fault surfaces or the plate boundaries. Offset geologic lines and contacts, tilted originally horizontal beds, rotated magnetization vectors in rocks, and seafloor magnetic anomalies are examples of geologic information bearing on rigid-body motions. For the remainder of our discussion, we shall consider strain without the encumberance of rigid-body motion; thus we will be studying homogeneous deformation of the sort that might be recorded within a single homogeneously deformed rock sample—for example, Figure 3-1.

From our knowledge of vectors (Figure 3-2), we know we can choose a coordinate system such that a given vector is parallel to one of the coordinate axes with the other two components zero. For example $U_i = (U_1, 0, 0)$ is displacement solely in the 1 direction. Similarly, we can choose a coordinate system such that for a specific stretch S_{ij}, all the components are zero except S_{11}, S_{22}, and S_{33}; that is, the three axes of the ellipsoid that can be used to represent a tensor are parallel to the three coordinate axes. In this case,

$$S_{ij} = \begin{bmatrix} S_{11} & 0 & 0 \\ 0 & S_{22} & 0 \\ 0 & 0 & S_{33} \end{bmatrix} \tag{3-8}$$

Then S_{11}, S_{22}, and S_{33} are called the *principal stretches,* generally written S_1, S_2, and S_3. From now on, to simplify equations, we shall consider only situations in which the coordinate axes are parallel to the *principal axes* of the stretch (for changing coordinate axes, see Box 3-2).

If during some deformation S_{ij}, a small-volume element dV is deformed to a new volume dV', the *volume ratio dV'/dV* is a measure of this change in volume. The volume ratio, a scalar, is just the product of the principal stretches:

$$\frac{dV'}{dV} = S_1 S_2 S_3 \tag{3-9}$$

Let's review what we have done in the preceding paragraphs in developing a general concept of strain and deformation. We found that we could easily describe the shape, position, and orientation of any object with a set of line segments or vectors, $\langle dX_i \rangle$, as illustrated in Figure 3-6. The problem of describing the deformation of any object is then just the problem of how any line segment or vector is translated, rotated, and stretched during the deformation (Fig. 3-7). The deformation of any line segment or vector can be separated into a rigid-body translation and rotation, which involves only motion relative to an external reference frame, and distortion plus change in volume, which make up strain and involve relative motion of atoms in the object. The details of the mathematical separation of strain from rigid rotation differ depending upon whether the strains are infinitesimally small (about 1 percent), such as in elastic deformation of rocks, or whether they are large finite strain, such as in the metamorphic flow of rocks. The order in which an object is strained and rotated affects the end result in finite strain. Finite strains are most easily described with the stretch tensor (Eq. 3-7), whereas infinitesimal strains are best described with the infinitesimal-strain tensor (Eq. 3-5). The nine components of the stretch or strain tensors, six of which are independent, relate the size and shape of an object before and after deformation. A coordinate frame can be found such that for a given strain, all components are zero except for S_{11}, S_{22}, and S_{33}, which are called the principal stretches (S_1, S_2, S_3), and in this case the coordinate axes are the principal-strain axes. The

---

BOX 3–2   Change of Coordinate Axes

It is straightforward to change from one coordinate system to another and calculate the components of the stretch S_{ij} or any other tensor relative to any new coordinate system. The stretch, of course, exists apart from any coordinate system.

If the old coordinate system has axes X_1, X_2, and X_3 and the new coordinate system has axes X_1', X_2', and X_3', then we can determine a set of nine *direction cosines*, a_{ij}, which are the cosines of the angles between the two sets of axes. For example, a_{21} is the cosine of the angle between the X_1 axis and the X_2' axis. To change the components of a vector S_j in coordinate system X_i to its equivalent components S_i' in system X_i', we simply multiply by the direction cosines:

$$S_i' = a_{ij}S_j$$

The computation of this equation is explained in Box 3-1.

Similarly, any second-rank tensor S_{kl} may be transformed to its components in the new coordinate system by the analogous equation for tensors:

$$S_{ij}' = a_{ik}a_{jl}S_{kl}$$

For example, the equation for the S_{11}' component is

$$S_{11}' = a_{11}a_{11}S_{11} + a_{11}a_{12}S_{12} + a_{11}a_{13}S_{13}$$
$$+ a_{12}a_{11}S_{21} + a_{12}a_{12}S_{22} + a_{12}a_{13}S_{23}$$
$$+ a_{13}a_{11}S_{31} + a_{13}a_{12}S_{32} + a_{13}a_{13}S_{33}$$

It is also straightforward, but a more-involved process, to calculate the principal stretches and their orientations given the components S_{ij} in any arbitrary coordinate frame. For a discussion of how to do this, see books on matrix or linear algebra, or see Fung (1969), Nye (1957), or Malvern (1969).

---

expression for displacement due to strain (Eq. 3-7) may be used to determine the shapes of objects after some homogeneous deformation S_{ij}. We use this theory in the following section to measure the natural distortion of rocks.

Measurement of Strain in Deformed Rocks

We may now treat the homogeneous deformation of any object because we can describe the undeformed and deformed shapes as two sets of vectors, $\langle dX_i \rangle$ and $\langle dX_i' \rangle$, and because we have a functional relationship between the two sets involving the stretch tensor (Eq. 3-7). These results allow us to determine the orientation and magnitude of strains in rocks by measuring certain classes of deformed objects, such as:

1. initially spherical or circular objects,
2. initially ellipsoidal or elliptical objects,
3. linear objects,
4. angular objects.

Measurements of deformed objects, such as fossils, are usually of necessity made on two-dimensional rock surfaces. For a complete three-dimensional analysis, measurements are made on nonparallel and commonly perpendicular planes; then the orientations and magnitudes of the principal strains can be calculated from the two-dimensional data using methods found in Ramsay (1967).

Consider the deformation of a sphere of unit radius with center X_i in three dimensions or consider the deformation of a circle in two dimensions. All points dX_i on the surface of the sphere satisfy the equation

$$dX_i dX_i = 1 \qquad (3\text{-}10)$$

that is, $dX_1^2 + dX_2^2 + dX_3^2 = 1$. Any point or vector whose components do not obey this equation are not on the surface of the sphere of unit radius. The equation relating a deformed material line or vector to its undeformed equivalent, no longer considering rigid-body motion, is (Eq. 3-7):

$$dX_i' = S_{ij} dX_j \qquad (3\text{-}11)$$

We substitute this relationship into the equation for a unit sphere (3-10) to obtain the equation that describes the final shape of an object that was a unit sphere prior to any homogeneous stretch S_{ij}. Since the principal stretches are parallel to the coordinate axes, substituting gives, in terms of principal stretches,

$$\left[\frac{dX_1'}{S_1}\right]^2 + \left[\frac{dX_2'}{S_2}\right]^2 + \left[\frac{dX_3'}{S_3}\right]^2 = 1 \qquad (3\text{-}12)$$

The deformed equivalents $\langle dX_i' \rangle$ of all material vectors $\langle dX_i \rangle$ describing the original unit sphere must satisfy this equation, which you may recognize as the equation of a triaxial ellipsoid with semimajor axes of lengths S_1, S_2, and S_3 oriented parallel to the coordinate axes.

This fundamental result (Eq. 3-12) tells us that any homogeneous deformation S_{ij} transforms a sphere into an ellipsoid whose axes are parallel to the principal strains and whose axial lengths are proportional to the principal stretches S_1, S_2, and S_3. Because of this simple relationship, initially spherical objects such as carbonate ooids, tuff lapilli, and reduction spots in slates (Fig. 3-9) are the most straightforward strain markers in rocks. If these objects, which were originally nearly spherical, are viewed on a two-dimensional cut placed through them, they appear in general as ellipses (Fig. 3-9). The ellipse immediately provides the orientation of the principal stretch components in the two-dimensional plane.

In contrast with orientation, the shape of the ellipse does not immediately tell us the magnitudes of the stretch components because in most cases we do not know the original volume or area of a strain marker. For this fundamental reason, we can determine only the orientations of the principal stretches and their ratios (S_1/S_2) and (S_2/S_3). We cannot determine the volume ratio $dV'/dV = S_1 S_2 S_3$ without additional information or assumptions. For example, if we assume no volume change—$dV'/dV = 1$—we may immediately compute the actual stretches.

We could write equations analogous to that of the sphere (Eq. 3-10) for other simple, or not so simple, geometric objects and see how they are distorted in any homogeneous deformation. Some of the more interesting results of such an exercise are the following (Fig. 3-10):

1. Straight lines always remain straight and flat planes always remain flat.

2. Parallel lines remain parallel and are extended or contracted by the same ratio.

3. Perpendicular lines that are parallel to the principal strain axes remain perpendicular; no other lines do.

4. Upon deformation, circular and elliptical cylinders are transformed into elliptical cylinders.

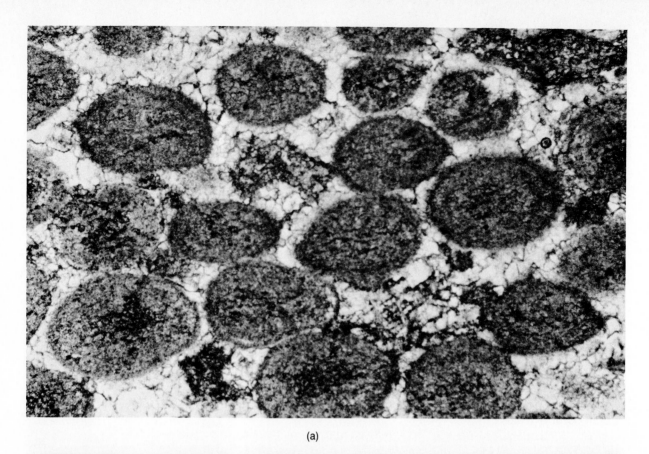

(a)

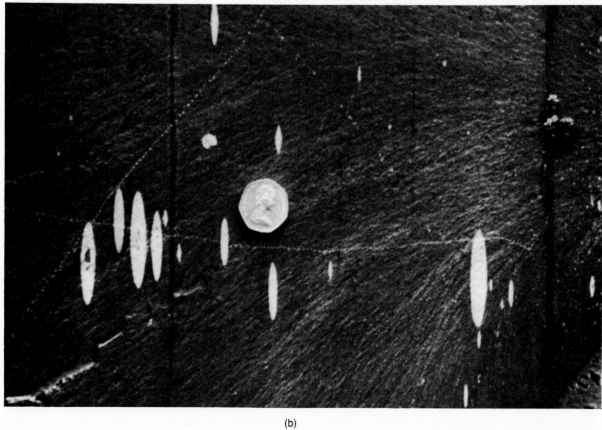

(b)

FIGURE 3–9 Deformed initially nearly spherical strain markers in rock. (a) Photomicrograph of now elliptical recrystallized carbonate ooids in a metamorphosed limestone, Appalachian mountain belt. (b) Elliptical reduction spots in Welsh slate. (Photograph by Jane Selverstone.)

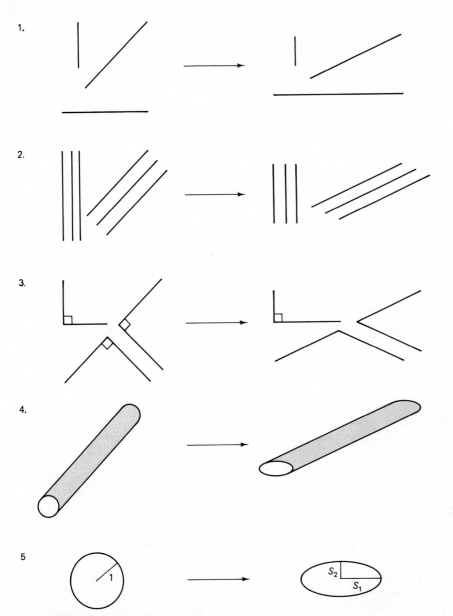

FIGURE 3–10 Examples of effects of homogeneous strain ($S_1/S_2 = 2.3$): (1) straight lines remain straight; (2) parallel lines remain parallel and undergo the same stretch; (3) perpendicular lines that are parallel to the principal stretch directions remain perpendicular, no other lines do; (4) circular cylinders deform to elliptical cylinders; and (5) circles and spheres become ellipses and ellipsoids.

5. An ellipsoid becomes a different ellipsoid and, in particular, a sphere becomes an ellipsoid. There is one ellipsoid for any specific strain that will transform into a sphere.

Properties 1 and 2 indicate that parallel beds of rock remain parallel upon homogeneous deformation. Most flat-lying sediments have deformed homogeneously; they have compacted from initial porosities of as high as 80 percent to present porosities of 30 percent or even much less.

Property 3 indicates that points joined by lines parallel to the principal stretches S_1, S_2, S_3 moved closer together or farther apart. In addition, all other points slide past one another; a line joining two such points rotates away from the S_3 direction and toward the S_1 direction (Fig. 3-10).

Property 4 indicates that a now-elliptical fossilized worm burrow has been deformed because the worm was a circular cylinder. Nevertheless, the strain generally cannot be determined from this distorted fossil because we generally do not know which section of the tube was originally circular; any section of a circular cylinder except for a right section is elliptical. Information on strain can be obtained from crinoid ossicles that were originally right-circular cylinders because the ends of the cylinders were originally circular (Fig. 3-11).

Property 5 poses difficulties for determining strain from distorted pebbles (Fig. 3-12). Pebbles are generally not spherical to begin with, but are approximately ellipsoids. Each pebble will be transformed into a different ellipsoid of shape and orientation dependent on the initial shape of the pebble and its orientation with respect to the strain, which is generally not known. These difficulties are not

FIGURE 3–11 Deformed bedding plane showing elliptical ends of crinoid ossicles from the flat-lying sediments of the Appalachian plateau, New York State. (Sampled by Terry Engelder.)

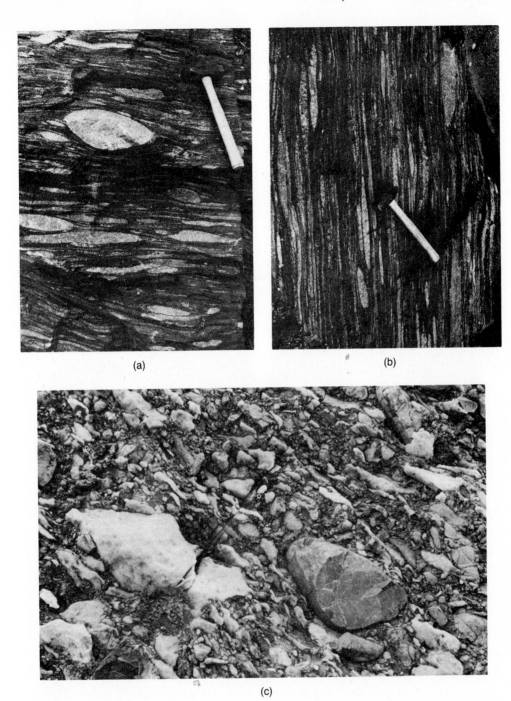

(a)   (b)

(c)

FIGURE 3–12 Deformed conglomerates. (a) and (b) are two faces of the same outcrop showing substantially different stretches, Grenville orogenic belt, Ontario. (c) is a conglomerate from the Canadian Appalachians showing very heterogeneous deformation; the quartzite cobble is undeformed whereas the numberous carbonate cobbles and pebbles are strongly and heterogeneously flattened.

entirely debilitating; in particular, the axes of the pebble ellipsoids become close to parallel with the strain axes at large strains even though the axial ratios are quite different (Fig. 3-13).

From the above discussion we see that objects of known original shape provide records of their own distortions; however, the completeness of these records depends strongly on details of the objects' geometries. Actual determina-

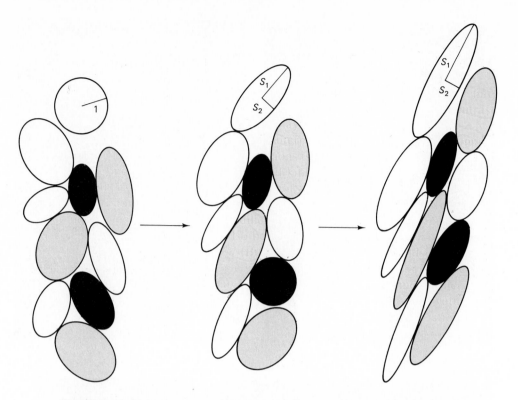

FIGURE 3–13 Convergence of orientations of principal axes of deformed pebbles (ellipses) with the orientations of the principal stretch axes at large strains. In contrast, the axial ratios of the pebbles do not converge to the ratios of the principal stretches, S_1/S_2.

tion of strain from measurements of distorted objects requires close attention to these geometric details. Special methods have been developed for many classes of geologic objects; for example, methods for spiral objects, such as ammonoids, were developed by Tan (1973). Some of the methods used for bilaterally symmetrical fossils, such as the trilobite in Figure 3-1, are given as an example in the following paragraphs. Comprehensive treatments of the methods of strain determination in rocks are given by Ramsay (1967) and Ramsay and Huber (1984).

Bilaterally Symmetrical Fossils. Many organisms are bilaterally symmetrical—for example, trilobites and brachiopods. Upon deformation this symmetry is generally lost. Bilateral symmetry implies that a line joining equivalent points on either side of the plane or line of symmetry, is perpendicular to the plane (or line), thereby defining a right angle. This initial right angle is the basis for determining distortion in several situations.

The two-dimensional distortion of a rock slab containing a number of fossils of original bilateral symmetry—for example, the trilobites in Figure 3-14—may be determined using a simple graphical method due to Wellman (1962). Wellman's construction is based on the following property of a circle. The two lines joining the ends of any diameter of a circle, D and D', to any other point on the circumference P_i meet in a right angle (see Fig. 3-15). Thus as soon as we choose an arbitrary diameter of a circle, DD', the points along the circumference, such as P_1, P_2, or P_3, may be used to represent the predeformational orientations of a set of right angles lying in a plane. After a homogeneous deformation, this circle and the slab of fossils it represents will be distorted into an ellipse. The points along the circumference of the circle will then lie on the circumference of the ellipse and their lines connecting to D and D' will be distorted in general to some new angle (Fig. 3-15).

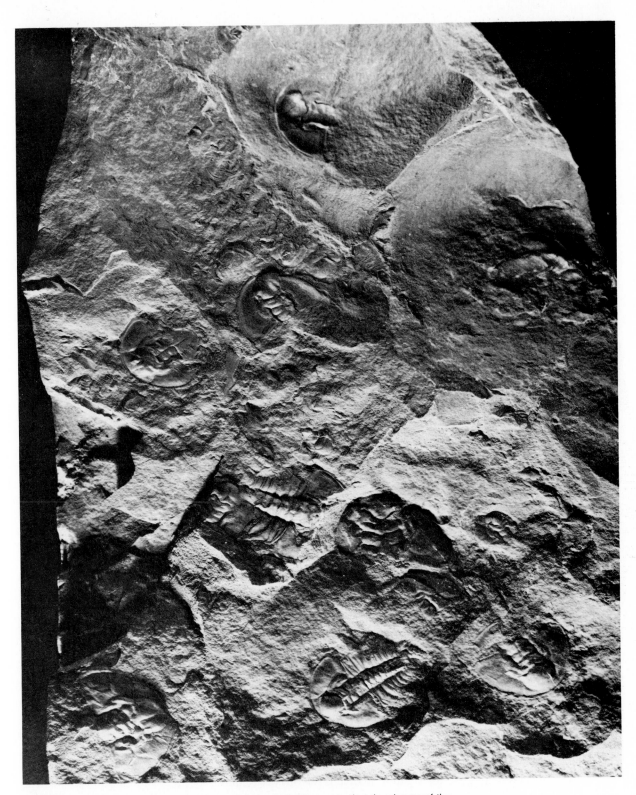

FIGURE 3–14 Bedding plane showing many deformed trilobites; note that the shapes of the trilobites depend on their orientations.

Wellman's method chooses some arbitrary line segment DD' and locates points on the circumference of the ellipse by plotting the present angles that were the original right angles. This is done keeping the rock slab in its proper

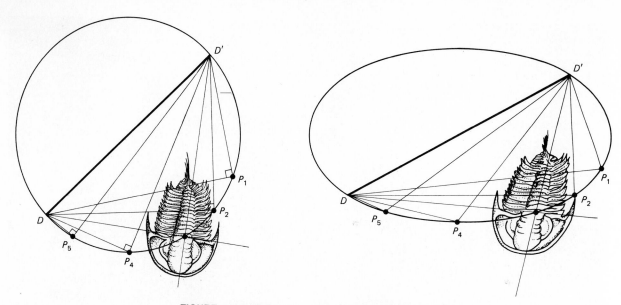

FIGURE 3–15 Wellman's construction for determining the finite strain directions and ratio of principal stretches from a set of deformed originally bilaterally symmetrical fossils, such as the trilobites in Figure 3-14. See discussion in text.

orientation relative to the arbitrary line segment DD' (see Fig. 3-15). Once all the fossils are plotted, the best-fitting ellipse may be determined.

Wellman's construction works best when there are many fossils and orientations. The minimum number of nonparallel fossils is two; in this case the orientation and shape of the strain ellipse is not very accurately located by graphical means; a nomogram is more convenient (see Ramsay, 1967). If only one distorted fossil is available, the strain cannot be determined without additional information. For example, if the orientation of the axis of principal stretch in the plane is known, then the stretch ratio can be determined.

Objects of known original shape record their own distortion, but they may or may not provide an accurate record of the overall distortion of the rock containing them. For example, the deformation of the cobbles in the conglomerate of Figure 3-12 is by no means homogeneous; some cobbles were stiffer and some were softer. As another example, a study of a deformed conglomerate in Saxony showed the axial ratios of pebbles to be strongly dependent on rock type: quartz, 1:1 to 2:1; quartzite, 8:1; graywacke, 10:1 to 12:1; schistose graywacke and shale, 18:1. The best strain indicators for a rock as a whole are objects whose mechanical properties are the same as the rest of the rock—for example, molds of fossil in which the shell material has dissolved away.

People have generally used strain indicators to help understand the origin of specific structures or types of structures; this will be our principal use of them. Regional studies of strain in rocks have been less common because of the great amount of work involved and the rarity of good strain indicators in some rock types. A well-known regional study is by Ernst Cloos (1971); it involved tens of thousands of measurements in a 15,000-km^2 region comprising much of the central Appalachian fold belt in Pennsylvania and the Virginias. Cloos showed that the strain throughout this large region was remarkably uniform in orientation, as well as magnitude.

Natural strains within the earth vary enormously; for example, the stretch ratios of a large number of deformed objects from many parts of the world are shown in Figure 3-16. On this graph, stretch ratios, S_1/S_2 or S_2/S_3, in the range 1 to 5 are most common. Pancake-shaped flattened strain ellipsoids are more common

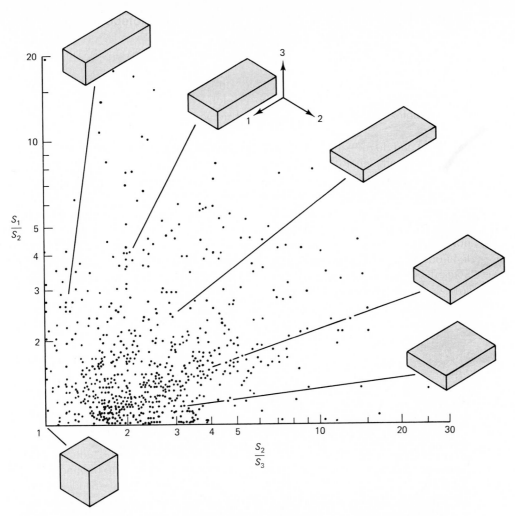

FIGURE 3–16 Graph of measured stretch ratios from throughout the world. The parallelepipeds illustrate the deformed shape of an original cube, corresponding to several points on the graph. Note that flattening deformations are much more common in nature than stretching deformations. (Data from Pfiffner and Ramsay, J. Geophysical Research, v. 87, p. 311–321, 1982, copyrighted by American Geophysical Union.)

than cigar-shaped stretched ellipsoids. Very large stretch ratios beyond 5 to 10 probably also exist, particularly at high metamorphic grade, but they involve such severe changes to the rock that identifiable original objects are largely destroyed.

Deformation Paths

We have seen in the previous sections that strain is a description of the net change in shape and size between an initial and a final state. We now ask the question: What is the history of strains and rotations that constitute the intermediate steps (recall Fig. 3-8)? In principle there is an infinity of geometrically possible strain histories or deformation paths that lead to the same final shape of a strain indicator, and its final shape tells us nothing that helps us decide which occurred. There are, however, small-scale textures and structures that record some aspects of this deformation path.

The net strain is the sequential sum of all the instantaneous strains. For example if a line of length L_0 is extended to L_n, we might envisage the longitudinal strain as the sum of a number of small increments $\delta L_i/L_i$:

$$\sum_{L_0}^{L_{n-1}} \frac{\delta L_i}{L_i} = \frac{L_1 - L_0}{L_0} + \frac{L_2 - L_1}{L_1} + \ldots + \frac{L_n - L_{n-1}}{L_{n-1}} \qquad (3\text{-}13)$$

We note in passing that if the increments are made arbitrarily small, this sum becomes an integral:

$$\bar{\epsilon} = \int_{L_0}^{L_n} \frac{dL}{L} = \log_e \frac{L_n}{L_0} = \log_e(1 + \epsilon) = \log_e S \qquad (3\text{-}14)$$

which is the *logarithmic*, or *natural, strain* mentioned earlier. Logarithmic strain is used sometimes in discussions of large strains and deformation paths; for example, Figure 3-16 has a log scale. It has an advantage for making comparisons between measured strains because the numerical values of the logarithmic strains are proportional to the relative changes in length. For example, if one line is contracted to half its original length and another is stretched to twice, the logarithmic strains are $-\log_e 2 = -.693$ and $\log_e 2 = .693$; in contrast, the stretches, $S = (L'/L)$, would be 0.5 and 2.

The most important concept in the study of deformation paths is the distinction between coaxial and noncoaxial deformation. If two successive increments of strain have the same principal axes with respect to the material, then this part of the strain history or deformation path is *coaxial*. If the successive principal axes are not parallel, but rotate with respect to the material, then the deformation is *noncoaxial*. The order of application of successive strains is important to the net strain in noncoaxial deformation, but not in coaxial deformation because it involves no rigid-body rotation of the material with respect to the instantaneous strain axes.

Of all the possible deformation paths, there are two simple geometric types that are commonly discussed and have special names. They are useful in illustrating the effects of coaxial and noncoaxial strain. *Pure shear* (Fig. 3-17) is used to denote coaxial strain and, strictly speaking, implies no change in volume as well. *Simple shear* (Fig. 3-17) is a noncoaxial rotational strain in which the amount of strain is directly related to the amount of rotation in a specific way, namely, the affine shear that we can easily perform on a deck of cards or telephone book. Simple shear, as defined, is a constant-volume, two-dimensional deformation, with no displacement in the third direction and no flattening perpendicular to the plane of slip.

In simple shear the principal strain axes start out oriented 45° to the plane and direction of slip. The principal elongation, S_1, rotates toward the plane of slip as deformation proceeds; the principal shortening, S_2, rotates toward perpendicular. This can easily be verified by tracing a coin on the side of a stack of cards or on the edge of the pages of this book and then shearing the stack homogeneously. The principal elongation, S_1, reaches the plane of slip only at infinite slip. It is important to note that some directions within the material are always lengthening (for example, aa'; Fig. 3-17), whereas other directions first shorten and then, once they have rotated past perpendicular to the slip plane, begin to lengthen (for example, bb'; Fig. 3-17).

The example of simple shear shows that a simple overall deformation can lead to some complexities in the history of the deformation. As a more-geological example, if a large volume of rock changed shape by noncoaxial strain, we can imagine that an initial shortening parallel to bedding—for example, oriented parallel to bb' in Figure 3-17—might produce folds that would later undergo extension and either unfold or boudinage, which is the breaking of stiffer rock layers into segments as they are stretched (Fig. 9-53). Many deformation histories

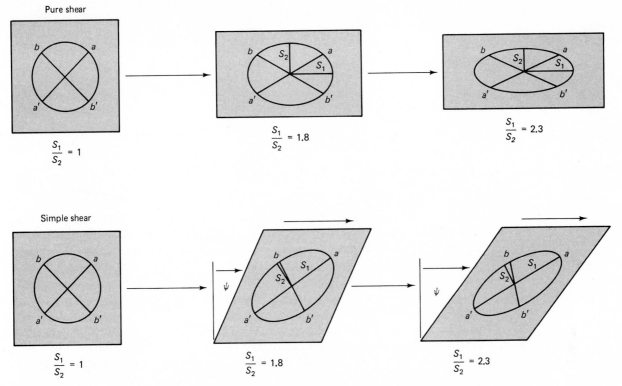

FIGURE 3–17 Examples of coaxial and noncoaxial deformations. Pure shear is a coaxial deformation with no change in volume; note that material lines *aa'* and *bb'* always lengthen and rotate away from the S_2-direction toward the S_1-direction. Simple shear is a noncoaxial deformation of the sort produced by shear of a deck of cards; note that material line *aa'* always lengthens, but *bb'* shortens until it rotates past vertical and then begins to lengthen. The angle ψ is called the *angle of simple shear.*

in rocks are probably more complicated than either pure shear or simple shear even for a single deformational event; for example, rather than simple shear, there might be flattening perpendicular to the slip plane as well.

Important information bearing on the deformation path is recorded in some rocks. For example, fibrous crystals in veins apparently grow parallel to the instantaneous principal elongation so that the curving fibers may record changes in the direction of elongation during the deformation. Another type of texture bearing on the deformation path is rolled metamorphic porphyroblasts, well known for garnets and albites. Porphyroblasts are large crystals that grow around adjacent minerals and thereby contain a record of the schistosity and metamorphic mineralogy during growth of each part of the crystal. Some porphyroblasts have rotated with respect to the schistosity during or after crystallization, while in other rocks they have not (Fig. 10-15). Cases are known in which the axis of rotation has changed during the course of deformation.

Strain Rate

The rate at which strain accumulates in deforming rocks is of particular importance in structural geology because the strength of rocks under conditions of high temperature flow is directly related to the strain rate, as we shall see in the next chapter. In this section we briefly investigate the strain rates at which deformation takes place within the lithosphere.

The most direct method of measuring strain rates is to set up an array of benchmarks and observe the deformation of the earth's surface by precise

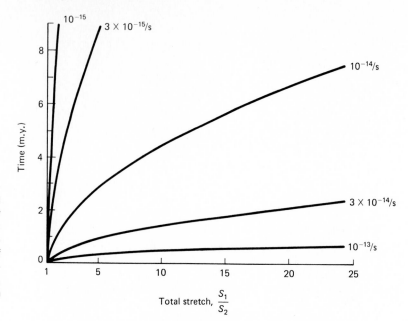

FIGURE 3–18 Time required to accumulate a total stretch S_1/S_2 at various instantaneous strain rates ($\dot{\epsilon} = \delta L/Lt$) for pure shear. The strain rate must be $\dot{\epsilon} = 10^{-14}$/s or faster to accumulate the observed typical geologic stretch ratios of 2–10 (Fig. 3-16). (After Pfiffner and Ramsay, J. Geophysical Research, v. 87, p. 311–321, 1982, copyrighted by American Geophysical Union.)

triangulation over a period of years. For example, geodetic measurements over a period of 20 years around the active San Andreas strike-slip fault in California indicate that any initially rectangular block is being changed into a parallelogram at a rate of 0.1 s of arc per year, which corresponds to a simple-shear rate parallel to the fault of 1.5×10^{-13}/s. This measurement is an average over the width of the fault zone at the surface. The deformation at depth may be more or less concentrated in a narrow zone, leading to larger or smaller actual strain rates. Similarly, we may estimate average strain rates along plate boundaries from plate motions if we assume how the deformation is distributed through the plate-boundary zone.

We may also estimate strain rates from finite-strain measurements in rocks if we know the time involved in the deformation. For example, some Miocene rocks that were deformed and metamorphosed in the active mountain belt of Taiwan are already at the surface, displaying stretch ratios S_1/S_2 of 3 to 5; the deformation took less than 2 to 3 m.y. However, we cannot simply divide the finite strain by the time to get the instantaneous strain rate; rather the instantaneous strain rate, $\dot{\epsilon}$, is equal to the natural, or logarithmic, strain, $\bar{\epsilon}$, divided by the time. Thus from Equation 3-14 we see that stretch accumulates exponentially with time t at constant instantaneous strain rate

$$S = e^{\bar{\epsilon}} = e^{\dot{\epsilon}t} \tag{3-15}$$

as is illustrated in Figure 3-18. Therefore, the stretch ratios of 3 to 5 produced in less than 2 to 3 m.y. in Taiwan indicate a strain rate of 10^{-14}/s or faster. If the deformation had taken only a hundred thousand years, the strain rate would have been about 10^{-13}/s. Estimates of duration of deformation for some of the measured stretches shown in Figure 3-16 indicate that strain rates of 10^{-13}/s to 10^{-14}/s are typical (Pfiffner and Ramsay, 1982). Later we will make use of these rates to estimate the strengths of rocks under metamorphic conditions.

STRESS

A body within which one part exerts a force on its immediately adjacent parts is said to be in a state of stress. These internal forces act on the scale of interatomic

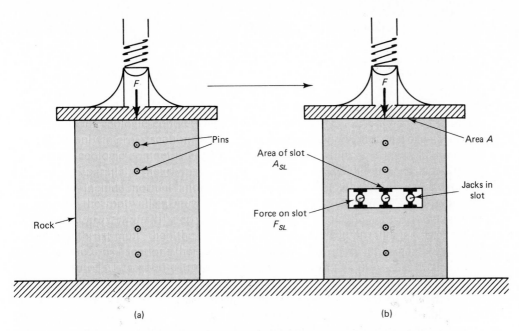

FIGURE 3–19 Schematic diagram of a block of rock placed under a compressive force **F** acting over an area **A**. In (b), a slot has been cut in the sample and jacks have been pumped up until the survey pins have returned to their original positions prior to cutting the slot (a). The jacks now support the load that was supported by rock prior to cutting the slot.

distances and arise during the deformation due to forced stretching or compressing of bonds from what otherwise would be their equilibrium positions.*

Stress is the total force exerted by all the atoms of one side of any arbitrary plane within a body on the atoms immediately on the other side, divided by the area of the plane. The concept of stress is closely related to elasticity because bonds are basically elastic in behavior (namely, stress is proportional to strain). The atomic basis of elastic behavior and stress is introduced in the next chapter.

Let's look more closely at the properties of stress. Suppose we place a cylinder or prism of rock under a uniform compression by applying a force **F** axially with a screw mechanism or hydraulic jack. What are the internal forces, or stresses, acting within the rock? We can perform a "thought experiment" to answer this question (Fig. 3-19).

We could first very accurately locate the positions of some points in the rock using metal pins. Next we cut a slot between them. The pins will move a very small distance closer together because the rock is elastic and the compression is no longer supported across the slot, but by the surrounding material. Next, we can put hydraulic jacks into the slot and force the walls of the slot and the pins back to their original positions. The jacks now support the load that was supported by the rock in the slot before it was removed. By adding up the loads on the jacks, we find the force, F_{sl}, that was acting perpendicular to the slot before it was cut. If we actually did this experiment, we would find that the force on the specimen divided by its cross-sectional area equals the force on the slot divided by the area of the slot: $F/A = F_{sl}/A_{sl}$, as we would expect.

If, instead, we cut an inclined slot, its walls would tend to slip or shear past each other, as well as to come together. In this case we need cross braces or

---

*In addition to these short-range forces, there are long-range forces, or *body forces*, the most important of which is gravity. Gravitational force g is exerted by a volume on all other volumes in inverse proportion to the square of the separation of their centers of mass r and directly proportional to the product of their masses ($g = Gm_1m_2r^{-2}$, where G is the proportionality constant).

parallel jacks, in addition to the perpendicular jacks, to restore the survey pins to their original positions (Fig. 3-20). The more the slot is inclined to the specimen axis (increasing θ), the less total force acts across the slot. This is true because the slot subtends less of the cross-sectional area of the sample, by a factor of cos θ, as shown in the inset of Figure 3-20. At θ = 90° no force acts across the plane of the slot in this experiment. A second important effect of inclination is that the portion of the total force on the slot supported by the perpendicular jacks depends upon the inclination as cos θ, whereas that supported by the parallel jacks depends on inclination as sin θ, as shown in the inset of Figure 3-20. The force per unit area resolved normal to the slot is the *normal stress,* σ_n, which in this thought experiment is $(\mathbf{F}/\mathbf{A})\cos^2\theta$. The force per unit area resolved parallel to the dip of the slot is $(\mathbf{F}/\mathbf{A})\cos\theta\sin\theta$, called the *shear stress,* σ_τ; it is a maximum when the slot is inclined 45° to the specimen axis. We now know the state of stress across any plane in our sample as a function of its orientation.

In general, another independent component of shear stress would be required to define fully the state of stress on a plane. In the case of our inclined slot, the third independent stress component would be a horizontal shear stress, perpendicular to the page in Figure 3-20, which is zero because we have only axially loaded our sample. The *state of stress on a plane* (Fig. 3-21) is given by one normal-stress component and two shear-stress components (σ_n, $\sigma_{\tau1}$, $\sigma_{\tau2}$).

We could consider a more-complex thought experiment with a more-complex set of externally applied forces—both compressions and shears—and write the analogous set of three equations that define the state of stress on an internal plane or slot of arbitrary orientation. However, let's turn the problem around; instead of defining the forces on the outside of a body and then calculating state of stress on a plane of arbitrary orientation, we shall make measurements to define the state of stress in some region below the surface of the earth. We could

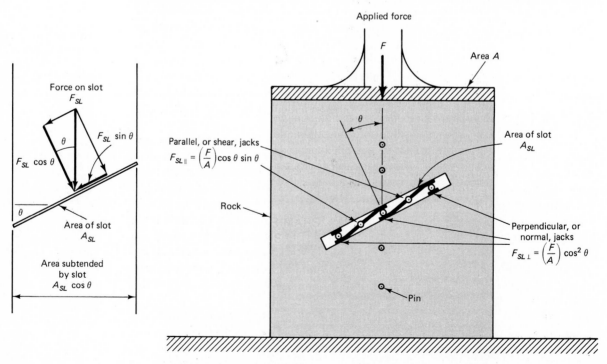

FIGURE 3–20 Schematic diagram of a block of rock under a compressive force **F**, similar to Figure 3-19 but with an inclined slot. Under these conditions both normal and shear jacks are required to return the survey pins to their positions prior to cutting the slot.

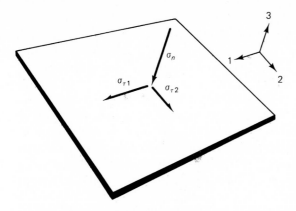

FIGURE 3–21 The state of stress at a single point on a plane is usually specified in terms of a normal-stress component, σ_n, and two perpendicular shear-stress components, $\sigma_{\tau1}$ and $\sigma_{\tau2}$.

go down into a deep mine and use our slot method; similar methods are in fact used in civil engineering.

With a single slot we can measure one normal-stress component and two mutually perpendicular shear-stress components. These three stress components completely define the state of stress on the plane. However these measurements are not sufficient to define fully the state of stress at this place within the earth; for example, we know nothing about the normal stress that is acting on any plane perpendicular to our slot. For this reason we might decide to measure the state of stress on three mutually perpendicular slots and arrive at nine mutually perpendicular stress components, three normal-stress components and six shear-stress components. With these nine measurements we can calculate the state of stress across any arbitrary plane using equations similar to those we worked out before. We will do this later for two dimensions using the Mohr diagram.

In matrix or tensor form, the nine measured values of the stress components relative to our coordinate system (coordinate axes perpendicular to the three slots) are called:

$$\sigma_{ij} \equiv \begin{bmatrix} \sigma_{11} & \sigma_{12} & \sigma_{13} \\ \sigma_{21} & \sigma_{22} & \sigma_{23} \\ \sigma_{31} & \sigma_{32} & \sigma_{33} \end{bmatrix} \qquad (3\text{-}16)$$

Figure 3-22(a) illustrates the meaning of this notation. The first subscript, i, indicates the plane perpendicular to the i direction; the second subscript, j,

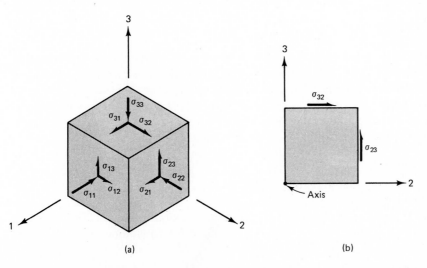

(a)   (b)

FIGURE 3–22 (a) The state of stress at a point, measured in any Cartesian coordinate system, is specified by the state of stress on the faces of an infinitesimally small cube, with the faces perpendicular to the axes of the coordinate system. Thus the state of stress at a point is specified by three normal-stress components and six shear-stress components (Eq. 3-16). (b) Only three of the shear-stress components are independent because there must be no unbalanced torques at equilibrium; for example, balance of torques about the 1-axis requires that $\sigma_{23} = -\sigma_{32}$. In general, $\sigma_{ij} = -\sigma_{ji}$.

indicates the direction in which the force acts. For example, σ_{12} is a shear stress acting on the 1 plane in the 2 direction and σ_{33} is a normal stress acting on the 3 plane. The quantity σ_{ij} is the *state of stress at a point,* in contrast with (σ_n, $\sigma_{\tau 1}$, $\sigma_{\tau 2}$), which is the state of stress on a particular plane passing through the point. If we know σ_{ij}, we can compute the state of stress on a plane of any orientation.

We did too much work in measuring the nine stress components on our three slots because not all the components are independent. This fact may be illustrated if we consider that the cube in Figure 3-22(a) is in equilibrium; therefore it has no unbalanced torques acting on it. Since the sum of the torques about any axis through the body must be zero, we have $\sigma_{ij} = \sigma_{ji}$ (for example $\sigma_{12} = \sigma_{21}$), as illustrated in Figure 3-22(b). Thus only six independent measurements are needed to define stress at equilibrium.

We noted earlier while discussing strain that a coordinate system can be found such that all components of the stretch are zero except for S_{11}, S_{22}, and S_{33}, which in this case are called the *principal stretches,* and are often labeled S_1, S_2, and S_3. The three principal axes of the strain ellipsoid are parallel to the three coordinate axes in this case. Similarly, we may find a coordinate system for stress such that all shear-stress components are zero:

$$\sigma_{ij} = \begin{bmatrix} \sigma_{11} & 0 & 0 \\ 0 & \sigma_{22} & 0 \\ 0 & 0 & \sigma_{33} \end{bmatrix} \tag{3-17}$$

where σ_{11}, σ_{22}, σ_{33} are called the *principal stresses* and are generally designated by σ_1, σ_2, σ_3, or simply σ_i. In this coordinate system, the three *principal axes* of the stress ellipsoid are parallel to the coordinate axes. Note that no shear stress acts across planes perpendicular to the principal axes, a fact that is important for understanding the origin of some structures (Chapters 6, 7, and 8).

The stress tensor may be factored or decomposed into deviatoric and mean, or isotropic, parts. The *mean stress,* σ_m, is

$$\sigma_m = \left[\frac{\sigma_{11} + \sigma_{22} + \sigma_{33}}{3} \right] \tag{3-18}$$

which represents the isotropic part of the stress. The *deviatoric stress,* Δ_{ij}, is

$$\Delta_{ij} = \begin{bmatrix} (\sigma_{11} - \sigma_m) & \sigma_{12} & \sigma_{13} \\ \sigma_{12} & (\sigma_{22} - \sigma_m) & \sigma_{23} \\ \sigma_{13} & \sigma_{23} & (\sigma_{33} - \sigma_m) \end{bmatrix} \tag{3-19}$$

or, in terms of the principal stresses, $\Delta_i = (\sigma_i - \sigma_m)$. We shall later see that most deformation, except for the dilation, is a result of the deviatoric stress—that is, the difference between the principal stresses rather than their absolute magnitudes. The mean, or isotropic, stress exerts an important control on which strain mechanisms operate, the strength of brittle materials, and the dilational strain. For example, fracture is inhibited with increasing mean stress.

The dimensions of stress are force per unit area. The unit of stress, or pressure, in the International System of Units (SI) is the *pascal* (Pa), which is one newton per square meter (N/m^2). Until recently it has been more common in geology and geophysics to use the *bar,* which is 10^5 Pa and is almost equal to atmospheric pressure at sea level (0.987 standard atmospheres). The pressure ($\sigma_1 = \sigma_2 = \sigma_3$) at the bottom of the ocean (about 5 km) is about 500 bars or 50 *megapascals* (1 MPa = 10^6Pa). The pressure at the bottom of continental crust (about 30 km) is about 10 *kilobars* (1 kb = 10^3 b) or one *gigapascal* (1

GPa = 10^9Pa). In engineering, tensile stresses are considered positive, whereas in earth science compression is normally considered positive because most stresses in the earth are compressive. We shall consider compression positive.

Mohr Diagram

Given the magnitudes of the principal stresses and their orientations, we can calculate the shear and normal stresses acting across a plane of any orientation. We now develop, for two dimensions, a way of graphically representing the state of stress (σ_n and σ_τ) on planes of every orientation in a region of uniform stress (principal stresses σ_1 and σ_2). This graphical representation, the *Mohr diagram,* will be very useful, for example when we consider fracturing of rock and sliding along faults and joints.

Consider a plane within a homogeneously stressed cube (square) whose normal makes an angle θ with the 1 direction (Fig. 3-23). At equilibrium, the forces acting in the 1 and 2 directions across the inclined internal plane must equal the forces acting across faces A and B, which are normal to the 1 and 2 directions:

$$\sigma_1 A = \frac{A}{\cos \theta} (\sigma_n \cos \theta + \sigma_\tau \sin \theta) \tag{3-20}$$

$$\sigma_2 B = \frac{B}{\sin \theta} (\sigma_n \sin \theta - \sigma_\tau \cos \theta) \tag{3-21}$$

These are the equations of equilibrium in terms of force for the 1 and 2 directions. We may solve these simultaneously for σ_n and σ_τ in terms of σ_1, σ_2 and θ:

$$\sigma_n = \sigma_1 \cos^2\theta + \sigma_2 \sin^2\theta$$
$$\sigma_\tau = (\sigma_1 - \sigma_2) \sin \theta \cos \theta \tag{3-22}$$

Substituting several trigonometric identities ($\sin 2\theta = 2\sin \theta \cos \theta$, $\cos^2\theta = (1 + \cos 2\theta)/2$, $\sin^2\theta = (1 - \cos 2\theta)/2$), we obtain a more useful form of the equations:

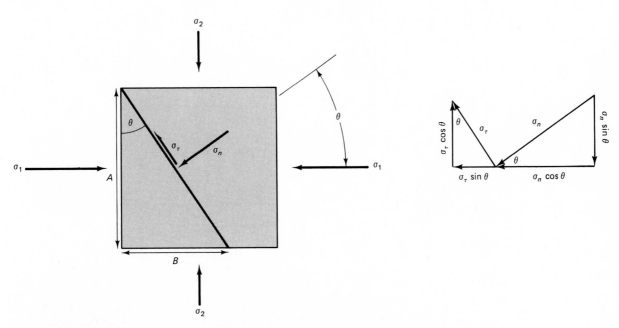

FIGURE 3–23 State of stress σ_n and σ_τ on a plane within a cube with normal stresses σ_1 and σ_2 acting on the faces. The normal to the plane has an inclination θ to the 1-axis.

$$\sigma_n = \left(\frac{\sigma_1 + \sigma_2}{2}\right) + \left(\frac{\sigma_1 - \sigma_2}{2}\right) \cos 2\theta \tag{3-23}$$

$$\sigma_\tau = \left(\frac{\sigma_1 - \sigma_2}{2}\right) \sin 2\theta$$

We may now graph σ_n versus σ_τ for all values of θ (that is, planes of all orientations θ) for a given set of principal stresses, σ_1 and σ_2 (Fig. 3-24). The stress (σ_n, σ_τ) on a single plane plots as a point. The state of stress on all possible planes graphs as a circle of radius $(\sigma_1 - \sigma_2)/2$, whose center is $(\sigma_1 + \sigma_2)/2$ from the origin. This graph of the state of stress is called the *Mohr circle*, named for Otto Mohr, who developed it. The radius of the circle is the deviatoric stress and its center is the mean stress. Values of 2θ can be measured directly on the diagram, as shown in Figure 3-24. The Mohr diagram illustrates some important properties of stress and has some important applications, as outlined below.

Planes perpendicular to the 1 and 2 directions have no shear stresses acting across them; these planes plot along the σ_n axis, $(\sigma_1, 0)$ and $(\sigma_2, 0)$. In contrast shear stress is a maximum at $2\theta = 90°$—that is, for planes oriented 45° to the principal directions. We also see that the larger the difference between σ_1 and σ_2 (the deviatoric stress), the larger the shear stress on any given plane. A hydrostatic, or isotropic, state of stress graphs on the Mohr diagram as a single point $(\sigma_n = \sigma_1 = \sigma_2, \sigma_\tau = 0)$ and no shear stresses are present. Compressive stresses are graphed to the right of the origin and tensile stresses to the left.

Given measurements of the state of stress on any two nonparallel planes, we may determine the principal stresses and their orientations graphically because the line connecting the two states of stress on a graph of σ_n versus σ_τ is a chord of the Mohr circle. Recall that the perpendicular bisector of a chord passes through

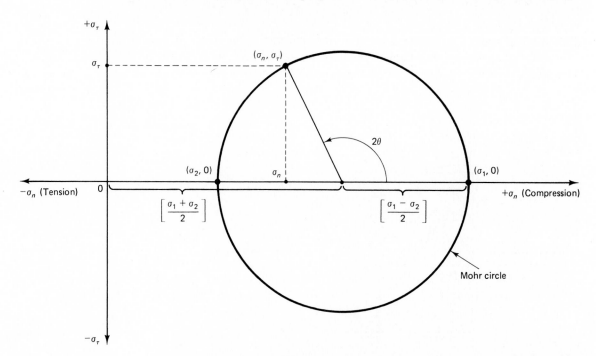

FIGURE 3–24 The $\sigma_n - \sigma_\tau$ stress space, which is called the *Mohr diagram for stress*. The state of stress at a point on a plane (σ_n, σ_τ) graphs as a point on this diagram. The state of stress on all planes passing through a point with principal stresses σ_1 and σ_2 graphs as a circle, called the *Mohr circle*, whose center is the mean stress and whose radius is the deviatoric stress. The definition of θ, specifying the orientation of a plane with respect to the principal axes, is shown in Figure 3-23.

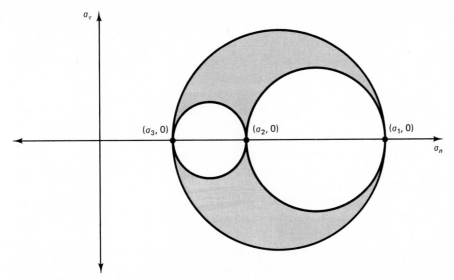

FIGURE 3–25 The Mohr diagram for three dimensions is composed of the Mohr circles for the three principal planes. The states of stress on any other planes plot within the shaded region.

the center of the circle. Therefore, we may locate the center of the circle where the bisector intersects the σ_n axis of the Mohr diagram.

Later we shall see that equations describing the state of stress required to fracture or frictionally slide along a plane in a rock may be plotted on these same $\sigma_n - \sigma_\tau$ coordinates; we shall then be able to see which planes are capable of slip or fracture for a given state of stress (σ_1 and σ_2). The effects of fluid pressure within a porous rock will also be illustrated with this graph. Similar equations may be used to develop a three-dimensional Mohr diagram (Fig. 3-25). This diagram has three circles corresponding to the σ_1–σ_2, σ_2–σ_3, σ_3–σ_1 planes; the state of stress on any arbitrary plane plots in the area between the circles.

Let's review the main points of the preceding paragraphs dealing with stress. The short-range forces acting within a body may be described in terms of forces per unit area, or stresses, acting across planes of various orientations and positions. The state of stress across a single plane may be described in terms of one normal-stress component, σ_n, which is the force per unit area acting perpendicular to the plane, and two shear-stress components, σ_τ, which describe the force per unit area acting parallel to the plane (Fig. 3-21). The sum of these three components is, in general, inclined to the plane. The state of stress on planes of all possible orientations through a single point may be described in terms of nine stress components, σ_{ij}, only six of which are independent (Fig. 3-22). A coordinate frame always can be found, in which all components are zero except for the normal stresses σ_{11}, σ_{22}, and σ_{33}, which are the principal stresses σ_1, σ_2, and σ_3. The axes of this coordinate frame are the principal stress axes. Given the principal stresses and their orientations, the state of stress on a plane of any orientation may be determined. The state of stress in two dimensions (σ_1 and σ_2) may be shown graphically on the Mohr diagram, in which states of stress (σ_n, σ_τ) on all planes graph as a circle whose radius is the deviatoric stress and whose center is the mean stress (Fig. 3-24).

EXERCISES

3–1 Show that $dV'/dV = S_1 S_2 S_3$ (Eq. 3-9).

3–2 Measure the ratio of principal stretches (S_1/S_2) for the deformed oolitic limestone in Figure 3-9 by measuring the lengths (V, W) of the principal axes of each ellipse and then plotting each set of measurements as a point on a cumulative graph of minimum axis V versus maximum axis W. For example, the first measurement, (V_1, W_1), is plotted at point (V_1, W_1), the second measurement, (V_2, W_2), at point $(V_1 + V_2, W_1 + W_2)$, and so on. A good estimate of the stretch ratio is then the slope of the line connecting the final point with the origin. What are the principal stretches if we assume a constant-area deformation and if we assume a 10 percent decrease in area?

3–3 Measure the ratio of the principal stretches for the slab of deformed trilobites in Figure 3-14 using Wellman's method. What is the orientation of the direction of principal elongation relative to the vertical edge of the photograph?

3–4 If a sedimentary rock (dry density, 2400 kg/m^3) whose grains have a density of 2650 kg/m^3 originally had a porosity (volume percent pores) of 50 percent and gained its present density by compaction due to burial and removal of pore fluid (no horizontal strain), what are the principal stretches (S_i) that describe this compaction? What would be the final ellipticity (ratios of principal axes of ellipse) of an originally spherical object that deformed with the sediments?

3–5 The distorted fossils observed in the bedding plane of Figure 3-11 are in flat-lying, unfolded rocks northwest of the Appalachian fold belt. Approximately the same level of strain is observed in a 100-km-wide zone adjacent to the fold belt with the principal shortening oriented perpendicular to the fold belt. The strain therefore records the outer limits of Appalachian deformation. Estimate the amount of horizontal shortening perpendicular to the fold belt recorded in this zone of flat-lying, but strained, rocks.

3–6 **(a)** Derive the equations for the Mohr's circle:

$$\sigma_n = \frac{\sigma_1 + \sigma_2}{2} + \frac{\sigma_1 - \sigma_2}{2} \cos 2\theta$$

$$\sigma_\tau = \frac{\sigma_1 - \sigma_2}{2} \sin 2\theta$$

(b) Let $\dfrac{\sigma_1 + \sigma_2}{2} = P$ and $\dfrac{\sigma_1 - \sigma_2}{2} = \Delta$. Show that these two parametric equations

can be combined into a single equation of the form

$$x^2 + y^2 = C^2$$

(showing explicitly that Mohr's circle is a circle).

3–7 **(a)** Suppose that the density of rock is 2.6 times that of water. Graph the vertical stress (normal to the earth's surface) as a function of depth for rock down to a 50-km depth. Compare this with the equivalent pressure of water. Suppose that no other stresses are applied and that the system is in equilibrium. What is the stress tensor as a function of depth? What is the mean stress?

(b) Now suppose in addition that there is a normal force applied by a colliding island arc in the east-west direction (call this the 2 axis). The force is constant with depth and of intensity 100 MPa. Now graph the horizontal stress on your graph of stress versus depth in part **a**. Draw the Mohr's circle for stress at 30 km and 3 km. What is the angle of the plane of maximum shear stress in each case (angle between normal to plane and vertical)?

3–8 Given the principal stresses 45 MPa and 80 MPa with σ_1 horizontal and σ_2 vertical, what is the state of stress on a bedding surface dipping 30°?

3–9 The state of stress (two-dimensional) is measured on two slots in a mine. One slot dips 32°E and has a normal compressive stress of 57 MPa and a shear stress of 12 MPa downdip. Another slot dips 84°E and has a normal compressive stress of 40 MPa and a shear stress of 3 MPa downdip. What is the state of stress in this region (what are the principal stresses and their orientations)? Recall that the perpendicular bisector of a chord passes through the center of the circle.

3–10 Given the principal stresses of magnitudes 63 MPa and 94 MPa, what is the state of stress on a joint face whose normal is oriented 53° to the σ_1 direction? The joint face will slide if the stress exceeds the static friction criterion $\sigma_\tau = 0.4$ MPa $+ \sigma_n\mu$, where μ is the coefficient of friction, 0.63 in this case. Will it slide?

3–11 Design a set of six independent measurements that will uniquely determine the state of stress in a mine assuming a homogeneous state of stress.

3–12 Suppose the strain recorded in Figure 3-9(b) took 2 m.y. to accumulate; what is the strain rate?

SELECTED LITERATURE

ELLIOTT, D., 1972, Deformation paths in structural geology: Geol. Soc. Amer. Bull. v. 83, p. 2621–2638.

MALVERN, L. E., 1969, *Introduction to the Mechanics of a Continuous Medium*, Prentice-Hall, Englewood Cliffs, N.J., 713 p.

MEANS, W. D., 1976, *Stress and Strain—Basic Concepts of Continuum Mechanics for Geologists;* Springer-Verlag, New York, 336 p.

NYE, J. F., 1957, *Physical Properties of Crystals*, Oxford, London, 322 p.

PFIFFNER, O. A., AND RAMSAY, J. G., 1982, Constraints on geological strain rates: arguments from finite strain states of naturally deformed rocks: J. Geophysical Research, v. 87, p. 311–321.

RAMSAY, J. G., 1967, *Folding and Fracturing of Rocks,* McGraw-Hill, New York, 568 p.

RAMSAY, J. G., AND HUBER, M. I., 1984, *The Techniques of Modern Structural Geology,* Vol. 1: Strain Analysis, Academic Press, London, 307 p.

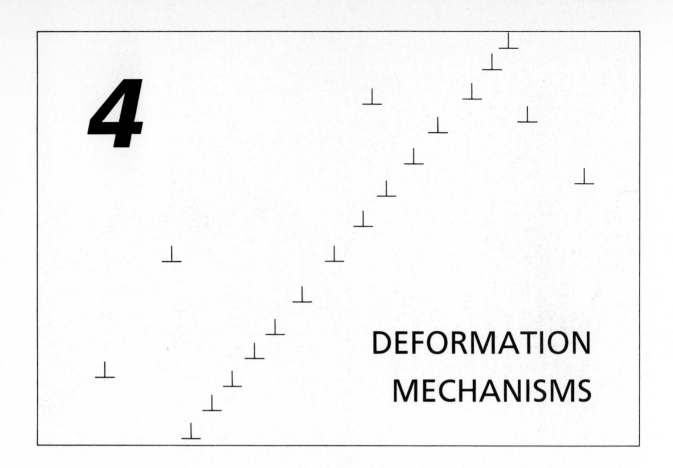

4

DEFORMATION MECHANISMS

INTRODUCTION

Rocks deform in response to the stresses acting locally within them, whatever the cause. Our concern in this chapter is the physical mechanisms by which rocks respond to these stresses on a microscopic or submicroscopic scale—that is, the *deformation,* or *strain, mechanisms.*

Only a few important deformation mechanisms exist; they may be divided into two categories: recoverable and permanent. The *recoverable mechanisms* are elastic deformation and thermal expansion; they do not involve breaking of atomic bonds—just their lengthening, shortening, and bending. If the cause of the deformation is removed, the body will return to its original size and shape. Recoverable strains in earth materials are generally less than a few percent.

If, however, the stress exceeds a critical value, called the *yield stress,* or *elastic limit* (Fig. 5-3), the rock will undergo *permanent deformation* by such mechanisms as fracture, sliding along grain boundaries, or motion of linear defects within crystals (dislocations). If the cause of the deformation is removed, much of the strain will remain. Permanent deformation can also be produced at stresses below the short-term yield stress given sufficiently long time and high-enough temperature because of *time-dependent mechanisms* such as diffusion of material from points of high stress to those of low stress or compaction by removal of pore fluid.

In this chapter we consider four important classes of distributed deformation: (1) elasticity, (2) plastic deformation within crystals by motion of linear crystal defects (dislocations), (3) flow by pressure solutions and (4) compaction by removal of pore fluid. Several other deformation processes that can be important on a microscopic scale will be treated elsewhere: frictional sliding in the chapter

FIGURE 4–1 Photomicrograph of deformed marble, El Paso Mountains, eastern California. The parallel bands are sets of deformational twin bands in calcite.

on faulting, buckling in the chapter on folding, and thermal expansion in the chapter on jointing, as well as in the following section. Fracture is the best-studied deformation mechanism because of its importance in engineering geology and rock mechanics; it warrants a separate chapter (Chapter 5).

Several deformation mechanisms may operate simultaneously. For example, the marble in Figure 4-1 has deformed by twinning of calcite, as shown by the sets of parallel twin bands. The crystal structure between the twin bands is distorted in some areas by motion of linear crystal defects (dislocations), as shown by the bending of the twin bands. Calcite has dissolved at points of high stress along grain boundaries and precipitated along cracks as veins, which are out of the field of view. The calcite in the veins also shows deformation by twinning. Thus twinning, motion of linear defects, fracture, and diffusion through pore fluid have operated, apparently simultaneously, to produce a deformation of this rock.

The dominant deformation mechanism of a rock varies with temperature, mean stress, deviatoric stress, pressure of the fluid in the pores, strain rate, chemical conditions, and history of the material. For example, near the surface of the earth, at low pressure and low temperature, many rocks deform by brittle fracture; at greater depths and higher temperatures, these same rocks flow by motion of linear crystal defects (dislocations) and diffusion.

ELASTICITY AND THERMAL EXPANSION

Elasticity and stress are closely related because the atomic or molecular bonds that transmit the stress are commonly elastic; that is, the relationship between force on the bond and bond distance is linear. The force on a bond is the atomic equivalent of stress and the percent change in bond length is the equivalent of strain.

We can get some appreciation for the physical origin of stress, elastic strain energy, and elastic constants by considering the changes in potential energy, U, and bonding force, F, that occur while a crystal is deformed. Bonds bend and twist, as well as compress and extend, during the deformation of actual crystals. In spite of such complexities, we can look at the deformation of a crystal qualitatively by considering the force on the bond and the potential energy of just a single pair of atoms in an ionic crystal such as halite (NaCl), while varying just bond length, r.

The potential energy of a pair of oppositely charged ions decreases as they are allowed to fall together due to their electrostatic attraction. At any distance r, the potential energy due to their attraction may be expressed as $-A/r$, where A is a constant related to the geometry of the crystal and the charges of the ions (Fig. 4-2). When the interatomic distance gets very small, however, there is a repulsion due to electron-cloud overlap. The ions must be forced to come closer together, causing an increase in potential energy. This repulsive increase may be approximated as B/r^{12}, where B is another constant. Note that the increase in potential energy is very rapid, as is shown by the dependence on interatomic distance r to approximately the -12 power.

By summing the attractive and repulsive contributions to the potential energy, we get the total potential energy, U, of the bond as a function of its length, r:

$$U = \frac{-A}{r} + \frac{B}{r^{12}} \tag{4-1}$$

The most-stable bond length ($r = d_0$ in Fig. 4-2) is the one that has the lowest potential energy, U_0. Electrostatic attraction dominates for bond lengths larger than d_0, whereas repulsion dominates for smaller bond lengths.

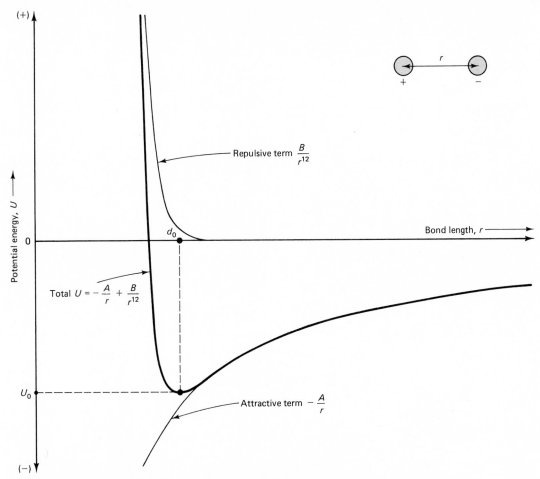

FIGURE 4–2 Schematic relationship between bond length, r, and potential energy of the bond, U. The equilibrium bond length, d_0, has the minimum energy, U_0.

In stressing a crystal there is a forced displacement of ions from their equilibrium separation (d_0). Whether bonds are lengthened or shortened, an increase in potential energy is produced (Fig. 4-2), which is called the *elastic energy*, or *elastic strain energy*. This energy is released when the atoms are allowed to return to their equilibrium, or stress-free, position (d_0). Alternatively, the elastic energy is the local source of energy for breaking bonds in permanent deformation if the yield stress is exceeded or if time-dependent deformation mechanisms are possible.

Now consider the force between the ions instead of their potential energy. From physics we recall that the force between the two ions is equal to the negative of the rate of change of potential energy with distance. That is, the force at each distance is equal to the slope of the potential energy curve (Fig. 4-2), which is the derivative of the potential energy equation (4-1):

$$F = \frac{dU}{dr} = \frac{-A}{r^2} + \frac{12B}{r^{13}} \qquad (4\text{-}2)$$

where compressive force is defined to be positive. The force equation is similar in form (Fig. 4-3) to the potential energy equation (Fig. 4-2) because the difference in the exponents of the two terms is so large. It is important to note that at equilibrium $(r = d_0)$, the force between the two ions is zero because the attractive and repulsive forces balance. A negative, or tensile, force is required to lengthen the bond and a positive, or compressive, force is required to shorten it.

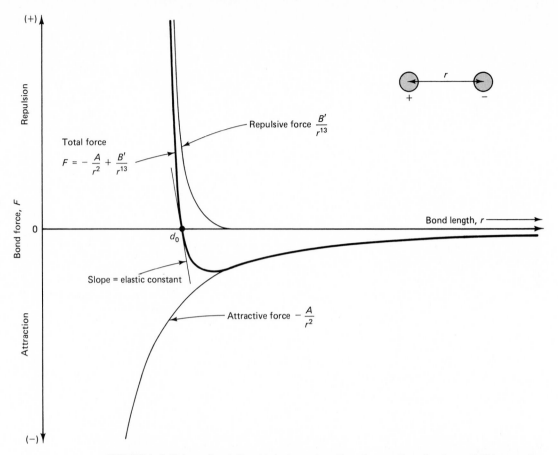

FIGURE 4–3 Schematic relationship between bond length, *r*, and bonding force, *F*. This curve is the derivative of the potential energy curve in Figure 4-2. The equilibrium bond length, d_0, has zero net force on the bond. The slope of the curve at d_0 is the elastic constant for the bond.

The force-distance curve is approximately a straight line near d_0 (Fig. 4-3), which implies that any forced change in bond length δr is equal to the change in force, δF, divided by a constant E, or

$$\delta F = E\delta r \qquad (4\text{-}3)$$

where E is the slope of the force–bond-length curve. On a macroscopic scale, in terms of stress σ and strain ϵ, Equation (4-3) becomes *Hooke's law*, the elastic equation ($\sigma = E\epsilon$). The elastic constant, E, is the first derivative of the force equation and the second derivative of the potential energy equation.

To have any permanent nonelastic deformation, the applied force must be large enough to break bonds. The critical force to break a bond is equal to the depth of the bonding force–bond-length curve. It is never reached simultaneously over most of the crystal; normally, it is only reached near imperfections that are sites of stress concentration or are weaker. If it were not for imperfections, solids would be orders of magnitude stronger and the surface of the earth would be a different world.

Before leaving the subject of stress and elasticity on an atomic scale, let us also consider the origin of thermal expansion. At absolute zero, T_0, and no stress, the interatomic distance has an equilibrium value d_0 with no vibration about that position. Increasing temperature increases the thermal vibration about the equilibrium position. As illustrated on the potential-energy–bond-length curve (Fig. 4-4), the thermal energy of the ion pair is $(U_2 - U_0)$ at temperature T_2. The bond

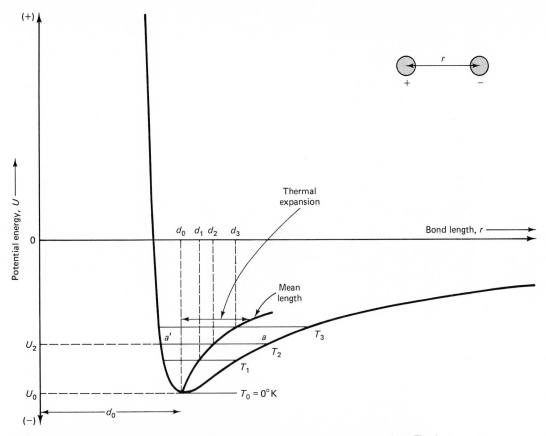

FIGURE 4–4 Schematic relationship between bond length, r, and bond energy, U (from Fig. 4-2), illustrating the effect of temperature on bond length. The mean length of the bond increases with temperature giving rise to thermal expansion.

vibrates along the curve with mean displacements a and a' to either side of the T_0-position (d_0, U_0). The mean interatomic distance at temperature T_2 is d_2, which is larger than d_0 because the trough of the potential energy curve is asymmetric, $(a - d_0) > (d_0 - a')$. This is the origin of thermal expansion. With an increase in temperature to T_3, the bond length will expand to d_3.

Because of the randomness of thermal vibration, any given bond at any instant may have a length and energy considerably greater or less than the mean values for that temperature (d_2 and U_2 for T_2 in Fig. 4-4). A few bonds may at times have enough energy to break and form different bonds. The higher the temperature, the more often there will be bonds with sufficient energy to break. For this reason the deformation mechanisms that involve diffusion, such as Herring-Nabarro creep and dislocation climb (see the next section) are more rapid at higher temperatures.

Finally, most minerals have different bonds in different directions and therefore have anisotropic thermal expansion and elasticity. If a crystal within a rock is not fully free to expand or contract upon change in temperature, *thermal stresses* develop, which can be important in rock deformation (see Chapter 6).

We have seen that stress on an atomic scale takes the form of forced elastic displacement of atoms from what otherwise would be their equilibrium positions. The increased potential energy associated with such forced lengthening or shortening of bonds is either used in permanent deformation or is released if the deforming forces are released and the atoms return to their stress-free positions.

Bulk elastic properties of rocks are relatively insensitive to irregularities on an atomic scale, such as missing atoms or linear defects, because the strain is distributed throughout the stressed material. Elastic strain is a bulk property to which all the bonds contribute. In contrast, most permanent deformation, except fluid flow, is concentrated in restricted zones controlled by irregularities, or flaws, in the material that are sites of local weakness or stress concentration. Thus yield stress depends very much upon flaws, whereas elastic deformation depends on the types of bonding, crystal structure, and preferred orientation of crystals in a rock. Cracks and grain boundaries also affect bulk elastic properties at low pressures.

PLASTIC DEFORMATION WITHIN CRYSTALS

Crystals are highly regular and symmetrical arrays of atoms. Nevertheless, they all contain large numbers of imperfections, both grown in and resulting from subsequent deformation. The crystal structure is orders of magnitude weaker because of these defects, and for this reason most plastic deformation within crystals is associated with the motion of defects through the crystal in response to applied stresses. We shall primarily consider point defects and linear defects that can be made to move when stressed.

Point Defects

A variety of point defects have been recognized in crystals (see Fig. 4-5(a)), including *substitutional impurities*, both self- and impurity *interstitials*, in which atoms occupy sites not usually occupied, and *vacancies*, which are regularly occupied sites that are left unfilled. Large, open structures like fluorite (CaF_2) have *Frenkel imperfections*, in which an atom leaves its regular site and occupies a nearby interstitial site; thus it is a vacancy–self-interstitial pair. Ionic crystals, such as halite (NaCl), may have *Schottky imperfections*, which are pairs of nearby vacancies with missing atoms of opposite charge, thereby satisfying the local charge balance. We primarily consider vacancies for purposes of illustration, but more-complex point defects (such as Frenkel and Schottky imperfections) are probably more important in rock-forming minerals.

A crystal containing a vacancy has more internal energy than a perfect crystal. This fact is illustrated in Figure 4-5(b), which shows the formation of a vacancy by taking an atom from inside a crystal and removing it to the crystal face. In this example it was necessary to break a total of ten bonds and form seven bonds to produce the vacancy. A total of three bonds were left broken, which is the energy required to form the vacancy.

Crystals contain more vacancies at higher temperature and fewer vacancies at higher pressure, as is observed in thermal expansion experiments. Thermal expansion of crystals can be measured in two ways: (1) The outside length of a crystal may be measured as a function of temperature, and (2) lattice spacing may be measured by X-ray diffraction as a function of temperature. The two methods do not yield exactly the same expansion; the bulk thermal expansion is greater than the lattice expansion. The difference is a result of the increased number of vacancies with increasing temperature. The equilibrium number of vacancies will form almost instantaneously within a crystal upon heating. It is possible to have more than the equilibrium number by cooling the crystal quickly (quenching), by stressing the crystal, or by bombarding it with protons or neutrons.

Vacancies are quite important for diffusion in solids; as vacancies diffuse in one direction, atoms in effect diffuse in the opposite direction. For example, when atom *a* in Figure 4-5(c) jumps into the vacancy, the vacancy in effect jumps to the

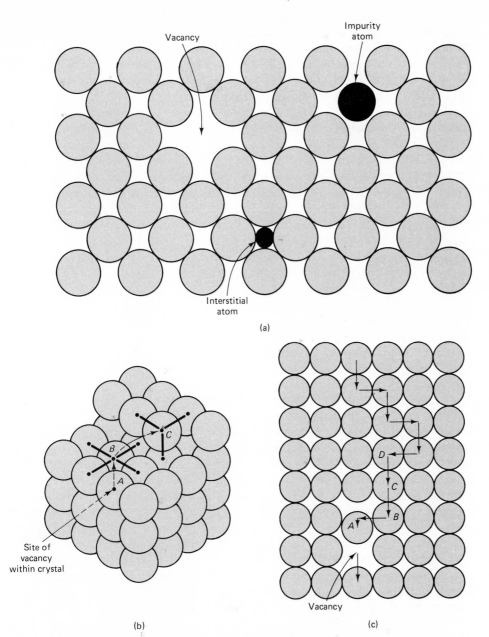

FIGURE 4–5 (a) Simple kinds of point defects. (b) Creation of a vacancy at *A*, one layer below the surface of the crystal, requires breaking five bonds of the atom at *B*, moving it to *C* on the surface of the crystal and forming three bonds, breaking five bonds of the atom at *A*, and moving it to *B*, forming four bonds. Thus the energy required to form the vacancy is the energy of the net three bonds broken. (c) Vacancies play an important role in solid-state diffusion. For example, atom *A* is moving into an adjacent vacancy, atom *B* may then move into the hole left by *A*, atom *C* may then move into the hole left by *B*, and so on, producing a flux of one atom downward across the crystal and one vacancy upward across the crystal.

former position of atom *a*. Diffusion without vacancies requires the atoms to move in interstitial paths, which commonly require higher thermal energy. It has been suggested that diffusion of vacancies might be an important deformation mechanism in some parts of the mantle. This process, called Herring-Nabarro creep, involves diffusion of vacancies toward crystal faces that are under compression and concurrent diffusion of atoms in the opposite direction toward less-stressed faces. Herring-Nabarro creep may be an important deformation mechanism at temperatures near the melting temperature (about 0.9 T_m K), small grain size, and at very low levels of differential stress. At higher stresses other mechanisms dominate, as is discussed later. Diffusion of vacancies is probably more important in less direct ways in rock deformation—for example, in the motion of line defects by the mechanism of climb, as we shall see shortly.

Translation and Twin Gliding

It is possible to stress crystals so that they slip along some internal crystallographic plane. This process of stress-induced slip within crystals is called *translation gliding* and is an important deformation mechanism in rocks, ice, metals, and other crystalline solids. For example, Figure 4-6(a) schematically shows that one-half of a crystal of halite (NaCl) has slipped past the other half along a single internal plane *A-A'*. Translation gliding is observed along crystallographic planes and directions that have a short distance between equivalent atoms, requiring a relatively small disturbance of the crystal structure for unit slip. For this reason, slip planes and slip directions have a relatively close packing of atoms; we would expect it to be more difficult to slip along internal plane *B-B'* than plane *A-A'* in Figure 4-6(a). Also, slip generally occurs in directions that do not juxtapose ions of like charge; for example, ionic crystals such as halite slip parallel to planes and directions of constant charge (for example, plane *A-A'* in Figure 4-6(a), but not plane *C-C'*). A particular crystallographic plane and direction along which slip may occur is called a *slip system.*

A second mode of translation within crystals is stress-induced twinning, or *twin gliding.* For example, Figure 4-6(b) schematically shows a band in the calcite lattice that is sheared 38.2° into a twinned state, the mirror image of the undeformed structure. Note that the total amount of shear a crystal may undergo is crystallographically fixed in twin gliding—38.2° in calcite; once the entire crystal is sheared to the twinned orientation, further deformation by this mechanism is impossible. In contrast, shear by translation gliding may assume any value. Twin gliding is a common deformation mechanism in plagioclase feldspar, dolomite, and calcite (Fig. 4-1). However, not all twinned crystals are formed by deformation; formation of twins during crystal growth is also common.

Early in the study of plastic deformation of crystals, it was learned that there is a critical value of the shear stress (acting on the slip plane and in the slip direction), below which translation gliding will not occur for a given temperature and strain rate. This behavior in crystals is given the name *Schmidt's law of strength,* which states that there exists a critical shear stress required for slip on any slip system. It is important to note that this critical shear stress is independent of the normal stress acting across the slip plane. This is in contrast with frictional behavior, which is observed across interfaces of low contact area or atomic cohesion. The critical shear stress for frictional slip is equal to the normal stress times the coefficient of friction (see Chapter 7).

If a crystal will slip along a certain plane and direction, it will also slip along all symmetrically equivalent planes and directions by virtue of the crystal's symmetry. For this reason crystals of high symmetry have many slip planes and directions. Slip starts first on the systems with the lowest critical shear strength. As slip proceeds, the different slip systems may start to interfere, for reasons we shall see, and increased stress is required for continued slip. The stress may get high enough to activate other slip systems of higher critical shear stress. This increase in stress required for continued deformation is called *work hardening,* well known in metals.

The individual crystals in a rock, or in any polycrystalline material, have many different orientations, so that their slip systems have many different orientations with respect to an applied stress. A slip system may operate in one

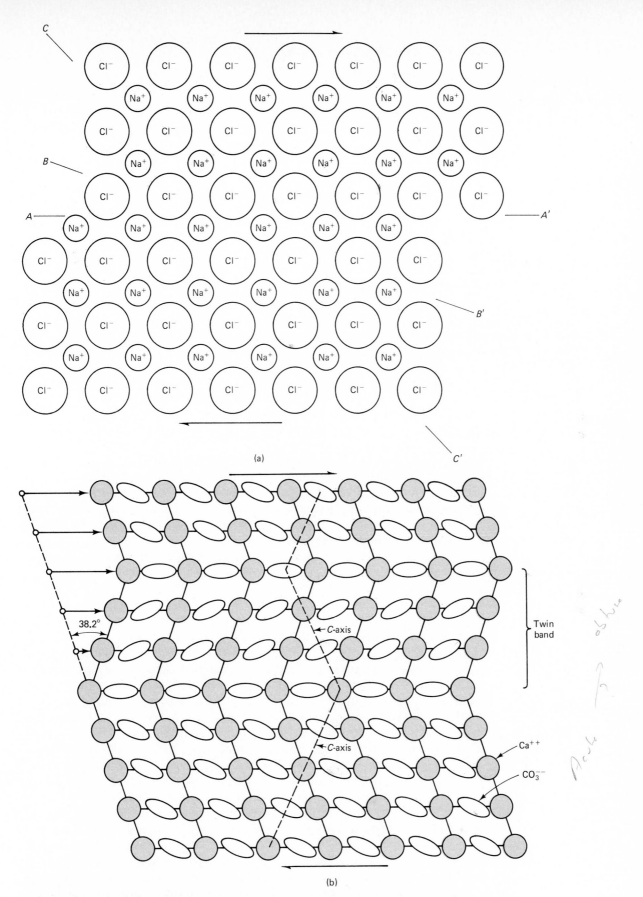

FIGURE 4–6 (a) Translation gliding of halite (NaCl) along the plane *A-A'*. (b) Twin gliding of calcite (CaCO₃). Note that the amount of slip is crystallographically controlled to be *2h* tan(38.2°/ 2), where *h* is the thickness of the twin band. In contrast, the amount of slip is unlimited crystallographically in translation gliding.

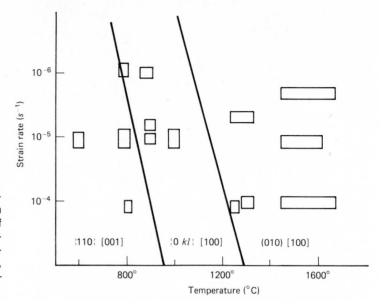

FIGURE 4–7 Illustration of the dependence of the dominant slip system in olivine on experimental conditions of temperature and strain rate. Boxes represent experimental conditions. (Simplified from Carter and Ave'Lallement, Geol. Soc. Amer. Bull., v. 81, p. 2181–2202, 1970.)

grain, but not in another. If the rock is to deform homogeneously at constant volume, there must be at least five independent active slip planes and directions. However, only four active planes may be required for heterogeneous deformation. If there are fewer active slip systems, the material will not be able to deform without voids opening along grain boundaries, which are in effect cracks that lead to fracture (see Chapter 5). Which slip systems operate at a given stress varies with temperature and strain rate; for example, Figure 4-7 shows the dominant slip systems observed in experiments on olivine as a function of temperature and strain rate. In some cases the transition between fracture and flow may be entirely due to an increase in the number of slip systems; for example, there are fewer slip systems in iron at lower temperatures, which accounts for some cold-weather failures of bridges and ships.

Motion of Dislocations

The stress required for simultaneous slip of all the atoms along an atomic plane by translation gliding apparently should be a periodic function due to the periodic crystal structure because all the bonds must first be broken and then reformed as the atoms reach the next equivalent lattice site. In 1926 Frenkel estimated that the maximum stress to produce the simultaneous slip should be about $\frac{1}{30}$ the shear elastic modulus of the crystal (about 10^5 MPa). More elaborate estimates and subsequent experiments with perfect-crystal whiskers and perfect analog cystals of bubbles have borne out this estimate. In contrast, the observed critical shear stress, in all but these special experiments, is about 10^{-5} of the shear modulus. This observation that crystals are orders of magnitude weaker than predicted for a perfect crystal led to the idea that slip starts at local imperfections.

In 1934 Taylor, Orowan, and Polanyi proposed the existence of *dislocations*, a type of line defect that would be the important imperfection for translation gliding. A dislocation in its simplest form is the termination of an extra plane of atoms (Fig. 4-8). This linear defect in the crystal structure can be made to move in response to shear stress, thereby producing slip within the crystal. For example, in Figure 4-9 bond *a* is broken and bond *b* is formed, thereby shifting the dislocation one lattice dimension to the right with respect to the atoms of the crystal. Continued slip will move the dislocation out to the edge of the crystal, thereby producing a step of one lattice dimension on the grain boundary and

down; the loop does not close because of the dislocation. The closure vector, **b**, is the Burger's vector. For normal dislocations the length of the Burger's vector is an integral number of lattice spacings, usually one, and its orientation and magnitude is a constant along the dislocation line. The plane containing the tangent vector and the Burger's vector is called the *slip plane*.

There are two simple types of dislocations: (1) the *edge dislocation* (Fig. 4-8), with the tangent vector perpendicular to the Burger's vector, and (2) the *screw dislocation*, with the tangent vector parallel to the Burger's vector. The screw dislocation has no intrinsic slip plane and does not need to move in the slip plane of the related edge dislocation. It is called a screw dislocation because the atomic planes spiral around the dislocation (Fig. 4-12). Except when dislocation loops are strongly controlled crystallographically, as in olivine (Fig. 4-10), a dislocation will, in general, have some mixture of edge and screw components.

The two opposite points of the dislocation loop with pure edge component have their extra half-planes on opposite sides of the slip plane, one pointing up and one pointing down. Similarly, the two opposite points with pure screw component spiral in the opposite sense. These opposing segments of the dislocation loop have

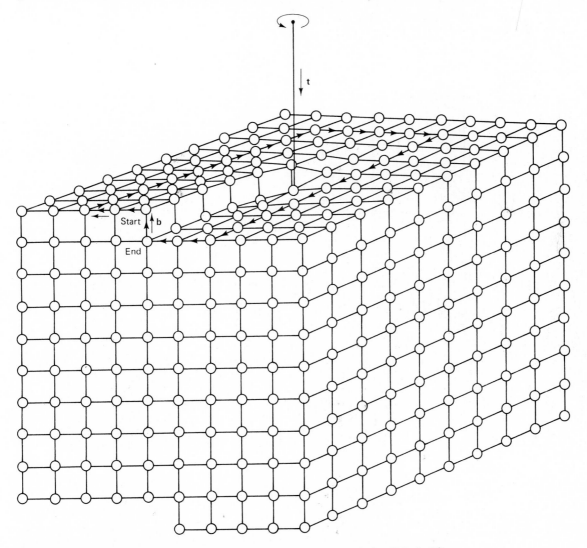

FIGURE 4–12 Schematic block diagram of a screw dislocation in a cubic lattice. Note that the Burger's and tangent vectors are parallel for a screw dislocation in contrast with edge dislocations (Fig. 4-8); therefore, it has no slip plane.

tangent vectors of opposite sign, and if two dislocations of opposite sign are brought together, they will merge to form a perfect crystal. For example, if a dislocation loop were shrunk down to zero size, no trace of it would be left, just perfect crystal.

When a dislocation is swept out of a crystal in response to an applied shear stress (Fig. 4-9), a step is left on the crystal face one Burger's vector wide—for example, 7 Å in olivine. Clearly, a very large number of dislocations must be moved to produce a noticeable deformation. How many dislocations are there in a crystal—are there enough to produce the strains we observe? We can measure the dislocation density, for example, by counting the number of etch pits per unit area, which is equivalent to the length of dislocation line per unit volume. Undeformed crystals have a density of about 10^3 to $10^4/cm^2$, whereas densities may be as high as 10^8 to $10^{12}/cm^2$ in the same crystal after plastic deformation at relatively low temperature. Clearly, there are mechanisms that produce new dislocations during deformation; crystals are not at a loss for dislocations.

One of the simplest dislocation sources is the *Frank-Read source,* an example of which is a mobile segment of an edge dislocation that is pinned by immobile segments at each end (Fig. 4-13). Given the critical shear stress on the slip plane, the mobile segment will bow out like a rubber band and finally loop back on itself. The two opposite sides of the dislocation—where they touch—are screw segments of opposite sign, so they merge to form a complete dislocation loop plus the original Frank-Read source back again. The process continues,

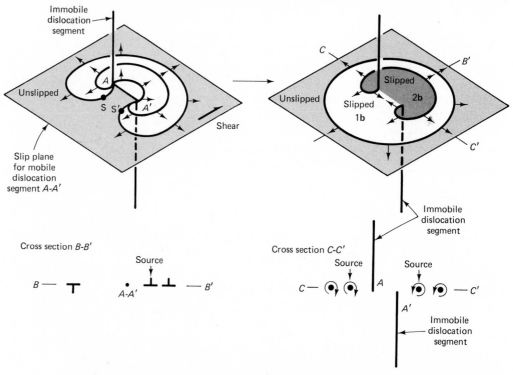

FIGURE 4–13 Schematic operation of a dislocation source composed of a mobile segment of an edge dislocation pinned by immobile segments at its end points *A* and *A'*. As a shear stress is applied to the slip plane, the dislocation segment bows outward, being pinned at points *A* and *A'*, as shown on the left. The arrows show the direction of motion of the dislocation line. At the last stage shown on the left the two opposite sides of the loop *S* and *S'* are about to touch, creating a complete new dislocation loop and preserving the source to create more loops, as shown on the right. *B-B'* is a cross section in which the edge dislocations are shown with the symbol T; an upright T has a downward-extending half-plane, whereas an upside-down T has an upward-extending half-plane. Cross section *C-C'* shows the opposite spiral sense of screw dislocations on opposite sides of the dislocation loop.

forming thousands of loops; the area inside each successive loop has slipped one Burger's vector more than the area outside. The stress needed to activate a dislocation source is thought to be the critical shear stress for translation gliding (Schmidt's law). Therefore, dislocation sources are the most important crystal imperfections for flow by motion of dislocations.

As deformation proceeds, the dislocations from all the different slip systems get tangled up, making further deformation more and more difficult. In the following paragraphs, we consider three mechanisms by which dislocations become more difficult to move and result in work hardening.

1. *Self-Stress Field.* Dislocations induce a *self-stress field* because they disturb the normal crystal structure. For example, the end of the extra half-plane in an edge dislocation (Fig. 4-8) causes a compressive stress field above the end of the half-plane and a tensile stress field below. This self-stress field dies away in proportion to one over the square of the distance from the dislocation. Self-stress fields of nearby dislocations interact and cause some dislocations to repel each other and others attract. For example, two edge dislocations of the same sign lying on the same slip plane repel each other, whereas two edge dislocations of opposite sign attract each other. In general, the more dislocations within a crystal, the higher their total repulsion and the higher the stress required to move them.

Therefore, as deformation proceeds—producing more dislocations—the stress required to continue deformation increases. There is, however, a limit to the number of dislocations because some are always leaving the crystal or are being annihilated by dislocations of opposite sign. At any temperature a steady state is soon reached at which the density of dislocations, ρ_d, does not change with time. At steady state there is a regular relationship between dislocation density and deviatoric stress, Δ, squared:

$$\rho_d = C\Delta^2 \qquad\qquad (4\text{-}4)$$

where C is a constant nearly independent of temperature. Figure 4-14 illustrates this relationship for experimentally deformed quartz and olivine. This experimental relationship has been used to estimate paleostresses in naturally deformed samples for which dislocation density is known; for example, typical stress differences estimated from deeply eroded plastic fault zones in a number of regions are 100 to 200 MPa, based on dislocation densities in quartz (Kohlstedt and Weathers, 1980).

2. *Dislocation Pileups.* A second mechanism by which dislocations become difficult to move involves immobile impurities. If an immobile, very tightly bonded impurity lies in the slip plane, an edge dislocation will not be able to slip through it (Fig. 4-15). As more dislocations encounter the obstacle, they tend to pile up because they are repelled by the other dislocations in addition to the impurity atom. The obstacle can be sidestepped if the temperature is high enough for diffusion of atoms or vacancies to the dislocation, thereby moving the edge of the extra half plane and the slip plane down or up (Fig. 4-15); this important process is called *dislocation climb*.

Glide (Fig. 4-9) and climb (Fig. 4-15) are the main mechanisms by which dislocations move through crystals. The velocity v at which dislocations move by this combination of mechanisms is observed experimentally to be proportional to the deviatoric stress Δ:

$$v = k\Delta \qquad\qquad (4\text{-}5)$$

where k is a constant that is a strong function of temperature because diffusion is involved in climb,

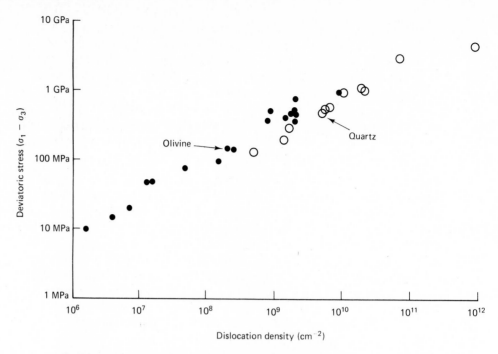

FIGURE 4–14 Experimentally determined relationship between dislocation density and deviatoric stress for quartz and olivine. (Kohlstedt and Weathers, J. Geophys. Research, v. 85, p. 6269–6285, 1980, copyrighted by the American Geophysical Union.)

$$k = K\exp\left(\frac{-Q}{RT}\right) \tag{4-6}$$

where K is another constant, R is the gas constant, T is the absolute temperature, and Q is the activation energy. The climb velocity of dislocations, v, becomes large as temperature, T, increases. As a result work hardening is less important at higher temperatures because pile-ups are circumvented.

3. *Jogs.* A third mechanism of work hardening is the production of immobile *jogs* in dislocation lines when dislocations of different slip systems move through one another. For example (Fig. 4-16), if an edge dislocation d_1 moves through a perpendicular edge dislocation d_2, it will offset, or jog, it by one Burger's vector, $\mathbf{b}_1$. If the two dislocations have perpendicular Burger's vectors, as is the case in Figure 4-16, the jogged segment of the second dislocation (d_2) will also be an edge dislocation, but with a slip plane perpendicular to the main segment of dislocation. Recall that the slip plane is the plane containing the dislocation line and Burger's vector. The second dislocation (d_2) will now be more difficult to move because the jog will not, in general, glide at the same stress as the rest of the dislocation; climb is required. Under these conditions, the jog acts as a drag on the dislocation (Fig. 4-16) and only can be made to move by diffusion of vacancies to it. Motion of jogs and climb around obstacles are favored by increased temperature because they involve diffusion.

In the previous paragraphs we have seen that temperature exerts an important control on plastic deformation because of the strong temperature dependence of climb (Eq. 4-6). For this reason we can qualitatively divide plastic deformation into (1) low-temperature plastic deformation displaying important work hardening, sometimes called *cold working,* and (2) high-temperature deformation, or *hot working,* displaying relatively little work hardening. These two modes of deformation produce significant differences in the microstructure of the

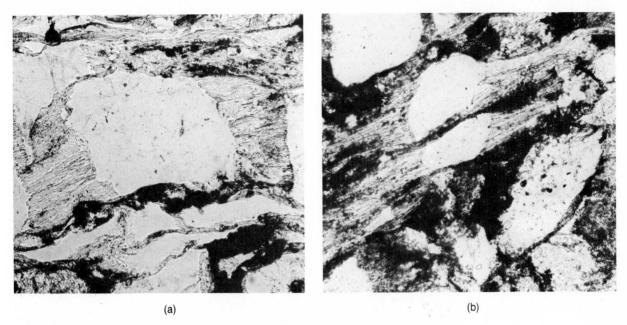

(a)   (b)

FIGURE 4–22 Pressure solution of quartz sand grains. The grain contacts parallel to the foliation show flattening by dissolution. Grain contacts perpendicular to the foliation have opened up with the space filled by fine-grained fibrous crystals of quartz and chlorite oriented parallel to the extension direction.

newly precipitated fibrous quartz and chlorite. The long axis of the fibers records the direction of elongation of the rock. As a result of this systematic dissolution and precipitation, the rock as a whole shortens perpendicular to the foliation and stretches parallel to the foliation in a relatively homogeneous manner.

Some pressure solution is much less homogeneous. For example, many limestones undergo substantial low-temperature deformation by dissolution along bedding planes and preexisting fractures that are subject to compression. Many centimeters of rock can be dissolved along these surfaces; therefore, they are commonly lined with a layer of insoluble residue. Because one or the other wall of the fracture is locally more soluble, the surface of dissolution develops a highly irregular, toothed shape; these irregular surfaces of dissolution are called *stylolites* (Fig. 4-23(a)). The teeth of the stylolites are normal to preexisting fracture surfaces that are perpendicular to the principal compression, whereas the axes of the teeth are inclined to fracture surfaces that experience large shear stress. The axes of the stylolite teeth are quite regular in orientation—probably parallel to the σ_1 direction. For example, Figure 4-23(b) shows the regular pattern of orientation of stylolite teeth through the Pliocene Jura fold belt; the teeth are oriented roughly perpendicular to the fold axes, which is the predicted orientation of σ_1. Most of these stylolites formed at depths of 1 km or less.

The precise mechanisms of pressure solution are not well documented; nevertheless, a qualitative explanation of the relationship between stress and dissolution can be given as follows. The mineral grains in a rock are more highly compressed near grain boundaries oriented perpendicular to the maximum compression or at point contacts of sand grains in sediments than along less-compressed grain boundaries or along the walls of fluid-filled cracks and pores. The highly stressed point contacts of the grains have a higher mean stress and therefore have more internal energy than the same mineral elsewhere in the rock because the bonds are more highly compressed and thus contain more elastic energy. Because of this greater internal energy, the stressed solid is less stable and more soluble in the pore fluid than the less compressed solid. The increase in

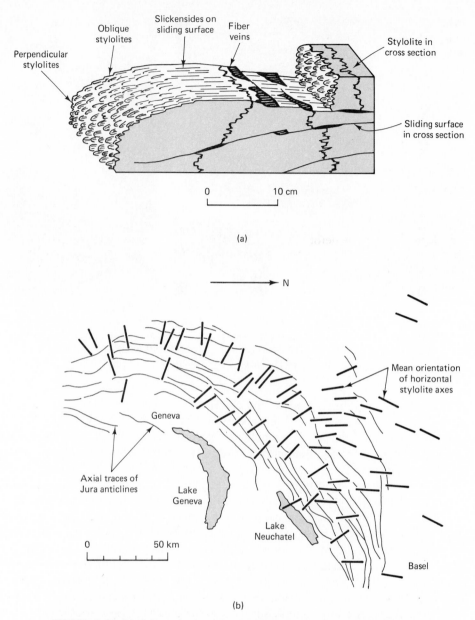

FIGURE 4–23 (a) Schematic drawing of relationship among perpendicular and oblique stylolites, sliding surfaces, and veins in deformation of limestones by pressure solution. (Modified after Arthaud and Mattauer, 1969.) (b) Map pattern of horizontal stylolite axes in the Pliocene Jura foldbelt of Switzerland and France. This map may approximate the orientation of the maximum compressive stress σ_1. (Simplified from Plessmann, 1972.)

solubility, δC, caused by an increase in mean stress or pressure, $\delta\sigma_m$, is approximately

$$\delta C = C_0(e^{\delta\sigma_m v/RT} - 1) \tag{4-7}$$

where C_0 is the initial solubility, V is the molar volume, R is the gas constant, and T is the absolute temperature. The diffusion of material from points of high stress to points of low stress takes place because of the concentration gradient, dC/dX_j, the diffusion rate, or flux, J_i, being

$$J_i = -D_{ij}\frac{dC}{dX_j} \qquad i, j = 1, 2, 3 \tag{4-8}$$

which is *Fick's first law of diffusion*. D_{ij} is the diffusion coefficient. The concentration gradient, which is produced by the stress difference, drives the diffusion. For this reason, pressure solution is sometimes called *grain-boundary diffusion flow*.

The rate of diffusion should decrease with grain size because the larger the grain the greater the distance to diffuse. This prediction agrees with the observation that pressure solution is more important in finer-grained rocks. Pressure solution also is developed more strongly in rocks having fine-grained clay between the sand grains.

COMPACTION

Compaction, the final mechanism of distributed deformation that we will consider, is important in the deformation of sediments, either due to loading by overlying sediment or to tectonic loading. Compaction is apparently also important to the mechanics of extraction of magma from partially melted rock in the mantle or lower crust. Most geologically important compaction involves the removal of a fluid phase from a porous solid; however, the fluid is negligible or absent in some cases, as in compaction of hot volcanic ash or lunar soil. We consider here the role of fluid flow and fluid pressure in compaction.

When sediment is first deposited, it contains substantial interstitial water. The volume percentage occupied by the fluid, or voids, is called the porosity, ϕ:

$$\phi = \frac{V_f}{V_f + V_s} \tag{4-9}$$

where V_f and V_s are the volumes of the fluid and solid, respectively. Another useful measure is void ratio θ_v, defined as

$$\theta_v = \frac{V_f}{V_s} \tag{4-10}$$

As more and more sediment is piled on top, the sediment in question undergoes *compaction*—reduction in porosity—in response to the overlying load. The granular framework cannot support the additional load and therefore deforms to a more-stable, less-porous configuration that can support the overburden.

Some muds at the sea floor have porosities as high as 80 percent, whereas Paleozoic shales buried to a depth of a kilometer or two may have a porosity less than 10 percent. Sandstones may have initial porosities as high as 45 percent, but soon after burial are reduced to 30 percent or even much less if they have a wide variety of grain sizes. Typical petroleum reservoir sands have porosities in the range of 10 to 30 percent.

The more deeply a sediment is buried, the lower its porosity, as a general rule. Figure 4-24(a) shows some observed porosity-depth relationships for shales from the Paleozoic of Oklahoma and the Tertiary of the Texas Gulf Coast and Venezuela. Observations such as these lead to an empirical relationship between porosity and depth z in well-compacted sediments, called *Athy's law:*

$$\phi = \phi_0 e^{-az} \tag{4-11}$$

where ϕ_0 and a are constants that vary with the sediment type and geologic province. Furthermore, there may be some relationship between age and state of compaction. Older sediments are more fully compacted, suggesting that compaction may be a time-dependent process. Compaction cannot occur instantaneously because it takes time for the pore fluid to leave the sediments and because time-dependent mechanisms like pressure solution are important in some compaction.

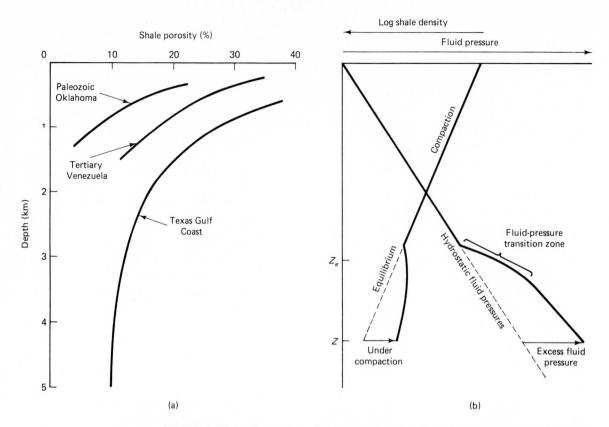

FIGURE 4–24 (a) Compaction of shale as a function of depth in several areas. (Data from Hedberg, 1936.) (b) Schematic relationship between degree of undercompaction and excess fluid pressure.

A complete theory of compaction would require consideration of the geometric details of the granular framework and all the deformation mechanisms operating to deform it—for example, grain fracture, plastic deformation, pressure solution, and slip of grains past one another. This understanding is largely unavailable, although different deformation mechanisms and granular geometries will in general lead to different constants in the porosity-depth equation (4-11). The best understood aspect of compaction is its relationship to pore-fluid pressure.

The behavior of a fluid-filled sediment undergoing compaction can be illustrated with Terzaghi's model, widely used in soil mechanics. The schematic model (Fig. 4-25(a)) consists of a chamber containing perforated plates separated by elastic springs. The assemblage of springs and plates is meant to model the solid granular framework of a sediment and the perforations would model its permeability. If we suddenly apply a force or weight W to the top, the structure will immediately compact so that the load supported by the springs S is equal to the applied weight,

$$W = S \tag{4-12}$$

The load S supported by the springs represents the stress supported by the solid granular framework of a sediment, called the *framework stress*.

If, instead, we fill the apparatus with water, it behaves in a substantially different way (Fig. 4-25(b)). Initially, the water is at an equilibrium fluid pressure, P_e, determined by its density and depth:

$$P_e = \rho_{H_2O}\, gz \tag{4-13}$$

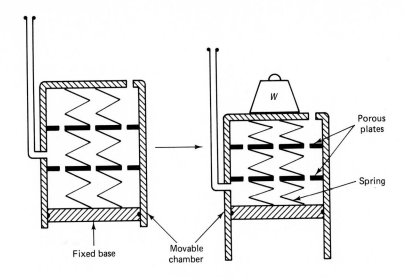

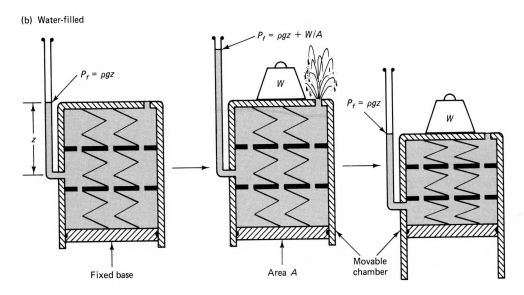

FIGURE 4–25 Terzaghi's analog model of sediment compaction, consisting of a movable chamber containing springs, porous plates, and a tube to measure the fluid pressure. (a) The dry apparatus compacts almost instantaneously in response to the load, W. (b) The water-filled apparatus responds to the load with a sudden increase in water pressure. As the water leaks out, the apparatus compacts until the pressure returns to the hydrostatic value (ρgz).

Immediately after the load W is applied to the top, the fluid pressure, P_f, is

$$P_f = P_e + \frac{W}{A} \qquad (4\text{-}14)$$

where A is the area of the plate. In contrast with the dry case (Fig. 4-25(a)), the springs support none of the applied load initially and do not compress. It is only as the fluid is allowed to leak out and the excess fluid pressure ($P_f - P_e$) decays away that the apparatus is able to compress and support some of the applied load. At this stage the weight is supported partly by the springs and partly by the fluid pressure:

$$W = (P_f - \rho_{H_2O}gz)A + S \qquad (4\text{-}15)$$

The apparatus will finally stop compressing when the fluid pressure returns to equilibrium (Eq. 4-13), at which point the load is supported entirely by the solid framework.

Compacting weak sediments, such as mudstones, behave in a manner that is very similar to Terzaghi's model, although most of the deformation is nonelastic. Sediments that are less compacted than the equilibrium porosity for their depth, described by Equation 4-11, have fluid pressures that are higher than the equilibrium or hydrostatic fluid pressure. The relationship between shale density and fluid pressure in a well is shown schematically in Figure 4-24(b). The amount of excess fluid pressure is directly related to the degree of undercompaction of the sediment. If an undercompacted sediment at depth z has the same porosity as completely compacted sediment at a shallower depth z_e, then the fluid pressure at depth z is

$$P_f = \rho_{H_2O}gz_e + \rho_r g(z - z_e) \qquad (4\text{-}16)$$

where ρ_r is the density of the rock.

Excess fluid pressures are commonly encountered in petroleum exploration of thick stratigraphic sequences of low permeability, in which disequilibrium compaction can persist for tens to hundreds of millions of years. These excess fluid pressures exert an important control on fracturing and faulting, as discussed in Chapters 5, 6, and 8.

STRESS-STRAIN RELATIONS

Having discussed the main physical mechanisms of distributed deformation in rocks, we may now consider the net effect of these mechanisms on the overall behavior of a rock. That is, what is the functional relationship between stress and strain, or strain rate? Such relations, called *stress-strain relations, rheologic models,* or *constitutive relations,* may be theoretical or empirical in origin. They will be useful in considering the origins of specific structures—for example, folds (Chapter 9).

Stress-strain relations in their simplest form may easily be visualized using conceptual mechanical models. Elastic behavior, for which stress is proportional to strain ($\sigma = E\epsilon$), can be visualized with a spring model (Fig. 4-26(a)). Viscous behavior, for which stress is proportional to strain rate ($\sigma = \eta\dot{\epsilon}$) may be visualized with a dash pot, which is a leaky piston and cylinder with fluid in both sides of the piston (Fig. 4-26(b)). Elastico-viscous (Maxwell) behavior ($\dot{\epsilon} = \dot{\sigma}/E + \sigma/\eta$) may be visualized as a dash pot and spring in series (Fig. 4-26(c)). Firmo-viscous (Kelvin or Voigt) behavior ($\dot{\sigma} = \eta\dot{\epsilon} + E\epsilon$) may be visualized as a dash pot and spring in parallel (Fig. 4-26(d)). For each of these mechanisms, a strain-time and a stress-time curve are shown for a sudden application of a load F at time t_1 and a removal of the load at time t_2. The elastico-viscous material shows exponential relaxation of the stress at constant strain. The firmo-viscous material shows transient creep (strain at constant stress).

Some materials approximate each of the simple model behaviors discussed above under certain conditions, but natural stress-strain relations of rocks are commonly more complicated. The equations relating stress and strain in rocks normally have more than one constant. For example, the elastic equation relating stress and strain, Hooke's law,

$$\sigma_{ij} = C_{ijkl}\,\epsilon_{kl} \qquad i, j, k, l = 1, 2, 3 \qquad (4\text{-}17)$$

will formally have $9 \times 9 = 81$ constants, although not all will be independent. Since both stress and strain have 6 independent components, the elastic stiffness

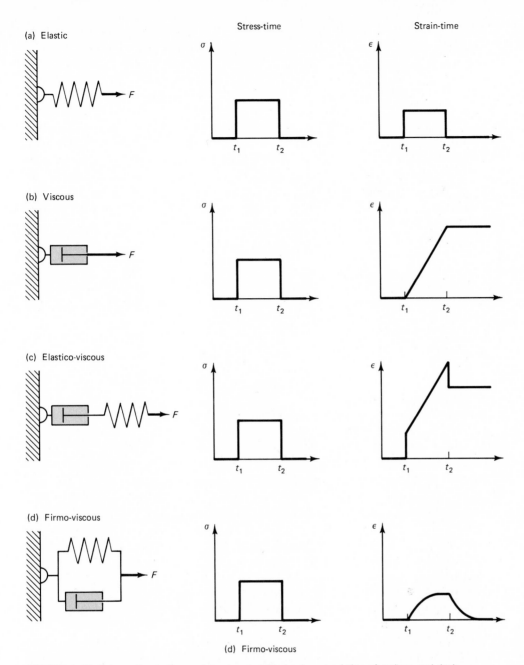

FIGURE 4–26 Schematic analog models of material behavior, consisting of springs and dash pots (fluid-filled piston and cylinder). The strain history produced in response to the sudden application of a force at time t_1 and removal of the force at time t_2 is shown.

(C_{ijkl}) can have no more than $6 \times 6 = 36$ independent components. A thermodynamic argument involving conservation of energy reduces this number to 21. The actual number of independent components may be obtained by applying *Neumann's principle,* which states that the symmetry of a physical property of a crystal or other material is no less symmetric than its point-group symmetry, that is, the symmetry of its external crystallographic form. Table 4-1 gives the number of independent elastic stiffness constants for each crystal-symmetry class, which ranges from 21 for the least symmetric to 2 for isotropic materials.

For isotropic materials, *Young's modulus, E,* and *Poisson's ratio, v,* are commonly used instead of the two stiffness coefficients (C_{1111} and C_{1122}), to which they are related as follows:

TABLE 4–1 Effect of Crystal Symmetry on Number of Independent Elastic Stiffness Constants C_{ijkl} and Thermal Expansion Coefficients α_{ij}

Crystal Class	Number of Independent Elastic Constants	Number of Independent Thermal Expansion Coefficients
Triclinic	21	6
Monoclinic	13	4
Orthorhombic	9	3
Tetragonal		
Classes 4, $\bar{4}$, 4/m	7	2
Classes 4mm, $\bar{4}2m$, 422, 4/mmm	6	2
Trigonal		
Classes 3, $\bar{3}$	7	2
Classes 32, $\bar{3}m$, 3m	6	2
Hexagonal	5	2
Cubic	3	1
Isotropic	2	1

$$C_{1111} = E \quad \text{and} \quad C_{1122} = \frac{-E}{\nu} \qquad (4\text{-}18)$$

For a simple axial compression of a cylinder (Fig. 4-27), Young's modulus is the ratio of stress to the longitudinal strain:

$$E = \frac{\sigma}{\epsilon_l} \qquad (4\text{-}19)$$

Poisson's ratio is the ratio of transverse to longitudinal strain:

$$\nu = \frac{\epsilon_t}{\epsilon_l} \qquad (4\text{-}20)$$

It should be noted that many rocks, perhaps the majority, are elastically anisotropic because they have a strong preferred orientation of anisotropic minerals such as quartz, feldspar, and olivine. Thus any schistose or foliated rock will have anisotropic elasticity. Large volumes of crust and upper mantle are elastically anisotropic; for example, there is commonly a 5 to 10 percent anisotropy in compressional-wave seismic velocity* in oceanic upper mantle, measured parallel and perpendicular to oceanic ridges. This difference is due to a strong preferred orientation of olivine, which is produced by solid-state flow during the formation of oceanic lithosphere. Olivine itself has a 20 to 25 percent anisotropy.

Thermal expansion is described by a simpler constitutive relationship than elasticity and has fewer independent coefficients (Table 4-1) because temperature is a scalar:

$$\epsilon_{ij} = \alpha_{ij}\Delta T \qquad i, j = 1, 2, 3 \qquad (4\text{-}21)$$

The principal strain directions must correspond to the principal crystallographic axes except for the triclinic and monoclinic (in part) cases.

The stress-strain relations for permanent deformation of rocks are considerably more complicated than the mechanical models of Figure 4-26. Stress-strain relations corresponding to important mechanisms of plastic deformation have been derived by a combination of empirical and theoretical considerations. These equations are somewhat tentative, subject to modification, and apply strictly to

---

*Seismic velocity is a simple function of the adiabatic elastic constants and density ρ. For the isotropic case the compressional-wave velocity is $V_p = [E(1 - \nu)(1 - 2\nu)\rho]^{1/2}$ and the shear-wave velocity is $V_s = [E/2\rho(1 + \nu)]^{1/2}$.

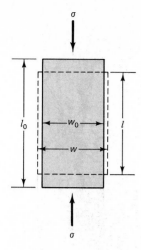

Longitudinal strain, $\epsilon_l = (l_0 - l)/l_0$

Transverse strain, $\epsilon_t = (w_0 - w)/w_0$

Young's modulus, $E = \sigma/\epsilon_l$

Poisson's ratio, $\nu = \epsilon_t/\epsilon_l$

FIGURE 4–27 Definition of Young's modulus, E, and Poisson's ratio, ν, for the axial loading of a cylinder of initial length l_0 and width w_0 and final length l and width w.

special geometric cases; nevertheless, they give considerable insight into the bulk behavior of rocks. The equations describing deformation by dislocation glide and climb are the best documented theoretically, experimentally, and in applications to naturally deformed rocks. In contrast, the equations describing diffusion flow are less well-documented and more tentative; nevertheless, they also provide insight into the behavior of rocks. For comparison, a list of important deformation mechanisms at high pressure and temperature is given in Table 4-2.

The physical basis for the stress-strain rate relationship corresponding to dislocation motion is illustrated as follows. Consider the crystal of height L and width W that is sheared parallel to its base by dislocation glide and climb, as shown in Figure 4-28. The shear-strain rate is

$$\dot{\epsilon}_{shear} = \frac{\Delta W}{Lt} \tag{4-22}$$

where t is the time required to produce the deformation and ΔW is the displacement of the top of the crystal relative to the base, which is equal to the

TABLE 4–2   Deformation Mechanisms at Elevated Pressure and Temperature*

A. Steady-state deformation mechanisms
   1. Diffusional creep by diffusive flow of simple ions
      a. Herring-Nabarro creep: lattice diffusion dominant
      b. Coble creep: grain boundary diffusion dominant
      c. Fluid phase transport: mass transport by diffusion through a fluid phase
   2. Dislocation creep involving climb as well as glide of dislocations
   3. Viscous flow of a liquid or liquid-solid mixture
B. Nonsteady deformation mechanisms
   1. Elastic and anelastic deformation
   2. Transient component of dislocation glide and creep: work hardening
   3. Twinning and kinking
C. Phenomena that do not directly contribute to strain but do influence the deformation
   1. Recovery and polygonization
   2. Grain boundary migration and recrystallization
   3. Grain growth
   4. Dissolution or precipitation of phases

*Modified from Stocker and Ashby (1973)

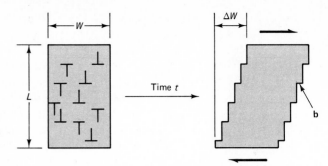

FIGURE 4–28 Schematic deformation of a crystal by sweeping out dislocations of Burger's vector, *b*; used in the development of Orowan's equation (Eq. 4-25).

number of dislocations swept out of the crystal, *n*, times the magnitude of their Burger's vector, *b*

$$\Delta W = n \, b \qquad (4\text{-}23)$$

The number of dislocations swept out in time *t* will be considered to equal or be proportional to the dislocation density, ρ_d, times the area of the crystal:

$$n = \rho_d(LW) \qquad (4\text{-}24)$$

Combining Equations 4-22, 4-23, and 4-24, we obtain *Orowan's equation,*

$$\dot{\varepsilon} = \rho_d \, vb \qquad (4\text{-}25)$$

where $v = W/t$ is the velocity of dislocation motion. Orowan's equation is the underlying basis of the main constitutive equations describing dislocation motion.

Let us apply Orowan's equation to steady-state deformation; under these conditions dislocation density is proportional to the square of the deviatoric stress Δ^2 (Eq. 4-4):

$$\rho_d = C\Delta^2 \qquad (4\text{-}26)$$

as is shown experimentally. The velocity of dislocation motion by glide and climb is strongly temperature-dependent and proportional to deviatoric stress Δ (Eqs. 4-5 and 4-6):

$$v = K\Delta \exp\!\left(\frac{-Q}{RT}\right) \qquad (4\text{-}27)$$

where *K* is some constant. The exponential term containing temperature describes the diffusion involved in climb, *Q* being the activation energy. Substituting these expressions (4-26 and 4-27) for dislocation density and velocity into Orowan's equation (4-25), we obtain:

$$\dot{\varepsilon} = (CKb)\Delta^3 \exp\!\left(\frac{-Q}{RT}\right) \qquad (4\text{-}28)$$

which is in close agreement with experiments on olivine, quartz, and ice. Experimental results on other minerals or rocks indicate a dependence on stress to a different power *n* of the stress Δ^n; for example, $n \simeq 8$ and 4 for various deformation regimes of calcite. A more general form of Equation 4-28 is

$$\dot{\varepsilon} = C_0 \, \Delta^n \exp\!\left(\frac{-Q}{RT}\right) \qquad (4\text{-}29)$$

where C_0 is a constant. Deformation obeying an equation of this form is called *power-law creep* and is a common mode of rock deformation at elevated temperatures under laboratory strain rates. Once the activation energies, *Q*, preexponential constant, C_0, and power dependence, *n*, are determined in the laboratory (see Appendix B), the flow law may be quantitatively extrapolated to geologic strain rates. The extrapolation of these experimental results obtained at high strain rates, such as 10^{-5}/s, to geologic strain rates of 10^{-14}/s is strictly justified only in cases in which it can be shown that the rate-limiting deformation mechanisms are identical geologically and experimentally. Otherwise the plastic strengths obtained by extrapolation should normally be considered maximum strengths for geologic applications because other deformation mechanisms may exist that operate at lower stresses than the mechanisms studied experimentally.

The plastic strengths of a number of common rocks and rock-forming minerals under a typical geologic strain rate of 10^{-14}/s are shown as a function of temperature in Figure 4-29, which was produced by extrapolating experimental results using equation 4-29 and similar equations. The first feature to note is that plastic strengths are very strongly dependent on temperature and mineralogy. In contrast, brittle strengths of rocks are to the first approximation independent of temperature and mineralogy, as we shall see in the next chapter. The resistance to plastic flow of any particular mineral or rock type drops from an essentially infinite value to a very low resistance over a very narrow temperature interval because of the negative exponential in Equation 4-29. For example, at 250° quartzite has a very high plastic strength of over 1000 Mpa, but at 300° the strength has dropped to a moderate value of 100 MPa. The critical temperature interval for the onset of plastic deformation is little affected by strain-rate changes within the normal tectonic range of 10^{-15}/s to 10^{-13}/s (Figure 3-18); an order-of-magnitude change in strain rate shifts the plastic-strength curve by only about 25°. Therefore, we can assign a narrow temperature interval to the onset of plastic deformation in any mineral, particularly at stresses above 50 MPa. The onset of plastic deformation in common rock-forming minerals is predicted experimentally (Figure 4-29) to be in the sequence halite $\rightarrow$ anhydrite $\rightarrow$ calcite $\rightarrow$ quartz $\rightarrow$ feldspar $\rightarrow$ pyroxene $\rightarrow$ olivine, which is in general agreement with field observations.

Rocks composed of mixtures of strong and weak minerals will have the plastic strength of the weaker mineral if it makes up more than about 25 to 30 percent of the rock. For example, in Figure 4-29 the aplite, which is composed of feldspar and quartz, has about the same strength as quartz. The diabase, which is composed of pyroxene and feldspar, has a strength closer to feldspar than diopside. Nevertheless, the plastic deformation of polyphase rocks is not well studied.

For mechanisms of diffusion flow, three different paths of diffusion may be considered: (1) crystal-lattice diffusion (Herring-Nabarro creep), (2) grain-boundary diffusion (Coble creep), and (3) diffusion in a grain-boundary fluid phase. Some combination of the last two may approximate pressure solution. The equations describing each of these paths have similar forms. The strain rate is proportional to the concentration of vacancies, C, which is a function of temperature times the velocity at which they diffuse, v, divided by a geometric description of the diffusion path, d^n:

$$\dot{\epsilon} \propto \frac{Cv}{d^n} \qquad (4\text{-}30)$$

where d is the grain diameter. The exponent n is 2 for crystal-lattice diffusion and 3 for grain-boundary diffusion. Experimentally, it is found that the diffusion velocity is proportional to the deviatoric stress, Δ, times an effective diffusion coefficient, D:

$$v \propto D\Delta \qquad (4\text{-}31)$$

Combining proportionalities 4-30 and 4-31, we obtain a general form of the diffusion-flow equation:

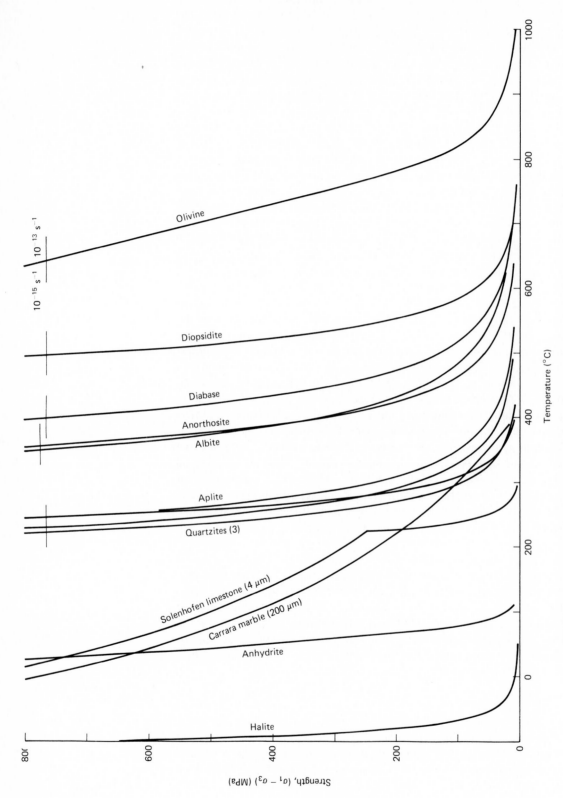

FIGURE 4-29 Plastic strengths as a function of temperature for a variety of rocks and minerals based on laboratory experiments. Strengths are computed for a strain rate of 10^{-14}/s with the bar at the top showing that the curves shift 30° to 40°C for an order of magnitude change in strain rate. Tectonic strain rates should be in the range 10^{-13}/s to 10^{-15}/s (see Fig. 3-18). (Computed based on data from Shelton and Tullis, 1981; Koch, Christie, and George, 1980; Schmid, Paterson, and Boland, 1980; Muller, Schmid, and Briegel, 1981; Brace and Kohlstedt, 1980; and Heard, 1972)

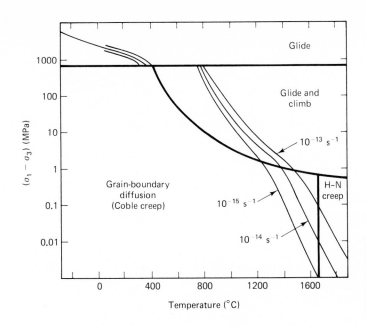

FIGURE 4–30 Deformation map for olivine, showing domains of dislocation glide and climb, and Coble and Herring-Nabarro creep. (Modified after Stocker and Ashby, 1973)

$$\dot{\epsilon} = C_0(T)\,\frac{D\Delta}{d^n} \qquad n = 2 \text{ or } 3 \qquad\qquad (4\text{-}32)$$

where $C_0(T)$ is a constant dependent on temperature. Note that strain rate is proportional to deviatoric stress; therefore, *diffusion creep* is viscouslike as opposed to the power-law dependence of dislocation creep (Eq. 4-29). A second important point is that strain rate in diffusion flow is very sensitive to grain size— inversely proportional to grain size squared or cubed. Therefore, diffusion flow should be much faster in fine-grained rocks, which is in qualitative agreement with observations in some naturally deformed rocks. Fine-grained slates that have deformed by pressure solution are generally more strained than interbedded coarse-grained slates.

The effect of grain size may be also seen experimentally in Figure 4-29. The Carrara Marble ($d = 200\ \mu$m) and the Solenhofen Limestone ($d = 4\ \mu$m) show similar plastic strengths for dislocation motion below about 200°, but at about 230° the much finer grained Solenhofen limestone suddenly exhibits a strength drop that reflects a diffusion creep mechanism.

Each constitutive relationship describes the stress-strain or stress-strain rate relation when a certain deformation mechanism or combination of mechanisms is dominant. No single mechanism will dominate over the entire range of earth conditions. Therefore it is useful to construct *deformation maps* showing the conditions under which each flow law is rate limiting. A deformation map for dunite (olivine rock) under upper mantle pressure conditions, at which brittle behavior is inhibited, is shown as an example in Figure 4-30.

EXERCISES

4–1 At what depth would you expect that quartz-rich rocks would begin to deform plastically at a deviatoric stress of 100 MPa or less, based on experimental data? Estimate this depth for average geothermal gradients of 10°, 20°, 30°, and 50°C/km.

4–2 How does the maximum horizontal stress at the onset of plastic deformation in Exercise 4-1 depend on the geothermal gradient? Assume as one case that σ_3 is vertical (compressional tectonics) and as another case that σ_1 is vertical (extensional tectonics).

4–3 How much will a 100-m length of basalt shrink as it cools from 1000°C to room temperature (25°C) given a thermal expansion coefficient of 2.5×10^{-6}/C?

4–4 Suppose a cylinder of diabase 10 cm in diameter and 25 cm long is placed under an axial compression of 10 MPa; how much will it shorten axially and how much will it expand transversely, given a Young's modulus of 10^{11} Pa and a Poisson's ratio of 0.25?

4–5 How much should the strength of a rock deforming by grain-boundary diffusion theoretically change for a decrease in grain size by a factor of ten?

4–6 Estimate the velocity of dislocation motion in quartz at 350°C and a steady-state strain rate of 10^{-14}/s. The Burger's vector is about 5 Å.

SELECTED LITERATURE

CARTER, N. L., 1976, Steady-state flow of rocks: Rev. Geophys. and Space Phys., v. 14, p. 301–360.

CARTER, N. L. AND RALEIGH, C. B., 1969, Principal stress directions from plastic flow in crystals: Geol. Soc. Amer. Bull., v. 80, p. 1231–1264.

ELLIOTT, D., 1973, Diffusion flow laws in metamorphic rocks: Geol. Soc. Amer. Bull., v. 84, p. 2645–2664.

KIRBY, S. H., 1983, Rheology of the lithosphere: Rev. Geophys. and Space Phys., v. 21, p. 1458–1487.

MAGARA, K., 1978, *Compaction and Fluid Migration,* Elsevier, New York, 318 p.

NICOLAS, A. AND POIRIER, J. P., 1976, *Crystalline Plasticity and Solid State Flow in Metamorphic Rocks,* Wiley, New York, 444 p.

WENK, H. R., ed., 1976, *Electron Microscopy in Mineralogy,* Springer-Verlag, New York, 564 p.

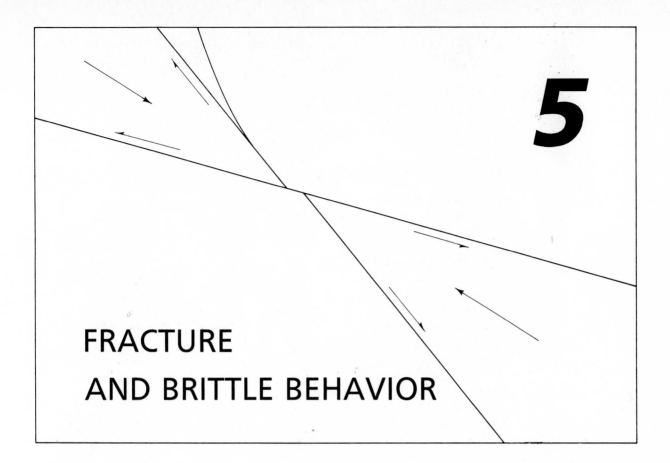

FRACTURE
AND BRITTLE BEHAVIOR

INTRODUCTION

Rocks at the surface of the earth—the only place we see them—exhibit brittle-elastic behavior. Rocks shatter, crack, split, or in some way fracture if we subject them to high-enough deviatoric stresses. They are elastic until they fracture. In contrast, the once deeply buried rocks we see in outcrop show signs that they once flowed, a behavior further from our direct experience, yet typical of the earth's interior. Fracture plays a major role in many natural structures of the upper crust, including joints, faults, and igneous intrusions. Much of our understanding of fracture in structural geology comes from the engineering field of rock mechanics, which developed because of the enormous engineering and economic importance of brittle-elastic behavior in rocks. For example, about 5 percent of the energy generated in the United States is said to be consumed in the fracture of rocks.

Fracture is the process of breaking to pieces either macroscopically or microscopically. If the pieces fit together without having changed shape, as with porcelain, it is called *brittle fracture* in everyday language (for example, Fig. 5-1). If the pieces will not fit, as with a wire broken after much bending, it is called *ductile fracture*. Considerable permanent deformation accompanies the slow growth of ductile fractures, but brittle fracture is sudden and most of the deformation is elastic, except for the crack.

This everyday distinction between brittle and ductile behavior is useful, but can be misleading; therefore, for our purposes we must define them—as well as the word *plastic*—more precisely. Brittle and plastic are used here to refer to groups of related deformation mechanisms; in contrast, *ductile deformation* is used as a general term for macroscopic flow of rock, without regard to the

FIGURE 5–1 Brittle fracture in a dark grey dolomitic limestone. The spaces that opened up during fracture are filled with white calcite. The fractures are confined to a bed of brittle dolomitic limestone; the overlying and underlying calcitic limestone, also dark grey, was plastic and flowed during the same deformation, which stretched the formation parallel to the bedding. The photograph shows one of the places where the dolomitic layer necked by brittle fracture. Lower Ordovician, Rheems, Pennsylvania Appalachians.

underlying deformation mechanisms. Ductile deformation may be brittle, plastic, or a combination of the two. If the ductile deformation is dominated by microscopic fracturing and frictional sliding, the deformation is called *cataclastic flow*. If the ductile deformation is dominated by the mechanisms of crystalline plasticity (Chapter 4), the deformation is called *plastic flow*.

Our grouping of mechanisms of permanent deformation into brittle and plastic reflects their dependence on the two most important intensive variables in the earth: pressure and temperature. *Brittle deformation* is defined as strongly pressure-dependent deformation involving an increase in volume, as a result of cracking. It includes fracture and frictional sliding. Brittle strengths are relatively insensitive to temperature and time, but increase with pressure, as we shall see. In contrast, *plastic deformation* is defined for our purposes as strongly temperature- and time-dependent, constant-volume deformation. It includes the mechanisms of dislocation climb and glide, twinning, and diffusion flow (Chapter 4). Plastic strengths are insensitive to pressure, but decrease exponentially with temperature (Fig. 4-29). Therefore, as a general rule plastic deformation dominates the deeper parts of the lithosphere, whereas brittle deformation dominates the upper lithosphere (Fig. 1-1).

Macroscopic brittle fracture depends to the first approximation only on the applied stress. In contrast, macroscopic ductile fracture depends on the stress history, temperature, and other factors. Therefore, brittle fracture is much simpler than ductile fracture. Nevertheless, the critical stress for brittle fracture, the *fracture strength,* does not have a unique value. It depends, as we shall see, on the mean stress (pressure) and the sign of the stress (tension or compression). Furthermore, at very high mean stress, high temperatures, and slow rates of loading, rocks are not brittle, but plastic. Ductile fracture is a transitional behavior between brittle fracture and plastic flow.

THE MOHR ENVELOPE—AN EMPIRICAL VIEW OF FRACTURE

The fracture strengths of many rocks have been measured to help design structures involving rocks; this information also provides much insight for structural geology. Strengths are commonly measured by subjecting cylindrical samples to an *axial load,* σ_a, applied by a piston or hydraulic ram, and a *confining pressure,* σ_c, applied radially around the side by a fluid that is isolated from the rock by a deformable rubber or copper jacket (see Fig. 5-2). Two distinct states of stress are possible: *uniaxial compression,* which is $\sigma_a = \sigma_1 > \sigma_2 = \sigma_3 = \sigma_c$ and *uniaxial extension,* which is $\sigma_c = \sigma_1 = \sigma_2 > \sigma_3 = \sigma_a$. In some tests there may be fluid within the pores of the rock; the pressure P_f of this fluid may be controlled and is important, as we shall see. The size of the cylinder is measured, and displacement and strain gauges are used so that stresses and strains can be calculated from the forces, pressures, and extensions measured. Very accurate results require more complex procedures and equipment, but the basic methods are essentially as shown.

Let's follow an experiment performed on dry rock at a constant confining pressure, graphing the course of the experiment on both a stress-strain diagram and a Mohr diagram (Fig. 5-3). First, we increase the confining pressure and axial load essentially together until the value of our desired confining pressure—for example, 50 MPa—is reached. The state of stress in the specimen plots as a single point on the Mohr diagram because the axial load equals the confining pressure. The stress difference, plotted on the stress-strain diagram, is zero. In the process

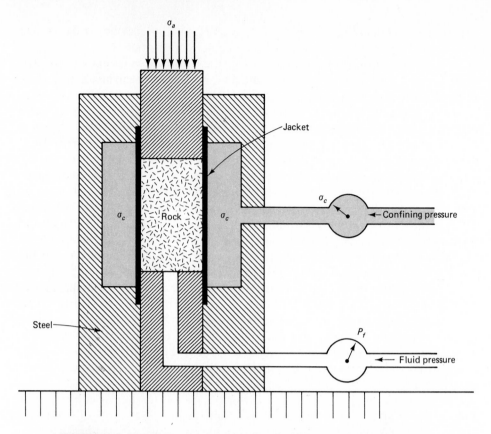

FIGURE 5–2 Simplified schematic drawing of triaxial testing apparatus. The cylindrical rock sample is subjected to an axial load, σ_a, and a radially applied confining pressure, σ_c. The fluid that exerts the radial load, σ_c, is separated from the rock by a weak but impermeable jacket so that the pore-fluid pressure, P_f, within the rock might be controlled independently from confining pressure.

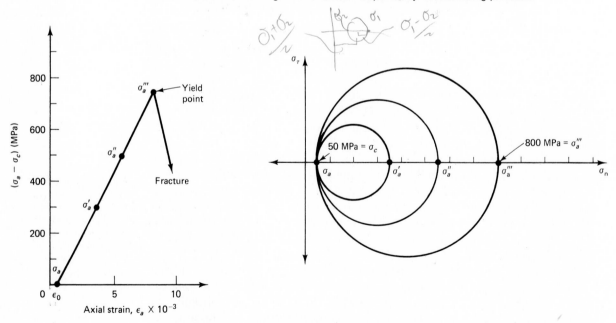

FIGURE 5–3 Stress-strain diagram and corresponding Mohr diagram for a fracture experiment at a confining pressure σ_c of 50 MPa.

of reaching this isotropic state of stress ($\sigma_a = \sigma_c$), both the length and volume of the sample has decreased; the initial strain is ϵ_0.

We will now increase the axial load, keeping the confining pressure constant and measuring the change in length of the sample as we go. The state of stress appears on the Mohr diagram as successively larger circles of diameter $\sigma_a - \sigma_c$, sharing only the confining pressure σ_c as a common point; the maximum shear stress, equal to the radius of the circle $|\sigma_\tau| = (\sigma_a - \sigma_c)/2$, becomes progressively larger. On the stress-strain diagram, the experiment graphs in a nearly linear fashion, reflecting the elastic behavior of the sample; the area under the line is a measure of the elastic energy stored in the sample due to the differential stress ($\Delta = \sigma_a - \sigma_c$)*. The sample finally fractures at an axial load σ_a of 800 MPa and the stress drops. Some of the elastic energy is expended in making the fracture, some in sound, and some in the frictional heating due to sliding. If the results are reproducible, we know that the fracture strength, $\sigma_a - \sigma_c$, of our sample is 750 MPa at a confining pressure of 50 MPa. When we remove the sample, we see that the fracture lies about 24° to the axis of the cylinder.

Several experiments will show that the fracture strength increases with confining pressure. The circles in Figure 5-4 show the state of stress at fracture for a homogeneous set of samples of the same rock, a diabase. These experiments define the states of stress that the rock is capable and incapable of supporting. The

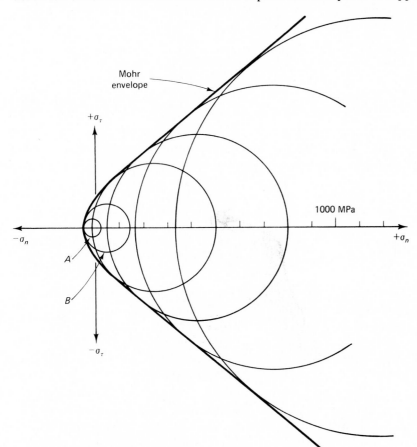

FIGURE 5–4 Mohr diagram for Fredrick Diabase (dry) at room temperature. Each circle represents the state of stress at failure at a different mean stress, showing that rock strength increases with mean stress. The locus of stress states that bounds the fields of stable and unstable stresses is called the Mohr envelope. (Data from Brace, 1964.)

---

*There is also elastic-strain energy stored in the sample due to the confining pressure, applied at the start of the experiment. Furthermore, the apparatus may contain considerable elastic-strain energy, which cannot be ignored in an actual experiment.

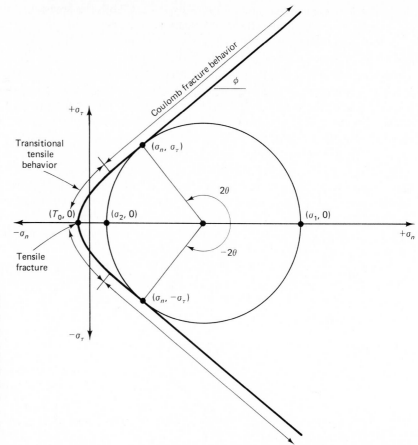

FIGURE 5–5 Three fields of the fracture, or Mohr, envelope. The tensile field shows a fixed strength, T_0. The Coulomb fracture field shows a linear increase in shear strength, σ_τ, with normal stress, σ_n, and a slope $\tan \phi$. The two points of tangency of the Mohr circle represent the states of stress on the two planes whose normals are θ from the σ_1-axis (Fig. 3-23). Transitional-tensile behavior is transitional between tensile and Coulomb fracture.

boundary on the Mohr diagram between these two fields of stress space is called the *fracture, failure,* or *Mohr envelope.* If the state of stress on any plane in the rock plots on the fracture envelope, the rock will begin to fracture. The final macroscopic fracture will generally be oriented close to the plane that is tangent to the fracture envelope ($\pm 2\theta$ in Fig. 5-5), barring any preexisting flaws or zones of weakness.

Four fields of fracture behavior are exhibited along the Mohr envelope (see Fig. 5-5):

1. *Tensile fractures* form perpendicular to the direction of the maximum tensile stress and the tensile strength is independent of the other principal stresses (for example, experiments A and B in Fig. 5-4). The Mohr circle is tangent to the fracture envelope at only one point, and only one direction of potential fracture exists. Typical values of the tensile strength T_0 are -5 to -20 MPa (see Table 5-1).

2. *Transitional tensile behavior* is exhibited above a critical value of the least tensile stress, σ_1. Beyond this value, which is approximately $\sigma_1 = |3T_0|$, the Mohr circle is tangent to the envelope at two points, rather than one, at the time of fracture. Two directions of potential fracture exist. The transitional tensile field is characterized by a rapid nonlinear increase in strength ($\sigma_1 - \sigma_3$) with increasing confining pressure, as well as a change in both the observed and predicted [$\theta = (90° + \phi)/2$] fracture orientation from parallel to roughly 30° to the maximum compression. Most joints are probably formed in the tensile or transitional tensile field.

TABLE 5–1   Compressive, C_0, Shear, S_0, and Tensile, T_0 Fracture Strengths and Coulomb
Coefficients μ and K of Some Dry Rocks*

	Fracture Strengths in MPa (or 10^{-2} kb)					
	C_0	S_0	T_0	$|C_0/T_0|$	$\mu = tan\phi$	K
Cheshire Quartzite	461	103	−28	16.5	0.9	5.0
Westerly Granite	229	37	−21	10.9	1.4	9.7
Frederick Diabase	487	114	−40	12.1	0.8	4.6
Gosford Sandstone	50	16	−3.6	13.5	0.5	2.6
Carrara Marble	90	23	−7.0	13.0	0.7	3.7
Blair Dolomite	507	112	−35	14.5	0.9	5.1
Webatuck Dolomite	148	44	−8	18.5	0.5	2.8
Bowral Trachyte	150	31	−14	10.9	1.0	5.8
Witwatersrand Quartzite	194	40	−21	9.3	1.0	5.8

*Modified from Jaeger and Cook, 1979, and Brace, 1964

3. *Coulomb behavior* is exhibited at intermediate confining pressure $\sigma_1 \geq \sim |5T_0|$ for many rocks. The strength increases linearly with increasing confining pressure. The fracture envelope in this range of behavior is described by the equation

$$|\sigma_\tau| = S_0 + \sigma_n tan\,\phi = S_0 + \sigma_n\mu \qquad (5\text{-}1)$$

which is commonly called the *Coulomb fracture criterion.* Here, S_0 is the *cohesive shear strength* of a plane with no normal stress σ_n acting across it and $tan\,\phi = \mu$ is the rate of increase of shear strength with increasing normal stress, sometimes called the *Coulomb coefficient,* or the *internal friction.* The angle ϕ (Fig. 5-5) is called the *angle of internal friction.* Typical values of S_0 and μ are given in Table 5-1 and Figure 5-6. The Coulomb fracture criterion may also be written in terms of principal stresses:

$$\sigma_1 = C_0 + K\sigma_3 \qquad (5\text{-}2)$$

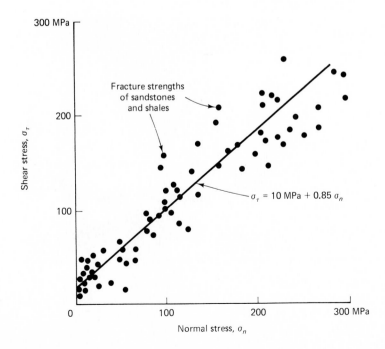

FIGURE 5–6 Summary of laboratory measurements of fracture strength of sandstone and shale. (Data from Hoshino and others, 1972.)

where C_0 is the *compressive strength*—$C_0 = 2S_0[(\mu^2 + 1)^{1/2} + \mu]$—and K is the *earth-pressure coefficient*—$K = [(\mu^2 + 1)^{1/2} + \mu]^2$. Typical values of C_0 are 500 to 5 MPa, with the tensile strengths T_0 of the same rock about an order of magnitude less (Table 5-1). It should be noted that, in addition to rock, the Coulomb fracture criterion describes the behavior of many soils, unconsolidated sand, concrete, and highly fractured, unsound rock, but with different parameters than the unfractured equivalent, under a wide range of conditions.

4. At very high confining pressure and increasing temperature, rock becomes distinctly nonbrittle on a macroscopic scale, and its strength increases more slowly with increasing confining pressure. This is the realm of *ductile fracture* discussed briefly near the end of this chapter.

ROLE OF CRACKS IN THE FRACTURE OF ROCKS

What are the underlying physical mechanisms that determine the shape and magnitude of the fracture envelope (Fig. 5-4) for a given rock? Why are rocks far weaker in tension than compression (Table 5-1)? The answers in large part lie in the existence of cracks or holes that produce large stress concentrations near their edges. If the cracks were not there, a homogeneous state of stress would exist, but with the cracks, adjacent regions have to assume the additional load. We can see this increase in stress near the tip of a crack in Figure 5-7 by virtue of the photoelastic effect; the crack is in a sheet of photoelastic plastic under applied tension.

In 1920, A. A. Griffith developed a theory of fracture strength based on the assumption that cracks exist throughout a material. For mathematical simplicity he approximated cracks as very flat elliptical holes. He reasoned that a small overall stress field, σ_0, would be locally enormously magnified at the tip of each crack. This intensification depends upon the orientation, length ($2l$), and width d

FIGURE 5–7 Isochromatic fringes (lines of constant deviatoric stress) illustrating the concentration of stress around the tip of a crack in a sheet of photoelastic plastic under tension applied normal to the crack. When this photograph was taken, the crack was propagating at about 400 m/s; a similar but slightly lower stress concentration would be present with a static crack. The dimension of the grid is 2.5 cm. (Photograph by School of Engineering, University of Washington, courtesy of A. S. Kobayashi.)

of each crack; thus some cracks grow in preference to others. For example, the tensile stress, σ, at the tip of an elliptical crack in an elastic plate under a regional tension, σ_0, normal to the crack is approximately:

$$\sigma \cong \frac{2}{3} \sigma_0 \frac{(2l)^2}{d} \qquad (5\text{-}3)$$

The theoretical cohesive strength of a material—that is, the stress required to break bonds to form a free surface—can be exceeded at the tip of the crack at a fairly low applied tensile stress, σ_0, because of the factor $(2l)^2/d$. For example, with a ratio $l : d$ of 30 : 1, the stress intensification is about 2400 times.

Next Griffith considered that elastic-strain energy is released as a crack forms or grows because most bonds in the vicinity of the crack are able to return to their equilibrium positions. In the case of a crack forming in a thin elastic plate, the energy release per unit thickness is

$$U_e = \frac{-\pi l^2 \sigma_0{}^2}{E} \qquad (5\text{-}4)$$

where E is the elastic constant, Young's modulus. On the other hand, energy is required to produce the crack surface; this energy expended per unit thickness is

$$U_s = -4l\gamma \qquad (5\text{-}5)$$

where γ is the energy expended in making the fracture surface per unit area (roughly 0.1 to 10 J/m^2), which includes the specific surface energy of the material, heat, acoustic energy, and energy consumed in plastic deformation.

The important point to note in Equations 5-4 and 5-5 is that the elastic energy released, U_e, is proportional to l^2, whereas the energy expended, U_s, is only proportional to l. Thus the system will become unstable and catastrophic brittle fracture will occur if the surface energy expended per increment of crack growth is less than the elastic energy released. The critical applied stress, σ_c, for catastrophic fracture in a thin plate under tension may be obtained from Equations (5-4) and (5-5) and is

$$\sigma_c \geq \sqrt{\frac{4E\gamma}{\pi l}} \qquad (5\text{-}6)$$

where $2l$ is the crack length. If we measure the surface energy, elastic constants, and length of the longest cracks of various orientations, we can predict the fracture strength of a material. Griffith did this for glass and compared the actual fracture strength with the predicted one to verify his theory. Similar work has been done with rock. The lengths of cracks in sound rocks are of the order of the grain size; therefore, strength increases with decreasing grain size in otherwise equivalent rocks.

The *Griffith theory of fracture* leads to a prediction of the size and shape of the fracture envelope:

$$\sigma_\tau^2 - 4T_0\sigma_n - 4T_0^2 = 0 \qquad (5\text{-}7)$$

where T_0 is the tensile strength. The theory is successful in predicting the shape of the fracture envelope in the tensile and transitional-tensile region, as well as the approximate orientation of fractures. Cracks oriented perpendicular to the principal tensile stress are the most likely to grow in the tensile region according to the Griffith theory, which is in agreement with the experimental observation that tensile fractures form perpendicular to σ_3.

In contrast with the success of the Griffith theory in the tensile and transitional-tensile field, there is a wide discrepancy between theory and observation in the Coulomb field. This fact led to a modified-Griffith theory, which assumed that cracks are closed under compression; thus their growth is affected by friction. The predicted fracture envelope of the modified Griffith theory in the compressive field is similar to the Coulomb criterion. Nevertheless, it is now known that the fracture process in the compressive field is much more complex than the growth of a single unstable crack, as was assumed by the Griffith theory. This added insight has come from precise experimental studies of compressive fracture of rocks.

MICROSCOPIC VIEW OF THE FRACTURE EXPERIMENT

Some experimentalists have taken a close look at the behavior of brittle rocks preceding catastrophic fracture and found that these rocks, particularly at low confining pressures, show important deviations from ideal elastic behavior. In addition to observing the details of the stress-strain curves, they placed sensitive microphones near the deforming samples in order to hear the opening, closing, and growth of cracks, as well as to locate precisely the elastic shocks within the sample, just as one might locate earthquakes with a set of seismometers.

The prefracture behavior of a rock may be divided into four main fields (see Fig. 5-8(a)). Stage I exhibits both axial and volumetric stress-strain curves that are concave toward the stress axis because loose grain boundaries are compacting and cracks are closing. This behavior is exhibited only in compressive tests and is very pronounced in altered, unsound, and unconsolidated rock. Stage II exhibits nearly linear stress-strain curves, that is, nearly linear elastic behavior. Stage III is marked by an increase in the nonelastic part of the volumetric strain and, in contrast with stages I and II, the volume of the sample begins to decrease less rapidly and then to increase. The increase in volume, or *dilation,* is due to the growth of cracks, as shown by a much-higher incidence of elastic shocks and by changes in seismic-wave velocities and electrical resistivity. The microscopic

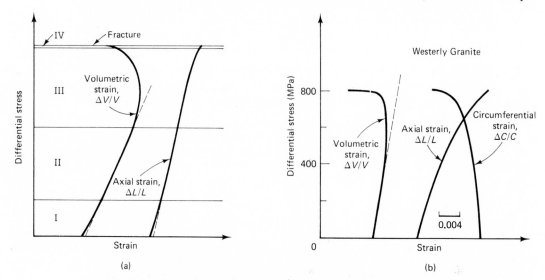

FIGURE 5–8 Stress-strain diagrams for axial, circumferential, and volumetric strain during prefracture deformation. (a) Schematic, showing four fields of prefracture behavior discussed in text. (b) Data for Westerly Granite. (From Brace, Paulding and Scholtz, J. Geophys. Research, v. 71, p. 3939–3953, 1966, copyrighted by the American Geophysical Union.)

fractures (microfractures) are located randomly in space but are oriented perpendicular to σ_3. Stage IV just precedes fracture and exhibits a great increase in nonelastic strain. Microfractures occur in rapid succession. They begin to develop in limited regions and then coalesce to form the macroscopic fracture surface, at which moment the sample becomes mechanically unstable—that is, it breaks.

These experiments illustrate the importance of cracks in fracturing. The macroscopic fracture is due to many coalescing cracks, however, rather than growth of a single crack, as was the simplifying assumption of the Griffith and modified Griffith theories.

ROLE OF FLUID IN THE FRACTURE OF ROCKS

Virtually all rocks contain pores and cracks filled with water, hydrocarbons, and other fluids. These fluids support part of the load that would otherwise by supported across grain contacts; thus the state of stress in the solid grains is modified. The mechanical behavior of dry and wet rocks, while closely related, is different.

Rocks exhibit a wide range of porosities (percent voids), from as high as 40 to 80 percent for unconsolidated sediments to as low as 0.01 to 0.001 percent in many coarse-grained igneous rocks. A rock with high porosity can be envisaged as loose grains in point contact, whereas rock with low porosity might be considered a solid containing a partially interconnected capillary network. The geometry of this network affects the ease with which the fluid may flow through the rock, that is, its permeability, k_{ij}. The rate of flow, V_i, a vector, is equal to the permeability divided by the viscosity of the fluid η, times the excess pore-pressure gradient that drives the flow:

$$V_i = \frac{k_{ij}}{\eta} \frac{dP_p}{dx_j} \qquad i, j = 1, 2, 3 \tag{5-8}$$

which is called *Darcy's law*.

The brittle behavior of fluid-filled rocks is well described by the concept of *effective stress*, which was introduced to the field of soil mechanics by K. Terzaghi in 1923. Although it has some theoretical basis, its status has been largely empirical. According to this concept, if we use effective stress, σ_{ij}^*, rather than stress, σ_{ij}, wet rocks behave the same as dry. In effect, the material acts as though the normal stresses are reduced by the amount of the pore pressure, P_p, while the shear stresses are unaffected:

$$\sigma_{ij}^* \equiv \begin{bmatrix} (\sigma_{11} - P_p) & \sigma_{12} & \sigma_{13} \\ \sigma_{21} & (\sigma_{22} - P_p) & \sigma_{23} \\ \sigma_{31} & \sigma_{32} & (\sigma_{33} - P_p) \end{bmatrix} \tag{5-9}$$

The fluid pressure supports part of the normal stress that would otherwise act across grain boundaries. The shear stress across grain contacts remains unchanged because the pore pressure is isotropic and fluids cannot support shear stress. Both fracture and frictional sliding obey the effective-stress law.

Equations 5-1 and 5-2 for dry Coulomb fracture become, in terms of effective stress, σ_n^*,

$$|\sigma_\tau| = S_0 + \sigma_n^* \tan \phi = S_0 + (\sigma_n - P_f)\tan\phi \tag{5-10}$$

and

$$\sigma_1^* = C_0 + K\sigma_3^*$$

or

$$(\sigma_1 - P_f) = C_0 + K(\sigma_3 - P_f)$$

(5-11)

The tensile strength becomes, in terms of effective stress, σ_n^*,

$$\sigma_3^* = (\sigma_3 - P_f) = T_0$$

(5-12)

The effective stress law has been shown to hold for rocks, many soils, and concrete. It holds for different fluids, including water, acetone, silicone oil, and kerosene. It is also valid for the very low porosities (0.01 to 0.001 percent) present in crystalline rocks such as granite and dunite.

Effective stress may be illustrated with the Mohr diagram (Fig. 5-9). The Mohr circle is in effect translated to the left by the amount of the fluid pressure; the fracture envelope and the size of the circle (deviatoric stress) are unaffected. This effective-stress behavior is illustrated with experimental data on fracture strength of a sandstone in Figure 5-10; the Mohr circles for effective stress at a variety of fluid pressures lie very close to the Mohr envelope for dry rock. Figure 5-9 also illustrates the basis of *hydraulic fracturing* of rocks. Given an initial state of stress within the rock, the pore-fluid pressure may be pumped up until the Mohr

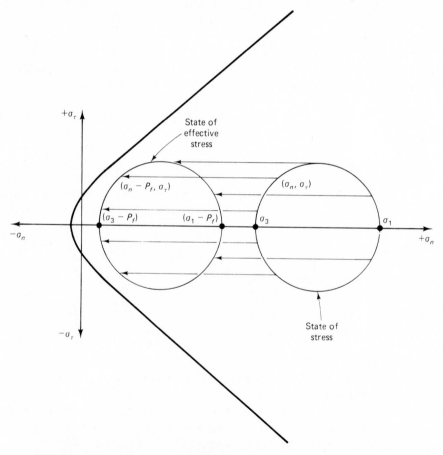

FIGURE 5–9 Mohr diagram showing the relationship between state of stress and state of effective stress. The Mohr circle for effective stress is the same size as for the stress, but it is shifted to the left by the amount of the pore-fluid pressure, P_f.

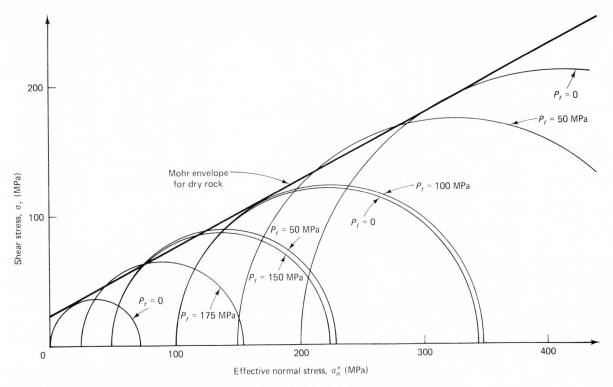

FIGURE 5–10 Mohr diagram (upper half) showing the state of effective stress at failure (ultimate strength) for various experiments with Berea Sandstone at 24°C. The Mohr envelope for dry rock is identical, within experimental uncertainties, to the Mohr envelope for effective stress at the various fluid pressures shown. (Data from Handin, Hager, Friedman, and Feather, 1963.)

circle for effective stress is translated far enough to the left to touch the fracture envelope. Hydrofracturing is done in oil wells to increase the permeability of the reservoir rocks.

An increase in pore-fluid pressure reduces the effective mean stress σ_m^*:

$$\sigma_m^* = \frac{\sigma_1^* \sigma_2^* \sigma_3^*}{3} = \frac{(\sigma_1 + \sigma_2 + \sigma_3 - 3P_p)}{3} \qquad (5\text{-}13)$$

Since strength and ductility increase with mean stress, the pressure of pore fluids makes rocks weaker and more brittle. Brittle fracture therefore takes place to greater depths in the earth than would otherwise be expected.

We recall that rocks undergo dilation before fracture because cracks open (Fig. 5-8). For slow rates of deformation, relative to the permeability, the fluid pressure will remain constant because of the inflow of additional fluid. In contrast, rapid rates of deformation will produce a drop in fluid pressure and an increase in effective normal stress; that is, the load supported across grain contacts increases because the fluid supports less of the load. This produces an apparent increase in strength, sometimes called *dilation hardening*. The interaction between dilation and pore-fluid pressure is apparently important in a number of geologic situations. Whether or not it is important in a specific case depends on the rate of loading versus the permeability and the size of the deforming region. The larger the deforming region, the farther the fluid must flow. Large dilatant regions on the order of kilometers across have been observed in the vicinity of active faults through the study of water wells and seismic-wave velocities. This phenomenon has played a major role in attempts to predict earthquakes.

EFFECT OF PREEXISTING FRACTURES

Laboratory experiments are normally carried out on intact specimens. In contrast, rocks in the field contain important preexisting fractures in the form of joints and faults. Once we have a fracture, what is the requirement for slip along it as opposed to forming a new fracture? To answer this question, we must consider the static friction acting across the preexisting fracture and determine the state of stress required to overcome the friction and allow sliding.

We shall consider the physical nature of friction in some detail in Chapter 8, but for the moment it is sufficient to note that the critical shear stress, σ_τ, required to overcome the friction across an interface is equal to the effective normal stress times the coefficient of friction:

$$|\sigma_\tau| = (\sigma_n - P_p)\mu_f \tag{5-14}$$

or, more generally,

$$|\sigma_\tau| = \tau_0 + (\sigma_n - P_p)\mu_f \tag{5-15}$$

where τ_0 is the cohesive shear strength of the interface, which is very small—about 1 MPa. The coefficient of friction μ_f ranges between about 0.6 and 0.85 for many rock interfaces. For the Weber sandstone from Rangely, Colorado, $\mu_f = 0.81$. The friction equation may be graphed as two lines on the Mohr diagram (see Fig. 5-11); if the stress acting across any interface within the rock lies on this line, slip will occur.

On the other hand, the same rock has a fracture strength described by the Coulomb-fracture equation

$$|\sigma_\tau| = S_0 + (\sigma_n - P_p)\mu \tag{5-16}$$

where in the case of the Weber sandstone, $S_0 = 70$ MPa and $\mu = 0.6$. This equation is also graphed in Figure 5-11. It is important to note that the friction equation applies only to preexisting interfaces, whereas the fracture equation applies to potential fracture planes of all orientations. Of course, both equations are of the same form and pertain to effective stress.

Suppose the Weber sandstone is subjected to the state of stress indicated by the Mohr circle shown in Figure 5-11. This circle is tangent to the fracture line at point f, so that the sample is about to fracture along a plane of orientation θ_f, where θ is the angle between the normal to the plane and the orientation of the σ_1 axis. On the other hand, any preexisting fracture of orientation between θ_1 and θ_2 should already have slipped, since the state of stress along them exceeds that of the friction law. If any fractures lie outside the range θ_1 to θ_2, they will remain frictionally locked and a new fracture will form of orientation θ_f.

Potential sliding along preexisting fractures is an important consideration in many engineering applications of rock mechanics. If the rock is unsound and contains a large number of preexisting fractures of many orientations, however, it will once again obey the fracture equation, but with constants different from those that apply to its unfractured equivalent.

Before leaving Figure 5-11, we observe that it deals with only a two-dimensional state of stress, for purposes of illustration. In many actual problems, the complete three-dimensional state of stress should be considered—for example, using the methods given by Jaeger and Cook (1979)—because preexisting fractures could have any three-dimensional orientation relative to the stress axes. Furthermore, the value of the intermediate principal stress, σ_2, has a significant effect on the slip of preexisting fractures, particularly on the orientation of the slip. For example, Figure 5-12 is a spherical projection showing a dipping fracture

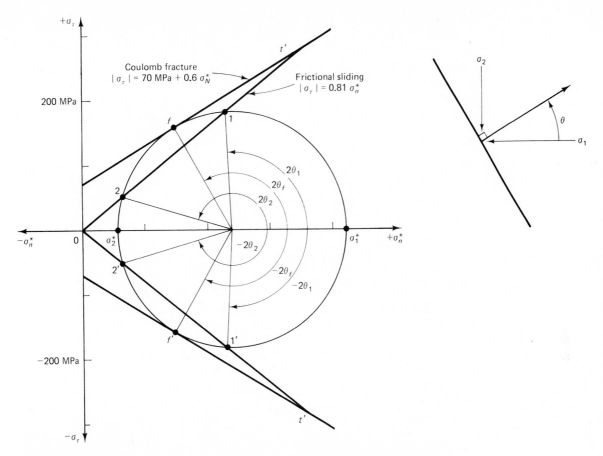

FIGURE 5–11 Mohr diagram showing the Coulomb fracture and frictional sliding criteria for the Weber sandstone from Rangely, Colorado. The state of stress shown is just sufficient for the rock to fracture with orientations $\pm\theta_f$. The Mohr envelope for fracture is tangent to the Mohr circle at points f and f'. The stress required for frictional sliding on preexisting fractures is significantly less than the fracture strength at low effective normal stress; therefore frictional sliding should occur on preexisting fractures with orientations between θ_1 and θ_2 and $-\theta_1$ and $-\theta_2$. (Data from Byerlee, 1975.)

plotted as a great circle and as its pole n. The maximum principal stress, σ_1, is vertical in Figure 5-12(a) and σ_2 and σ_3 are horizontal. As the magnitude of the intermediate stress, σ_2, varies between σ_3 and σ_1, the orientation of the maximum shear stress on the preexisting fracture, which is the slip direction, varies between σ_τ and σ_τ'. Similarly, in Figure 5-12(b), the slip directions vary between σ_τ and σ_τ' with σ_2 vertical. Thus we see that the relative magnitude of the intermediate principal stress exerts an important control on the direction of slip. This relative magnitude is sometimes represented by

$$\Phi = \frac{\sigma_2 - \sigma_3}{\sigma_1 - \sigma_3} \qquad (5\text{-}17)$$

which takes on values between 0 and 1. In uniaxial compression $\sigma_1 \geq \sigma_2 = \sigma_3$, $\Phi = 0$, and in uniaxial extension $\sigma_1 = \sigma_2 \geq \sigma_3$, $\Phi = 1$.

Slip on preexisting fractures, as well as on newly formed ones, is an important process in natural brittle deformation in the earth. By measuring the orientations of a family of fractures and their directions and sense of slip, it is possible to compute the orientations of the principal stresses and Φ, which is important structural and tectonic information (Angelier, 1979a, b).

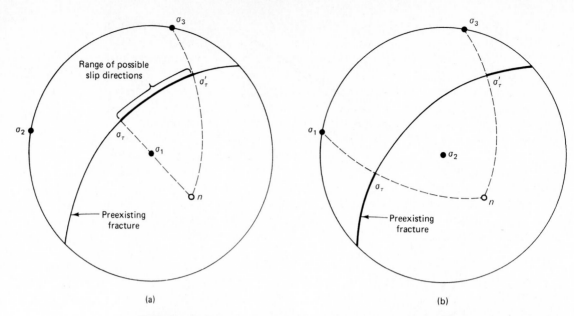

FIGURE 5-12 Spherical projections showing the effect of the magnitude of the intermediate effective stress σ_2 on the orientation of the maximum shear stress (slip direction) on a preexisting fracture. In (a), σ_1 is vertical; if $\sigma_2 = \sigma_3$, then the slip direction is oriented σ_τ, whereas if $\sigma_2 = \sigma_1$, the slip direction is oriented σ'_τ. A similar diagram is given in (b) for σ_2 vertical. (After Angelier 1979a, b.)

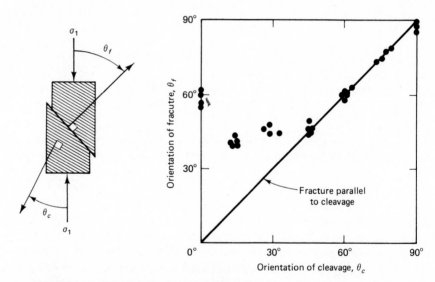

FIGURE 5-13 Effect of the orientation of the plane of slaty cleavage on a fracture orientation in compressive fracture experiments on Martinsburg Slate, a strongly anisotropic rock (Figs. 10-1 and 10-5). (Data from Donath, 1961.)

FRACTURE OF ANISOTROPIC ROCKS

As we might expect, fracture strength is not isotropic in rocks exhibiting stratification, strong schistosity, slaty cleavage, or other metamorphic or igneous foliation. As an example, Figure 5-13 illustrates the experimental fracture behavior of the Martinsburg Slate, a strongly anisotropic rock (Figs. 10-1 and 10-5). For all orientations of the cleavage relative to the principal compression, σ_1, the fractures dipped in the same direction as the slaty cleavage. Furthermore, for inclinations of the cleavage to the σ_1-direction of as much as 45°, the fracture always lay parallel to the slaty cleavage. Only in cases for which the plane of anisotropy was nearly perpendicular to the σ_1-direction was it unimportant in localizing fractures. The fracture strength of the Martinsburg Slate is a minimum when the cleavage plane is roughly parallel to the preferred orientation of fracture in isotropic rocks (30°).

Fracture anisotropy is one of the oldest subjects in structural geology and rock mechanics. Quarrymen have long known that many rocks—including many without obvious schistosity or foliation, such as some granites—exhibit three perpendicular directions of preferred fracture. The easiest and smoothest direction is commonly called the *rift* (or the cleavage in slate); the *grain* is somewhat more difficult; and the *hard way* is still more difficult. An arbitrary direction is hardest of all. Several studies have shown a strong preferred orientation of microscopic cracks, particularly in quartz, parallel to the three fracture directions. Rocks free of quartz, such as gabbro, diabase, and anorthosite, generally do not exhibit this phenomenon.

BRITTLE-PLASTIC TRANSITION

We have seen that brittle-fracture strength under compressive conditions is strongly dependent on solid pressure and fluid pressure (Figs. 5-4 and 5-10). Experiments also show that brittle fracture has a relatively small sensitivity to temperature and strain rate (Paterson, 1978; Brace and Jones, 1971); to the first approximation, brittle-fracture strength is independent of temperature and strain rate. In contrast, strength in the ductile-fracture regime shows more substantial temperature and strain-rate dependence, but less dependence on pressure. This change in dependence on environmental parameters is expected because the mechanisms of plastic deformation that begin to operate in the ductile regime are strongly temperature- and strain-rate sensitive (Figure 4-29), but are independent of pressure.

The regime of ductile fracture is not as well explored in the laboratory or the field as are the purely brittle or plastic regimes. For this reason it is convenient to view ductile fracture first in terms of a simple model of the brittle-plastic transition. In this model there is no ductile fracture, but instead a sudden change from brittle to plastic behavior is encountered with increasing temperature and pressure. This model is illustrated for diabase on the Mohr diagram shown in Figure 5-14. The brittle strength is taken directly from the Mohr diagram for diabase at room temperature shown in Figure 5-4, with the assumption that it will be valid for brittle fracture at all temperatures and strain rates. The plastic strengths of diabase at various temperatures are taken directly from Figure 4-29 and plotted as horizontal lines, showing their independence of effective normal stress, σ_n^*. According to this simple model, the brittle-plastic transition takes place as a sudden change in deformation mechanism when the brittle and plastic strengths are equal. Thus the model Mohr envelope at a fixed temperature would follow the brittle Mohr envelope in Figure 5-14 until it crosses the plastic strength for that temperature; then it would follow the plastic strength.

We should realize that the actual brittle-plastic transition will be more complex than our simple model. The actual strength in the transition may be greater or less, as is shown schematically in Figure 5-14. For example, the strength in the brittle-plastic transition for the Solenhofen Limestone is a little less than would be predicted by the abrupt theory, as is shown by experimentally determined Mohr envelopes in Figure 5-15. Two phenomena that may contribute to the complexity of deformation near the brittle-plastic transition are as follows: (1) Each mineral of a polyphase rock will undergo its brittle-plastic transition in a different temperature range, causing the rock to deform by a combination of

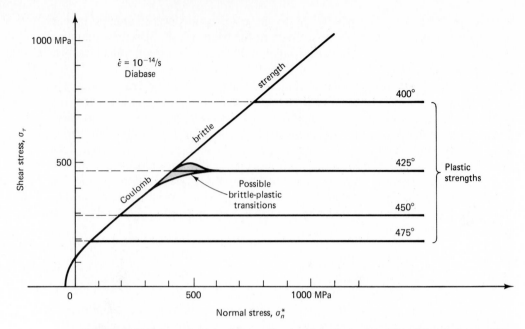

FIGURE 5–14 Schematic Mohr diagram showing a simple model of the brittle-plastic transition in diabase. The diagram is simply a combination of the low temperature brittle strengths taken from Figure 5-4 and the plastic strengths at various temperatures taken from Figure 4-29 for a strain rate of 10^{-14}/s. This simple model of rock strength assumes that brittle strength is completely independent of temperature and plastic strength is completely independent of pressure. Actual brittle-plastic transition might lie above or below the strength predicted, as shown schematically by the shaded region.

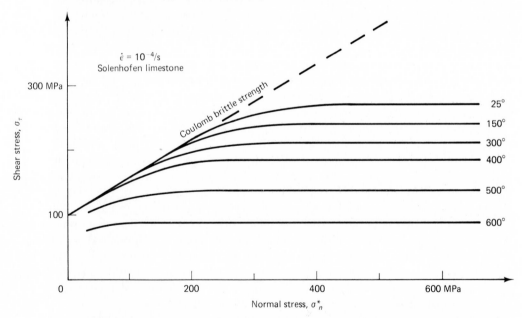

FIGURE 5–15 Experimentally determined Mohr envelopes for Solenhofen Limestone at a strain rate of 10^{-4}/s and various temperatures. The observed strengths in the brittle-plastic transition are less than would be predicted by the simple model of Figure 5-14. (Based on data of Heard, 1960.)

fracture and flow, particularly if the brittle phase is abundant. For example, in the deformed granite of Figure 5-16(a), the feldspar has deformed by brittle fracture, whereas the quartz has flowed. In contrast, if the bulk of the rock is plastic, the minor brittle phases will behave as rigid particles. (2) Healing of cracks is an

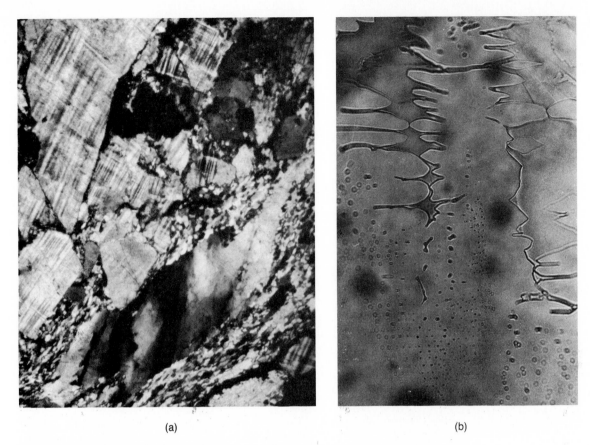

(a)   (b)

FIGURE 5–16 (a) Photomicrograph of deformed granite from near Grenville Front, Sudbury, Canada. The potassium feldspar in the upper left had deformed by brittle fracture, whereas the quartz in the lower center has deformed plastically. (b) Partially healed crack in quartz viewed normal to the plane of the crack. In the top upper right, the crack is segmented into planar polygonal fluid inclusions. Below we see progressive segmentation into tubular fluid inclusions and then point fluid inclusions. All the fluid inclusions lie on the plane of the crack. (Photomicrograph courtesy of David Smith.)

important process in rocks that causes them to be stronger under geologic strain rates than we might predict from laboratory experiments of brittle fracture at moderate temperatures. Healed cracks are commonly observed under the microscope as planar arrays of fluid inclusions, particularly in quartz grains (Fig. 5-16(b)). Crack healing involves diffusive mass transfer that first segments the crack into planar fluid inclusions, which then segment into tubular inclusions and finally into point inclusions, all of which lie in the plane of the original crack.

EXERCISES

5–1 Construct a Mohr diagram showing the Coulomb fracture criterion and tensile fracture strength for the Westerly Granite (Table 5-1). Plot the state of effective stress at a depth of 5 km, assuming an isotropic rock stress equal to the pressure of the overburden ($\rho g z$, where $\rho = 2620$ kg/m^3 and $g = 9.8$ m/s^2), for three different fluid pressures: (a) hydrostatic = $\rho_{H_2O} g z$, where $\rho_{H_2O} = 1030$ kg/m^3; (b) fluid pressure = $0.8\ \rho g z$; and (c) fluid pressure = $\rho g z$.

5–2 Continuing from 5-1, consider that the region begins to undergo horizontal tectonic compression. Plot the successive states of stress as Mohr circles, finally plotting the Mohr circles for fracture. What is the rock strength and horizontal stress at fracture for each of the three fluid pressures?

5–3 Continuing from 5-1, consider that the region begins to undergo horizontal tectonic extension. Plot the successive states of stress as Mohr circles, finally plotting the Mohr circles for fracture. What is the rock strength and horizontal stress at fracture for each of the three fluid pressures? Which are Coulomb and which are tensile fractures?

5–4 What are the dips of the fractures for each of the fluid pressures in 5-2 and 5-3? What is the predicted sense of displacement on each fracture?

5–5 What is the ratio of compressive to extensile fracture strengths for each of the fluid pressures in 5-2 and 5-3? Why is the rock stronger in horizontal compression?

5–6 Estimate the temperature conditions during the deformation of the granite shown in Figure 5-16(a), assuming a stress of less than 100 MPa and a strain rate of 10^{-14}/s.

5–7 Making use of the geometry of the Mohr construction for Equation 5-1, derive Equation 5-2 for the Coulomb fracture criterion in terms of principal stresses. Also show that $K = (1 + \sin \phi)/(1 - \sin \phi)$.

SELECTED LITERATURE

JAEGER, J.C., AND COOK, N. G. W., 1979, *Fundamentals of Rock Mechanics,* 3rd ed., Methuen, London, 593 p.

LAWN, T. R., AND WILSHAW, T. R., 1975, *Fracture of Brittle Solids,* Cambridge, London, 204 p.

PATERSON, M. S., 1978, *Experimental Rock Deformation: the Brittle Field,* Springer-Verlag, New York, 254 p.

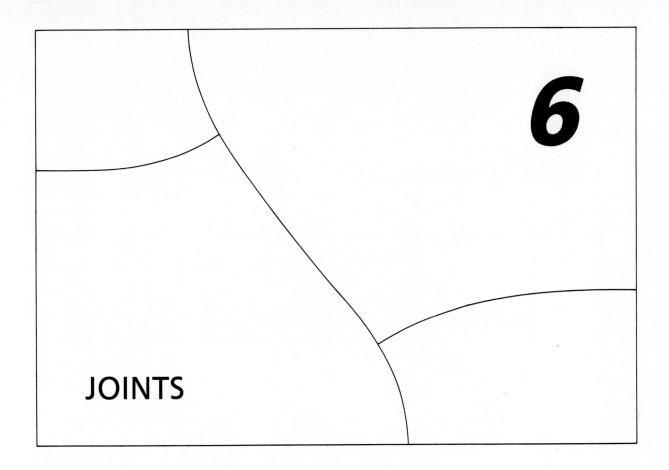

JOINTS

6

INTRODUCTION

Rock at the surface of the earth is cut by a variety of fractures and cracks, into which roots force their way and water seeps. Deeper within the earth these fractures become less and less common, as is sometimes noted in deep mines and quarries. Nevertheless, some cracks still exist at intermediate crustal depths, particularly in plutonic igneous rocks that have cooled substantially since crystallization. Cracks of deep origin are normally healed with vein minerals. Cracks and fractures are a widespread structural feature of the brittle upper part of the crust.

Any thin natural planar crack that is not a fault, bedding, or cleavage and is larger than the grain size of the rock is a *joint* in the broadest sense of the word. The word is said to have originated with British coal miners who thought the rocks were "joined" along these fractures, just as bricks or building stones are joined together in building up a wall.

Joints are of considerable practical importance. They are a widespread plane of potential slip and therefore must be considered for safety and economics in quarrying, mining, and civil engineering. The orientation and spacing of joints can significantly affect the ease of mining and subsequent handling of coal and some ore. Joints are important to groundwater hydrology and the design of dams because they affect porosity and permeability. Joints are paths for circulation of hydrothermal ore-forming solutions.

Most joints show no visible displacement parallel to their walls and are therefore not nascent faults, as can be seen by correlating details of the rock across the fracture. In addition, many joints are composed of closely spaced fractures arranged *en échelon,* contain ramp-shaped steps or, in unweathered exposures, may exhibit a variety of delicate surface undulations, all of which are incompatible with frictional sliding along the joint face (for example, Fig. 6-1). Surface structures characteristic of faulting, such as scratches, grooves, gouge, or brecciation, are not seen along joint faces, except if they have been subjected to later stresses sufficient for frictional sliding. Many joints may have formed as narrow open fissures because they contain vein minerals. These observations suggest that most joints are probably tensile or transitional tensile fractures (see Fig. 5-5) and did not form with significant compressive normal stress acting across the fracture surface.

DESCRIPTION AND CLASSIFICATION OF JOINTS

Joint faces, or even the traces of joints, are rarely seen in their entirety because the outer surfaces of most rock exposures are composed of a number of intersecting and abutting joint faces, as well as bedding planes, schistosities, faults, eroded surfaces, and man-made fractures. For this reason relatively little is known of the complete three-dimensional shapes of joints. Instead, most studies of joints have focused on the more readily observed features, particularly orientation and spacing.

What most readily attracts one's attention of what is visible of joints is their remarkable smoothness and their existence in nearly parallel sets that cut across other sets with no apparent interaction or offset (see Fig. 6-2). Joints may be symmetrically oriented with respect to other structures in the outcrop; for example, they may be perpendicular to bedding, fold axes, or planar and linear fabrics such as slaty cleavage. Some joints are closely parallel or perpendicular to the surface of the earth. In other cases they are seemingly unrelated to any other structures or surfaces.

The joints that have been most studied and are best understood are those that are in some way systematic or regular in their arrangement; these are called *systematic joints*. The term *joint set* is applied to all systematic joints within a region that are parallel to one another and may be distinguished from other sets of different orientation. Joint sets are said not to affect one another, but rather crosscut without deflection. Two prominent joint sets in flat-lying sandstone may be seen in the vertical aerial photograph shown in Figure 6-2.

In practice, the recognition of individual joint sets may be aided by plotting a representative sample of measured joint orientations as poles on a spherical projection (Chapter 2) or by plotting strikes of vertical joints on a rose diagram, which is a circular histogram of orientations. Figure 6-2 has a rose diagram of the joints seen in the aerial photograph. Spherical projections and rose diagrams of joint orientations are useful in recognizing potential planes of slip in mining and civil engineering.

The term *joint system,* in contrast with joint set, is applied to two or more joint sets that are thought to be genetically related—for example, conjugate sets of *shear joints,* which systematically maintain acute dihedral angles of about 5° to 60° between each other. Another example of systematic joint systems is columnar jointing, well known in lava flows, dikes, and sills, which is an effect of inhomogeneous thermal contraction during cooling of the lava (Fig. 6-27). Systematic joints of the same set or system are sometimes characterized by a distinctive wall-rock alteration, one of the few properties of joints that allow their relative ages to be determined.

Not all joints are developed in systematic sets and systems; there are also many less regular fractures, which are called *nonsystematic joints*. Nonsystematic joints usually meet, but do not cross, other joints; many examples may be seen in Figure 6-2. As a group, they are less smooth and planar than the systematic joints.

FIGURE 6–1 Joint face in graywacke sandstone showing upward radiating plumose structure perpendicular to conchoidal undulations; Mesozoic Franciscan complex, northern California Coast Range.

FIGURE 6–2 Vertical aerial view showing joints in the flat-lying Cedar Mesa Sandstone (Permian) in Canyonlands National Park, Utah. The two sets of systematic joints intersect with an angle of 70° and crosscut each other without deflection. A rose diagram of the orientations of the systematic joints is shown. The nonsystematic joints do not crosscut the systematic joints, but abut them at a high angle, approaching 90°. The area of the photograph is about 500 m by 700 m. (Photograph by George E. McGill.)

JOINT-SURFACE MORPHOLOGY

Subtle textures and structures may be seen on joint surfaces, particularly with proper lighting (for example, Fig. 6-1). Joint-surface morphology provides some insight into the fracture mechanisms responsible for the joints. Joint-surface structures are generally best seen in artificial exposures such as road cuts and quarries because the structures are delicate and easily destroyed by weathering. It is important, however, to distinguish between the many fractures created during the quarrying or road construction and those existing within the rock beforehand. Artificial fractures are fresher and free of staining, wall-rock alteration, and vein deposits. Many natural fractures are composed of a series of subparallel, *en échelon* fractures (Fig. 6-3). In order to expose an entire joint surface, cross fractures must be formed between the *en échelon* fractures and are often created during excavation, so are fresher. Many joint surfaces are extremely smooth, with most of their roughness due to the later cross fractures connecting individual *en échelon* fractures (Figs. 6-1 and 6-3).

The best-known joint-surface structures are plumose structures, conchoidal structures, and the joint fringes (Fig. 6-4). These three structures are not developed in every exposed joint surface; nevertheless, they are widespread. They are seen to be closely related geometrically where they are present together.

FIGURE 6–3 Joint surface composed of a set of closely spaced *en échelon* fractures in the Triassic Lockatong Formation, New Jersey.

The *fringe* is a discrete band of *en échelon* fractures along the edge or termination of the main joint surface (see Figs. 6-5 and 6-6). *Plumose structure* is composed of very gentle linear undulations and *en échelon* fractures on the surface of the joint that fan outward from a single point or line and terminate at the fringe (see Figs. 6-1 and 6-7). Individual fringe joints may have their own plumose structure (Fig. 6-6). Some artificial fractures in rock and brittle metal exhibit plumose structure and show that the fracture starts in the center of the plume and propagates along the plume trajectories to the fringe. *Conchoidal structure* consists of discrete changes or steps in the orientation of the joint surface (Fig. 6-1). The steps are oriented perpendicular to the plumose structure. Conchoidal structure represents a discrete discontinuity in the propagation of the fracture. Occasionally you observe a single joint face in nearly its entirety that has developed without interference from other structures; such joints are seen to be elliptical in plan, with the axis of the plumose structure parallel to the long axis of the ellipse. We return to joint-surface structures when we consider the origin of joints.

JOINTS IN RELATION TO OTHER STRUCTURES

Regional Patterns of Jointing

Many systematic joints exhibit regionally consistent patterns of orientation. Furthermore, these consistent patterns are observed to persist throughout the

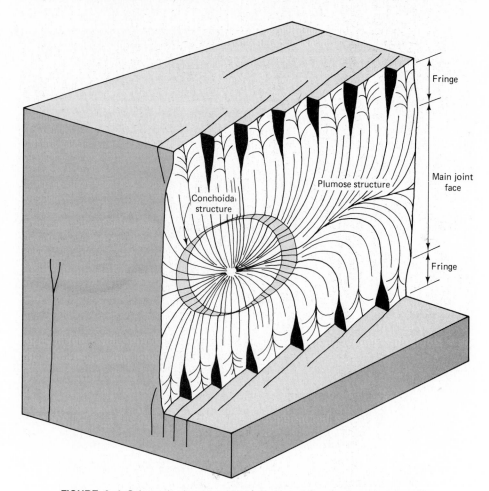

Fringe

Main joint face

Fringe

Plumose structure

Conchoidal structure

FIGURE 6–4 Schematic drawing showing some of the common morphologic features of joint surfaces. The black cross fractures connecting between the *en échelon* fractures of the fringe commonly are man made, formed during excavation of the rock mass and exposure of the joint face.

stratigraphic section, even though the spacing and degree of development of the fractures varies with rock type and bed thickness. Thus many joints have formed in response to stresses that are regionally consistent in orientation.

Several significant regional studies have been made of jointing in flat-lying sedimentary rocks of the cratonic foreland just beyond the edge of fold-and-thrust mountain belts. For example, Nickelsen and Hough (1967) and others measured the orientations of joints in the central Appalachian Plateau just northwest of the strongly folded Valley-and-Ridge Province (see Fig. 6-8). This region exhibits smooth vertical joints in nearly flat-lying sedimentary rocks (Fig. 6-9). The strata are, nevertheless, not undeformed; a train of broad, gentle folds exists with limb dips of less than a degree and axes parallel to the edge of the Appalachian Fold Belt. Furthermore, about 10 percent regional horizontal shortening perpendicular to the fold axes is recorded in deformed fossils (Fig. 3-11). It is notable, then, that the major systematic joint sets in sandstone and shale are approximately perpendicular to the gentle folds and parallel to the direction of principal shortening of the fossils. The gentle folds and the systematic joints in sandstone and shale both disappear toward the craton in northwesternmost Pennsylvania and Ohio. It thus appears, from these regional patterns, that the systematic joints are an integral part of Appalachian foreland structure. Similar patterns of systematic joints are reported for the cratonic foreland of the Ouachita fold-and-thrust belt in Oklaho-

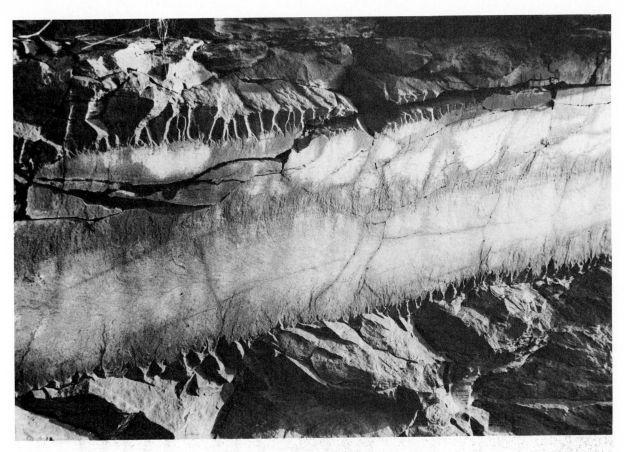

FIGURE 6–5 Joint surface confined to a single carbonate-rich bed; the edge of the joint surface is marked by a fringe of *en échelon* fractures; Triassic Lockatong Formation, New Jersey.

ma (Melton, 1929) and of the Cordilleran fold-and-thrust belt in Alberta (Babcock, 1973).

The regional pattern of jointing in the Appalachian Foreland is, nevertheless, complex. Whereas the broad sweep of the regional jointing suggests a unified regional stress field for their origin, this stress field does not bear a fixed orientation to the fold belt. The joints are broadly perpendicular to the Appalachian fold belt in central Pennsylvania, but are parallel to the fold belt in eastern New York. The strikes of the major joint sets change about 90°. This change is associated with a major swing and sharp bend in the Appalachian fold-and-thrust belt (Fig. 6-8).

The pattern of jointing shown in Figure 6-8(a) is that measured in siltstone and shale. Interbedded coal exhibits a regional pattern of joints (Fig. 6-8(b)) that is different from the shale joints, even though joints in different coal seams are parallel. The coal joints and shale joints formed in different stress fields, apparently at a different time.

The coherence of the orientations of joints over large regions indicates that the scale of the processes that control joint orientations is regional. Some of the possible processes are discussed in a later section.

Not all regional studies of joints have uncovered a systematic relationship between joint orientation and other structures in the same rocks, as may be illustrated with a study of joints in the Grampian Ranges of western Victoria, Australia (Spencer-Jones, 1963) (Fig. 6-10). Sandstones and conglomerates of the Upper Paleozoic Grampians Group are broadly folded and intruded by slightly

FIGURE 6–6 Fringe of *en échelon* fractures at the edge of the upward-extending joint surface. The rough fracture extending below and to the right of the fringe is man made. Siluro-Devonian metamorphosed sandstone, New Hampshire.

younger granite and granodiorite. The joints within the Grampians Group are steeply dipping, and their orientations are independent of rock type. The joints are not genetically related to the folding because the patterns of jointing bear no systematic or symmetrical relationship to the fold axes. What is seen is widespread parallelism of joint sets over distances of at least 50 km (Fig. 6-10), suggesting that the joints were produced by a reasonably homogeneous regional stress field. The Victoria Valley Granodiorite, which has intruded the Grampians Group, has a significantly different pattern of jointing dominated by two orthogonal joint sets parallel to the long and short horizontal dimensions of the intrusion (Fig. 6-10).

Regional studies of jointing in the flat-lying sedimentary rocks of the Colorado Plateau of Arizona, New Mexico, and Utah (Hodgson, 1961; Kelly and Clinton, 1960) also uncovered no systematic angular relationship between regional joint systems and the changing trends of major folds. The regional joint pattern of the Colorado Plateau, which passes through the entire stratigraphic section, is composed of overlapping regions dominated by independent joint sets and systems such as those shown in Figure 6-2. Studies at the Grand Canyon in Arizona, where the Precambrian basement of the Colorado Plateau is exposed in the canyon bottom, have given the insight that major joint trends are parallel to schistosity and other old structures in the basement. Thus the anisotropy of the basement can in some way influence the fracture of the overlying cover, perhaps through differential compaction or epeirogenic warping.

FIGURE 6–7 Plumose structure on a set of subparallel joint surfaces in Eocene metamorphosed sandstone, southern Taiwan.

Regional patterns of jointing in basement rocks are commonly composed of overlapping regions dominated by independent joint sets and systems similar to those of the sedimentary cover of the Colorado Plateau. Systematic joints in the Precambrian gneisses and granites of the Beartooth uplift in Montana in the western United States are nearly vertical and are dominated by four major regional joint sets, each of which is best developed in its own overlapping subregion (Spencer, 1959). Two of the sets form an orthogonal system. Igneous dikes of several Precambrian and Tertiary ages are exposed within the uplift and intruded parallel to the main joint systems. Thus parts of the fracture patterns have persisted since the Precambrian.

Fracture anisotropy, such as rift and grain (Chapter 5), may exhibit distinctive regional patterns similar to those exhibited by systematic joint sets. Furthermore, some—but not all—systematic joint sets may be parallel to the fracture anisotropy. In some cases fracture anisotropy is known to be an immediate effect of preferred orientation of microcracks or microjoints, which have a spacing of a few millimeters (Chapter 5). The microjoints at each locality in the Beartooth Mountains form a pattern of two nearly orthogonal, vertical, perpendicular sets (Wise, 1964). The microjoint sets and parallel rift and grain

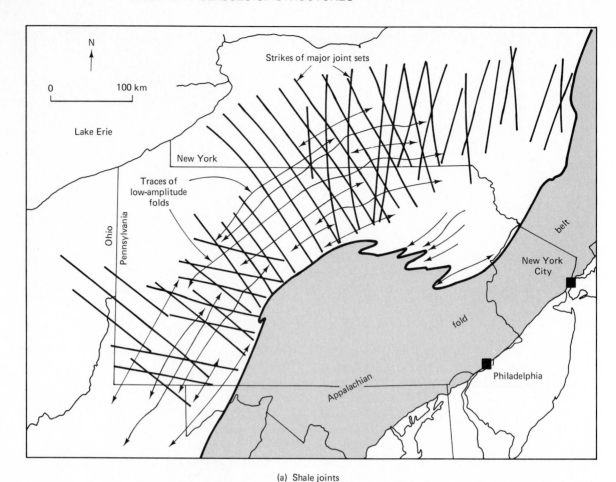

(a) Shale joints

FIGURE 6–8 Strikes of major joint sets in the central Appalachian foreland, west and north of the strongly deformed Appalachian fold belt. (a) Joints in shale and siltstone. (b) Joints in coal. (Compiled from Nickelsen and Hough, 1967; Engelder and Geiser, 1980; and Rodgers, 1970.)

simulate the orientations of some sets of ordinary joints and present a much simpler, more well-defined regional pattern. Some ordinary joints are parallel to the rift, grain, and microjoint directions, but other unrelated joint directions exist as well. The parallelism of fracture anisotropy with some joints is also observed in sedimentary rocks.

In spite of the insights outlined in the preceding paragraphs, geologists have not been notably successful in determining the tectonic significance and ultimate origin of regional joint patterns. This incomplete success in part reflects the difficulty of dating joints. Only if two joints form at the same moment do they record the same stress field. Therefore, it is uncertain to what extent systematic regional patterns of jointing, such as those shown in Figures 6-8 and 6-10, can be equated with stress fields. Even if the joint patterns reflect a stress field, its age is generally unknown. Furthermore, as discussed in a later section, a variety of subtle distortions of the rock are capable of producing joints.

Joints in Relation to Local Structures and Topography

On an outcrop scale, many joints bear little systematic relationship to anything else. Nevertheless, some do and are therefore the joints most susceptible to explanation. Some of the commonly observed systematic relationships between joints and local structure and topography are listed below.

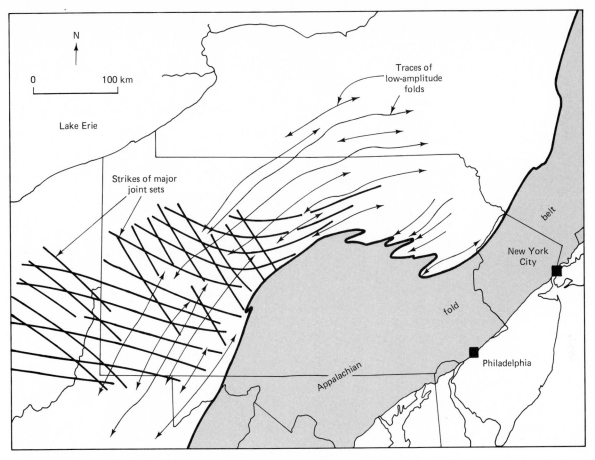

N

0   100 km

Lake Erie

Strikes of major
joint sets

Traces of
low-amplitude
folds

belt

New York
City

fold

Appalachian

Philadelphia

(b) Coal joints

FIGURE 6–8 Continued

Joints in bedded or otherwise layered rocks are very commonly perpendicular to layering and display a spacing that is a function of rock type and bed thickness. Joint spacing increases with increasing bed thickness. Many joints are roughly parallel or perpendicular to the local surface of the earth, whether or not the surface is horizontal. Many early-formed joints in igneous rocks are perpendicular to the outer surface of igneous intrusions. A set of joints commonly forms perpendicular to fold axes or linear fabrics. Joints that are perpendicular to a linear fabric in the rock are sometimes called *cross joints*. Joint sets may be parallel to the fracture anisotropy.

These and other systematic relationships between joints and local structure and topography provide some basis for developing theories of jointing, as we do later. However it is appropriate to consider first the various mechanisms by which rocks are stressed and consider the state of stress in the earth's crust because joints are a response to these stresses.

STATE OF STRESS IN THE EARTH'S CRUST

We now consider, as an interlude in our discussion of joints, the various mechanisms by which rocks are stressed and how this stress may be released by permanent deformation of the rock. These considerations will give some understanding of why rocks may deform by jointing in some situations, but deform in others by faulting and still others only elastically. We shall also consider the results of·actual measurements of stress made in deep mines and boreholes.

FIGURE 6–9 Vertical systematic joint set in flat-lying Devonian calcareous siltstones, central New York.

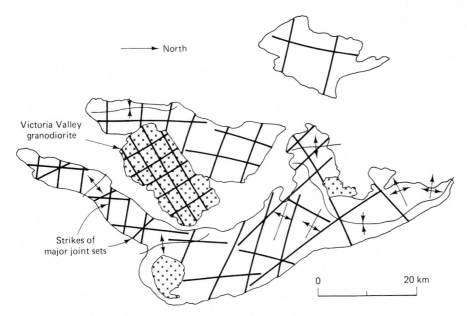

→ North

Victoria Valley
granodiorite

Strikes of
major joint sets

0   20 km

FIGURE 6–10 Map illustrating regional patterns of steeply dipping systematic joints in the sandstones and conglomerates of the Upper Devonian and Lower Carboniferous Grampians Group, western Victoria, Australia. (Data from Spencer-Jones, 1963.)

Loading History

Let us make a distinction between stress and loading. For this discussion, we define *loading* to be the history of applied forces, displacements, and temperature changes that produces the history of stress fields experienced by a body. In contrast, recall that stress is a property of a single point in a body and a single time (Chapter 3). The net deformation of a body of rock is determined by the history of stress fields that it experiences and is therefore the result of the loading history.

There are three especially important mechanisms of loading: (1) gravitational, (2) thermal, and (3) displacement loading.

1. *Gravitational Loading.*   By far the most important control on the state of stress in the earth is the weight of the overlying rocks. The vertical compressive stress is caused primarily by the gravitational load. This fact is illustrated in Figure 6-11(a), which is a graph of the measured vertical compressive stress, σ_z, as a function of depth in Norwegian mines. The observed vertical normal stress, σ_z, is very close to that predicted from the weight of the overburden:

$$\sigma_z = \rho g z \tag{6-1}$$

where ρ is the density of the overburden (about 2700 kg/m^3 in this case), g is gravity (9.8 m/s^2), and z is the depth (Fig. 6-11(a)). Vertical stresses that exceed the overburden are occasionally encountered, often because of material heterogeneities; nevertheless, Equation 6-1 remains a very good working approximation for the vertical stress in most situations.*

---

*The complete expression for the normal stress in the vertical direction σ_{zz} at a depth h is

$$\sigma_{zz} = \rho g z - \int_0^h \frac{\partial \sigma_{xz}}{\partial x}\, dz - \int_0^h \frac{\partial \sigma_{yz}}{\partial y}\, dz$$

where x and y are orthogonal horizontal directions. The last two terms describe the net contribution of the vertical shear-stress components to the vertical normal stress, which may be significant in some situations.

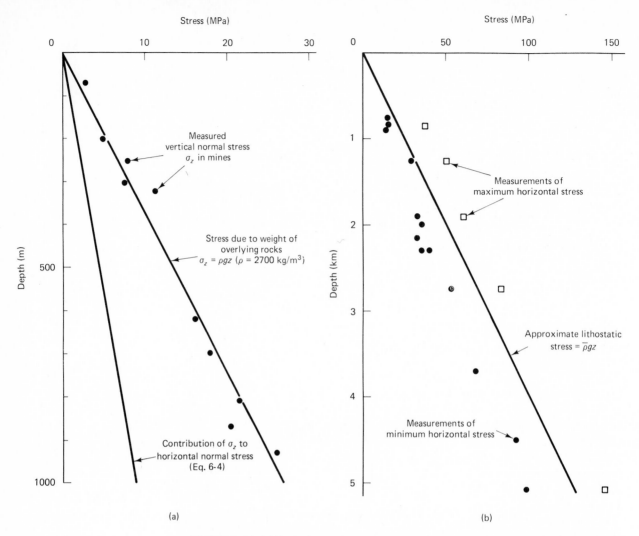

FIGURE 6–11 (a) Vertical normal stress measured in Norwegian mines in comparison with the predicted lithostatic overburden ρgz. (Data from Bjørn, 1970.) (b) Maximum and minimum horizontal stresses measured in the United States in comparison with the lithostatic overburden. (Data from Haimson, 1977; McGarr and Gay, 1978.)

Gravitational loading not only dominates the vertical stress, but also affects the horizontal stress because rocks tend to expand horizontally in response to the gravitational load. Elastic materials, including rocks, tend to expand in the directions perpendicular or transverse to the applied compressive stress (Fig. 4-27).

The transverse expansion can be described by Poisson's ratio, v, which is the ratio of the transverse strain ϵ_t, to the longitudinal strain, ϵ_l, when the material is free to expand transversely (Eq. 4-20):

$$v = \frac{\epsilon_t}{\epsilon_l} \qquad (6\text{-}2)$$

If the rock is not free to expand transversely ($\epsilon_t = 0$), a transverse stress, σ_t, is created, which we may calculate from the elastic stress-strain relation

$$\epsilon_x = \epsilon_y = \frac{1}{E}\left[\sigma_x - v\left(\sigma_y + \sigma_z\right)\right] \qquad (6\text{-}3)$$

where z is the vertical (longitudinal) direction, x and y are the horizontal directions, E is Young's modulus, and σ_z is the applied (longitudinal) stress—that is, the gravitational load. If the transverse strain is zero ($\epsilon_t = \epsilon_x = \epsilon_y = 0$), then the induced transverse stress ($\sigma_t = \sigma_x = \sigma_y$) follows from Equation 6-3:

$$\sigma_t = \frac{v}{1 - v}\sigma_z \tag{6-4}$$

This horizontal stress, induced by the gravitational load, is about one-third the vertical axial stress, σ_z, because Poisson's ratio is commonly about one-fourth. Thus the gravitational load makes a significant, if modest, contribution to the horizontal stress if the rock is not free to expand horizontally (see Fig. 6-11(a)).

   2. *Thermal Loading.*   If a homogeneous rock is slowly heated or cooled, it will homogeneously expand or contract. The relationship between strain, ϵ, and temperature change, ΔT, is (Eq. 4-21)

$$\epsilon = \alpha\Delta T \tag{6-5}$$

where α is the linear coefficient of thermal expansion.

   If, however, the rock is not free to expand or contract, stress will be generated. This mechanism of stress generation is called *thermal loading*. An everyday example of thermal loading comes from dropping a very cold ice cube into a glass of water. The ice cracks audibly because the warmed ice on the outside of the cube expands relative to the still-very-cold inside. Stress is thereby generated, placing the inside in tension and the outside in compression.

   Thermally generated stresses may be described using the elastic stress-strain relationship (Eq. 6-3) with thermal effects (Eq. 6-5) included:

$$\epsilon_t = \frac{1}{E}\left[\sigma_t - v\left(\sigma_t + \sigma_l\right)\right] + \alpha\Delta T \tag{6-6}$$

If a confined rock ($\epsilon_t = 0$) is cooled through a temperature interval ΔT, the induced thermal stress, ignoring the applied longitudinal stress σ_l, is then

$$\sigma_t = \frac{\alpha E\Delta T}{1 - v} \tag{6-7}$$

For example, consider a confined rock ($\epsilon_t = 0$) that cools 100°C but is not allowed to shrink ($\alpha = 10^{-6}/°\text{C}$, $E = 10^5$ MPa, $v = 0.25$). The induced tensile stress, by Equation 6-7, is -13 MPa, which is approximately the tensile strength of rock (Table 5-1). Therefore, cooling of confined bodies of rock can cause joints to form.

   The confinement of rock that is essential for thermal loading is provided by two mechanisms.

   (a) *Spatial variation of* ΔT.   Inhomogeneous heating or cooling has already been illustrated with the cracking ice cube. A geologic example is the rapid, and therefore inhomogeneous, cooling of a lava flow, which places its upper surface in tension because the top has initially cooled more than the inside. It is the difference in temperature change between two parts of the rock that are joined together that gives rise to the stresses. If the lava were cooled slowly and therefore homogeneously, no stress would be generated and no joints would form. Another example of inhomogeneous cooling is provided by a dike intruded into a cool country rock; the dike eventually cools to the regional temperature, whereas the country rock undergoes no net change in temperature. Thermal stresses are generated because the dike is welded to the country rock and the net temperature change is heterogeneous.

(b) *Inhomogeneous material properties.* Even with a homogeneous temperature change, thermal stresses will develop if the body is composed of materials joined together that have different thermal expansion or elastic coefficients α, ν, and E. For example, thermal stresses develop on the scale of individual crystals in rocks composed of interlocking minerals of contrasting thermal-elastic properties. This last process is discussed in more detail under the heading "Residual Stresses."

3. *Displacement Loading.* A third major mechanism by which rocks are loaded in the earth is the forced displacement of their adjacent surroundings, which is of major importance in tectonic deformation. For example, rocks at a convergent plate boundary are placed under horizontal compression because the adjacent lithospheric plates are moving together continuously. Other examples of displacement loading include epeirogenic warping of the crust in response to inhomogeneous heating or cooling of the bottom of the lithosphere (Chapter 1) or warping of strata due to differential compaction of adjacent rocks in a heterogeneous sedimentary section. What is common to all these examples is the externally forced displacement of one boundary of a body of rock relative to another. The stresses in the body can be thought of as arising in response to the applied displacements.

State of Stress in the Upper Crust

The actual state of stress in a rock will vary with time through the interplay of the loading mechanisms that stress the rocks and the deformation processes that may dissipate the deviatoric stress. The loading mechanisms include gravitational, thermal, and displacement loading, discussed earlier, and the dissipative processes include fracture, frictional sliding, buckling, and flow.

It is important to realize that deformation dissipates the deviatoric stress largely through changes in the horizontal components of the stress because the vertical stress is generally fixed by the mass of the overburden (Eq. 6-1). The vertical stress cannot be dissipated by mechanisms of permanent deformation unless these mechanisms affect the mass of the overburden. The gravitational load varies with time, largely through erosion, deposition, and tectonic thinning and thickening—for example, by thrust faulting.

Processes of creep, such as plastic deformation, diffusion flow, and compaction, are all capable of reducing the deviatoric stress in time, largely through changes in the horizontal components of the stress. In the limit of very weak rocks, the deviatoric stress would be reduced through deformation to zero; the state of stress would then be isotropic and equal to the vertical stress, which is controlled by the weight of the overburden. This fluidlike, isotropic state of stress is called a *lithostatic state of stress,* by analogy with the hydrostatic state of stress in a fluid at rest.

A lithostatic state of stress may be approximated in weak unconsolidated sediments, deep-seated metamorphic terrains at high temperature, and within the aesthenosphere (Fig. 4-29). Petrologists normally assume the state of stress in rocks to be lithostatic. The prediction, or rule of thumb, that the state of stress within the earth should be approximately lithostatic is often called *Heim's rule.* Most direct measurements of stress within the earth do not agree with Heim's rule. The horizontal stresses are generally significantly greater or less than the vertical stress; for example, measurements from North America are shown in Figure 6-11(b). Heim's rule is violated because measurements have been made only at depths of less than 5 km, where most rocks have a long-term finite strength.

The maximum and minimum possible value of the horizontal stress at any depth is limited by the strength of the rock. Processes that may fix the long-term strength of a body of rock at any temperature and pressure include plastic deformation, diffusion flow, faulting, buckling, compaction, and brittle fracture. Of these, by far the best known and most important in the upper 5 km of the crust is brittle fracture; the fracture strengths of many rocks have been measured (for example, Table 5-1). If the deviatoric stress exceeds the fracture strength, the rock must fracture and dissipate some of the stress. Therefore, we may use measurements of fracture strength as maximum limits on the deviatoric stress within the earth. We shall now determine what these limits are using Coulomb fracture theory (Chapter 5).

It is important to recall that the fracture strength of rock is not controlled by stress, but by effective stress. Therefore, we must consider the pore-fluid pressure, P_f, in addition to the stress. It is convenient to define a dimensionless quantity λ, the *Hubbert-Rubey fluid-pressure ratio,* which is the ratio of the fluid pressure to the vertical stress due to the rock overburden:

$$\lambda = \frac{P_f}{\bar{\rho}gz} \tag{6-8}$$

where $\bar{\rho}$ is the mean rock density, z is the depth, and g is gravity. The fluid-pressure ratio, λ, is useful because it is relatively constant over large rock volumes; it may range from 0 for dry rocks to 1 for fluid pressure equal to overburden stress. A *hydrostatic fluid pressure,* also called *normal fluid pressure,* is the static fluid pressure fixed by the depth below the water table or sea level, z_f, and the mean density of the fluid, $\bar{\rho}_f$ (about 1000 kg/m^3 for water):

$$P_f = \bar{\rho}_f g z_f \tag{6-9}$$

Hydrostatic values of λ range from about 0.37 to 0.47, depending on the mean density of the overlying rock and the salinity of the water; $\lambda = 0.465$ is often used for young sediments. Nonhydrostatic fluid pressures are called *abnormal fluid pressures,* even though fluid pressures well in excess of hydrostatic ($\lambda = 0.5$ to more than 0.9) are widely encountered in deep drilling for petroleum (Chapter 7). Fluid pressures that are equal to the total pressure ($\lambda = 1$) are commonly assumed for convenience in metamorphic petrology, but are probably unusual. Fluid pressures less than hydrostatic are relatively uncommon.

The Coulomb fracture criterion (Eq. 5-2) in terms of effective stress (Eq. 5-11) is

$$(\sigma_1 - P_f) = C_0 + K(\sigma_3 - P_f) \tag{6-10}$$

where C_0 is the cohesive strength and K is the earth-pressure coefficient. If the vertical normal stress is a principal stress and is gravitational, then the maximum horizontal stress at any depth z, limited by the fracture strength of the rock, is

$$\sigma_1 - \lambda\bar{\rho}gz = C_0 + K\bar{\rho}gz(1 - \lambda)$$

or

$$\sigma_1 = C_0 + \bar{\rho}gz\,[\lambda + K(1 - \lambda)] \tag{6-11}$$

and the minimum horizontal stress is given by

$$\rho gz\,(1 - \lambda) = C_0 + K(\sigma_3 - \lambda\bar{\rho}gz)$$

or

$$\sigma_3 = -\frac{C_0}{K} + \bar{\rho}gz[\lambda + \frac{1}{K}(1 - \lambda)] \tag{6-12}$$

Using the appropriate values of the compressive cohesive strength, C_0, and the earth-pressure coefficient, K, we may calculate the limits on the horizontal stress for a given rock as a function of the fluid pressure ratio, λ. For example, Figure 5-6 shows laboratory measurements on many Tertiary sandstones, siltstones, and shales from Japan that give a mean value of $C_0 = 15$ MPa and $K = 5$. Using Equations 6-11 and 6-12, we compute the maximum and minimum horizontal stresses that are possible in the Japanese Tertiary basins at various fluid-pressure ratios, as shown in Figure 6-12. It should be noted that as the fluid-pressure ratio,

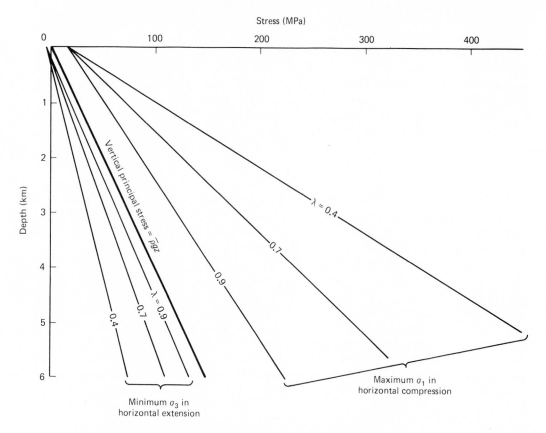

FIGURE 6–12 Maximum and minimum possible horizontal stresses at various fluid-pressure ratios λ (Eq. 6-8) based on Coulomb fracture strengths (Eqs. 6-11 and 6-12). Computed for typical sandstones, siltstones, and shales ($C_0 = 15$ MPa, $K = 5$).

λ, approaches 1, the rocks progressively weaken, the maximum stress approaches $C_0 + \bar{\rho}gz$, and the minimum stress approaches $\bar{\rho}gz - (C_0/K)$.

At present it is difficult to directly compare predicted limits on the stress based on rock strength with actual measurements of stress because fluid pressures are commonly not known. Nevertheless, a comparison of the *in situ* stress measurements in the United States (Fig. 6-11(b)) with the brittle-fracture strengths of Figure 6-12 suggests that the stress may be controlled by rock strength with fluid-pressure ratios, λ in the range of 0.4 to 0.7, σ_2 vertical, and $\sigma_1 - \sigma_2 = C_0$. In the future, actual stress measurements in deforming regions will probably play an important role in structural geology. At present it is known that the deviatoric stress is a linear function of depth in several regions (for example, Fig. 6-11(b)) and regionally consistent patterns of orientation of horizontal stress components have been observed in North America and Europe (Fig. 6-13).

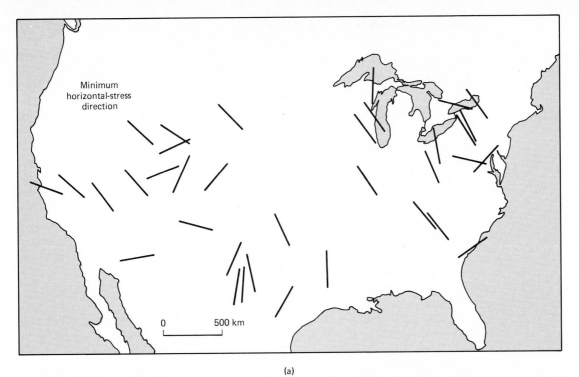

(a)

(b)

FIGURE 6–13 Regional patterns of present-day stress showing the orientation of the horizontal maximum or minimum stress component in (a) the United States and (b) western Europe. (Data compiled from Zoback and Zoback, 1980, and other sources.)

State of Stress in the Lower Crust and Upper Mantle

With increasing depth in the earth, we expect rocks to undergo a transition from a brittle to a plastic deformation mechanism because of the increase in brittle strength with pressure and the decrease in plastic strength with temperature (Chapter 5). We can combine these two effects to predict the maximum and minimum horizontal stresses in the lithosphere, as limited by rock strength. For example, if we combine the stresses just predicted for brittle sedimentary rocks in Figure 6-12 with the plastic strength of quartz from Figure 4-29, we can obtain an estimate of the maximum and minimum possible horizontal stresses in quartz-rich lithosphere, as shown in Figure 6-14(a). It should be noted that the maximum deviatoric stress is predicted to be at the brittle-plastic transition. Furthermore, an increase in fluid-pressure ratio produces an increase in the depth of the brittle-plastic transition, whereas it reduces the maximum deviatoric stress in the lithosphere.

The predicted stress-depth curves are substantially different for different rock types because of the large differences in plastic strength with mineralogy (Fig. 4-29). For example, we would predict oceanic lithosphere, which is rich in olivine to shallow depths, to be substantially stronger than quartz-rich lithosphere. Furthermore, if the Moho at the base of a quartz-rich crust lies at a depth intermediate between the brittle-plastic transitions for quartz and olivine, we would predict a lithosphere composed of a strong upper crust and upper mantle with an intervening weak lower crust—in essence it is like a jelly sandwich (Fig. 6-14(b)). It is apparently for this reason that the upper crust commonly deforms independently from the rest of the lithosphere in continental deformation.

Residual Stresses

Before returning to the subject of joints, one further aspect of stress generation and dissipation in the earth should be addressed, that of *residual, or locked-in, stresses*. Residual stresses are stresses locked into a body even when no forces act on the outside of the body. For example, a sample of granite sitting on a table will have deviatoric stresses locked into its individual mineral grains. These residual stresses are caused by changes in temperature and pressure in the heterogeneous material. Most rocks are heterogeneous intergrowths of minerals of different thermal expansion coefficients and elastic contents; for example, a quartz grain may be surrounded by a garnet crystal. On a larger scale, entire bodies of rock of contrasting thermoelastic properties are interlayered and interlocking within the crust; for example, a granitic intrusion may be imbedded in a schist terrain. As a consequence of this thermoelastic heterogeneity, stresses are generated because each material expands or contracts differently in response to temperature and pressure changes. The residual stresses produced by pressure release and cooling during uplift and erosion play an important role in the formation of some joints.

If we scribe two arbitrary lines of equal length on different minerals or rocks, they will generally have different and unequal lengths at some other temperature and pressure. If the two materials are intergrown, elastic distortion and stresses will result. Most rocks are observed in outcrop at a pressure and temperature quite different from their crystallization or consolidation; therefore, they will in general contain residual stresses unless the distortions in the heterogeneous material are relaxed by fracture or flow.

Consider a garnet crystal that grows around a 1-cm spherical quartz grain at 450°C and 500 MPa. The quartz sphere fits precisely into the spherical hole in the garnet with no mismatch. After some change in temperature or pressure, the

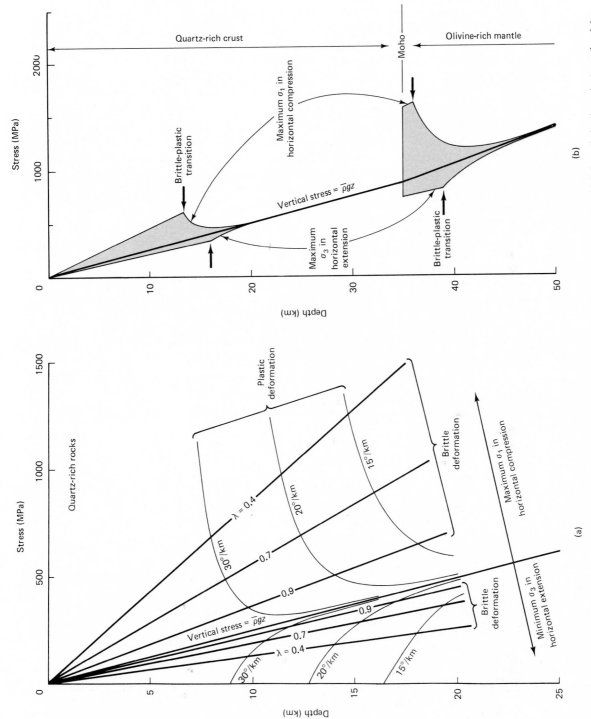

FIGURE 6-14 Maximum and minimum possible horizontal stresses based on laboratory measurements of brittle and plastic rock strengths. (a) Strengths of quartz-rich rocks based on Figures 6-12 and 4-29. (b) Example of strength of continental lithosphere composed of 35 km of quartz-rich crust overlying an olivine-rich upper mantle ($20°C/km$ and $\lambda = 0.8$). Note that the lithosphere in this case is composed of a strong upper crust and upper mantle and a weak lower crust.

quartz inclusion in general will no longer fit because quartz and garnet have different properties of thermal expansion and compressibility. If the 1-cm quartz sphere were free to expand and if it were brought to an earth-surface temperature (25°C), it would change to a uniaxial ellipsoid with lengths of the principal axes of 0.9927 and 0.9953 cm, because quartz has anisotropic thermal expansion (Eq. 4-21). In contrast, if the 1-cm spherical hole in the garnet were free to shrink without the quartz inclusion, it would become a 0.9905-cm spherical hole when brought to earth-surface temperature. The hole would be spherical because garnet has isotropic thermal expansion. Thus cooling makes the quartz inclusion an ellipsoidal peg in a smaller spherical garnet hole; the result is elastic deformation and stress in and around the quartz inclusion.

Elastic distortion produces photoelastic effects in crystals. For example, Figure 6-15 is a photograph of a quartz inclusion of one crystallographic orientation in a host quartz crystal of a different orientation; photoelastic effects of the distortion can be seen around the inclusion. Furthermore, two fractures that radiate from the inclusion have dissipated part of the residual stress.

ORIGIN OF JOINTS

Jointing in Light of Mohr Diagram

In the introduction to this chapter we reviewed the evidence that many, perhaps most, joints are tensile or transitional tensile fractures. Joints show little or no displacement parallel to their walls and show no scratches, slickensides, or gouge indicative of frictional sliding. Some fresh, unweathered joints display delicate surface markings that are incompatible with frictional sliding (for example, Figs. 6-1 and 6-4). Many joints apparently formed as narrow open fissures because they contain vein minerals that grew outward from the walls of the fissure, although the opening of some joints is later than the fracture. For these reasons joints appear to form without significant compressive stress acting across the fracture surface; the effective normal stress, σ_n^*, appears to be generally tensile ($\sigma_n^* \leq 0$).

From the perspective of the Mohr diagram, the principal domain of joint formation ($\sigma_n^* \leq 0$) appears to be the domain of tensile and transitional-tensile fracture (Figs. 6-16 and 5-5). Some shear joints have orientations suggesting they might be Coulomb fractures ($\phi \cong 60°$, Fig. 6-16,); alternatively, they could be transitional-tensile fractures ($\sigma_n^* \cong 0$) because fracture orientation, 2θ, is relatively insensitive to effective normal stress over the transition between Coulomb fracture and transitional-tensile fracture. Joint-surface morphology is the main observational criterion for distinguishing shear joints as Coulomb fractures or transitional-tensile fractures.

The shape and size of the Mohr diagram places an upper limit on the deviatoric stress for joint formation, which in turn fixes a maximum depth for joint formation. The maximum deviatoric stress for a tensile effective normal stress tangent to the Mohr envelope ($\sigma_n^* \leq 0$) is limited by the shape of the Mohr envelope to about six times the tensile strength, T_0 (Fig. 6-16):

$$(\sigma_1^* - \sigma_3^*) \leq -4\sqrt{2}T_0 \qquad (6\text{-}13)$$

based on Griffith theory of fracture (Eq. 5-7). For joints that are true tensile fractures, the maximum deviatoric stress is further limited to four times the tensile strength (Fig. 6-16):

$$(\sigma_1^* - \sigma_3^*) \leq |4T_0| \qquad (6\text{-}14)$$

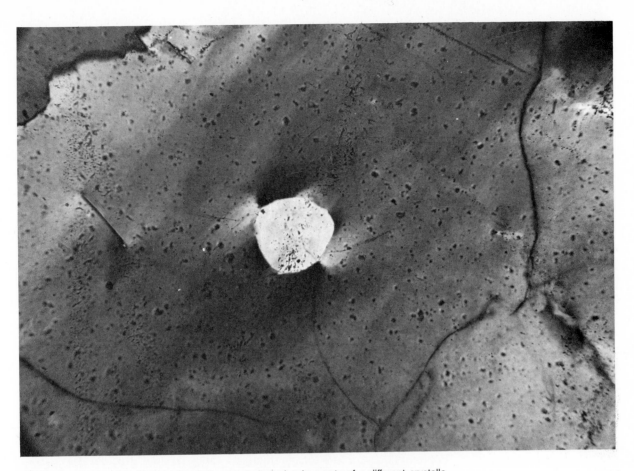

FIGURE 6–15 Photomicrograph showing a quartz inclusion in quartz of a different crystallographic orientation. Residual stresses have produced opposing light photoelastic fringes surrounding the inclusion and several small cracks radiating from it.

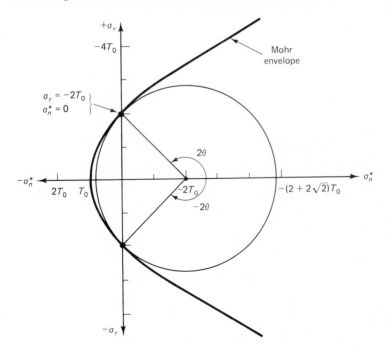

FIGURE 6–16 Mohr diagram showing the maximum effective stress for noncompressive normal stress on the fracture surface. Based on Equation 5-7.

The vertical stress due to the gravitational load is compressive (Eq. 6-1), whereas joints require tensile effective normal stress ($\sigma_n^* \leq 0$). Therefore, the requirement for deep formation of joints is both small deviatoric stress and high fluid-pressure ratio, λ (Eq. 6-8). Equations 6-13 and 6-14 may be used to estimate the maximum depth of jointing as a function of fluid-pressure ratio λ and tensile strength T_0. It follows from Equation 6-13 and Figure 6-16 that the maximum effective stress, σ_1^*, for jointing is limited to about five times the tensile strength:

$$\sigma_1^* \leq \ | \ (2 + 2\sqrt{2})T_0 \ |$$ (6-15)

If σ_1 is vertical, then (Eqs. 6-1, 6-8)

$$\sigma_1^* = \rho g z(1 - \lambda)$$ (6-16)

Equating 6-15 and 6-16,

$$\rho g z(1 - \lambda) \leq \ | \ (2 + 2\sqrt{2})T_0 \ |$$

and finally,

$$z \leq \left| \frac{(2 + 2\sqrt{2})T_0}{\rho g(1 - \lambda)} \right|$$ (6-17)

which gives the maximum depth, z, of jointing with σ_1 vertical as a function of fluid-pressure ratio, λ. For example, the maximum depth of jointing in the Gossford Sandstone ($T_0 = -3.2$ MPa, Table 5-1) is about 1 km at hydrostatic fluid pressure ($\lambda = 0.4$), whereas the maximum depth is about 7 km at $\lambda = 0.9$. Very strong rocks, such as the Cheshire Quartzite ($T_0 = -28$ MPa), are capable of tensile fracture throughout much of the crust. If σ_2 or σ_3 is vertical, the maximum depth of jointing is less than for σ_1 vertical. This analysis ignores any contribution of residual stresses to the vertical stress.

For joints that are true tensile fractures, the maximum effective stress (σ_1^*) is limited to three times the tensile strength (see Eq. 6-14 and Fig. 6-16):

$$\sigma_1^* \leq \ | \ 3T_0 \ |$$ (6-18)

The maximum depth of true tensile joints is therefore (equating 6-16 and 6-18):

$$Z = \left| \frac{3T_0}{\rho g(1 - \lambda)} \right|$$ (6-19)

which is about 60 percent of the maximum depth of transitional tensile joints (Eq. 6-17). A graph of this equation is given in Figure 6-17, which illustrates the principal domain of joint formation.

The Mohr diagram provides one further insight into the origin of joints. We recall from Chapter 3 that planes perpendicular to the principal stresses (σ_1, σ_2, σ_3) are planes with no shear stress acting across them. The surface of the earth and open fluid-filled joint fractures are examples of such planes of no shear stress because they are fluid-solid interfaces, and fluids cannot support shear stress. Therefore, the principal stresses have three possible orientations at the surface of the earth or near open joints; σ_1, σ_2, or σ_3 must be perpendicular to the interface. We know from the Mohr diagram and fracture experiments (Chapter 5) that tensile fractures form perpendicular to the least principal stress ($\theta = 90°$, Fig. 6-16). Therefore, we may expect that joints forming in a near-surface environment will be either perpendicular (σ_1 or σ_2 vertical) or parallel (σ_3 vertical) to the surface of the earth. Analogous theories of dike, sill, and fault orientations are given in Chapters 7 and 8. Similarly, tensile joints propagating near a preexisting open joint fracture will bend into perpendicular or parallel orientation with respect to preexisting open joints (for example, see the nonsystematic joints of Fig. 6-2). Crosscutting joints that are not orthogonal (for example, the systematic joints of

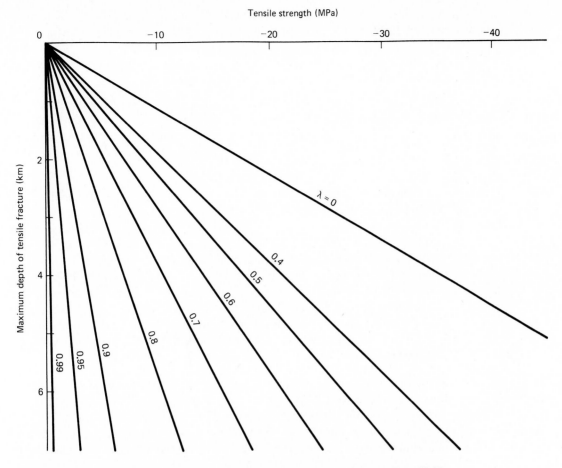

FIGURE 6–17 Maximum depth of tensile fracturing as a function of tensile strength T_0 and fluid-pressure ratio $\lambda = P_f/\rho gz$ (Eq. 6-19, σ_1 vertical, and $\rho = 2700$ kg/m³).

Fig. 6-2) apparently must be preexisting joint sets that were frictionally locked or cemented by vein material when the second joint set formed.

Uplift and Denudation in Jointing

It is generally thought that many joints are formed by the release of tensile stresses generated during uplift and denudation. Uplift and denudation modify the preexisting state of stress in three ways: (1) horizontal stretching through the geometry of uplift, (2) expansion through the release of the gravitational load, and (3) contraction through cooling. The net effect of these processes is horizontal tension in many situations.

1. *Initial State of Stress and Strain.* We must first consider the initial state of stress and elastic strain before uplift. We then determine how it is modified in any episode of uplift and denudation to a new state of stress that may cause jointing or other deformation. Widely different initial states of stress and elastic strain are theoretically possible, limited only by the strength of the rock. In order to emphasize how uplift and denudation modify the initial state, we shall first consider an isotropic or lithostatic initial state of stress. For example, suppose a body of homogeneous rock has crystallized or lithified at some depth h in the earth and is subjected to an isotropic stress P:

$$P = \sigma_z = \sigma_x = \sigma_y = \bar{\rho}gh \qquad (6\text{-}20)$$

where $\bar{\rho}$ is the mean density of the overburden and g is gravity. After some uplift and erosion Δz (negative), the state of stress is modified to a new value:

$$\sigma_z = \bar{\rho}g(h + \Delta z)$$

$$\sigma_x = \sigma_y = \bar{\rho}gh + \Delta\sigma_s + \Delta\sigma_g + \Delta\sigma_t \qquad (6\text{-}21)$$

where $\Delta\sigma_s$ is the change in stress caused by horizontal stretching during uplift, $\Delta\sigma_g$ is the change in stress caused by expansion upon release of the gravitational load, and $\Delta\sigma_t$ is the change in stress caused by thermal contraction upon cooling. The new horizontal stress is determined by the sum of these three effects, which are considered as follows.

2. *Horizontal Stretching.* As segments of the crust undergo uplift, they may be stretched horizontally. For example, the beds at the crest of the large anticline in Figure 2-1 have stretched because of the bending associated with the uplift during folding. In a similar way the epeirogenic uplift of the Guyana Highlands (Fig. 1-4) produced stretching of the top of the lithosphere. The amount of stretching and associated change in stress, σ_s, caused by bending is inversely proportional to the radius of curvature of the flexure. For this reason the more strongly curved parts of folds formed under brittle upper crustal conditions commonly display a higher concentration of joints and small faults. We shall not consider further the stress produced by stretching because it depends so specifically on the tectonic setting.

3. *Release of the Gravitational Load.* The release of the gravitational load by erosion causes expansion of the compressed rock. However, the rock is not free to expand in the horizontal directions; therefore, much of the horizontal compressive stress remains after erosion. The change in horizontal stress, $\Delta\sigma_g$, is related to the change in gravitational load, $\bar{\rho}g\Delta z$, through Poisson's ratio, v (Eq. 6-4):

$$\Delta\sigma_g = \frac{v}{1 - v}\, \bar{\rho}g\Delta z \qquad (6\text{-}22)$$

Considering that v is about one-fourth, the decrease in horizontal stress is only about one-third the decrease in gravitational load.

4. *Thermal Contraction.* A change in horizontal stress, $\Delta\sigma_t$, is produced during erosion as a result of cooling. Thermal stresses are generated because the rock is not free to contract horizontally, as described by Equation 6-7:

$$\Delta\sigma_t = \frac{\alpha E}{1 - v}\, \Delta T \qquad (6\text{-}23)$$

Thus if we can specify the cooling history during uplift and erosion, we can compute the thermal stress. In the case of slow uplift, the change in temperature, ΔT, is simply determined by the equilibrium thermal gradient dT/dz:

$$\Delta T = \frac{dT}{dz}\, \Delta z \qquad (6\text{-}24)$$

However, during very rapid uplift temperatures may be substantially greater (for example, Alberede, 1976), therefore Equations 6-23 and 6-24 may be used to estimate the maximum thermal stress:

$$\Delta\sigma_t = \frac{\alpha E}{1 - v}\, \frac{dT}{dz}\, \Delta z \qquad (6\text{-}25)$$

5. *Net Effect of Uplift.*   The stresses generated during uplift and erosion are the net effect of release of the gravitational load (Eq. 6-23) and thermal contraction (Eq. 6-25), ignoring possible horizontal stretching. Combining with Equation 6-21, we obtain the horizontal stress, σ_x, at depth $z = (h + \Delta z)$) after some erosion Δz:

$$\sigma_x = \bar{\rho}g \left[z - \frac{1 - 2v}{1 - v} \Delta z \right] + \frac{\alpha E}{1 - v} \frac{dT}{dz} \Delta z \qquad (6\text{-}26)$$

The resulting stress is a function of several material properties ($\bar{\rho}$, v, E, α) and the thermal gradient, dT/dz. The stress is in fact quite sensitive to the exact values of these quantities, especially because the effects of the gravitational load (Eq. 6-22) and cooling (Eq. 6-25) are normally of similar magnitude, but of opposite sign. Therefore, horizontal extension is produced in some cases, but horizontal compression occurs in others.

An example of the change in state of stress upon uplift and erosion is given in Figure 6-18(a). Consider a sandstone lithified at a depth of 5 km under an isotropic

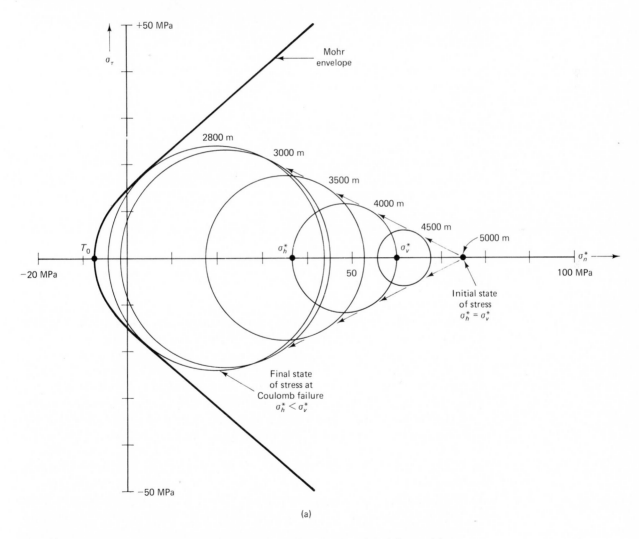

(a)

FIGURE 6–18 Modification of an initial state of stress as a result of erosion of the overlying rocks and associated cooling (Eq. 6-26, 20°C/km, $\lambda = 0.4$, $\rho = 2550$ kg/m^3). (a) An initial lithostatic state of stress at 5 km depth is modified to a Coulomb fracture stress in horizontal extension at 2.2 km of erosion. (Figure continued on next page.)

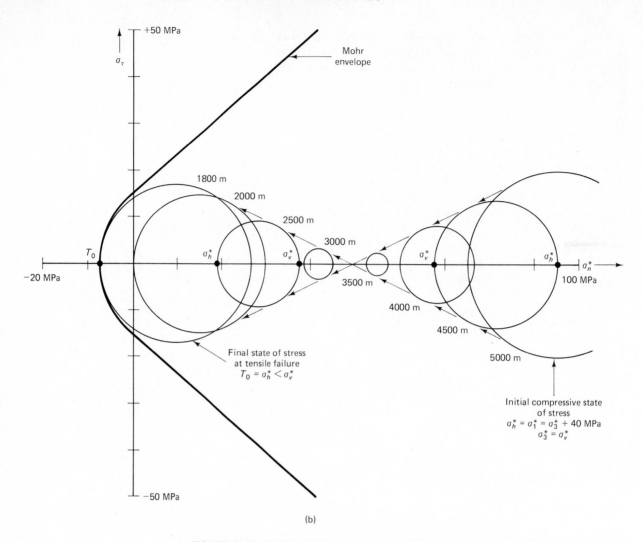

FIGURE 6–18 Continued. (b) An initial horizontal compressive state of stress ($\sigma_h - \sigma_v = 40$ MPa) at 5 km depth is modified to an isotropic state of stress at about 1.7 km of erosion and is brought to tensile fracture at about 3.2 km of erosion.

state of stress ($\sigma_h^* = \sigma_v^*$) with a 20°/km geothermal gradient and a hydrostatic fluid pressure ($\lambda = 0.4$). With typical elastic properties and thermal expansion coefficient, the cooling will dominate over the release of the gravitational load; therefore, successive states of stress upon erosion will show increasing horizontal extension ($\sigma_v^* > \sigma_h^*$). Finally, after about 2.2 km of erosion, the rock will be at a state of failure—in this case, by Coulomb fracture with a very small normal stress.

The initial state of stress in general will not be isotropic, which has a significant effect. Consider the same sandstone being uplifted from a depth of 5 km, but this time with an initial horizontal compressive stress ($\sigma_h^* > \sigma_v^*$), as shown in Figure 6-18(b). In this case the deviatoric stress decreases with erosion because of the dominance of thermal contraction over gravitational load until the stress is isotropic at about 1.8 km of erosion. The state of stress then shows increasing horizontal extension ($\sigma_v^* > \sigma_h^*$) until the rock fails by tensile fracture at about 3.2 km of erosion.

We have seen the importance of uplift and erosion on state of stress in these two examples (Fig. 6-18). A considerable variety of stresses is possible, depending on the rock types and the tectonic conditions of uplift. In many cases, uplift and erosion lead to formation of joints by tensile fracture, as shown in Figure 6-18(b).

FIGURE 6–19 Joints in an Ordovician shale and sandstone sequence passing through many beds, Topman's Bluff, Jefferson Co., New York. (G. D. Walcott.)

Joints in Bedded Sedimentary Rocks

Joints perpendicular to bedding are the most obvious structural feature of flat-lying sedimentary rocks (for example, Fig. 6-9). It is observed that joint spacing in interbedded sediments is a function of rock type; for example, joints are much more closely spaced in coal than in sandstone or shale. Furthermore, joint spacing is a function of bed thickness; joints are much more numerous in thin beds than in interlayered, thick beds of the same rock type. A related observation is that some joints are confined to a single bed (for example, Fig. 6-5), whereas other joints pass through many beds (Fig. 6-19).

To understand these observations, let us consider a set of interbedded layers $1, 2, \ldots, n$ of thickness $d_1, d_2, \ldots, d_n$ (Fig. 6-20). Each layer of different rock will in general have different elastic constants $E_1, E_2, \ldots, E_n$ and tensile strengths $T_1, T_2, \ldots, T_n$. To start with, assume that the beds have undergone compaction and lithification, that they are all at approximately the same lithostatic state of stress, and that the vertical dimension between the top and bottom bed is small. The horizontal stress, σ, and strain, ϵ, in each layer is then $\sigma = E_1\epsilon_1 = E_2\epsilon_2 = \ldots = E_n\epsilon_n$.

If the beds are now uniformly stretched (Fig. 6-20) by some strain ϵ_x—for example, as a result of epeirogenic warping (Chapter 1)—the new horizontal stress σ_n generated in each of the n layers will be now quite different:

$$\sigma_1 = E_1(\epsilon_1 - \epsilon_x)$$

$$\sigma_2 = E_2(\epsilon_2 - \epsilon_x)$$

$$\vdots \qquad\qquad\qquad (6\text{-}27)$$

$$\sigma_n = E_n(\epsilon_n - \epsilon_x)$$

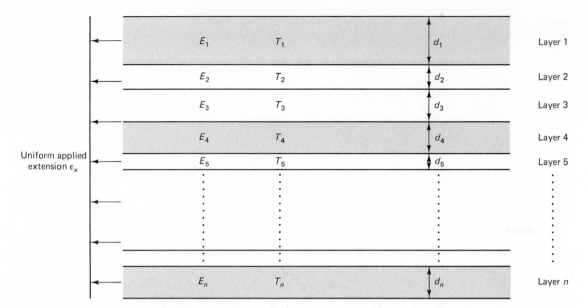

FIGURE 6–20 Mechanics of jointing in bedded sediments. Each layer, 1, 2, . . . , n, has different elastic properties, E, and tensile strength, T; therefore, they respond differently to a homogeneous stretching ϵ_x.

Thus a uniform applied strain will generate quite different layer-parallel normal stress in layers of contrasting elastic properties. The new stress may reach the tensile strength in some layers, but not in other layers, each layer having its own strength. It is the layers of high stiffness, E, and low tensile strength, T, that will fracture first. The fractures in general will be confined to the layer in which they initiate unless the adjacent layer is also near its own fracture strength.

Thus we have an explanation of joint spacing as a function of rock type in terms of contrasts in elastic stiffness and tensile strength, although we have not spoken specifically of spacing. To do this we must consider the problem of joint spacing as a function of layer thickness.

As the applied stretching ϵ_x increases—for example, as epeirogenic warping continues—a joint will eventually form at the weakest point in the system. The propagation of a joint in a weak, highly stressed bed does not necessarily precipitate rupture in neighboring beds, although the stress will, of course, increase to some extent to the neighboring beds near the joint. We recall from Griffith theory (Eq. 5-4) that a crack releases significant stress only within a radius of about one crack length. Therefore, the joint has little effect on the state of stress in its own bed more than one bed thickness away; the rest of the bed is still very close to the stress for tensile failure. With a small increase in strain, other joints will form, each only relaxing the stress within about one bed thickness; soon the entire bed will be jointed with a spacing roughly equal to the bed thickness. Closer spacing requires a significant increase in strain (Hobbs, 1967) or healing of joints by vein precipitation. If all the layers have very similar strengths and elastic properties, then joints may propagate through many layers (for example, Fig. 6-19).

Sheet Structure

Sheet structure, or *sheeting,* also called *exfoliation,* is the development of joints that divide rock into broad, gently curved shells, approximately parallel to the topographic surface (for example, Fig. 6-21). Sheeting is best displayed on convex

FIGURE 6–21 Sheeting of granitic rock conformable to the topographic surface, Half Dome, Yosemite National Park, California. (F. C. Calkins.)

topographic forms, but also develops below concave forms. Sheeting is well developed in many bodies of granitic rock, especially quartz monzonite and granite, and less-commonly developed in granodiorite, massive sandstone, and other rock that is relatively free of other joints. Sheeting joints are largely independent of earlier formed structures because they cut across igneous inclusions, septa of metamorphic country rock, pegmatitic and aplitic dikes, mineralogic foliation and layering, cross bedding, and some earlier joints. Sheet structure is evidently a near-surface phenomenon because the number of sheeting joints increases regularly toward the surface of the earth and because the joints are closely parallel to the topography (Fig. 6-22).

The nature of the relationship between sheeting and topography has been controversial. Some workers have felt that sheeting controls the development of topography and fixes the location of domes and valleys. By this theory the sheets were present before the topography. In contrast, most workers have felt that sheets form roughly parallel to the topography. This latter view is probably correct in light of the observed decrease and eventual disappearance of sheeting with depth (Fig. 6-22).

Sheeting has been most extensively studied in the granite quarries of New England (Jahns, 1943; Johnson, 1970); the sheets range in thickness from less than a meter near the original ground surface to between 5 and 10 m at a depth of 30 to 40 m. At the Fletcher Quarry in Massachusetts, the sheeting is independent of the mineralogic layering and the foliation; they intersect at a variety of angles from 0 to 90°. Individual sheeting fractures are roughly elliptic in plan; some are as long

FIGURE 6–22 Sheet structure in granitic rock extending to a depth of 30 m with decreasing number of fractures, Norfolk County, Massachusetts. (T. N. Dale.)

as 70 m. A well-developed rift, grain, and hardway fracture anisotropy (see Chapter 5) exists within the granite and is controlled by preferred orientation of microcracks. The sheeting and most of the microcracks are parallel to the rift, which is the easiest plane of splitting.

It is widely observed that many sheets are in compression parallel to their length. This compression is documented by the lateral expansion of sheets after they are cut out of a quarry. Furthermore, sheets are observed to have buckled outwards at the ground surface (Fig. 6-23). This buckling of sheets may occur suddenly in quarries.

Most modern theories have proposed that sheeting joints are caused by compression parallel to the topographic surface, based on the evidence that the sheets are in compression parallel to the topographic surface.

There has been considerable speculation and controversy on the origin of the compression parallel to the topographic surface. It has been thought that denudation by the Poisson's ratio effect (Eq. 6-4) would produce horizontal compression; however, as shown in a previous section, the net effect of uplift and denudation is commonly horizontal tension rather than compression, if thermal expansion is also considered (Eq. 6-26). A very low geothermal gradient is commonly required for Poisson's ratio to dominate over thermal expansion. An alternative origin has been tectonic compression, but this is unsatisfactory because the best development of sheeting is predicted for valleys, as valleys have the greatest stress concentration in horizontal compression. In contrast, moun-

are probably younger. Early thermal contraction joints can play an important role in some mineral deposits—for example, the skarn iron ore of Iron Springs, Utah (Mackin, 1947).

EXERCISES

6–1 Determine which joints are systematic and which are nonsystematic in Figure 6-2, using tracing paper. Which criteria were useful?

6–2 Given an interlayered sequence of sandstone ($E = 30$ GPa, $T_0 = -3$ MPa), shale ($E = 30$ GPa, $T_0 = -4$ MPa), and dolomite ($E = 70$ GPa, $T_0 = -8$ MPa), which will first begin to form joints upon stretching of the stratigraphic sequence parallel to bedding? Why?

6–3 Derive Equations 6-13 and 6-17 for the maximum depth of jointing given Equation 5-7 for the Mohr envelope in the tensile regime, which is based on the Griffith theory of fracture (Fig. 6-16).

SELECTED LITERATURE

Joint-Surface Morphology

BAHAT, D., 1979, Theoretical considerations on mechanical parameters of joint surfaces based on studies in ceramics: Geological Magazine, v. 116, p. 81–92.

HODGSON, R. A., 1961, Classification of structures on joint surfaces: Amer. J. of Sci., v. 259, p. 493–502.

KULANDER, B. R., BARTON, C. C., AND DEAN, S. L., 1979, *The Application of Fractography to Core and Outcrop Fracture Investigations,* U.S. Department of Energy, Morgantown Energy Technology Center, Report METC/SF-79/3, Morgantown, West Virginia, 174 p.

Regional Studies of Joints

HODGSON, R. A., 1961, Regional study of jointing in Comb Ridge–Navajo Mountain area, Arizona and Utah: Amer. Assoc. of Petroleum Geologists Bull., v. 45, p. 1–38.

KELLY, V. C., AND CLINTON, N. J., 1960, Fracture systems and tectonic elements of the Colorado Plateau: Publication University New Mexico Geology, no. 6, 104 p.

NICKELSON, R. P., AND HOUGH, V. D., 1967, Jointing in the Appalachian Plateau of Pennsylvania: Geol. Soc. Amer. Bull., v. 78, p. 609–629.

State of Stress in the Earth

BRACE, W. F., AND KOHLSTEDT, D. L., 1980, Limits on lithospheric stress imposed by laboratory experiments: J. Geophys. Research, v. 85, p. 6248–6252.

HAXBY, W. F., AND TURCOTTE, D. L., 1976, Stresses induced by the addition or removal of overburden and associated thermal effects: Geology, v. 4, p. 181–184.

McGARR, A., AND GAY, N. C., 1978, State of stress in the earth's crust: Ann. Rev. Earth Planet. Sci., v. 6, p. 405–436.

VOIGHT, B., AND ST. PIERRE, B. H. P., 1974, Stress history and rock stress: *Advances in Rock Mechanics,* Proceedings 3rd Congress Internat. Soc. Rock Mech., vol. 2A, National Academy of Sciences, Washington, D.C., p. 580–582.

Theories of Jointing

HOBBS, D. W., 1967, The formation of tension joints in sedimentary rocks: an explanation: Geological Magazine, v. 104, p. 550–556.

PRICE, N. J., 1974, The development of stress systems and fracture patterns in undeformed sediments: *Advances in Rock Mechanics,* Proceedings 3rd Congress Internat. Soc. Rock Mech., vol. 1A, National Academy of Sciences, Washington, D.C., p. 487–496.

MUEHLBERGER, W. R., 1961, Conjugate joint sets of small dihedral angle: J. of Geol., v. 69, p. 211–219.

SECOR, D. T., JR., 1965, Role of fluid pressure in jointing: Amer. J. of Sci., v. 263, p. 633–646.

Sheeting

BRUNNER, F. K., AND SCHEIDEGGER, A. E., 1973, Exfoliation: Rock Mechanics, v. 5, p. 43–63.

JOHNSON, A. M., 1970, *Physical Processes in Geology* (Chap. 10, "Formation of Sheet Structure in Granite"), Freeman, Cooper and Co., San Francisco, 577 p.

Twidale, C. R., 1973, On the origin of sheet jointing: Rock Mechanics, v. 5, p. 163–187.

Columnar Joints

LACHENBRUCH, A. H., 1962, Mechanics of thermal contraction cracks and ice-wedge polygons in permafrost: Geol. Soc. Amer., Special Paper 70, 65 pp.

PECK, D. L., AND MINAKAMI, T., 1968, Formation of columnar joints in the upper part of the Kilauea Lava Lake: Geol. Soc. Amer. Bull., v. 79, 1151–1169.

RYAN, R. A., AND SAMMIS, C. G., 1978, Cyclic fracture mechanisms in cooling basalt: Geol. Soc. Amer. Bull., v. 89, p. 1295–1308.

SPRY, A., 1962, The origin of columnar jointing, particularly in basalt flows: Geol. Soc. of Australia J., v. 8, p. 191–216.

TOMKEIEFF, S. I., 1940, The basalt lavas of the Giants Causeway district of northern Ireland: Bulletin Volcanologique, v. 6, p. 89–143.

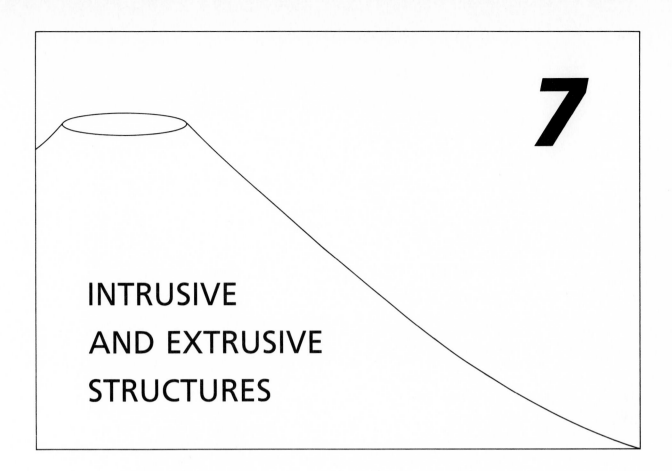

INTRUSIVE AND EXTRUSIVE STRUCTURES

7

INTRODUCTION

Magma, mud, and salt are united in their ability to pierce into, displace, and form conduits through their overburden and to flow upward toward the surface of the earth. This behavior involves fluid or fluidlike flow of the intruding material, which by virtue of its low strength and higher pressure is able to displace and deform the stiffer country rock. It is this fluid or fluidlike pressure-driven flow and deformation that unites the diverse phenomena treated in this chapter, ranging from the slow rise of cylindrical intrusions of salt *(salt domes)* through the overlying strata at rates measurable by the timespan of deposition of overlying sedimentary formations, to the high-velocity, even supersonic, rocket-enginelike flow of magma-gas mixtures in volcanic necks and diamond pipes. We begin by considering small shallow-level igneous intrusions such as dikes and sills and generally proceed toward deeper-level and larger intrusions, making a short detour to consider the deformational aspects of extrusive structures and eruptive mechanics. The end of the chapter is devoted to sedimentary intrusive and extrusive structures.

SHEET INTRUSIONS

Many intrusions are tabular or sheetlike (for example, Fig. 7-1); as a group they are called *sheet intrusions,* or *sheets,* which are general terms that subsume dikes, sills, and cone sheets, all discussed in the following sections.

A *dike* (Br. *dyke*) is a discordant sheet intrusion that cuts across bedding or foliation in the country rock at a high angle. The word is also used to mean any

FIGURE 7–1 Resistant dike standing in relief above more erodable shale and sandstone country rock, Ship Rock, New Mexico. This dike is one of a set that radiates from the volcanic neck outcropping as the distant peak (compare Fig. 7-11).

approximately vertical intrusive sheet. The name *dike* is derived from the word for *wall* or *fence* in the north of England and in Scotland because igneous dikes are sometimes seen projecting like a wall above the softer strata on both sides, which have been eroded away (Fig. 7-1). In contrast, a *sill* is a concordant sheet intrusion injected parallel to the layering of the host rock (Fig. 7-2). The word is also used to mean any approximately horizontal intrusive sheet. The term *sill* apparently is derived from the common architectural word for a horizontal member, as in a window sill or the sill (threshold) of a house. Miners in the north of England commonly called horizontal bands of rocks "sills," not distinguishing between sedimentary strata and igneous intrusions or extrusions. Sheet intrusions range in thickness from millimeters to kilometers, with most being in the decimeter to decameter range.

Shapes of Sheet Intrusions

Dikes and sills are not perfectly planar intrusions. Sills commonly intrude close to a single stratigraphic horizon and then crosscut sharply to a higher horizon; for example, the sill in Figure 7-2 abruptly changes stratigraphic horizon. In this example the two segments of the sill have different thicknesses; the difference is taken up in the country rock by a fault, which emanates upward from the top of the crosscutting segment of the intrusion. Faults that are produced through differential intrusion are called *intrusive faults*. Ridge-ridge transform faults are intrusive faults on a grand scale.

Some major sills are observed to be regionally bowl-shaped; as they intrude they propagate outward and upward through the stratigraphic section from their

FIGURE 7–2 Precambrian gabbroic sill stepping abruptly from one horizon to another in Proterozoic strata. Note that the two segments of the sill differ in thickness with the difference taken up along an intrusive fault, which emanates from the top of the crosscutting segment of the intrusion. Banks Island, Northwest Territory, Canada. (Geological Survey of Canada, Ottawa.)

feeder conduit. This common property of major sills is illustrated in Figure 7-3 with a map of intrusions in the Connecticut Valley Mesozoic sedimentary basin. If the map is viewed down structure (Chapter 2), the sills are seen in cross section to step generally upward and outward to the north and south from near the base of the stratigraphic section. Other bowl-shaped sill complexes that have not been tilted may display ring-shaped map patterns.

Individual dikes display a variety of departures from ideally planar morphology that reflect nonplanar aspects of the fractures into which the magma is emplaced. Some fractures may be preexisting, but many dikes apparently lie along hydraulic tensile fractures that were propagated by the magma itself, as discussed in the next section. Therefore, it should come as no surprise that dike morphology has some features in common with joint morphology because many joints also appear to be hydraulic tensile fractures (Chapter 6). Some dikes display lineated surface textures parallel to their direction of propagation similar to plumose structure of joints. Many dikes apparently intruded *en échelon* fractures and may display sharp jogs in their walls (Fig. 7-4) that are analogous to cross fractures connecting *en échelon* fractures in joints (Fig. 6-4). The progressive joining of *en échelon* segments of dikes (Fig. 7-5) leaves a ragged torn fragment of wall rock at the jog (Fig. 7-6). Some dikes and sills near their termination are composed of *en échelon* sheets that do not connect in map or outcrop view (for example, Fig. 7-7); they are apparently the equivalent of *en échelon* fringe joints (compare Fig. 6-4).

It has been commonly assumed that dikes and sills have knife-edged, tapered terminations. Nevertheless, observation shows that blunt, bulbous terminations are apparently the rule. This shape reflects a plasticlike behavior of the country rock at the tip of the propagating sheet intrusion (Pollard, 1973).

Dikes need not be injected vertically from a major conduit, but can be injected obliquely or even horizontally, as shown by plumose surface markings and other evidence. For example, detailed mapping of dikes near Spanish Peaks,

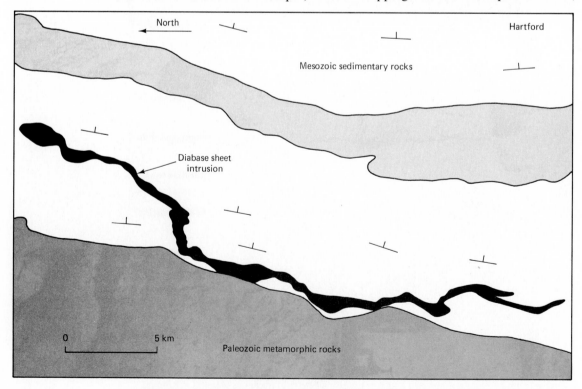

FIGURE 7–3 Map of saucer-shaped sill complex intruding the east-dipping sedimentary rocks of the Connecticut Valley Mesozoic graben. (Simplified from Davis, 1898.)

FIGURE 7–4 Jog in light-colored Miocene dike intruding siltstone, Point Magu, California.

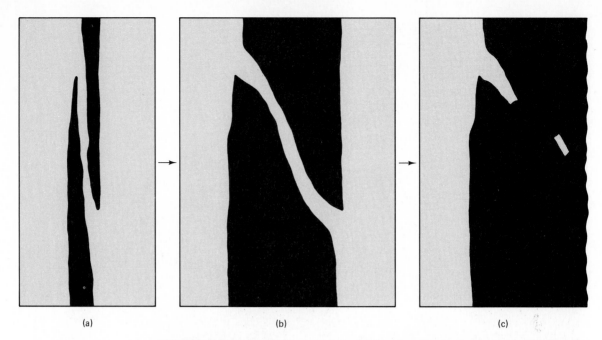

(a)   (b)   (c)

FIGURE 7–5 Diagram showing progressive joining of *en échelon* segments of expanding dike (shown in black). The dike propagates perpendicular to the page.

Colorado, shows that many of the vertical dikes have blunt, rounded terminations with shallow rather than steep plunges and exist as a series of offset *en échelon* segments; therefore, the present ground surface is close to the tops of the dikes. The direction of the flow of the magma in the dike and, to some extent, the direction of dike propagation, may be estimated from elongated and aligned vesicles, which were developed by shear of magma near the walls of the dike during intrusion. These estimated flow directions show that predominately horizontal flow occurred at Spanish Peaks, especially in the longer dikes. Detailed patterns of flow for several dikes suggest that magma was emplaced upward and outward from the central Spanish Peaks intrusion (Fig. 7-12) and that the dike initiation zone was at or above the Paleozoic-Precambrian unconformity, approximately 9 km deep.

Sheet intrusions, both dikes and sills, may exist in great swarms of subparallel sheets (for example, Fig. 7-8). In fact, the sheets may be so numerous that little, if any, intervening country rock is present. The most extreme case is *sheeted-dike complexes,* which are composed of dikes that have intruded the centers of preexisting dikes. Thus a dike will have its own chilled contacts, but will contain within it a chilled dike, which may itself contain other dikes. Sheeted-dike complexes are apparently a major constituent of oceanic lithosphere. Sheeted sill and cone-sheet complexes are also known. The interiors of dikes are a preferred site for new dikes to be injected; the reason for this may be that the interiors of dikes are coarser-grained and therefore have a lower fracture strength than finer-grained rocks of the same material (Chapter 5).

Sheet Intrusions as Fluid-Filled Cracks

Anderson (1942) proposed that sheet intrusions in homogeneous rock are injected most easily along preexisting fractures that are perpendicular to the least principal stress, σ_3, because this orientation requires the least work (force times displacement). If a preexisting fracture is not present, a propagating sheet intrusion can be

FIGURE 7–6 Disrupted wall of black Proterozoic diabasic dike, remaining after the joining of *en échelon* dike segments (compare Fig. 7-5). East coast of Hudson Bay, Canada. (Geological Survey of Canada, Ottawa.)

treated as a very flat fluid-pressurized elliptical hole, that is, a Griffith crack (see Chapter 5). The tensile stress at the tip of a pressurized Griffith crack is at a maximum if the crack is oriented perpendicular to σ_3; furthermore, the dilation of the crack will be accomplished with the lowest magma pressure in this perpendicular orientation. Therefore, we predict that sheet intrusions in homogeneous rock

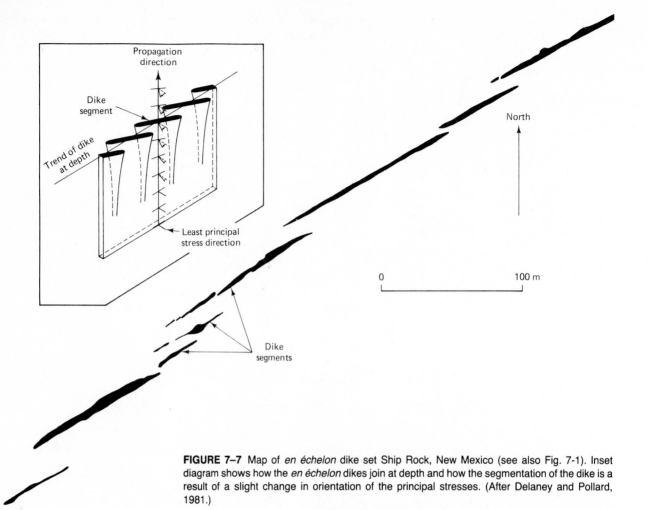

FIGURE 7–7 Map of *en échelon* dike set Ship Rock, New Mexico (see also Fig. 7-1). Inset diagram shows how the *en échelon* dikes join at depth and how the segmentation of the dike is a result of a slight change in orientation of the principal stresses. (After Delaney and Pollard, 1981.)

FIGURE 7–8 Swarm of dark Miocene andesitic dikes, Pt. Magu, California.

should propagate perpendicular to σ_3 in the manner of tensile fractures. This conclusion is supported by the fact that many individual dikes have surface structures similar to joints. Both sheet intrusions and most joints are considered to be natural hydraulic tensile fractures, one involving magma and the other water.

Therefore, we can expect the orientation in which sheets intrude to be controlled solely by the prevailing stress orientation, except for when the intrusion is affected by planar discontinuities such as bedding, faults, and joints. Sheet intrusions, especially dikes unaffected by previous structures, should approximate surfaces perpendicular to σ_3; that is, they should approximate the σ_1-σ_2 plane. For this reason we may use the near-surface orientations of stresses to predict the possible orientations of near surface sheet intrusions, by analogy with Anderson's theory of joint and fault orientation (Chapters 6 and 8). The surface of the earth is a fluid-solid interface, which can support no shear stress; therefore, the principal stresses must be oriented parallel and perpendicular to the earth's surface. Three distinct situations are possible, σ_1 vertical, σ_2 vertical, and σ_3 vertical. In homogeneous rock, vertical dikes are predicted if σ_1 or σ_2 is vertical; sills are predicted if σ_3 is vertical. (Table 7-1).

TABLE 7–1   Anderson's Theory of Dikes and Sills

Orientation of Stress	Orientation of Sheet Intrusions
σ_1 vertical	Vertical dike striking parallel to σ_2
σ_2 vertical	Vertical dike striking parallel to σ_1
σ_3 vertical	Horizontal sills

Relationship Between Dike Swarms and Regional Stress Patterns

If we follow the Anderson theory in considering vertical dikes to be magma-filled hydraulic tensile fractures that are oriented perpendicular to the least compressive stress, σ_3, then we may consider the map patterns of coeval dike swarms to be maps of the stress fields at the time of intrusion. In many cases the orientations of the deduced stress fields are quite homogeneous over great distances, a fact that is in agreement with present-day stress patterns in plate interiors (Fig. 6-13). Swarms of Precambrian diabase dikes are approximately parallel for hundreds of kilometers across substantial parts of the Canadian shield (Fig. 7-9). A Mesozoic diabase dike swarm stretches in a great arc over much of the length of Appalachian mountain belt, from Alabama to New England, part of which is shown in Figure 12-23. A northwest-trending early Tertiary dike swarm covers western Scotland, northwest Ireland and adjacent areas (Fig. 7-10).

In contrast with the simple parallelism of regional dike swarms, we find that dike swarms near central intrusive complexes and volcanic feeder pipes are more complex. This complex pattern of sheet intrusions arises because the pressure of the magma chamber affects the state of stress in the surrounding region. This phenomenon may be understood if we consider the country rock to be an elastic medium and the central intrusion to be a vertical cylindrical hole in the elastic medium. The edge of the hole is a fluid-solid interface, so the principal stresses are oriented radially and tangentially (Fig. 7-11(a)). If the magma in the hole is at a higher pressure than the medium, the radial stress is the maximum horizontal compression, σ_1, as would be expected from the magma pressure. The least

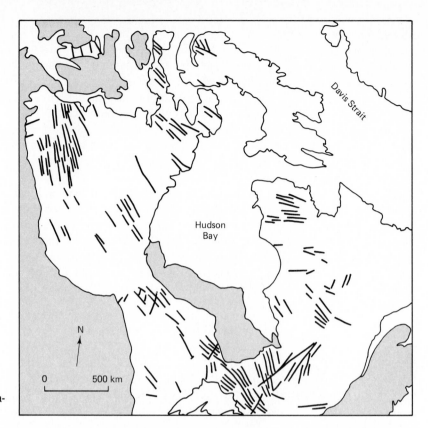

FIGURE 7–9 Map of Proterozoic diabase dike swarms, Canadian shield.

compressive stress, σ_3, is oriented concentrically about the hole. Therefore, if the pressure is high enough for dikes to be emplaced into the country rock, they are predicted to be radial to the central intrusion because the dikes propagate perpendicular to σ_3. Radial dikes are, in fact, commonly observed around central intrusive complexes and volcanic necks; for example, the dike in Figure 7-1 is one of a set of dikes radiating from the Shiprock volcanic neck, which is seen in the background. A radial dike pattern similar to the predicted one is shown in Figure 7-11(b).

The radial dikes do not propagate indefinitely out into the country rock because the stress field caused by the magma pressure decays away from the central intrusion. If the intrusion has radius r_0, then the radial and tangential principal stresses at a distance r from the center are (Figure 7-11(a))

$$\sigma_r = P_m \left(\frac{r_0}{r} \right)^2$$

$$\sigma_\theta = -P_m \left(\frac{r_0}{r} \right)^2$$

$$(7\text{-}1)$$

where P_m is the pressure of the magma in excess of the pressure of the country rock. Therefore, the radial and tangential stresses induced by the magma pressure decay away as the inverse square of the distance from the intrusion.

The map pattern of dikes about an intrusive center is more complicated if a regionally anisotropic stress field of tectonic origin is present in addition to the radial field generated by the magma pressure. The resulting pattern of dikes reflects a summation of the regional and the local magmatic stress fields; the classic example of this phenomenon is the Spanish Peaks igneous center of southeastern Colorado (Odé, 1957; Johnson, 1968; Muller and Pollard, 1977; Smith, 1978).

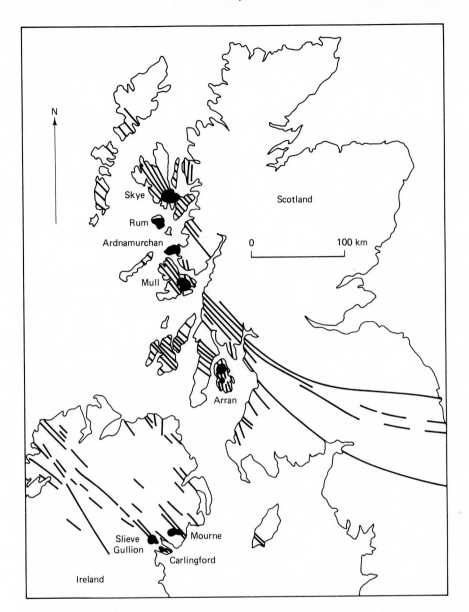

FIGURE 7–10 Distribution of early Tertiary northwest-trending basaltic-dike swarms and central intrusive complexes of the British Isles. (After Richey and Thomas, 1930.)

The Tertiary (22 to 27 m.y.) Spanish Peaks igneous center consists of two major central intrusions, each about 4 km across, and numerous dikes (Fig. 7-12(a)), which may be classified into several sets based on igneous petrology and orientation. The older, east-trending dikes are mostly lamprophyres and basalts, whereas the younger, approximately radial dikes are mostly andesitic porphyries. This correlation of dike pattern with petrology reflects a change in the stress field with time.

The approximately radial pattern of the andesitic-porphyry dikes has a distinct east-west axis of symmetry, which led Odé (1957) to suggest that the pattern reflects the superposition of a radial magmatic stress field on a rectilinear regional stress field. Odé modeled the sedimentary rocks as an elastic plate under a regional rectilinear stress with the intrusive center represented as a vertical pressurized circular hole. He modeled the stiffer Precambrian and Paleozoic rocks of the Sangre de Cristo Mountains to the west as a rigid undeforming boundary to

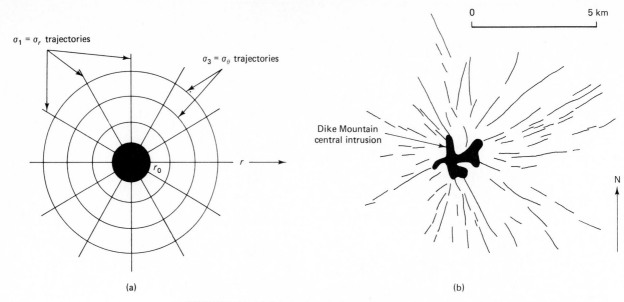

FIGURE 7–11 (a) Orientation of principal stresses around a pressurized hole; σ_r and σ_θ are the radial and tangential principal stresses. (b) Radial dike pattern around Dike Mountain intrusion near La Veta Pass, Colorado. (After Johnson, 1968.)

the elastic plate. Odé and Muller and Pollard (1977) showed that the resulting theoretical stress field (Fig. 7-12(b)) has a pattern quite similar to the dike pattern, assuming that the dikes are perpendicular to the least compressive stress, σ_3. Thus the theory that map patterns of vertical dikes represent $\sigma_1 - \sigma_2$ stress trajectories seems well established.

The dike pattern at Spanish Peaks exhibits a greater abundance of dikes in directions closely parallel to the regional direction of maximum compressive stress, σ_1, which is approximately east-west. Therefore, we might use preferred orientation of radial dikes as a qualitative indicator of the regional orientation of maximum compression, σ_1 (compare Figs. 7-12(a) and 7-12(b)). This idea has been extended to estimate regional stress fields near active volcanic complexes based on the arrangement of parasitic cones and craters around major central volcanoes because the parasitic cones are fed by radial dikes in many cases. Regional stress fields estimated from volcanic asymmetry are generally in good agreement with seismologically determined stress orientations.

SHEET INTRUSIONS OF SUBVOLCANIC INTRUSIVE COMPLEXES

Three principal classes of regular sheet intrusions are known to develop in the vicinity of central, subvolcanic intrusive complexes that are approximately circular in plan (Fig. 7-13). *Radial dikes,* already discussed, are suites of vertical dikes radiating from a common center (for example, Fig. 7-11(b)). *Cone sheets* are relatively thin sheets that occupy a suite of inverted conical fissures with a common vertical axis (Fig. 7-13). Cone sheets normally dip inwards. *Ring dikes* are approximately circular in plan and are vertical or are dipping steeply outward; they commonly consist of a circular segment rather than a complete ring. Ring dikes generally occupy a circular fault along which the central block, which is part of the roof of the central intrusion, has subsided into the magma chamber. If the subsiding block extends to the surface, the effects of the subsidence are pit craters, caldera collapse, and cauldron subsidence, to be discussed later. If the subsiding block does not reach the surface, the complete ring dike forms a bell-

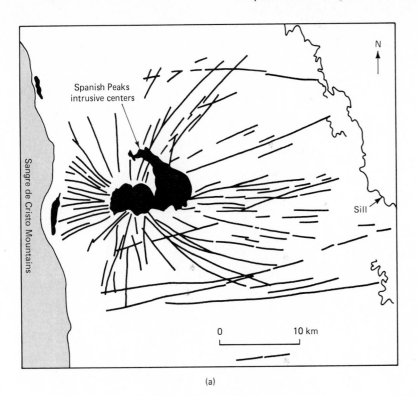

(a)

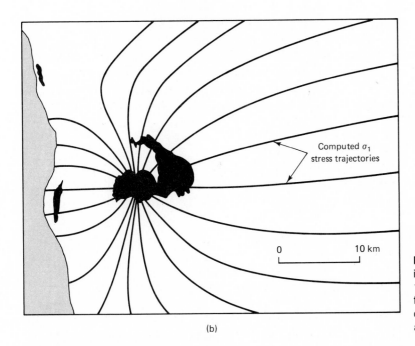

(b)

FIGURE 7–12 (a) Dike patterns at Spanish Peaks, Colorado. (After Johnson, 1968.) (b) Theoretical regional stress field at Spanish Peaks; compare with the observed dike pattern in (a). (After Muller and Pollard, 1977.)

shaped intrusion closed at the top. A principal distinction between cone sheets and ring dikes comes from the displacement of the central block; there is uplift with cone sheets but subsidence with most ring dikes.

British Tertiary Igneous Centers

A classic area of sheet intrusion around central subvolcanic complexes is the Tertiary igneous province of the northern British Isles (Fig. 7-10). The igneous

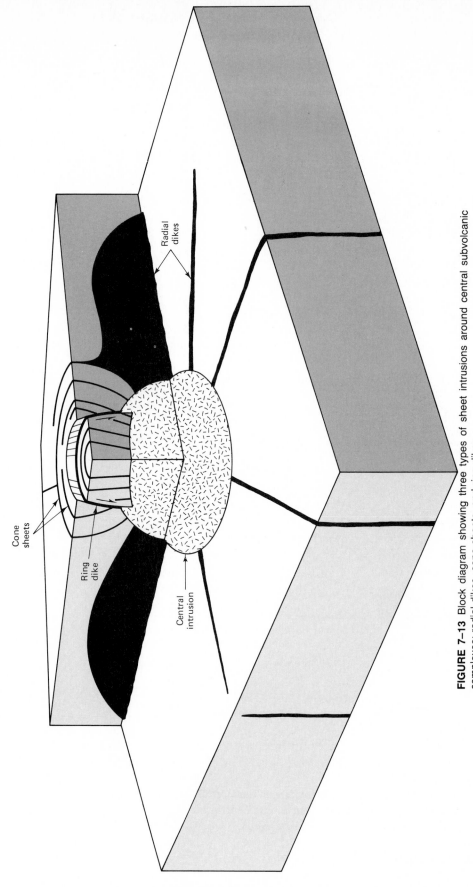

FIGURE 7-13 Block diagram showing three types of sheet intrusions around central subvolcanic complexes: radial dikes, cone sheets, and ring dikes.

Radial dikes

Cone sheets

Ring dike

Central intrusion

activity included the vast outpouring of basaltic lava flows, analogous to the flood basalts of the Deccan Plateau of India and the Columbia River Plateau of the United States (Figs. 13-2, 13-40). Approximately 6000 km^2 of plateau lavas are still preserved on land, with considerably more offshore. The magmatism was possibly associated with an early stage of the Iceland hot spot (Chapter 1). The lava flows are associated with northwest-striking swarms of feeder dikes, which are most concentrated along the west coast of Scotland (Fig. 7-10). The foci of igneous activity are the central volcanic vents and subvolcanic intrusions of Skye, Rum, Ardnamurchan, and Mull. It is there that cone sheets were first discovered and theories of their origin first developed.

Cone sheets were discovered by the petrologist Alfred Harker at Skye, where they have an average inclination of about 35° to 45° toward the intrusive center. Sheets outcropping near the center are more steeply inclined. Individual cone sheets are generally well separated from one another by country rock, but occasionally are so numerous as to be sheet upon sheet without intervening country rock, in the manner of sheeted-dike complexes. Individual sheets at Mull are 5 to 10 m thick.

The large number of cone sheets at some igneous centers account for considerable aggregate uplift of the material inside the cone. The aggregate thickness of cone sheets at Ardnamurchan is about a kilometer, with 300 m of sheets concentrated in an 800-m section. A similar aggregate thickness is present at Mull. The average inclination may be as low as 35°, yielding an estimated uplift of about a kilometer. It is the excess magma pressure that elevates the central block during intrusion.

Ring dikes are also particularly well displayed at Mull, where they group themselves about two centers aligned in a northwest-southeast direction (Fig. 7-14) parallel to the regional dike swarms. The dikes are generally steeply dipping—about 70° to 80° outward for the Loch Ba ring dike. Loch Ba is the most perfect ring dike at Mull with a diameter of 6 to 8 km and a width of about 100 m. The dike has the map pattern of a flat-sided ring, with a symmetry related to the northwest trend of the British Tertiary igneous complex (Fig. 7-10). The ring dike follows a line of faulting and outlines an area of central subsidence.

The age relationships between ring dikes and cone sheets at Mull are complex, but are generally consistent with a gradual northwestward migration of igneous activity, with the ring dikes generally being later than the cone sheets. The Loch Ba ring dike (Fig. 7-14) seems to be cut by no cone sheets at all. The transition in time from cone sheets to ring dikes apparently reflects changes in the state of stress associated with fluctuating magma pressure. Initially, the magma pressure apparently was high, forcing up the roof of the intrusion during cone-sheet emplacement. Later, the magma pressure apparently was lower, with a collapse of the roof of the chamber during ring-dike emplacement.

Origin of Cone Sheets and Ring Dikes

A theory of the formation of cone sheets, ring dikes, and radial dikes was first developed by E. M. Anderson, to explain the structures at Mull and the other Scottish intrusive centers. Similar structures have been found in many parts of the world. Anderson considered the possible patterns of stress orientation that might exist in the vicinity of a near-surface magma chamber lying at a depth roughly equivalent to its width. Once we know the patterns of stress orientation, we also know the patterns of sheet orientation because sheet intrusions lie perpendicular to σ_3, as already discussed.

Three considerations are important to understanding the orientation of sheet intrusions around a near-surface magma chamber: (1) the orientation of stresses at

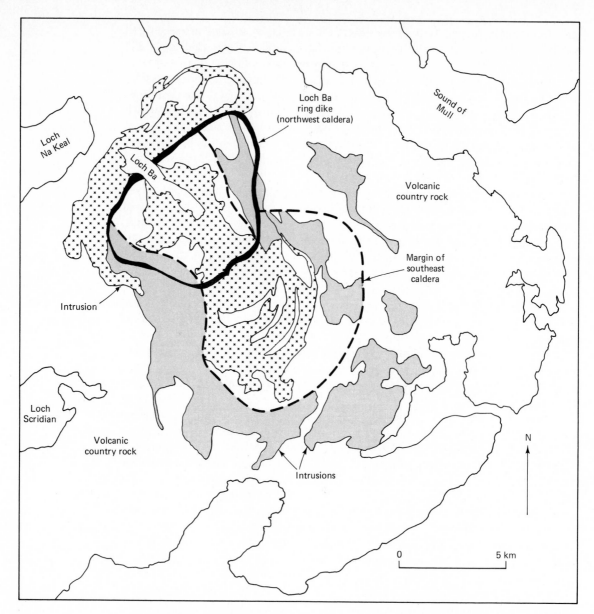

FIGURE 7–14 Map of Loch Ba ring dike and related caldera structure at Mull, Scotland. (After Bailey and others, 1924.)

the earth's surface, (2) the orientation of the regional stress field away from the igneous center, and (3) the orientation of stresses at the surface of the magma chamber. Whatever the orientation of the principal stresses at depth, they must bend to be horizontal and vertical at the surface of the earth because it is a fluid-solid interface that can support no shear stresses (Chapter 6). Furthermore, whatever the local stress orientations near the magma chamber, they must bend to become parallel to the regional stress field away from the magma chamber; the regional near-surface stress field may have σ_1, σ_2, or σ_3 vertical (Table 7-1). Finally, principal stresses must lie perpendicular to the surface of the magma chamber because this surface is also a fluid-solid interface. Based on the above considerations, we can deduce the patterns of sheet intrusions that may exist around subvolcanic igneous centers.

If the pressure in the magma chamber is increased above that of the overburden, then a system of tensile stresses will be imposed on the country rock, with σ_3 parallel to the wall of the magma chamber (Figure 7-11(a)). If the magma pressure is sufficiently high for hydraulic tensile fracturing, then the magma will be injected into the wall rock as a set of sheet intrusions. Two different intrusive situations are possible: (1) If the regional deviatoric stress is small relative to the excess magma pressure P_m (Eq. 7-1) during sheet intrusion, then the orientations of the sheets near the intrusive center are nearly independent of the regional stress. In this case vertical radial dikes are injected at depth and cone sheets are injected into the roof (Figs. 7-13, 7-15). At greater distance from the intrusive center, the sheet intrusions will change orientation to that of the regional stress field. (2) If the driving pressure of the magma during sheet intrusion is small relative to the regional deviatoric stress, the orientations of sheet intrusions will be more strongly affected by the regional stress. For example, if the regional stress is appropriate for sills (σ_3 vertical), cone-sheets will be emplaced above the magma chamber and sills will be emplaced to the sides of the chamber at greater depth.

If the conditions within the magma chamber are reversed and the pressure of the magma falls significantly below that of the surrounding country rock, then the roof of the magma chamber may collapse along outwardly dipping Coulomb shear fractures, along which dikes may intrude as the roof founders (Fig. 7-16). This is the classic explanation for ring dikes.

The pressure in the magma chamber changes greatly during an eruptive cycle. In general, the pressure builds up to higher and higher levels until the roof fails and magma erupts to the surface. Extrusion will lead to a drop in magma pressure. Therefore, we might expect radial dikes, cone sheets, or possibly sills to form first, depending upon the regional stress field. Late in an eruptive cycle we might expect ring dikes to form during collapse of the roof of the magma chamber during major extrusion. These deductions are in agreement with the observation at Mull and elsewhere that cone sheets of any eruptive episode are generally older than the ring dikes.

Volcanic Collapse Structures

Ring dikes were interpreted in the previous section as the result of collapse of the roof of the magma chamber. The primary evidence for this interpretation is

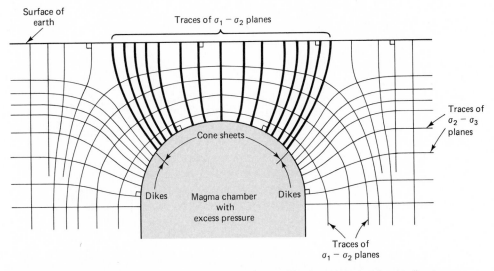

FIGURE 7–15 Theoretical stress trajectories around pressurized magma chamber leading to cone sheet emplacement, based on a regional stress field with σ_3 vertical and magma pressure $P_m > (\sigma_1 - \sigma_3)_{regional}$. (Based on Bailey and others, 1924; Roberts, 1970.)

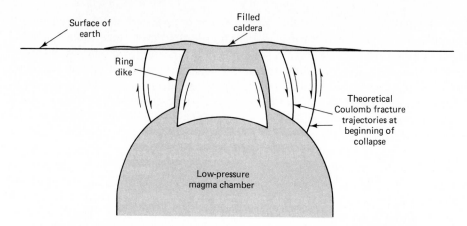

FIGURE 7–16 Schematic shear fracture trajectories around a low-pressure magma chamber leading to collapse of roof of chamber and formation of caldera and ring dike.

stratigraphic evidence that the block inside the ring dike has moved downward relative to the outside, sometimes even dropping volcanic rocks into the magma chamber. Nevertheless, the best evidence for collapse of the roofs of magma chambers is the widespread circular depressions in volcanic terrains. For example, the inner and outer craters of Kilauea Volcano, Hawaii (shown in Fig. 7-17), are both of collapse origin. The inner crater is a historic structure. The larger, outer crater displays numerous fault scarps subparallel to its wall that displace the topographic surface downward toward the center. Some of the eruptions have reached the surface along the wall of the crater; therefore, they are apparently the surface effect of ring-dike formation at depth.

FIGURE 7–17 Kilauea crater, Hawaii. (U. S. Geological Survey.)

Not all volcanic craters have a subsidence origin; some are a result of the extrusive process. Therefore, the term *cauldron* is used as an inclusive term for all volcanic subsidence structures, regardless of their size, shape, depth of erosion, or connection to the surface. The craters at Kilauea are therefore cauldrons. Very large roughly circular or oval volcanic depressions are *caldera,* which can be 10 km or more in diameter and contain enormous thicknesses of volcanic rock, including numerous individual volcanoes. The small Mihara caldera in Japan, 3 to 4 km in diameter, is shown in Figure 7-18. The caldera contains an inner composite cone.

Calderas appear to be surface expressions of shallow level stocks or batholiths. For example, geophysical studies show that the active Yellowstone caldera in the western United States, which is 50 km in diameter, overlies a still-molten batholith. The presence of magma at depth is shown by the lack of transmission of shear waves, as well as by gravity and magnetic data. Batholiths also apparently underlie the many ancient calderas that exist in western New Mexico, based on gravity data.

The subsidence of calderas appears to be directly linked with the tremendous outpourings of magma and occurs during the eruption. For this reason a single eruptive unit may be hundreds of meters thick in the caldera, but no more than tens of meters thick outside, even though the caldera may never have a large topographic expression. The Valles caldera in northern New Mexico formed during a Pleistocene eruption that scattered 200 km^3 of rhyolitic ash over a wide area of the western United States. As a result of this great eruptive volume, the roof of the magma chamber collapsed to form the caldera. The edges of the caldera floor tend to subside more than the center, apparently because of eruption

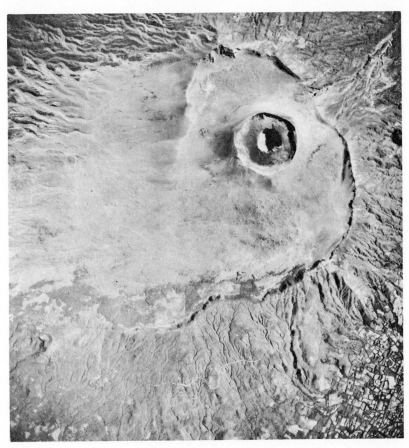

FIGURE 7-18 Mihara caldera, Japan. (U. S. Geological Survey.)

along ring dikes and curvature of Coulomb fracture trajectories (Fig. 7-16). The result is arching of the caldera floor with updoming and the formation of a moat that fills with volcanic materials coming up the ring dikes, including extrusion of viscous rhyolitic domes. The highly faulted arched caldera floor is called a *resurgent dome* (see Fig. 8-21).

Ancient deformed or eroded calderas may be recognized by their ring dikes or highly faulted resurgent domes. In addition, they are commonly recognized by the great thickness of individual ashflows.

LACCOLITHS

The simplest and most regular intrusions are the sheet intrusions, just considered. The next most complex intrusive types are perhaps *laccoliths,* which are sill-like bodies concordant to the stratigraphic layering, but in contrast with sills have bulged upward, folding the overlying strata into a dome. The floors of laccoliths are considered to be generally flat or planar, but the roofs are arched and broadly convex. Cross sections of two laccoliths in Montana are shown in Figure 7-19. They illustrate the general sill-like concordance of laccoliths with the presumed flat floor and the upward bending of the roof. Note, also, that the roof is faulted along the flank of some laccoliths in such a way that the roof is pushed upward as a plug or a flap. This faulting takes place after the bending.

The ideal laccolith is circular and domal in plan, such as the laccoliths of the Bearpaw and Judith Mountains, Montana; however, others are tonguelike projections from a central discordant intrusion, as in the Henry Mountains, Utah. A variety of actual shapes exist, but the essential character of a laccolith is a concordant, sill-like body displaying an upward bending of the overlying strata.

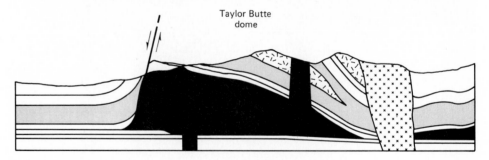

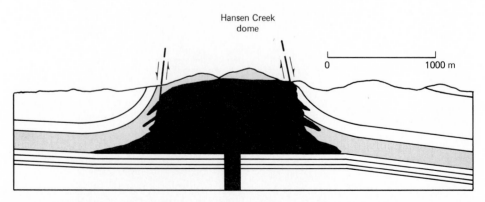

FIGURE 7–19 Cross sections of Taylor Butte and Hansen Creek laccoliths Bearpaw Mountains, Montana. (After Hearn, Pecora, and Swadley, 1963.)

Laccoliths are considered to begin as tabular sills that spread laterally. As they grow in diameter, the magma pressure gains enough leverage to bend the overlying strata to form a laccolith. The mechanics of bending will not be treated until Chapter 9 on folding; here it is sufficient to note that the amount of bending is proportional to the fourth power of the radius of a sill or laccolith in map view. For example, the vertical deflection, w, at the center of a circular intrusion is (Pollard and Johnson, 1973):

$$w = \frac{P_m}{64\,R}\,r^4 \qquad (7\text{-}2)$$

where P_m is the driving pressure of the magma, R is the bending resistance (flexural rigidity) of the overlying strata, and r is the radius of the sill or laccolith. Therefore, we see that a sill can grow in radius until it reaches a critical radius r_c, at which bending of the overlying strata suddenly begins. Once the critical radius is reached, the intrusion grows very little in radius; the influx of magma is accommodated by the vertical deflection. Furthermore, large stresses are generated at the edges of the laccolith in the overlying strata as a result of the bending and the stretching of the strata; these stresses often lead to faulting around the edge of the intrusion as seen in the Taylor Butte and Hansen Creek domes (Fig. 7-19).

The diameters of laccoliths are observed to increase with the thickness of the overlying strata at the time of intrusion. For example, the uppermost laccoliths in the Henry Mountains, Utah, have diameters of about 2 km, whereas those about a kilometer deeper have diameters of about 4 km (Johnson, 1970). Similar relationships are observed elsewhere.

This observed relationship between laccolith diameter and overburden thickness is predicted by a consideration of the bending resistance R of the overlying strata, which entered into Equation 7-2 for the deflection as a function of radius. The resistance to bending of a layer increases with its thickness; for example, it is much easier to bend a sheet of aluminum foil than to bend a centimeter-thick plate of aluminum. The bending resistance, or flexural rigidity, R, of a single layer increases in proportion to the third power of the layer thickness t:

$$R = \frac{Bt^3}{12} \qquad (7\text{-}3)$$

where B is the elastic modulus of the material. If we adopt a certain ratio of vertical deflection, w, to radius, r, as the definition of beginning of laccolith formation, then from Equations 7-2 and 7-3, the critical radius, r_c, is proportional to thickness, t, of the overlying layer

$$r_c \propto t \left(\frac{B}{P_m}\right)^{1/3} \qquad (7\text{-}4)$$

Therefore, laccoliths should increase in diameter with depth.

Actual stratigraphic sections are composed of numerous layers, and a stack of many thin layers has a smaller resistance to bending than a single layer of the same total thickness. For example, a centimeter-thick stack of sheets of aluminum foil has much less bending resistance than a centimeter-thick plate of the same aluminum. The bending resistance of a stack of freely sliding layers is the sum of the resistances of the layers. If there are n layers, the flexural rigidity, R, is

$$R = \sum_{i=1}^{n} \frac{B_i t_i^3}{12} \qquad (7\text{-}5)$$

Therefore, the effective thickness of a stratigraphic section for bending resistance is something less than its total thickness. For example, Pollard and Johnson (1973) estimate the effective thickness of the stratigraphic section to be one-half to one-third the true thickness during laccolith formation at Henry Mountains, Utah.

PLUTONIC INTRUSIONS

We use the word *pluton* as a general term for any large igneous intrusion of granitic texture that is not clearly a sheet intrusion; the name derives from Pluto, the Roman god of the underworld. These large intrusions generally display medium- to coarse-grained granitic textures because they have cooled slowly as a result of their large ratio of volume to surface area or because of crystallization deep in the crust. In contrast, dikes and sills generally display finer-grained textures unless they were emplaced into a high-temperature country rock.

Smaller plutons, less than 100 km^2 in outcrop area, are commonly called *stocks*. Larger plutons, greater than 100 km^2 in outcrop area, are called *batholiths;* the word has Greek roots meaning *deep rock*. The word *batholith* is not restricted to discrete plutons, but is also applied to vast terrains of thousands of square kilometers of granitic rock, composed of dozens of discrete plutons—for example, the Sierra Nevada batholith in California (Figs. 13-12, 13-13). Batholiths composed of multiple plutons are called *composite batholiths*.

Levels of Intrusion in the Crust

Plutons can be classified according to their depths of emplacement and are called epizonal, mesozonal, and catazonal plutons, from shallowest to deepest. The terms generally are used to refer to the depth of emplacement of the level presently exposed as a result of erosion. It is conceivable that a single intrusion might span several levels of the crust in its final solidified form. In any case, as the magma works its way toward the surface, the pluton in turn will be catazonal, mesozonal, and then epizonal; thus the study of the properties of plutons exposed by various depths of erosion helps us envisage the complete shapes of intrusions and the variety of forms they may have taken as they work their way to the surface.

Epizonal plutons are those emplaced into the uppermost levels of the crust, perhaps in the upper 5 to 7 km, with relatively little metamorphic flow in the surrounding country rocks. The deformation of the country rock during epizonal intrusion is dominated by brittle fracture and faulting; any folding or flow of the country rock commonly involves nonmetamorphic deformation mechanisms. Epizonal plutons are generally not strongly foliated.

Many epizonal plutons appear to be the magma chambers of large volcanic complexes because they intrude their volcanic cover. A principal distinction between epizonal subvolcanic plutons and volcanic necks or pipes is the large ratio of volume to surface area for subvolcanic plutons, which accounts for the relatively slow cooling and coarser-grained textures. The country rock of epizonal plutons generally displays only contact metamorphism. If the country rock is relatively porous and permeable, the high thermal gradient will drive extensive convection of the ground water with associated hydrothermal alteration, leaching, and—in some cases—important ore deposition.

Plutons emplaced into intermediate levels of the crust, with regional temperatures of the country rock in the range 300°–600°C, are called *mesozonal plutons*. Deformation of the country rock during intrusion may involve important metamorphic flow and recrystallization; however, brittle fracture and faulting are also important during intrusion. Brittle structures such as dikes, faults, and fractured blocks of country rock that have fallen into the magma all may show signs of subsequent plastic deformation.

Catazonal plutons are intrusions emplaced into country rock that was at or above the temperature of initiation of melting. There may be little distinction in mechanical properties between the partially melted country rock and the partially solidified intrusion. In many cases there are imperceptible gradations between plutonic bodies and their partially melted country rock; the boundaries between the two are arbitrarily defined in many practical cases of regional geologic mapping. In contrast, mesozonal plutons generally have clearly defined contacts with the country rock.

Stocks and Small Batholiths

The modes of emplacement of isolated stocks and small batholiths are commonly much better known than those of large batholiths because the deformation of the surrounding country rock is generally more easily observed, especially in mountainous areas of high relief. Plutons range from foliated *concordant* intrusions that have contacts roughly parallel to the deformed country rock to *discordant* intrusions that are commonly little foliated and sharply crosscut previous structure. We illustrate this variety of intrusive types by describing several well-studied small plutons.

Concordant Plutons.   The 10-km² Birch Creek pluton of late Cretaceous age in the White Mountains of eastern California (Nelson and Sylvester, 1971) is located within the nearly vertical east limb of a major south-plunging anticline involving Upper Precambrian and Lower Cambrian sedimentary rocks (Fig. 7-20). It should be noted that a few kilometers to the east is the edge of the very large

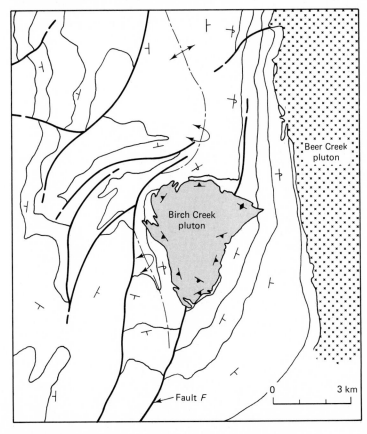

FIGURE 7–20 Map of Birch Creek pluton, Inyo Mountains, California (After Nelson and Sylvester, Geol. Soc. Amer. Bull. 1971, v. 82, p. 2891–2904.)

Jurassic Beer Creek pluton, which—together with its contact metamorphosed wall rock—apparently acted as a nearly rigid buttress during the intrusion of the Birch Creek pluton in the late Cretaceous. The strata on the east side of the Birch Creek pluton do not appear to have been deflected from their preplutonic orientations. The pluton has expanded westward into the core of the anticline as a great tumor, overturning the east flank of the anticline and displacing it as much as 3 km to the west. The pluton is relatively concordant and strongly foliated along its margins.

The history of intrusion of the Birch Creek pluton is shown schematically in map view and cross section in Figure 7-21(a) through (d). Prior to intrusion, fault *F* may have already existed on the east flank of the anticline (Fig. 7-21(a)). The pluton may have initially injected as a sheet intrusion along the fault (Fig. 7-21(b)) and then, as it gained leverage, expanded greatly toward the west but little toward the east (Figs. 7-21(c) and (d)). It appears that the Birch Creek pluton is somewhat like a laccolith in that the magma pressure caused the wall rock to bend outward, producing a bulbous, relatively concordant intrusion.

Concordant plutons that are strongly foliated near their contacts and bend the country rock outward are widespread. Some are somewhat laccolith-like, analogous to the Birch Creek pluton. Others have a shape and wall rock deformation similar to the salt domes discussed later in this chapter. An example is the Bidwell Bar batholith in the northern Sierra Nevada of California (Compton, 1955), which has shouldered aside its metamorphic country rock, dragging it upward and producing strongly foliated and vertically lineated gneissose walls.

Discordant Plutons.   Other stocks and small batholiths are dominantly crosscutting, displaying little folding or flow of the country rock in response to the magma pressure. Crosscutting plutons have deformed their country rock by fracture and faulting. The country rock can be displaced upward in a way somewhat analogous to the upward diplacement of cone sheets. Alternatively, the roof can be displaced downward as in caldera collapse and ring dike formation.

Another process by which magma moves up through the higher, more-brittle levels of the crust is *magmatic stoping*. Given a large magma chamber, above which roof and wall rocks are dislodged from time to time by some combination of thermal stresses, hydraulic fracture, and gravity, the magma body will gradually move upward by stoping if the blocks are able to be dislodged and sink at a rate

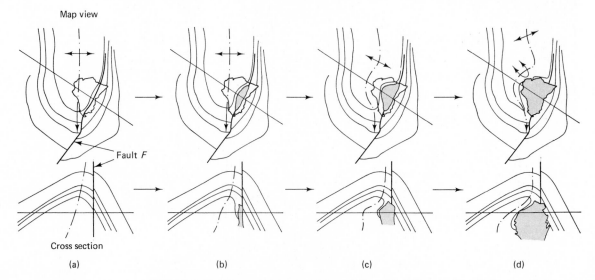

FIGURE 7–21 Schematic intrusive history of Birch Creek Pluton. (After Nelson and Sylvester, Geol. Soc. Amer. Bull. 1971, v. 82, p. 2891–2904.)

that is rapid relative to the rate at which the magma cools and crystallizes. The process is documented directly by the presence of fragments of the country rock, called *xenoliths* (foreign rocks), within many plutonic bodies (Fig. 7-22) and indirectly by irregular discordant contacts of some plutons. In some cases the fragments appear to be actively dislodged through the injection of ramifying systems of dikes and sills as a result of high magma pressure and roof collapse. Stoping becomes progressively more important higher in the crust for three reasons: (1) the increased temperature contrast, ΔT, between the magma and the wall rock increases the thermal stress (Eq. 6-7), (2) rocks become more brittle with decreasing temperature, and (3) fracture strengths decrease with decreasing pressure (Figs. 6-12 and 6-17).

An example of a discordant pluton is the Miocene Little Chief stock in the Panamint Range on the west side of Death Valley, California, which has an outcrop area of about 30 km^2. The intrusion is well-exposed with a relief of about 2 km and was emplaced into a stratigraphically well defined Upper Precambrian and Cambrian sedimentary sequence, which enables the deformation of the country rock to be studied with some precision. This is essentially the same sequence as the wall rock of the nearby concordant Cretaceous Birch Creek, discussed earlier (Fig. 7-20).

The Little Chief stock lies within a simple east-dipping homoclinal structure, as shown by a structure contour map giving the elevation of a well-defined stratigraphic horizon (Fig. 7-23). Mapping of the contacts of the stock shows that it has the form of a steep-sided dome, with the contacts dipping from vertical to 60° outward, although in a few areas the contacts dip inward as gently as 35 to 40°. The roof of the intrusion is partly exposed on the east side of the intrusion and is a nearly rectangular trapdoor of sedimentary rock bounded by vertical faults. A trapdoor has opened up to the west with vertical slip of almost 1 km on the western side and very minor slip on the east side, as shown by the structural contour map. A comparison of the volume created by the deformation of sedimentary rocks with the volume occupied by the stock indicates that the near-surface volume of the stock was accommodated by doming, flowage of the carbonate-rich wall rocks, and uplift of the trapdoor roof. Apparently little of the

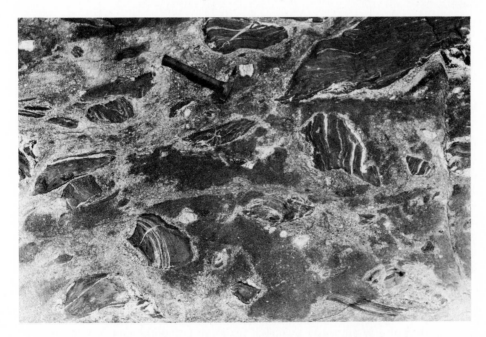

FIGURE 7-22 Xenoliths in granitic rock, northern Taiwan.

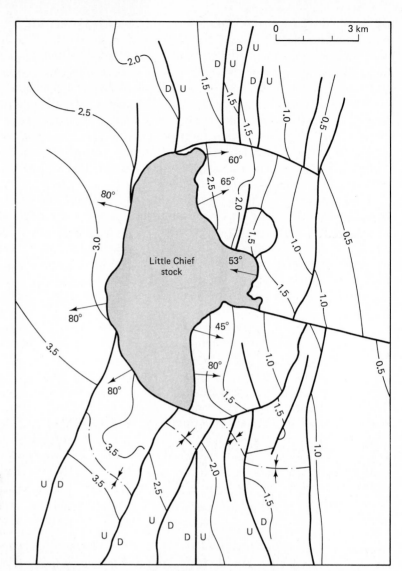

FIGURE 7–23 Structure contour map of stratigraphic horizon in the Eocambrian Johnnie Formation surrounding the Little Chief stock, Panamint Range, California. Elevations in kilometers. (After McDowell, Geol. Soc. Amer. Bull., 1974, v. 85, pp. 1535–1546.)

magma was emplaced by assimilation of country rock or by settling of blocks of country rock through the magma, that is, by stoping.

Large Batholiths and Deeper Igneous Bodies

As we turn to larger plutonic bodies, generally the great composite batholiths, or to plutonic bodies of any size in deep-seated metamorphic terrains, our specific understanding of the primary structural form and physical mechanisms of emplacement becomes rather meager in comparison with our understanding of small, shallow-level intrusive bodies. This fact is largely a result of our common inability in these terrains to specify the deformation of the country rock that took place during the emplacement of the igneous body. In the case of large composite batholiths, the country rock may simply be other plutonic rocks that contain insufficient primary structure to record the deformation. In the case of deep-seated metamorphic terrains, both the igneous and country rocks in many cases have undergone substantial plastic deformation after the formation of the igneous body. For these reasons only a somewhat tentative general understanding of igneous structure is presently available for large batholiths and deeper igneous bodies. Nevertheless, a substantial body of field experience exists around the world that leads to certain generalizations (for example, Pitcher, 1979; Hutchinson, 1970).

In those areas where large composite batholiths or deep-seated plutonic metamorphic terrains have been mapped in detail, the shapes of some igneous bodies are found to be similar to shallow-level and smaller intrusions, including large dike or sill-like sheet intrusions (Fig. 7-24), cone sheets, ring dikes, and laccolith, or balloonlike, plutons (for example, Larsen, 1948; Gastil and others, 1975; Hutchinson, 1970). Many other bodies simply have irregular shapes, whose origins are not presently understood. Nevertheless, we can conclude that many of the mechanisms of formation of igneous structures that are common for small, shallow-level bodies are probably also important at deeper levels.

Nevertheless, deeper igneous bodies also exhibit some important differences in appearance from shallow intrusions; these are the differences that were already emphasized in the depth classification of plutons as epizonal, mesozonal, and catazonal. Two important changes in mechanical behavior appear with increasing ambient temperature. The first change is the brittle-plastic transition representing the change from epizonal to mesozonal plutons. At mesozonal levels both the country rock and the crystallized magma are generally plastic. Before the magma has crystallized, however, brittle hydraulic fracture behavior can be a widespread response to rapid fluctuations in magma pressure, forming dikes and sills. These sheet intrusions commonly undergo later plastic deformation, including boudinage and folding (for example, Fig. 7-25).

The second important change in mechanical behavior with increasing ambient temperature comes with the initiation of melting of the country rock. This

FIGURE 7–24 Edge of complex sheet intrusion laterally injected into folded low-grade metamorphic lower Eleonore Bay Group (Proterozoic) north side of Nathorst Glacier, east Greenland. (Photograph courtesy of John Haller.)

FIGURE 7–25 Plastically deformed granitic sheet intrusions in 400-m vertical rock exposure, upper Kildedalen, Ostenfeld Land, east Greenland. (Photograph courtesy of John Haller.)

is represented by the change from mesozonal to catazonal plutons. Hydraulic fracture is still possible under these conditions (Shaw, 1980) and is represented by widespread dikes and sills, often plastically deformed. Nevertheless, catazonal plutons and their country rock commonly display important regions of *migmatite,* that is, "mixed rock" representing a mixture of magma and unmelted fragments or residue. These plutons display a broad irregular gradation between heterogeneous nebulitic or gneissic plutonic rock and surrounding migmatitic country rock that makes mapping of the interrelationships between plutons and country rock cartographically difficult. The structure of migmatitic zones is commonly concordant with metamorphic structure.

MAGMA PRESSURE AND THE ERUPTIVE MECHANISM

Our discussion so far has presumed that a body of fluid magma is maintained at some pressure, to which the brittle-elastic country rock responds. For example, if the magma pressure is relatively high, it may hydraulically fracture the country rock, producing dikes, sills, and cone sheets; or it may bend the country rock to produce laccoliths. If the magma pressure is sufficiently low, the roof of a magma chamber may collapse, producing ring dikes and cauldrons. It is the excess pressure that drives magma toward the surface to produce volcanic eruptions. Even the relatively passive Hawaiian eruptions display awesome fountainlike behavior. In this section we consider the origin of this magma pressure and its role in the eruptive mechanism.

Origin of Magma Pressure

Several mechanisms apparently can play significant roles in the development of the elevated magma pressures that dominate the mechanics of intrusion and eruption. The principal mechanisms are (1) volume increase on melting, (2) density difference between magma and overlying and surrounding rock (buoyancy effect), and (3) boiling of magma during crystallization and eruption.

Volume Increase on Melting. Rocks undergo a significant increase in volume upon melting, with the anomalous exception of ice. For example, many important rock-forming minerals undergo a 6 to 18 percent increase in volume upon melting at 1 atm (Table 8-2). If the rock that is melting is not completely free to expand, there will be an increase in pressure. This increased pressure may play an important role in the separation of the melt from the partially melted rock, which is an initial step in eruption (Shaw, 1980).

TABLE 7–2   Volume Change on Melting at 1 atm*

Mineral	$\rho_{solid}(kg/m^3)$	$\rho_{glass}(kg/m^3)$	Volume Change (%)
Forsterite	3223	3035	6.2
Fayalite	4068	3764	8.1
Clinoenstatite	3183	2735	16.4
Diopside	3275	2846	15.1
Pyrope	3582	3031	18.2
Anorthite	2765	2700	2.7
Albite	2605	2382	9.4
Sanidine	2597	2400	8.2

*After Yoder (1976)

Density Difference Between Magma and Rock.   Excess magma pressure can be produced by the density difference between magma and its rock. This mechanism can be illustrated by a thought experiment. Suppose an isolated sill of magma exists at a depth h (Fig. 7-26); then the pressure on the magma is the same as the solid overburden, which is $\bar{\rho}_r gh$, where $\bar{\rho}_r$ is the mean density of the overlying rock. Now suppose we place a pipe from the surface of the earth to the envelope of magma at depth h. How high could the magma rise in the pipe? The straightforward answer, of course, is that it will rise to a height, h_m, such that the pressure of the column of magma equals the pressure of the solid overburden:

$$\rho_m gh_m = \bar{\rho}_r gh \tag{7-6}$$

where ρ_m is the density of the magma. The elevation, Δh, of the top of the column of magma relative to the surface of the earth is then

$$\Delta h = h_m - h = \left(\frac{\bar{\rho}_r}{\rho_m} - 1\right)h \tag{7-7}$$

which is normally positive because the density of magma is less than the density of the same solid rock. Therefore, a static column of magma extending from the surface of the earth to a depth h below would have an excess pressure at the surface of

$$P_{\text{excess}} = \rho_m g\Delta h = gh(\bar{\rho}_r - \rho_m) \tag{7-8}$$

This excess magma pressure would be available to drive eruption.

Our thought experiment is actually somewhat artificial because columns of magma of great vertical extent are unstable because the country rock is not rigid. The base of the column would be pinched off and the magma squirted to the surface (Pollard, 1976; Pollard and Holzhausen, 1979).

Boiling of Magma.   The third important mechanism by which excess magma pressure is produced is the expansion during boiling of the magma as a result of crystallization or drop in solid-overburden pressure. Comprehension of this mechanism requires some understanding of the physical chemical behavior of magmas undergoing crystallization. A complete discussion is beyond our scope; however, it is important to mention a few salient properties of igneous crystallization because of their importance to the development of magmatic structures. The

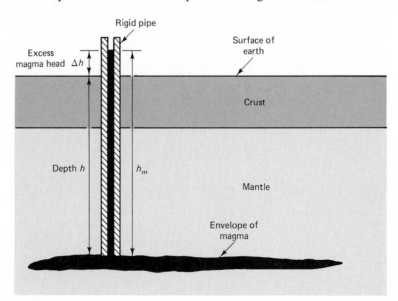

FIGURE 7–26 Origin of excess magma pressure by difference in density between magma and country rock. See text for discussion.

crystallization of magma depends on the pressure, temperature, and amount of dissolved water and other fluids, such as carbon dioxide.

If the melted rock—for example, basalt or granite—contains all the water, it can possibly dissolve; then its chemical behavior contains three fields of stability that can be shown on a temperature-pressure diagram (Fig. 7-27). At the highest temperatures and pressures, the material will be composed of two phases, a silicate-rich melt and a water-rich fluid. At lower temperature and pressure, there is a field in which the magma is partially crystallized; the material is composed of melt, water-rich fluid, and crystals. This field of partially crystallized magma is very broad in the case of water-saturated basalt, whereas it is quite narrow for a water-saturated granite (Fig. 7-27). Finally, at the lowest temperatures only crystals plus water-rich fluid are stable; the magma has completely crystallized. The boundary between crystals and melt plus crystals is called the *solidus*, shown as a line in Figure 7-27. The boundary between all melt and melt plus crystals is called the *liquidus*.

If the total amount of water present is less than the maximum amount that can be dissolved in the melt, then the liquidus line is displaced to higher temperatures. For example, at a pressure of 1000 MPa the liquidus for granite is

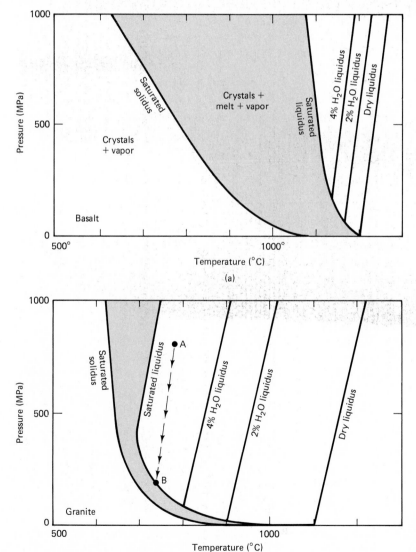

FIGURE 7–27 Solidus and liquidus curves for basalt and granite. See text for discussion of role of crystallization in generation of excess magma pressure.

displaced about 250°C to about 1010°C if two percent water by weight is present. Thus we see that undersaturated melts begin to crystallize at higher temperatures than saturated melts. Let us now consider the effects of these phase relations on the mechanical behavior of the magma and, in particular, its pressure.

As an example, suppose a granite magma that is at a pressure of 800 MPa (depth of 25 km) and a temperature of 775°C (*A* in Fig. 7-27) is brought to higher levels of the crust. In the process it will lose some heat to the wall rock by conduction. Suppose the rising magma encounters the liquidus at a pressure of 200 MPa (depth of 6 km) and starts to crystallize. The rate at which it crystallizes depends upon the rate at which the magma can lose heat to the wall rock because of the heat of crystallization. As the magma crystallizes, little of the water goes into crystallizing minerals, such as hornblende, biotite, or muscovite; therefore, the proportion of water to melt increases. The melt is water-saturated, so water in excess of the saturated amount is released on crystallization to form bubbles in the magma. Both the bubbles and the crystals greatly affect the mechanical properties of the magma. Viscosity is increased by the crystals and may be either increased or decreased by the water-vapor bubbles, depending on their concentrations. The bubbles may rise to the roof of the intrusion and attempt to expand, creating very high pressures in the intrusion. This increase in pressure may serve to inhibit crystallization, but as sufficient heat is released to the wall rock, the magma will continue to crystallize. Thus if the magma chamber has some strength, the pressure will gradually increase as the magma crystallizes and releases its dissolved gases. When the pressure exceeds the strength of the magma chamber, it will fail, causing injection of magma to higher levels of the crust. If the magma breaks through to the surface, it will erupt catastrophically because the boiling magma is able to expand without constraint, suddenly releasing its dissolved gas. The most violent volcanic eruptions are a result of release of pressure on vapor-saturated magmas.

Eruptive Mechanism

The three important pressure-producing mechanisms just discussed and the various mechanisms by which the wall rock deforms in response to the excess magma pressure combine to produce the specific eruptive mechanisms. As one example we outline a model of a kimberlite eruption developed by McGetchin and Ullrich (1973).

Kimberlites are volatile-rich ultramafic magmas derived from deep in the upper mantle. They commonly exist in dikes or pipes no more than a few hundred meters across, which are commonly filled with abundant fragments of crustal and upper mantle rocks. These fragments are the source of much of what we know about the upper mantle, especially under the continents. Diamonds are found in kimberlites. The dikes or pipes are often called *diatremes,* which is a word for volcanic conduits filled with fragmental materials and thought to be produced by gas-rich volcanic activity. A map and cross sections of the breccia-filled Moses Rock kimberlitic dike in southeastern Utah are shown in Figure 7-28. Most of the fragments are from nearby Upper Paleozoic sedimentary formations. Only a small percentage of the breccia is composed of crystalline fragments carried up in the dike from below the sedimentary cover, and of these, less than 1 percent are peridotites, eclogites, and pyroxenites of upper mantle origin. The fact that the mantle xenoliths are substantially more dense than the magma suggests that the eruption was high velocity from the upper mantle to the surface.

The beginning of a kimberlite eruption is apparently the propagation of a fracture upward from a pressurized magma chamber in the upper mantle. The origin of the magma pressure is the increase in volume on melting and the fluid

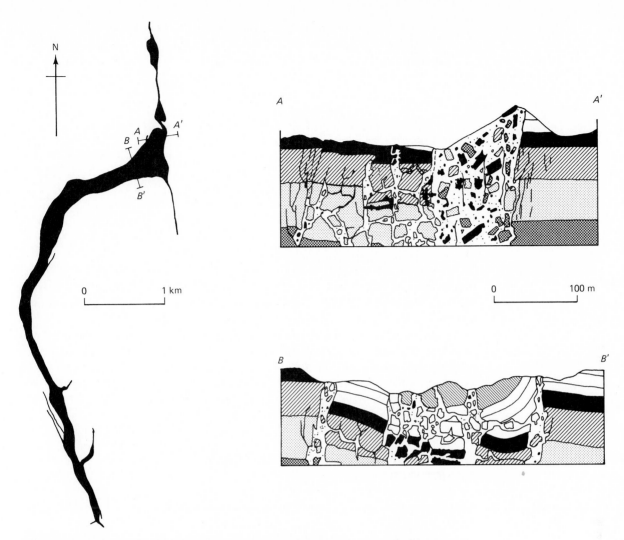

FIGURE 7–28 Map and cross sections of Moses Rock dike, Utah. Different layers are members of the Permian Cutler Formation, blocks of which are displaced both upward and downward. (From McGetchin and Ullrich, J. Geophys. Research, v. 78, p. 1833–1853, 1973, copyrighted by the American Geophysical Union.)

head resulting from the height of the magma chamber. Magma will flow into the fissure driven by the fluid head, which is proportional to the height above the reservoir (Eq. 7-8). Once the crack begins its upward propagation, it will reach the surface unless the magma supply in the reservoir is expended or unless the stress state in the crust dictates that it propagate as a sill. When the surface is breached, a decrease in pressure will result, which propagates downward as an expansion wave. As the walls erode the dike or pipe enlarges and the flow velocities increase because of decreased drag. Eruption continues until the reservoir is expended.

A numerical model of the velocity, density, and temperature of a water- or carbon dioxide-rich kimberlitic eruption originating at a depth of 100 km is shown in Table 7-3. The velocity is of the order of 25 to 50 m/s through most of the crust; however, within a few hundred meters of the surface it reaches supersonic velocities. Note that as velocity increases, the gas temperature and density greatly decrease. Kimberlites are noted to be cold eruptions, possibly explaining the survival of diamonds in these bodies.

High flow velocities are not limited to kimberlites, although kimberlite velocities are extreme. The 1968 eruption at Arenel, Costa Rica, ejected sizable

TABLE 7–3   Numerical Hydrodynamic Calculations for Erupting Kimberlite*

Depth	Gas Temperature (°C)	Density, ρ (kg/m^3)	Velocity, v (m/sec)	Mach Number	Dynamic Pressure, $\rho v^2/2$ (MPa)
Surface	−19	17	334	3.8	0.9
10 m	79	52	304	2.9	2.4
50 m	209	150	261	2.1	5.1
100 m	287	240	235	1.7	6.7
300 m	440	510	186	1.09	8.8
1 km	661	1050	116	0.49	7.1
3 km	871	1790	58	0.15	3.1
10 km	954	2660	33		1.5
50 km	1000	2960	25		0.9
90 km	1000	2960	25		0.9

*After McGetchin and Ullrich (1973)

blocks, producing impact craters to ranges of 5.5 km. Exit velocities were calculated at about 220 m/s. These very high velocities are to be expected only near the surface, where large gas expansions are possible.

Minimum rates of ascent of magma through the crust and upper mantle may be inferred from the settling rates of xenoliths. For example, Spera (1980) computes a settling rate of 5.1 m/s for a 30-cm ultramafic xenolith (ρ = 3450 kg/m^3) from the 1801 eruption of Hualalai Volcano, Hawaii (magma density 2800 kg/m^3). The actual magma ascent rate must have been somewhat greater than the settling rate because the xenolith obviously reached the surface.

SEDIMENTARY INTRUSIVE AND EXTRUSIVE STRUCTURES

Diapirism

A diapir is a body of rock that has moved upward by piercing and displacing the overlying strata; the word is derived from the Greek verb *diaperein,* meaning *to pierce.* The word *diapir* is usually applied to bodies emplaced largely as solids, such as salt and mud intrusions, and is somewhat less commonly applied to magma. Diapiric structures are driven upward by buoyant forces due to the contrast in density between the lighter rock or magma of the intruding diapir and the more-dense overlying and surrounding strata (Eq. 7-6). Diapirs are well known to form in evaporites (especially salt), mud, and shale. In deeper levels of the crust, diapirism may be the mode of formation of some gneiss domes. Igneous diapirism is thought to occur in the mantle and lower crust and is probably an important process in the vertical motion of magma through the aesthenosphere and lithosphere (Marsh, 1982). Some plutons are similar to salt domes in shape and wall-rock deformation. We emphasize salt domes here because they are by far the best-known diapiric structures; this knowledge is largely a result of their economic importance for petroleum, salt, and sulfur.

Hundreds of salt domes are known in the Gulf Coast of Louisiana, Texas, and Mexico; for example, Figure 7-29 shows the salt domes in part of offshore Louisiana. Salt domes are also well known in northwestern Germany (Fig. 7-30), the Mediterranean, Romania, the Ukraine of the Soviet Union, Iran (Fig. 1-31), Gabon, Senegal, and the Canadian Arctic. In addition to these areas of diapirism, there are many regions in which salt plays an important mechanical role in folding and faulting because of its weakness and buoyancy. For example, salt serves to

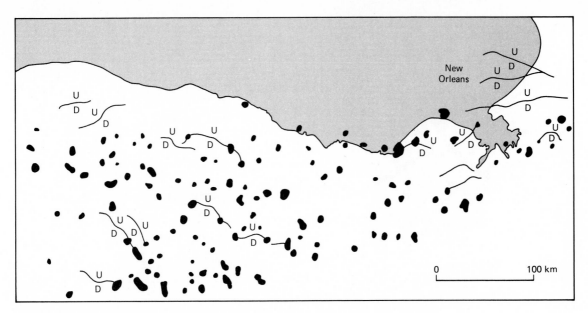

FIGURE 7–29 Map of distribution of salt domes and normal growth faults of offshore Louisiana. Black dots show the areas of the salt domes at 3-km depth. (Simplified after Lafayette Geological Society, 1973.)

detach the folded layers of the southeastern Zagros Mountains of Iran from their underlying basement; this is also a region of important diapirism in which many of the salt domes have reached the surface, as can be seen in the satellite photograph in Figure 1-31.

At their lowest amplitudes salt domes are gentle undulations of the upper surface of the salt, either as linear ridges and troughs or as gentle basins and domes (called *salt pillows*). At this stage the salt structures are not true diapirs because they have not pierced the overburden. As the domes or ridges grow, the overlying strata begin to updome and stretch by slip on sets of normal faults. If the regional stress field is nearly isotropic, the normal faults develop in outwardly radiating patterns; for example, see Figure 7-31(a), which is a structure contour map of the Clay Creek dome in Texas, drawn at a stratigraphic horizon that has not yet been pierced by the fault. If the regional stress field is more strongly anisotropic, the fault pattern may have a more pronounced preferred orientation, such as shown in the Hawkins dome (Figure 8-20). With growth of a dome, the

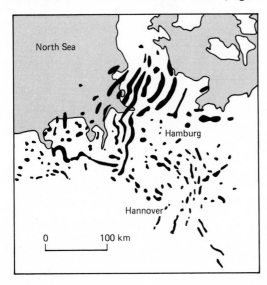

FIGURE 7–30 Map of distribution of salt structures in northwest Germany. (After Trusheim, 1960.)

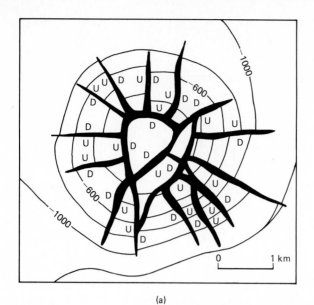

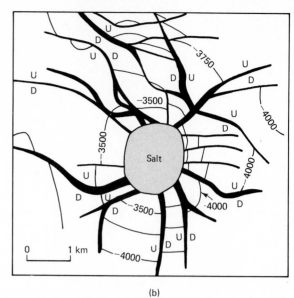

(a)   (b)

FIGURE 7–31 Structure contour maps of sediments above and adjacent to salt domes. Contours in meters. (a) Clay Creek dome, Texas, drawn at the top of the Wilcox Formation. (Modified after Parker and McDowell, 1951.) (b) Belle Isle dome, Louisiana. (Modified after O'Neill, 1973, Gulf Coast Association Geological Society Transactions, v. 23, p. 115–135.)

strata are progressively pierced, as is shown in Figure 7-31(b) for the Belle Isle dome, Louisiana. Thus we see that the overlying strata are first pushed upward, stretched, faulted, and finally pierced by the salt.

The deformation of the overlying and surrounding strata is rather gentle compared with the deformation of the salt. For example, Figure 7-32 gives structure contours on the outer surface of the salt, as well as a cross section, for the Cote Blanche Island dome. The dome is a nearly vertical walled cylinder with an overhang on the north flank, whereas the sediments that have been pierced are only gently upturned. The internal structure of the salt in well-developed diapirs consists of isoclinal, attenuated, vertically plunging, refolded, and faulted folds,

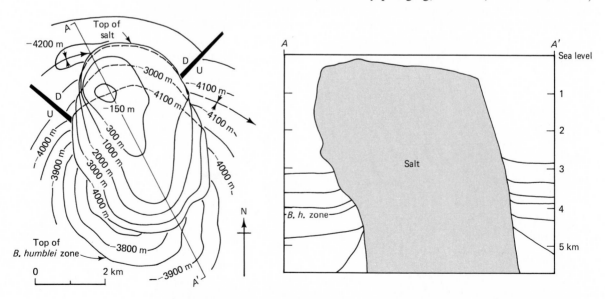

FIGURE 7–32 Cross section and structure contour map of Cote Blanche Island dome, Louisiana. The central contours are on the top and side of the salt whereas the outer contours are on the top of the biostratigraphic zone *B. humblei*. (After Atwater and Forman, 1959.)

which have been described as resembling those in a handkerchief drawn or pushed vertically from the center through a small ring. Thus folded salt layers are commonly observed on the ceilings of salt mines, which are cross sections normal to the fold axes, whereas the folds are not seen on the vertical walls except very obliquely. The deformation mechanisms that accompany flow of the salt are very similar to those of high-grade metamorphic rocks and involve plastic deformation, hot working, and annealing recrystallization (Fig. 4-29).

At these more-advanced stages of growth, salt diapirs are most commonly vertical cylinders having the shape of an upward-pointing thumb (Figs. 7-32 and 7-33). Alternatively, they form vertical sheets, which are well-developed in the deeper parts of the northwest German salt basin (Fig. 7-30). As the supply of salt becomes exhausted, domes may detach from their underlying source, first forming mushroom shapes and then inverted teardrops (Fig. 7-34), both of which are common in the German basin.

Several styles of deformation exist in the strata directly overlying and surrounding a salt dome. With a thicker cover, such as the Heidelberg dome shown in Figure 7-35, grabens develop directly over the dome because of stretching, and the formations thicken over the structure because of growth faulting. In other structures growing with a fairly thin sedimentary cover, such as the White Castle dome in Figure 7-33, a plug of sediment is perched passively on the crest of the dome and the formations thin over the crest of the structure. Finally, when the salt plug essentially reaches to the surface, as in the case of the Cote Blanche Island dome (Fig. 7-32), continued growth is accompanied by deposition of shallow-water sediments directly adjacent to the salt, or the sediments were pierced shortly after deposition. In this last case growth of the dome keeps pace with basin subsidence and the sediments move downward around the salt plug with little deformation (Fig. 7-32).

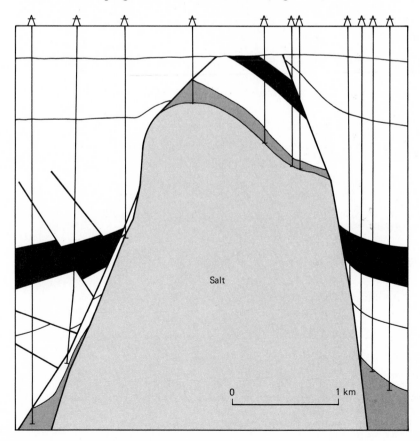

Salt

0   1 km

FIGURE 7-33 Cross section of the White Castle dome, Louisiana. Note the thinning of the stratigraphy over the dome. (Simplified after Smith and Reeve, 1970.)

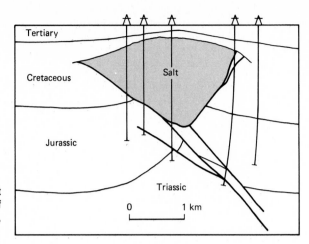

FIGURE 7–34 Cross section of Eilte salt stock, Germany. The stratigraphic age of the salt is Triassic. (After Trusheim, 1960.)

Mechanics of Salt Diapirism

Salt diapirism occurs because of several important physical differences between rock salt and the enclosing sediments. Evaporitic sediments have little porosity soon after burial and thus have a density very close to the constituent minerals, 2160 kg/m^3 in the case of halite (NaCl). In contrast, the mineral constituents of other sediments have a density greater than about 2600 kg/m^3; however, the sediments are generally less dense than rock salt when first deposited because of their high porosity. Clastic sediments compact slowly with burial because of the relatively greater strengths of silicate mineral grains, their low solubilities, and—in some cases—the low permeability of the rock. Thus rock salt is initially more dense than clastic sediments; however, the salt keeps approximately the same density with greater burial, whereas the clastic sediments become more dense, equaling rock salt at about 1 km in the Gulf Coast (see Fig. 7-36).

Once the salt is buried deeply enough to be less dense than the overlying sediments, there is a tendency for the salt layer to flow horizontally and thicken locally if there is some heterogeneity in the overburden pressure. This heterogeneity in the overburden pressure, P, may be due to horizontal variation in density,

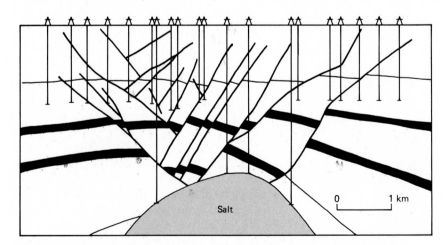

FIGURE 7–35 Cross section of the top of the Heidelberg dome, Mississippi. Note the thickening of the strata directly over the dome, but the thinning of strata toward the dome on the flanks. Compare with the White Castle dome in Figure 7-33. (Simplified after Hughes, 1960, Gulf Coast Association Geological Society Transactions, v. 10, p. 155–173.)

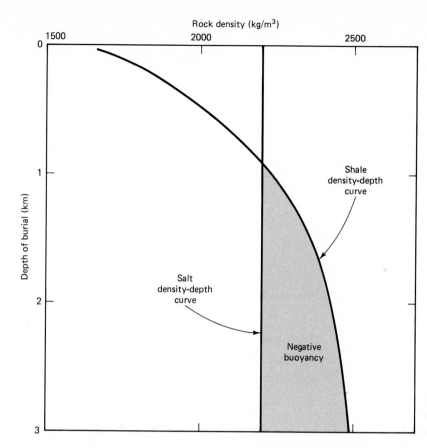

FIGURE 7–36 Comparison of shale density and salt density as a function of depth.

ρ_o, or in thickness, h, of the overburden. In Figure 7-37(a) this horizontal variation in overburden pressure is represented as an undulation of height Δh in the upper surface of the salt. The difference in overburden pressure, ΔP, between points below swells and troughs in the upper surface of the salt (points A and B, Fig. 7-37(a) is just the change in elevation of the upper surface Δh times the density contrast $(\rho_o - \rho_s)$:

$$\Delta P = \Delta h g \, (\rho_o - \rho_s) \qquad (7\text{-}9)$$

where ρ_s is the density of the salt and ρ_o is the density of the sedimentary overburden. If the salt is weak enough, it will flow horizontally from the trough (A) toward the swell (B) in response to the pressure difference, ΔP.

Note that horizontal flow of the salt from A to B (Fig. 7-37(a)) by itself produces a net increase in overburden thickness above point B and a decrease in thickness above point A, thereby reducing the horizontal pressure gradient that is driving the flow of salt from A to B. Substantial flow of the salt requires substantial redistribution of the overburden because the flow of the salt by itself tends to reduce the horizontal pressure gradient. If the overlying sedimentary strata were purely elastic layers, there would be an elastic restoring force that would inhibit the growth of any perturbation of the sediment-salt interface. Thus some initial perturbation of the interface exceeding the elastic limit of the overburden is necessary to initiate a diapir; mere burial of salt by more-dense sediments may never lead to diapirism. Some diapirs are known to have initiated along faults or other weak spots. The redistribution of the overburden may take place by flow of sediment from over the dome into the syncline, by sliding off the dome, or by a greater rate of deposition in the syncline than over the dome, as

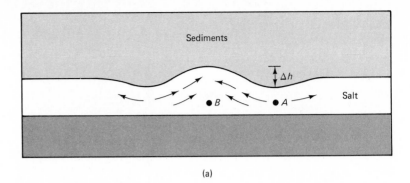

(a)

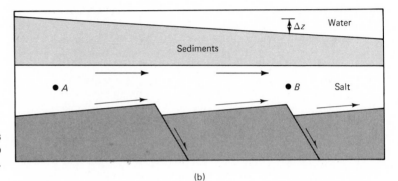

FIGURE 7–37 Schematic diagrams showing the flow of salt in response to differences in overburden pressure (Eqs. 7-9 and 7-10).

(b)

may be observed in detailed stratigraphic studies of sediments that were deposited over a growing dome.

The horizontal pressure difference in the salt may also be produced by the depositional slope of the overlying sedimentary load (Figure 7-37(b)). In this case the pressure difference is

$$\Delta P = \Delta z g \left(\rho_o - \rho_\omega \right) \tag{7-10}$$

where ρ_o is the density of the sedimentary overburden and ρ_ω is the density of water. Figure 7-37(b) also illustrates the effect of large topographic irregularities at the base of the salt. The flow direction is everywhere from A toward B, but there is a convergence of flow over the buried faults because of the thinning of the salt layer. This convergence produces a buildup of salt, the formation of salt pillows, and, eventually, salt domes if the strength of the overburden is exceeded.

Because of the circular symmetry of a salt dome in map view, the salt flows into the dome radially, creating a rim syncline. The depth of the syncline is less than the height of the dome because of the radial flow. The higher the dome grows, the larger the horizontal pressure gradient that is driving the flow of salt from the source bed into the dome (Eq. 7-9). If the pressure difference is maintained through redistribution of the overburden, the dome will continue to grow in volume until the layer of salt under the rim syncline is exhausted (Fig. 7-38).

As the dome rises it meets increasingly less dense sediments. Once the dome reaches the level at which the density of the salt and the surrounding sediments are equal, the salt may tend to spread horizontally if the sediments are weak and deeper salt continues to rise. In this way, an increasingly mushroomlike shape is produced. If, however, lateral spreading of the salt is sufficiently inhibited by the strength or density of the country rock, the salt may actually reach the surface and extrude, as is observed in dry climates—for example, along the Persian Gulf (Fig. 1-31).

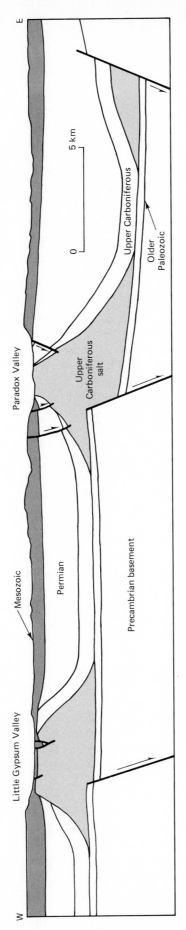

FIGURE 7–38 Cross sections of the Little Gypsum Valley and Paradox Valley salt ridges, Paradox basin, Colorado. Note that the growth of the salt structures was Permian during the drawdown of the adjacent synclines. A detail is shown in Figure 8-22. (After Cater, 1970.)

Mud Diapirism and Mud Volcanoes

The theory of salt diapirism requires some modification if it is applied to certain other diapiric materials. The major change is in the origin of the density inversion. In the case of mud diapirs, for example, the mud is less dense than the overlying sediments because of undercompaction due to low permeability of the mud, rather than low mineral density (as in the case of evaporites).

Undercompacted mud is an important diapiric material in many sedimentary basins that have been subjected to rapid sedimentary or tectonic loading. For example, near the mouth of the Mississippi River, older shelf and prodelta clays that have been loaded by rapidly deposited bar sediments form diapiric intrusions into and through the more permeable and compactable overlying sediments. These diapiric intrusions, called *mud lumps,* produce low ephemeral islands and submerged mounds 50 to 100 m in diameter, which are rapidly destroyed by erosion. The history of mud-lump formation during the last 100 years is well-documented for the mouth of the Mississippi River near South Pass, where numerous navigational surveys have been made because of the importance of the area as a shipping route (Morgan and others, 1968).

Mud diapirs along the south coast of Trinidad exist in an active tectonic environment. Several have formed ephemeral islands during this century in Erin Bay along a line of active folds and faults, which displays mud volcanic activity on the adjacent land (Higgins and Sanders, 1967). Islands appeared in 1911 and 1928 accompanied by large amounts of gas, including spectacular flames and a mushroom cloud in 1911. In August 1964 a new mud island appeared a few kilometers offshore; it rose to approximately 8 m above sea level and had an original diameter of about 230 m. The extrusion of soft plastic mud, displaying grooved plastic slip surfaces, took place largely during the first 2 days and ceased altogether a few days later. Compaction and erosion followed until the island disappeared below sea level 8 months later.

In the more extreme cases of undercompaction, mud can become fluidized and erupt to the surface in volcanic fashion. Such volcanoes are easily eroded and not widely noted in the geologic record; nevertheless, they are widespread phenomena in presently active tectonic areas associated with thick clastic sedimentary sequences. They are found in Borneo, Timor, Burma, Malaya, India, Pakistan, Iran, Iraq, the Soviet Union, Rumania, New Zealand, New Guinea, Taiwan, Trinidad, Venezuela, Colombia, Ecuador, and Peru. True mud volcanoes are unrelated to volcanic igneous or geothermal activity; they should not be confused with mud pots, geysers, or other thermally pressured flow that may in some cases involve mud. Mud volcanoes are almost always associated with anticlines or faults; such structures appear to play a role in allowing the mud to reach the surface.

Mud volcanoes are generally much smaller than their igneous counterparts, ranging from small mud cones 0.5 to 50 m high and 5 to 150 m in diameter (Fig. 7-39), up to truly impressive volcanoes as much as 500 m high and 6 to 10 km in diameter. They range from steep-sided conical shapes to broad shields. Some exhibit important cauldron subsidence.

In the Ramri Islands of Burma, mud volcanoes generally range from 50 to 100 m in diameter and are normally 5 to 7 m high. Thus they have a very low profile, which is due to the fluidity of the mud. They commonly have a small, nipplelike cone at the actual vent (compare with Fig. 7-39). The islands are formed largely of rock fragments and mud extruded by the volcanoes. One island, Cheduba, erupted historically in a violent fashion with gas that ignited to form a pillar of fire 300 m high.

FIGURE 7–39 Crest of mud volcano, southern Taiwan, with men standing on an active vent. The surrounding mudflows are a few days old.

Such violent eruptions with associated methane gas are fairly common. The local names of the mud volcanoes of the Baku Peninsula of the Caspian Sea include those that testify to their violent activity: Keireki (a forging furnace), Otman-Bozdag (a mountain emitting flame), Janadag (a burning mountain), Pilpilja (squelching of gas bubbles in the cone of mud), and Airantekjan (mountain issuing kefirlike liquid). Otman-Bozdag has erupted violently, forming a column of smoke 14 km high and throwing 1-m blocks of rock tens to hundreds of meters from the vent.

The nearly universal association of methane with mud volcanoes has caused some speculation that it plays a role in the eruptive mechanism. Methane-producing reactions directly increase the fluid pressure. Furthermore, when the methane produced exceeds that which will dissolve in the pore water, bubbles of gas exist that greatly reduce the effective permeability, leading to higher fluid pressures. Nevertheless, the primary requirement for mud volcanoes is mud that can be fluidized, that is, it is *thixotropic*. It thus requires relatively high porosity.

Sedimentary intrusive structures, particularly *sandstone dikes,* are sometimes observed in outcrop. They also require a fluidized sediment and may be emplaced by hydraulic fracturing.

EXERCISES

7–1   Compare and contrast the form and mechanics of salt intrusions and of magmatic intrusions.

7–2   Consider a salt dome of radius r and height $h;$ what is the approximate drawdown of the rim syncline if its width is $2r$? This consideration is important to constructing retrodeformable cross sections across salt domes.

7–3   Predict the pattern of present-day subvolcanic dike injection in the volcanic field of southern Italy and Sicily (compare with Fig. 6-13(b)).

7–4   The form and style of intrusion in the crust changes with depth in response to changes in mechanical properties of the country rock, as well as other factors. How will the intrusive styles differ between country rocks of quartz-rich rocks, carbonates, gabbros, and peridotites, just considering their different mechanical properties?

SELECTED LITERATURE

General

ANDERSON, E. M., 1951, *The Dynamics of Faulting and Dyke Formation, with Applications to Britain,* Oliver and Boyd, Edinburgh, 191 p.

HARGRAVES, R. B., ed., 1980, *Physics of Magmatic Processes,* Princeton University Press, Princeton, N.J. 585 p.

NEWALL, G., AND RAST, N., eds., 1970, Mechanisms of Igneous Intrusion, Liverpool Geol. Soc., Geol. J. Special Issue No. 2, 380 p.

Sheet Intrusions

GRETNER, P. E., 1969, On the mechanics of the intrusion of sills: Canadian J. Earth Sci., v. 6, p. 1415–1419.

DELANEY, P. T., AND POLLARD, D. D., 1981, Deformation of host rocks and flow of magma during growth of minette dikes and breccia-bearing intrusions near Ship Rock, New Mexico: U. S. Geol. Survey Professional Paper 1202, 61 p.

POLLARD, D. D., 1973, Derivation and evaluation of a mechanical model for sheet intrusions: Tectonophysics, v. 19, p. 232–269.

POLLARD, D. D., AND HOLZHAUSEN, G., 1979, On the mechanical interaction between a fluid-filled fracture and the earth's surface: Tectonophysics, v. 53, p. 27–57.

POLLARD, D. D., MULLER, O. H., AND DOCKSTADER, D. R., 1975, The form and growth of fingered sheet intrusions: Geol. Soc. Amer. Bull., v. 86, p. 351–363.

ROBERTS, J. L., 1970, The intrusion of magma into brittle rocks, p. 287–338 in Newall, G. and Rast, H., eds., *Mechanism of Igneous Intrusion,* Liverpool Geol. Soc., Geol. J. Special Issue No. 2, 380 p.

Volcanic Centers

SMITH, R. L., BAILEY, R. A., AND ROSS, C. S., 1961, Structural evolution of the Valles Caldera, New Mexico, and its bearing on placement of ring dikes: U. S. Geol. Survey Professional Paper 424-D, p. 145–149.

WILLIAMS, H., 1941, Calderas and their origin: University of California Publications, Geol. Sci., v. 25, p. 239–346.

Laccoliths

JOHNSON, A. M., AND POLLARD, D. D., 1973, Mechanics of growth of some laccolithic intrusions in the Henry Mountains, Utah, I. Field observations, Gilbert's model, physical properties and flow of magmas: Tectonophysics, v. 18, p. 261–309.

POLLARD, D. D., AND JOHNSON, A. M., 1973, Mechanics of growth of some laccolithic intrusions in the Henry Mountains, Utah, II. Bending and failure of overburden layers and sill formation: Tectonophysics, v. 18, p. 311–354.

Plutons

BUDDINGTON, A. F., 1959, Granite emplacement with special reference to North America: Geol. Soc. Amer. Bull., v. 70, p. 671–748.

HAMILTON, W., AND MYERS, W. B., 1967, The nature of batholiths: U. S. Geol. Survey Professional Paper 554-C, 30 p.

HUTCHINSON, W. W., 1970, Metamorphic framework and plutonic styles in the Prince Rupert region of the central Coast Mountains, British Columbia: Canadian J. Earth Sci., v. 7, p. 376–405.

PITCHER, W. S., 1979, The nature, ascent, and emplacement of granitic magmas: J. Geol. Soc. London, v. 136, p. 627–662.

Eruptive Mechanisms

MARSH, B. D., AND CARMICHAEL, I. S. E., 1974, Benioff zone magmatism: J. Geophys. Res., v. 79, p. 1196–1206.

MARSH, B. D., 1982, On the mechanics of igneous diapirism, stoping, and zone melting: Amer. J. of Sci., v. 282, p. 808–855.

SHAW, H. R., 1980, The fracture mechanisms of magma transport from the mantle to the surface, in Hargraves, R. B., ed., *Physics of Magmatic Processes,* Princeton University Press, Princeton, N.J., p. 201–264.

SPERA, F. J., 1980, Aspects of magma transport, in Hargraves, R. B., ed., *Physics of Magmatic Processes,* Princeton University Press, Princeton, N.J., p. 265–323.

Diapirs

BRAUNSTEIN, J., AND O'BRIEN, G. D., eds., 1968, *Diapirism and Diapirs,* Amer. Assoc. Petroleum Geol., Mem. 8, 448 p.

HALBOUTY, N. T., 1979, *Salt Domes Gulf Region, United States and Mexico,* 2nd ed., Gulf Publishing Company, Houston, Texas, 561 p.

RAMBERG, H., 1967, *Gravity, Deformation and the Earth's Crust as Studied by Centrifugal Models,* Academic Press, London, 214 p.

TRUSHEIM, F., 1960, Mechanism of salt migration in Northern Germany: Bull. Amer. Assoc. Petrol. Geol., v. 44, pp. 1519–1540.

Sedimentary Volcanism

HIGGINS, G. E., AND SAUNDERS, J. B., 1974, Mud volcanoes—their nature and origin: Verhandl. Naturf. Ges. Basel, v. 84, p. 101–152.

VAKUBOV, A. A., ALI-GADE, A. A., AND ZEIMALOV, M. M., 1971, *Mud Volcanoes of the Azerbaijan, SSR,* Acad. Sci. Azerbaijan, SSR, Baku, 258 p. (in Russian with English summary).

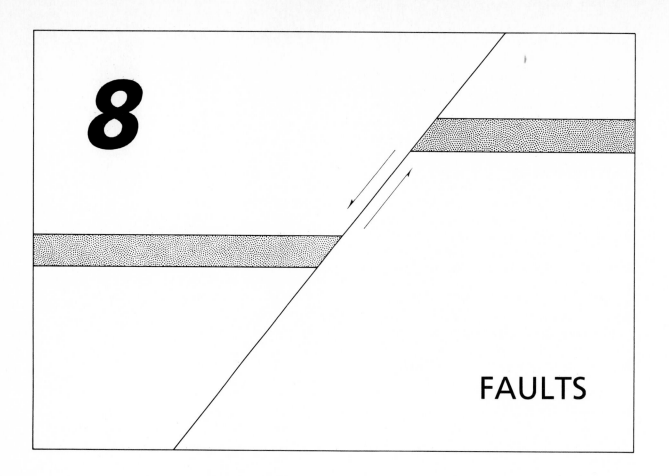

8

FAULTS

INTRODUCTION

There are three ways by which rocks undergo large deformation: (1) They flow, thereby experiencing a more- or less-distributed deformation (Chapter 10); (2) they buckle or bend, deflecting the rock layers, which allows considerable overall shortening with only moderate internal deformation (Chapter 9); or (3) they deform by slip of one body past another along discrete surfaces or zones, with little deformation away from the slip surfaces. These planes or zones of slip are called *faults,* which are the concern of this chapter.

The word fault was not initially a scientific term. According to Charles Lyell (Principles of Geology, 3rd ed., 1834):

> fault, in the language of miners, is the sudden interruption of the continuity of strata in the same plane, accompanied by a crack or fissure varying in width from a mere line to several feet, which is generally filled with broken stone, clay, etc.

This early definition, which reflects the practical problems of tracing coal beds in mining, could be construed to imply slip along the plane of the fault; nevertheless, the emphasis of the definition is not on slip, but on discontinuity. Lyell's definition points out to us the important fact that from an observational point of view, faults are first and foremost discontinuities across which there is generally some mismatch in the geology. The fact that the mismatch is a result of slip becomes clear only if equivalent strata can be correlated across the fault surface.

By their very nature as zones of concentrated slip, faults are commonly also zones of crushing, hot-water circulation, and, in some cases, initially weaker rock; these properties make fault zones generally more erodable and less visible

walls of some thrust sheets or gravity slide blocks. Some may be injected as dikes (for example, Pierce, 1979), whereas others may be dragged up along fault imbrications (see Gretener, 1977). A more-general fracturing, with brecciation and associated vein formation, is observed in brittle formations adjacent to some faults, particularly if the fault surface has substantial bends (for example, the left side of Fig. 8-1). Coarsely brecciated zones may be sites of hydrothermal circulation, alteration, and ore deposition.

Folding exists in and near some fault zones and reflects several distinct processes. Drag along the fault surface produces asymmetric folds of the fault-zone material, including gouge, fiber veins, and mylonite. Some crosscutting faults are initially folds that break across as slip proceeds; the process of folding is part of the mechanism of fault propagation (Chapter 9). Finally, as fault blocks slip past one another along a nonplanar fault surface (Fig. 8-1), at least one fault block must deform to conform to the fault surface; if the rocks are layered, they commonly bend in a process called fault-bend folding (see Chapter 9).

KINDS OF FAULTS

Numerous systems of fault classification have been proposed over the years and many are in current use. It is thus important to be aware of at least the more-widespread vocabulary and the associated concepts and phenomena, some of which were presented in Chapter 2. Some vocabulary has arisen because of the inferential nature of faults and is relatively noncommittal, describing only what is in fact known about a fault. *Strike faults* are faults whose strike, or trace, across the land surface is parallel to the strike of adjacent rock layers. Faults that are steeply dipping are *high-angle faults,* whereas *low-angle faults* are shallowly dipping.

If the sense of slip on a fault is known in addition to its dip, we have a more specific nomenclature. Faults with horizontal slip vectors are called *strike-slip faults*. Faults that slip up or down-dip are called *dip-slip faults* and include normal and thrust faults (Chapter 2). All others are *oblique-slip faults*. It is of interest in mechanical theories of faulting that most faults that have not been later deformed are approximately strike-slip, normal or thrust faults; we return to this fact later in the chapter.

Another distinction that may be made in fault classification concerns the contraction and extension of bed length as a result of faulting. *Contraction faults* lie at a low angle to rock layering and produce shortening parallel to layering. *Extension faults* lie at a high angle to rock layering and produce lengthening parallel to layering. The terms are independent of orientation of bedding and are useful in describing outcrop-scale faults in regions of strongly folded rocks.

In spite of these and other fault classifications, there remain for most purposes only three important classes of faults: normal, strike-slip, and thrust. These three fundamental classes of faults are analogous to the three major classes of plate boundaries: divergent, transform, and convergent. Plate boundaries, particularly in continental crust, are tens of kilometers wide or more and are composed of a zone of active faults. Normal faulting is dominant within divergent plate boundaries, although some strike-slip faults may be present striking at a high angle to the normal faults. Strike-slip faulting is dominant within transform zones, although some normal and thrust faulting is commonly present, oriented at an angle to the strike-slip faults. Thrust faulting is dominant within convergent plate boundaries, with minor transverse strike-slip faulting present as well.

Although plate-boundary slip is by far the most important setting for fault development, important faulting is not restricted to this setting. Faulting also develops in stress fields generated in response to folding, igneous and sedimentary intrusion, gravity sliding, and other processes not directly related to plate-boundary slip. In the following sections we describe the typical properties of normal, strike-slip, and thrust faults as they are developed in both plate-boundary and non-plate-boundary settings.

Normal Faults

A *normal fault* is a dip-slip fault in which the hanging wall drops relative to the footwall (Chapter 2). Normal faults in a given area may exist in two sets with parallel strikes but opposite dips, called *conjugate* normal faults. A set of conjugate normal faults is shown in Figure 8-14. Commonly the two fault sets have quite different aggregate slip. The fault set with relatively minor displacement is called *antithetic* with respect to the principal fault set. For example, the fault set dipping to the right in Figure 8-14 is antithetic to the set dipping to the left because it has much less aggregate slip.

FIGURE 8–14 Conjugate normal faults in Tertiary tuff, south-central Oregon. The height of the road cut is about 5 m.

Within active normal-fault terrains, the down-dropped blocks form topographic basins called *grabens* (German, *hole*). Grabens may be bounded either by a single set of faults on one side of a tilted fault block, called a *half-graben,* or by a conjugate pair on both sides of the basin, called a *full graben.* The term *graben* has several additional meanings in addition to this topographic one; it is also applied to ancient, sediment-filled normal fault basins that are no longer topographic basins involving recent fault slip. Structurally down-dropped blocks bounded by normal faults on any scale are also called grabens—for example, the down-dropped block in Figure 8-14. An intervening structurally high fault block between a set of grabens is called a *horst;* for example, a horst is present on the northwest side of Figure 8-10.

Normal faults are typically steeply dipping, usually in the range 55 to 70°. Their dips can be even greater, approaching 90° in some near-surface faulting. However, other normal faults dip substantially less than 50°, even as low as a few degrees in the extreme (see detachment faults later in this chapter). There are two principal reasons for this large variation in dip of normal faults: (1) Many normal faults pass downward into a relatively flat stratigraphic décollement of shale or salt, along which the hanging-wall fault block slides. More-deep-seated normal faults associated with crustal extension may flatten either abruptly or gradually into the zone of plastic flow in the lower crust or upper mantle (Fig. 1-27). Normal faults that gradually flatten with depth are called *lystric* normal faults, based on the Greek work *listron,* or *shovel,* because of their curved snow-shovel shape in cross section. The principal evidence for the lystric shape of many normal faults is the tilting of the hanging-wall block toward the fault, although slip on both gradually curved and abruptly flattening faults can produce block rotation. Figure 13-39 shows an example of this tilting in the Railroad Valley half-graben of eastern Nevada. The beds filling the graben show progressively greater tilting with increasing age, demonstrating the rotation of the fault block during graben formation. The exact shape of the fault system is not easily determined from the seismic section. (2) Some normal faults have low dips because of later rotation, commonly in association with progressive horizontal extension. For example, the older normal faults in the Yerington district of Nevada (Fig. 8-15) originally formed at 55° to 70° to the horizontal bedding; now the beds have rotated to steep dips and the faults have rotated to nearly horizontal. The younger normal faults are less rotated and have a dip closer to the 55° to 70° initial dip of new faults. Older faults become inactive because they rotate in the stress field until they are frictionally locked; then new faults must propagate for continued deformation.

There are two distinct kinematic situations in which normal faults develop in brittle or semibrittle rocks: (1) net horizontal extension of the rock mass that is undergoing faulting, and (2) collapse of the rock mass as a result of removal of

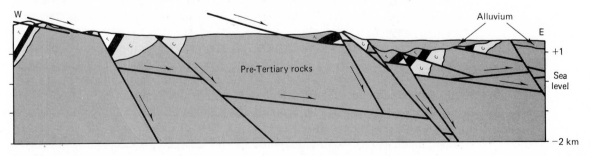

FIGURE 8-15 Normal-fault structure of the Yerington district, Nevada, based on extensive drilling data. Older normal faults have been rotated to near horizontal, maintaining their initial bedding cutoff angle. (Simplified after Proffett, 1977.)

material from below. The principal settings dominated by net horizontal extension include continental rifts, midocean ridges, and backarc basins associated with plate extension; pull-apart zones associated with upper-end, large-scale gravity slides, particularly on continental margins; and stretching of beds associated with flexural folding. The principal settings of normal faulting involving collapse without net horizontal extension include removal of magma from subvolcanic igneous centers, removal of underlying evaporites by dissolution, or flow, and differential compaction of underlying sediments. Examples of normal faulting in these situations are given below.

1. *Overall Horizontal Extension.* The most important tectonic setting of normal faulting is zones of plate divergence, including continental rifts, midocean ridges, and backarc basins. These normal-fault zones of plate divergence generally display voluminous basaltic or silicic volcanism, although the magmatism is minor in some continental rifts.

(a) *Continental rifting.* Continental rifts provide the best on-land exposure of structure dominated by normal faulting, particularly where horizontal extension has terminated without complete continental separation. Important examples include the East African rift, the Basin and Range of the western United States, the Rhine-Bresse graben of western Europe, the Oslo graben of Norway, the Reconcavoux basin in coastal Brazil, and the Jurassic-Triassic basins of the eastern United States and Canada.

A common result of active subaerial normal faulting is the formation of topographic grabens, commonly with internal drainage and associated lakes or playas. For example, the great lake basins of east Africa, including lakes Rudolf, Albert, Tanganyika, and Nyasa, lie along the graben systems of the East African rift. The main rift is generally 50 to 100 km wide and extends for several thousand kilometers.

Another example of active continental rifting is in the large area of internal drainage in the western United States called the Great Basin. The Great Basin contains many horsts and grabens, which divide the larger region of internal drainage into many smaller internal basins. Individual grabens or horsts are 10 to 20 km wide and typically extend for more than a hundred kilometers along strike; whereas the Great Basin or Basin and Range rift system as a whole is 300 to 800 km wide (Figure 13-38). The average spacing of major active normal faults in the Great Basin is about 15 km; therefore, the width of each active fault block is significantly less than the crustal thickness of 20 to 30 km and more nearly equal to the value of the depth of the brittle-plastic transition of 10 to 15 km that would be predicted from rock mechanics (Fig. 6-14). In fact, 98 percent of the earthquakes are shallower than 15 km, and 80 percent are shallower than 10 km. Therefore, the brittle fault behavior only exists in the upper crust. Furthermore, the rotation of the fault blocks with slip (Figs. 13-39 and 8-15) indicates that faults flatten with depth, either gradually or abruptly.

(b) *Oceanic rifting.* Rift valleys in oceanic areas exhibit some significant differences from those on the continents because the crust that is undergoing faulting in the oceanic realm has been created through igneous intrusion and extrusion synchronously with faulting, whereas continental faulting largely involves preexisting crust. The geometric relations of faults and offset layers must be somewhat complex in the ocean. Furthermore, it is suspected that crystallizing magma chambers, which are responsible for the layered gabbroic and ultramafic complexes exposed in on-land fragments of oceanic crust, are present at shallow depths of a few kilometers below active oceanic rifts. Therefore, some faulting in the oceanic realm may be analogous to the structures of caldera collapse associated with magma outflux (Chapter 7), whereas other faulting will be associated with horizontal tectonic extension and flexure of the lithosphere.

The most widespread and abundant data bearing on the structure of oceanic rifts is bathymetric, supplemented by seismic and magnetic profiles and bottom photographs. A few small areas have been surveyed in considerable detail, including the segment of the Mid-Atlantic Ridge southwest of the Azores surveyed under the FAMOUS Project (lat 37°N). In this region the ridge consists of a median valley and adjacent rift mountains. The inner floor of the rift valley is 2 to 3 km wide with a central zone of ridges and depressions (Fig. 8-16). These ridges are composed of the volcanic products of fissure eruptions, whereas the lows may be either grabens, volcanic collapse structures, or simply valleys between volcanic ridges. Volcanism dominates the morphology of the inner floor; faults at the surface are widespread but have small throws of the order 2 to 10 m and account for little of the relief. More-intense faulting and fracturing is observed on the flanks of the rift valley, with hundreds of small faults and fractures that trend parallel to the median valley. As the walls of the rift valley are reached, fault slip increases to hundreds of meters on individual faults and accounts for essentially all the relief between the inner floor and the rift mountains. The surface slope is reduced near the crest of the rift mountains through bending of the lithosphere and slip on normal faults that dip away from the inner rift. As a steady-state system, new crust appears to be produced largely in the inner rift, where faulting occurs through horizontal extension and through collapse of shallow magma chambers. The relief between the floor of the rift valley and the crest of the flanking rift mountains appears to be produced in steady state through the elevation of former rift-valley floor in escalator fashion as a result of patterns of flow in the mantle below the oceanic ridge. The brittle crust deforms by normal faulting as it rides up the escalator, first on faults dipping inward and then on faults dipping outward (Macdonald, 1982).

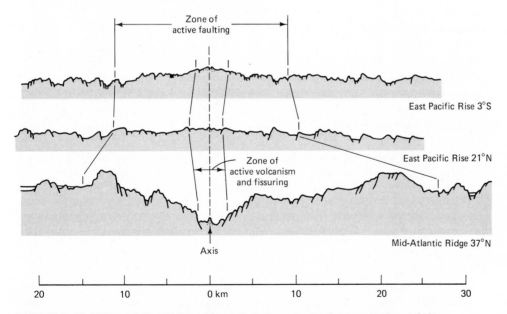

FIGURE 8–16 High-resolution bathymetric and shallow structural cross sections of three midocean ridges (vertical exaggeration about four times). The upper cross section, showing the smoothest bathymetry, is characterized by the fastest-spreading rate (more than 9 cm/year), whereas the bottom cross section, showing a pronounced axial valley, has the slowest rate (less than 5 cm/year). (After Macdonald, 1982. Reproduced, with permission, from the Annual Review of Earth and Planetary Science, v. 10, © 1982 by Annual Reviews Inc.)

Similar normal faulting and rift valley structure can be observed on land where the Mid-Atlantic Ridge is exposed in Iceland. In Iceland the ridge topography is dominated by massive volcanism because of the Iceland hot spot. This volcanism takes place as linear fissure eruptions along the rift and eruptions of volcanic mountain edifices. Numerous tension-crack fissures and small normal faults of a few meters to tens of meters slip, similar to the FAMOUS area, are present at the ground surface in the central rift. One well-known fault scarp at Thing Vallir in Iceland is marked by the collapse of the lava layer in a way that is suggestive of magma-chamber collapse rather than horizontal extension (Fig. 8-17).

(c) *Normal faults in gravity slides.* A third important setting of normal faulting is in major gravity slide structures. Large-scale landslides tens of kilometers across develop in a variety of settings as a secondary response to the relief created by tectonic processes, as is discussed under the heading of "detachment faults" later in this chapter. Normal faults develop in the head wall of the slides. For example, a system of normal faults exists in the sediments of the Gulf of Mexico continental margin, especially in Louisiana and Texas (Fig. 8-18). These normal faults are concave in plan, opening toward the south. There is an overall progression in age, with faults generally becoming younger toward the Gulf of Mexico, in the direction of sediment progradation. The master faults, which are those with the largest slip, dip toward the Gulf of Mexico and flatten markedly with depth until they are nearly horizontal within shale or salt décollement horizons. Numerous synthetic and antithetic secondary faults terminate at or above the master fault. Most of these faults are syndepositional and

FIGURE 8–17 Fault scarp of Mid-Atlantic Ridge exposed on land at Thing Vallir, Iceland. The rift valley is to the right.

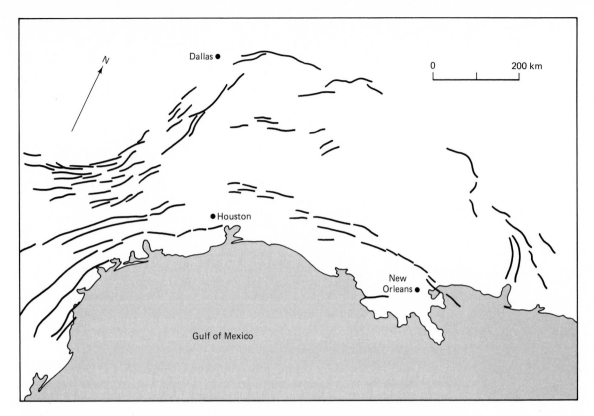

FIGURE 8–18 Map of normal faults cutting the sediments of the coastal plain of the U.S. Gulf Coast. Most of the slip is down toward the Gulf of Mexico. In addition a number of antithetic faults of smaller displacement exist. This pattern of faulting is the result of large-scale sliding of the continental-margin sediments out into the Gulf of Mexico. (Simplified from Tectonic Map of North America, 1969.)

exhibit significant increases in sediment thicknesses in the down thrown side. Anticlines, commonly called *rollovers,* may develop as a result of bending of the hanging-wall block as it conforms to the curved fault surface (see Fig. 9-41). Rollover anticlines are an important site of hydrocarbon entrapment. Some of the flattening normal faults in the Gulf Coast are closely associated with shale or salt diapirs.

   (d) *Normal faults in response to flexure.*   A fourth important situation for normal faulting in response to horizontal extension is the stretching of competent layers during flexure of anticlinal and sometimes synclinal folds. An example is the Kettleman Hills anticline of California, which displays numerous normal faults of minor slip (Fig. 8-19). Notice that the crestal faults strike parallel to the fold axis where the anticline is not plunging but transverse to the fold axis where it begins to plunge, implying a change in direction of stretching. Another example of secondary normal faulting as a result of stretching is the faulting of competent layers above salt diapirs prior to being pierced by the salt (Chapter 7), as is illustrated by the structure contour and fault map of the Hawkins field in Texas (Fig. 8-20).

   2. *No Overall Horizontal Extension.*   Normal faulting can also be produced through the collapse of rock into a space created through the removal of underlying rock wih no associated net horizontal extension. During outflow of magma from a near-surface magma chamber, the roof rock may collapse into the space vacated, with the associated formation of a cauldron, caldera, or collapse crater at the surface (Chapter 7).

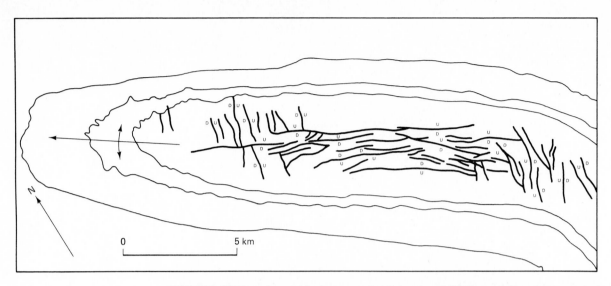

FIGURE 8-19 Map of crestal faults exposed at the surface in Kettleman Hills anticline, central California. All these faults have small displacement, from one to tens of meters, and have formed as part of the folding process. The strike faults are dip slip, mostly down-dropped to the crest. The transverse faults display oblique slip with the slickensides plunging toward the axial surface, which shows that the slip was synchronous with flexure. (Simplified from Woodring, Stewart, and Richards, 1940.)

The process of collapse is accomplished largely through normal faulting in and around the subvolcanic igneous center. Normal faulting in igneous centers is also produced through stretching and updoming during episodes of influx of magma (compare with salt diapirs, Fig. 8-20). The net effect of the cycles of magma influx, eruption, and collapse is the production of large numbers of normal faults in the immediate vicinity of the igneous center.

Faulting due to igneous collapse and updoming is often characterized by much closer spacing than tectonic normal faulting and displays a wide variety of strikes. For example, Figure 8-21 is a map of normal faults in the vicinity of the Timber Mountain caldera complex of Miocene age in southwestern Nevada. The region lies within the Basin and Range normal-fault province; away from the caldera center, the typical north-trending normal faults are well developed. Within the caldera complex, the density of faults is greatly increased in a roughly circular area with a wide variety of strikes, in contrast with the regional fault pattern. At Timber Mountain these faults outcrop within the central resurgent dome; the ring fractures of the moat are covered.

Many hydrothermal ore deposits are associated with subvolcanic igneous centers. They commonly display the complex fault structure typical of collapse and dilation of the country rock.

Normal faulting through the removal of material from below is not limited to igneous centers. Another important origin of collapse normal faulting is through the removal of underlying salt or other evaporites through dissolution or through the differential compaction of shale masses. For example, the deformation of the Paradox basin of eastern Utah and western Colorado is characterized by a series of long, linear northwest-trending folds displaying broad simple synclines, 15 to 20 km wide, and tight, narrow anticlines 4 to 5 km wide (Fig. 7-38). The anticlines are cored by a great thickness of Upper Carboniferous salt that has flowed from the adjacent synclines into the anticlines during Permian folding. The folded late Paleozoic succession is unconformably overlain by Jurassic and Cretaceous sediments that are little folded but are substantially down-dropped along normal faults into the cores of the salt anticlines (Fig. 8-22). This structural complexity is

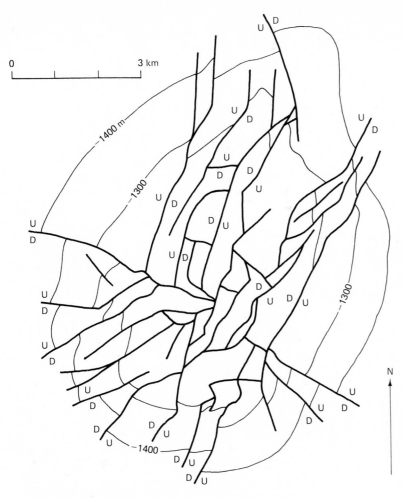

FIGURE 8–20 Structure contour and fault map of the Hawkins dome, Texas, drawn at the top of the Woodbine sand, which is a horizon above the present level of salt piercement. Only selected contours are shown. (Simplified after Wendlandt, Shelby, and Bell, 1946.)

a result of collapse of the crests of the anticlines in Tertiary time because of removal of underlying salt by dissolution.

Strike-Slip Faults

Strike-slip faults are characterized by nearly horizontal slip parallel to the strikes of vertical or steeply dipping faults. Strike-slip faults are traditionally thought of as vertical faults, but some have a discernible inclination; for example, the well-known San Andreas fault in California has a southwestward dip in a region south of San Francisco, as shown by accurate locations of earthquakes.

There are two principal tectonic settings of strike-slip faulting: (1) transform plate boundaries within both oceanic and continental crust and (2) secondary strike-slip faults that join together *en échelon* normal or thrust faults in areas dominated by normal fault or thrust tectonics.

Transform plate boundaries exist as both major segments many hundreds of kilometers long, such as the Romanche and Vema fracture zones in the South Atlantic, connecting widely separated spreading centers or subduction zones, and as minor, sometimes less clear-cut, strike-slip zones tens of kilometers long connecting minor offsets in spreading centers. Major strike-slip transform zones in the continents include the San Andreas zone in California, the Dead Sea system, and Anatolia fault in Turkey, the Philippine fault zone, the Denali fault in Alaska, the Alpine fault in New Zealand and the Great Glen fault of Scotland (Figs. 13-29, 1-29, 1-28, 2-29). Typical lengths are of the order of hundreds to thousands of kilometers.

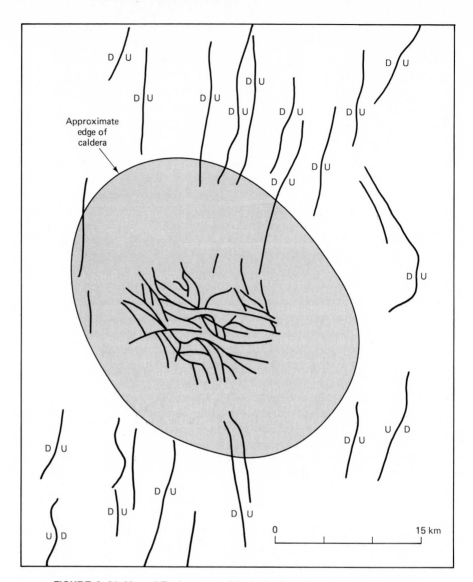

FIGURE 8–21 Map of Tertiary normal faults in the vicinity of Timber Mountain caldera, Nevada. The approximate region of caldera collapse is shown in grey. Faults within the caldera are exposed in a central resurgent dome. Faults in the outer moat are covered by volcanic and sedimentary rocks younger than caldera formation. (Data from Cornwall, 1972.)

Slip along continental, as well as oceanic strike-slip, faults terminates in zones of extension or compression. At the ends of simple oceanic ridge-ridge transforms, new crust is formed through magmatic activity at the ridge crests. Net slip along the active transform segment increases as the crust travels from its site of formation to the opposite rise crest, where the slip reaches a maximum. Once a segment of crust passes the opposite rise crest, the fault becomes an inactive fracture zone with no major continued slip; the net fault slip is equal to the separation of equivalent age crust across the inactive fracture zone. Minor continued vertical displacement may take place through differences in thermal subsidence across the fault (Fig. 1-10).

In contrast, net fault slip along a transform zone in continental crust or in pretransform oceanic crust is a constant along the fault and goes to zero where the

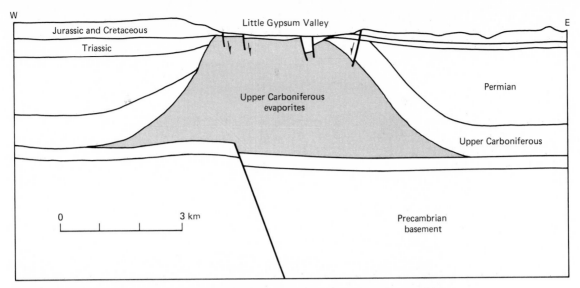

FIGURE 8–22 Cross section of salt-cored anticline showing late normal faulting caused by collapse of the crest through removal of salt by dissolution, Paradox basin, Colorado. A detail of Figure 7-38. (Simplified after Cater, 1970.)

fault terminates in zones of extension or compression. For example, slip along the left-lateral Garlock strike-slip fault in southern California terminates in a series of grabens north of the fault (Fig. 13-32). Net fault slip is not constant among strata deposited during the period of fault slip; slip decreases with decreasing age, leading to a complex paleogeographic history (for example, see the discussion of San Andreas fault system, Chapter 13).

A characteristic of major strike-slip fault zones in both continental and oceanic crust is the lack of a perfectly planar continuous fault surface; instead the fault zones are composed of *en échelon* fault segments that are joined kinematically by zones of compressive or extensional deformation, as shown diagrammatically in Figure 8-23(a). For example, if a right-lateral fault steps *en échelon* to the right, then extension must take place in the overlap zone. This extension commonly takes place by transverse normal faulting and the formation of a basin in the *en échelon* zone; examples are given in Figure 8-23 with fault maps of the Hope fault, South Island, New Zealand, and the San Jacinto fault in California. In contrast, if the right-lateral fault steps *en échelon* to the left, then compression must take place in the overlap zone. An important large-scale example is the zone of folding and thrusting in the Transverse Ranges of southern California associated with an *en échelon* step to the left of the right-lateral San Andreas fault zone (Figure 8-23(d)). Zones of *en échelon* faulted anticlines are observed along strike slip zones—for example, along the Rinconada fault in Figure 8-23(e). The anticlines are oriented roughly perpendicular to the expected maximum compression. These folded structures above strike-slip faults are sometimes called *flower structure* if they are associated with thrust faults that root into the strike slip zone.

The second important setting of strike-slip faults is in areas dominated by normal or strike-slip faulting. Normal and thrust faults may terminate in strike-slip faults called *tear faults* that absorb the slip and transfer it to another normal or thrust fault. For example, most of the slip at the northeast end of the Pine Mountain thrust is taken up along the Russell Fork tear fault (Fig. 8-24). A small amount of slip continues on the Pine Mountain fault to the northeast of the Russell Fork fault and is consumed in folding. A tear fault also exists at the southwestern end of the Pine Mountain thrust, the Jacksboro fault (Fig. 8-24).

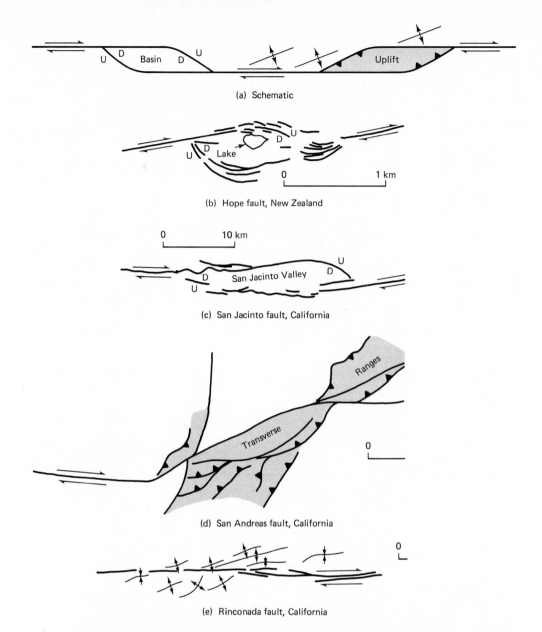

(a) Schematic

(b) Hope fault, New Zealand

(c) San Jacinto fault, California

(d) San Andreas fault, California

(e) Rinconada fault, California

FIGURE 8–23 Formation of extensional basins and compressional uplifts between *en échelon* segments of strike-slip faults and arrangement of *en échelon* folds. (After Sharp, 1975; Geologic Map of United States, 1974; Geologic Map of California, 1977.)

Thrust Faults

Thrust faults are dip-slip faults in which the hanging wall has moved up relative to the footwall. Thrust faults generally dip less than 30° during active slip and commonly dip between 10° and 20° at their time of formation. Many thrust faults lie along bedding planes over part of their length. Thrust faults dipping greater than 45° are sometimes given the special name *reverse faults*.

The structural settings in which thrust faults develop may be divided into two major classes: (1) compressive plate boundaries in both continental and oceanic settings and (2) secondary faulting developing in response to folding, flexure, igneous or sedimentary intrusion, or large-scale landsliding. Compressive plate boundaries are by far the more important.

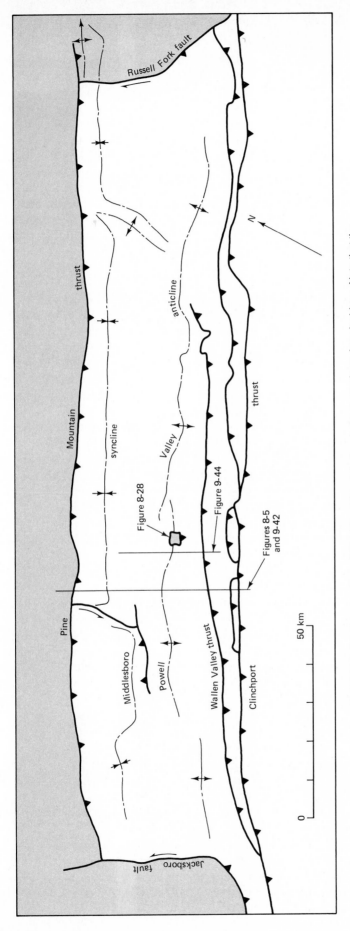

FIGURE 8–24 Structural map of the Pine Mountain thrust sheet, southern Appalachians. Note that the strike-slip faults function as tears that connect thrust faults. Locations of cross sections and maps seen in other figures are shown. Faults to the southeast of the Clinchport thrust are not shown. (Simplified from Harris and others, in Fischer and others, ed., Studies of Appalachian Geology: Central and Southern, © 1970, Wiley-Interscience, New York.)

1. *Thrusts at Compressive Plate Boundaries.*   Our detailed knowledge of thrust faulting along compressive plate boundaries is largely limited to foreland fold-and-thrust belts on land (Chapter 1), where there has been extensive geologic mapping, deep drilling, and seismic exploration in the search for petroleum and coal resources. Thrust faulting in accretionary wedges involving oceanic lithosphere is much less well studied but nevertheless undoubtedly is widespread. Here we emphasize the better-known continental thrusting.

There are basically two types of continental fold-and-thrust belts: (1) those in which thrust faulting is the dominant structural style at the surface, such as the southern Appalachians (Figure 12-4) and the Cordilleran fold-and-thrust belt in western Canada and the United States (Fig. 13-20), and (2) those in which folding is the dominant structural style at the surface, such as the central Appalachians (Fig. 12-18), the Sierra Madre Oriental of Mexico, and the Jura Mountains of Switzerland and France (Fig. 2-26). The structure of both types of fold-and-thrust belts is fundamentally dominated by thrust faults at depth that dip toward the interior of the mountain belt and run along bedding planes over much of their length. The deformed structure is detached from its substrate along a bedding-plane zone of slip called a *décollement* (French, *unglue*). The individual thrust sheets are typically displaced upward and outward toward the margin of the mountain belt (Fig. 12-4). The direction of overthrusting or overfolding is called the *vergence,* which is commonly toward the craton. For example, the Appalachian fold-and-thrust belt verges toward the craton throughout Canada and the United States (Fig. 12-1).

Individual thrust faults commonly cut progressively up through the stratigraphic layering toward the margin of the mountain belt in a sequence of discrete crosscutting segments called *ramps* or *risers* and separated by bedding-plane décollement segments (see Fig. 8-25). Ramps also develop along strike, associated with changes in décollement horizon along strike. The locations of some ramps are known to be fixed by interruptions of the décollement by normal faults (Fig. 8-25) or by facies changes along the décollement horizon. Ramps commonly have dips of less than 20°; however, many faults are now steeper because of rotation by *imbrication* or overlapping, of the thrust slices (Fig. 8-5). The bending of thrust sheets as they imbricate or ride over ramps is discussed more fully in Chapter 9, on folding.

Many thrust faults bifurcate in the ramp areas with the total displacement distributed among the many fault splays, as well as absorbed in folds (Fig. 8-26). Each fault splay apparently undergoes slip at a separate time, with one fault slipping until it locks, then a new fault propagating and slipping until it locks, and so on. If the imbricated fault slices are confined between two décollements, as in

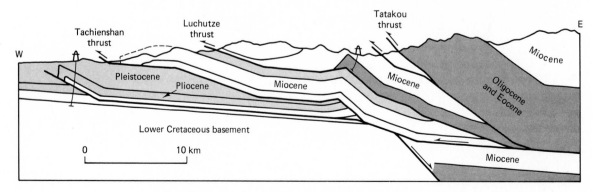

FIGURE 8–25 Cross section of active fold-and-thrust belt of western Taiwan, showing the influence of a preexisting normal fault on the locations of ramps.

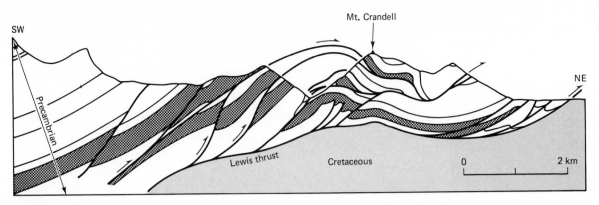

FIGURE 8–26 Imbrications of the Lewis thrust sheet, Montana. (After Boyer and Elliott, 1982.)

Figure 8-27, the compressed zone between the two décollements is called a *duplex structure*.

The widespread development of décollement fault segments along thrust faults reflects the profound control on the structural development that is imposed by the anisotropy and heterogeneity of the stratigraphic sequence. Décollement segments of thrust faults are confined to stratigraphic horizons that appear to be inherently weak either because of low rock strength, high fluid pressures, or both. The major décollement horizon in the Alps and Jura is a layer of Triassic evaporites, largely anhydrite (Fig. 2-26). Important décollement horizons in the

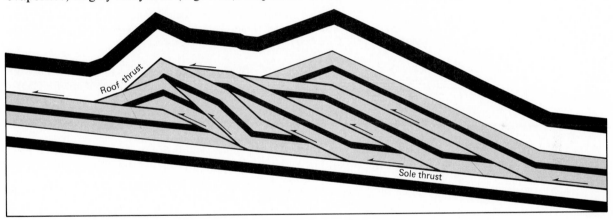

(a) Schematic duplex

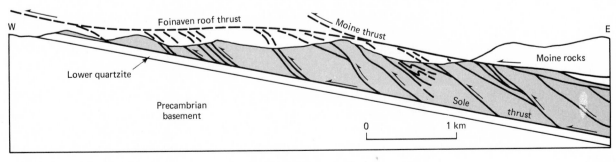

(b) Eroded duplex, Scotland

FIGURE 8–27 (a) Schematic drawing of a duplex structure. (b) Example of a duplex structure of the Moine thrust system, Scotland. (Cross section simplified after Elliott and Johnson, Trans. Roy. Soc. Edin., **71,** 69–96, 1980.)

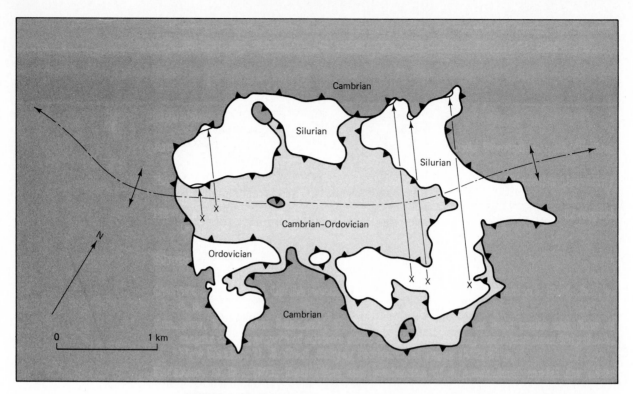

FIGURE 8–28 Chestnut Hill window of the Pine Mountain thrust, southern Appalachians, showing the net slip of five small fault slivers of the Pine Mountain system. See Figure 8-24 for location of window. (Simplified from maps of Miller and Fuller, 1954.)

Appalachians are Devonian black shale, Silurian salt, Ordovician shale and middle Cambrian shale and possibly salt. The major décollement in the Zagros Mountains of Iran is Cambrian salt. In some other mountain belts, less-strong stratigraphic contrasts in strength exist; in these areas décollement horizons have been selected for less-obvious reasons.

As thrust sheets slip over their bases, small areas may permanently stick from time to time, resulting in the propagation of a new branch of the main fault, above or below the locked region. This process produces numerous fault slivers along major thrust faults. An example of these fault slivers can be seen in the Chestnut Ridge window* through the Pine Mountain thrust fault in the southern Appalachians (Fig. 8-28). Individual slivers have been translated a shorter distance than the overlying hanging wall; for example some of the small slivers on the north side Chestnut Ridge window have moved only a kilometer or two (Fig. 8-28), whereas the large sheet of Cambrian and Ordovician rocks has moved 3 to 5 km and the overlying Cambrian rocks of the Pine Mountain thrust sheet have translated about 10 to 14 km. Fault slivers provide a record of the footwall geology that the thrust sheet rode over and therefore provide important information on the subsurface geology.

Occasionally an extensive record of the overriding thrust sheets is preserved as a result of simultaneous folding and thrusting, thereby producing a syncline filled with a stack of thrust sheets. For example, a stack of folded thrust sheets is displayed in the thrust structure of the Carboniferous Antler orogenic belt in

---

*A *window* is an erosional exposure of footwall rocks completely surrounded in map view by hanging-wall rocks; the equivalent German word *Fenster* is also often used in English. The German word *Klippe* is used for an erosionally isolated outlier of hanging-wall rocks (for example, see Fig. 12-9).

Nevada (Fig. 13-7). Each successively higher thrust slice has been translated farther relative to the footwall and is less strongly folded.

Many major thrust sheets display thin fault slivers along their base; nevertheless, the faults themselves typically display remarkably narrow zones of deformation considering the large translation, to the surprise of most observers. For example, Figure 8-29 is a photograph of the Copper Creek thrust fault in the southern Appalachians, which places Cambrian shales and sandstones of the Rome Formation over Ordovician carbonates. The strata of the hanging wall and footwall display relatively minor disruption at distances more than a meter from the contact. The zone of most intense deformation is generally a few centimeters thick. The fault has slipped several tens of kilometers. The great Champlain thrust of the northern Appalachians (Fig. 8-11) displays little deformation in the hanging wall of Cambrian dolomite, whereas the footwall zone of Ordovician shale displays extensive deformation over a thickness of several meters. These out-crops illustrate an important mechanical property of thrust faults, as well as faults in general; displacement is concentrated in relatively narrow zones. Apparently continued slip along a weakened surface in many situations offers less resistance than the development of a new surface of slip.

Thrust faults associated with compressive mountain belts along plate boundaries apparently root at depth through the lower continental crust and upper mantle into the aesthenosphere. As a result, thrust sheets in the interiors of mountain belts commonly involve deeper rocks, including older basement rocks, more-deeply metamorphosed rocks, and—in some cases—upper mantle. Great basement or metamorphic thrust sheets are well documented in Scandinavia, Scotland, the Austrian Alps, the Appalachians, the Himalayas, and other moun-tain belts. For example, the great East Alpine thrust sheet of Austria places Paleozoic basement rocks and their sedimentary cover over Mesozoic and Cenozoic sediments and metamorphosed sediments. Several major windows, the Tauern, Otztal, and Engadine, document the existence of this great flat thrust

FIGURE 8-29 Photograph of Copper Creek thrust of the southern Appalachians, near Knoxville, Tennessee. The fault is a décollement in both the hanging wall and footwall, placing shales and sandstones of the Cambrian Rome Formation over Ordovician carbonates.

sheet over a distance of 50 to 80 km across strike. Similar, but much larger, basement thrust sheets are documented with seismic data in the Appalachians (Figs. 12-21 and 12-22). Basement thrusts are also associated with the Laramide basement uplifts of the U.S. Rocky Mountains (Chapter 13) and late Cenozoic basement uplifts in the Venezuelan and Colombian Andes. For example, the Wind River thrust, which is responsible for the Wind River uplift in Wyoming, appears to flatten to a décollement in the lower crust or along the Moho (Fig. 13-26).

   2. *Thrust Faults in Secondary Settings.*   Thrust faults develop in secondary settings, in addition to their primary development along compressive plate boundaries. Thrust faults are subsidiary structures of major strike-slip fault zones, as discussed earlier in this chapter (for example, Fig. 8-23). In addition, thrust faults develop in the toe region of some major subaerial or submarine slides, as discussed in the following section on detachment faults. One additional setting of secondary thrust faulting is in response to folding. For example, as chevron folds undergo extreme shortening, thrust faults develop that are confined to bedding on one flank of the fold but are crosscutting on the other flank; these are called *limb thrusts*.

Detachment Faults

Most bedding-plane thrust sheets associated with compressive plate boundaries have slipped updip and they root or disappear at depth into the aesthenosphere, because they are shallow crustal expressions of plate boundaries. Nevertheless, décollement-bounded sheets in some geologic settings have slipped downdip over much of their length and are essentially colossal subaerial or submarine landslides, in some cases hundreds of square kilometers in area. The faults in this category of low-angle normal faults involving downhill sliding along a décollement are called *detachment faults*. A common property of detachment faulting, as with intact landslides in general, is a zone of extension in the upper part, or head, of the slide and a zone of compression in the lower part, or toe, of the slide. In some cases, detachment faults have been confused with orogenic compressive thrusts, but they may be distinguished by the fact that orogenic thrust sheets undergo overall compression, whereas detachment fault sheets must have zones of extension.

   Two well-studied examples of detachment faulting are in the western United States near the eastern edge of the Cordilleran mountain belt in the Bearpaw and Beartooth Mountains of Montana and Wyoming. These two examples will serve to illustrate several aspects of detachment faulting.

   The Bearpaw Mountains are an uplifted early-to-middle Eocene laccolithic and volcanic center approximately 50 km in diameter surrounded by flat-lying, undeformed sedimentary rocks of the Great Plains of northern Montana, approximately 200 km east of the eastern edge of the compressive Cordilleran mountain belt. However, the sediments immediately flanking the Bearpaw Mountains exhibit décollement thrusting involving slip directed away from the mountains, as well as normal faulting within the décollement sheets (Fig. 8-30). Frank Reeves mapped the belt of faulted folds around the Bearpaw Mountains in the early 1920s and proposed that the volcanic rocks and their floor of uplifted Tertiary and Upper Cretaceous sedimentary rocks slid plainsward on a weak layer of bentonitic shale within the Cretaceous sequence. This hypothesis has been essentially substantiated by extensive drilling, surface mapping, and theoretical analysis (Hearn, 1976; Gucwa and Kehle, 1978). The drill-hole data indicate that the complex and highly faulted surface geology is only surficial and that the deeper horizons are much less deformed, showing smooth dips off the Bearpaw Mountain uplift of 5 to 30 m/km.

which is Amontons's first law. The coefficient of friction, μ_f, is independent of the area of contact and is a material property:

$$\mu = \frac{S}{Y} \qquad (8\text{-}4)$$

We see from Bowden's theory, based on experiments with metals, that Amontons's second law is a result of the fact that the frictional resisting force is proportional to the microscopic—not the macroscopic—area of contact, which in turn is proportional to the normal force. Friction can be a function only of macroscopic contact area when the two surfaces are in perfect contact; then the frictional resistance of an interface is independent of normal force and is given by Equation 8-2 rather than the friction equation (Eq. 8-3).

In general, Bowden's concept of flattening asperities and increasing microscopic contact area with increasing normal stress appears to be qualitatively correct for rocks, although at low temperatures, the asperities undergo both fracture and plastic flow. The physical process of friction in rocks is probably more complex than those of Bowden's theory (Byerlee, 1978). Nevertheless, at moderate to high normal stresses, most rock interfaces display a frictional behavior close to Amontons's laws with a coefficient of friction that is essentially independent of rock type (Fig. 8-33). Writing Equation 8-3 in terms of stresses, it is found experimentally that

$$\sigma_\tau = 0.85\sigma_n^* \qquad (\sigma_n^* < 0.2 \text{ GPa})$$
$$\sigma_\tau = 50 \text{ MPa} + 0.6\sigma_n^* \qquad (0.2 \text{ GPa} < \sigma_n^* < 2 \text{ GPa}) \qquad (8\text{-}5)$$

where σ_τ is the peak frictional shear stress and σ_n^* is the effective normal stress. Effective normal stresses in the upper crust are generally less than 0.2 GPa. This

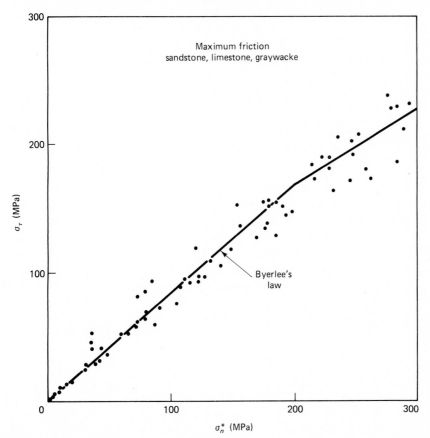

FIGURE 8–33 Measurements of maximum friction in sandstone, limestone, and graywacke. (Data from compilation of Byerlee, 1978.)

empirical relationship at moderate to high effective normal stress is sometimes called *Byerlee's law* of rock friction.

Fault surfaces that contain a layer of gouge also obey Byerlee's law and display a frictional behavior that is similar to clean rock interfaces involving either strong and weak rocks. The principal silicate materials that display substantially lower friction than Byerlee's law are montmorillonite, vermiculite, and illite; the first two contain loosely bound interlayer water in the clay structure, which has been suggested to act as a pseudo-pore pressure, thereby reducing the effective stress (Eq. 8-5). This explanation is not valid for illite however. Other platy minerals, such as chlorite, kaolinite, halloysite, and serpentine, display normal frictional properties. Therefore, most gouge probably has a normal coefficient of friction unless certain clay minerals, such as montmorillonite, vermiculite, or illite, are present.

The general independence of friction with rock type as given by Byerlee's law is an important observation, which for many purposes greatly simplifies considerations of frictional behavior. It should be noted, however, that this regular behavior holds only for moderate to high effective normal stresses. The frictional properties of rock are much more variable under the low normal stresses of the near-surface environment of interest to civil and mining engineering.

MECHANICAL THEORIES OF FAULTING

There exists no single comprehensive theory of mechanics of faulting. This reflects not only the mechanical complexity of the problem, but also the fact that several distinct classes of questions may be asked of faults. For example geologists have attempted primarily to understand the initial formation of faults and the major mechanical events that lead to their final disposition. In contrast seismologists are concerned primarily with the short-term behavior of faults, including the slip associated with single earthquakes, and the relationship between this slip and the seismic radiation recorded by seismometers. The behavior of faults over periods of up to a few hundred years has become common ground to both structural geology and geophysics. Theory of faulting is an important area for future research because there is much yet to be learned.

Anderson's Theory of Faulting

The simplest and in many ways the single most enlightening theory of faulting is that first presented by E. M. Anderson near the beginning of this century. *Anderson's theory* assumes as a simplification that shallow-level faults are Coulomb fractures and then explores the simple consequences of that assumption. We recall from Chapter 5 that there exist two possible orientations of Coulomb fractures with respect to the principal-stress directions (see Fig. 5-5). The line of intersection of the two fracture orientations is parallel to the intermediate principal stress, σ_2. The direction of maximum principal compression, σ_1, bisects the acute angle between the fractures, whereas the direction of least principal compression, σ_3, bisects the obtuse angle. The material shortens parallel to σ_1 and expands parallel to σ_3 as a result of slip along the fractures. Therefore, if faults are approximated as Coulomb fractures, we can predict the orientation and sense of slip of the faults given the orientation of the principal stresses and the shape of the Mohr envelope ($\mu = \tan \phi$ in Fig. 5-5). Alternatively, if we know the orientations of the faults and their slips, we can infer the orientations of the principal stresses.

The next essential step in Anderson's theory is to consider the possible orientations of the principal stresses. We recall from Chapter 3 that only planes that are perpendicular to the principal-stress axes are planes of no shear stress. The earth's surface is a plane of no shear stress because the surface is a fluid-solid interface, and fluids cannot support shear stresses. Anderson therefore assumed in the case of shallow-level faulting that one of the principal stresses would be perpendicular to the surface of the earth. There are thus three possible orientations of the principal stresses: σ_1-, σ_2-, and σ_3- vertical; the other two principal stresses must be horizontal in each case. The Anderson theory of faulting therefore predicts three possible arrangements of near-surface faults (Fig. 8-34). If σ_1 is vertical, the fractures will dip more than 45° (that is, at $45° + \phi/2$) and the slip will be downdip. If σ_2 is vertical, the fractures will be vertical and the slip will be horizontal. If σ_3 is vertical, the fractures will be inclined at less than 45° (that is, at $45° - \phi/2$) and the slip will be updip (Fig. 8-34). We are now in a position to compare theory with observation.

Anderson's theory of faulting is remarkably successful. It predicts that there should be three distinct types of near-surface faulting corresponding to normal, thrust, and strike-slip faults. Normal faults commonly exist in two opposing sets and dip more steeply than 45° as predicted by Anderson's theory. Strike-slip faults

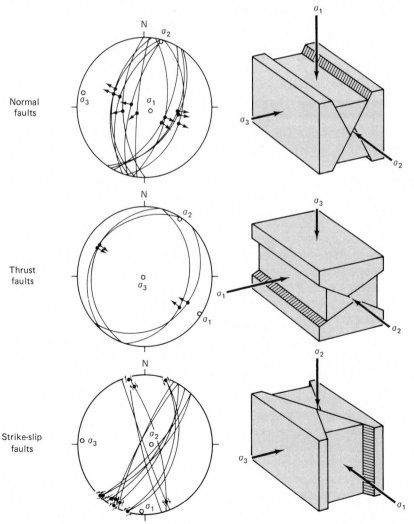

FIGURE 8–34 Three classes of conjugate faults according to Anderson's theory of faulting, together with typical data on fault and slickenside orientation from the Aegean. (After Angelier, 1979b.)

are normally nearly vertical. Thrust faults dip less than 45°. In other ways Anderson's theory is less successful; for example, high-angle reverse faults are not predicted. Thrust faults are dominated by bedding slip and commonly dip at less than the predicted angle (less than 20°). Furthermore, strike-slip and thrust faults do not exist in conjugate pairs as commonly as normal faults. Nevertheless, given the simplicity of the assumptions, Anderson's theory is quite successful, particularly in that it predicts three major classes of faults. Furthermore, it allows us to estimate the orientations of the principal stresses from the orientations and senses of slip on faults (Fig. 8-34).

Simple Extensions of Anderson's Theory

The Coulomb fracture criterion deals with fracture at a single point in a homogeneous isotropic material that is homogeneously stressed. It is to be expected that Anderson's theory, which equates faults with Coulomb fractures, will have some shortcomings, because rock is commonly layered, laterally variable, and anisotropic and contains preexisting fractures and faults. Furthermore, faults are not just fractures at a point; they are three-dimensional planes of recurring slip. Several of these complexities can be treated as simple extensions of Anderson's theory by direct application of some of the other aspects of brittle fracture that were presented in Chapter 5. These concern the effects of preexisting fractures and of anisotropy, which are treated in the following paragraphs. In a following section we consider the effect of some geologically reasonable stress fields on fault shape in light of Anderson's theory.

In Chapter 5 we considered the problem of Coulomb fracture in a body containing a preexisting fracture along which frictional sliding might occur (Fig. 5-11). We found that, in general, for preexisting faults or fractures to be capable of sliding and therefore not be frictionally locked, they must have an orientation rather close to the orientation in which a new fracture would form in unfractured rock. The larger the difference between the fracture strength and the frictional stress, the greater the range of possible orientations of preexisting faults that are able to slide rather than forming a new fracture (Fig. 5-11). These results can be applied directly to theory of faulting. They imply that in regions undergoing a second deformation, old faults will generally not be reactivated unless they are rather close to the orientation in which a new fault would form.

Most rocks in the crust exhibit some layering or planar anisotropy, such as bedding, gneissic banding, or foliation. This planar structure can exert an important control on the development and orientation of brittle fractures and faults. In Chapter 5 we considered the experimental brittle-fracture behavior of slate, which is a strongly anisotropic rock (Fig. 5-13); those experiments will give us some insight into the effects of layering on fault orientation. The fracture orientation in slate is relatively insensitive to the anisotropy if σ_1 is within about 30° of perpendicular to the plane of weakness anisotropy; the fracture orientation is roughly what it would be in a similar, but isotropic, rock. In contrast, if σ_1 is more inclined such that there is moderate shear stress on the plane of weakness, then slip is commonly along the anisotropy or very close to it (Fig. 5-13). If the principal compression is nearly parallel to the planar anisotropy, a fracture may form roughly parallel to it or the layers may buckle. These experimental observations suggest that layering or planar anisotropy will affect thrust faulting more commonly than strike-slip or normal faulting in first-generation deformation because bedding will be roughly horizontal at the start of deformation. In more-complex areas, the rock anisotropy and layering may exert important controls on fault orientation if the anisotropy and layering is in the proper orientation relative to the stress.

Fault Orientation and Localization within Stress Fields

Anderson's theory of faulting provides an estimate of the orientation of faults in a homogeneous near-surface stress field. The orientation of Coulomb fractures in other more-complicated, yet geologically reasonable, stress fields can also be considered. For example, in Chapter 7 we considered possible states of stresses around near-surface magma chambers as an indication of possible orientations of dikes and magma-chamber-related faulting (for example, Fig. 7-15). Here we review several other theoretical examples of stress fields related to faulting.

Hafner (1951) considered the stresses within a slab of rock being pushed from one end, a problem that is related to thrust-fault imbrication. The slab of rock is considered to be sitting originally at lithostatic pressure due to gravity and is then pushed from one side with balancing frictional forces along the base to satisfy the conditions of static equilibrium (no unbalanced forces or torques). A resulting stress field is shown in Figure 8-35, which exhibits a concentration of stress at the back end of the block and a curving of the stress trajectories in response to the frictional resistance of the base. If we consider faults to form in the Coulomb fracture orientation $\theta = (45° + \phi/2)$, then we may determine the predicted orientations of faults, as shown in Figure 8-35. Hafner's result predicts that if a conjugate set of thrust faults develops, the foreward set will be flatter, which may agree with natural observations. Figure 8-35 also predicts that the faults will step up from the décollement at a finite angle.

In Hafner's theory stresses were applied to the exterior of the body and an internal stress field was calculated. Alternatively, displacements may be applied to the exterior of an elastic slab and the resulting stress trajectories may be calculated. Sanford (1959) used this second approach to determine the stress field at the edge of a vertical block uplift to determine the possible orientations of related secondary faults.

Theories of Short-Term Fault Behavior

Fracture theory is also useful if we wish to consider seismic faulting. In this case, however, the pertinent theories of fracture mechanics are those dealing with the dynamics of crack propagation because they allow consideration of the emitted elastic radiation. There are also dynamic theories dealing with frictional sliding along a preexisting interface or fault. In both sets of theories the elastic waves emitted during fracture or sliding along one part of the fault modify the stress field as the waves travel through other parts of the fault. The details are beyond our scope, but dynamic theories provide considerable insight into the behavior of a fault during a single earthquake (for example, Rice, 1980).

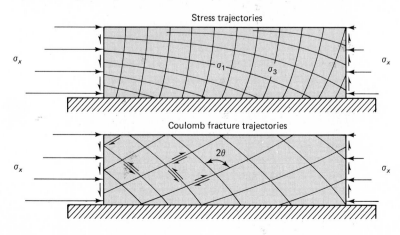

FIGURE 8–35 Schematic drawing of stress trajectories and corresponding Coulomb-fracture trajectories within a thrust block experiencing a net push from the left. (Based on models of Hafner, 1951.)

One of the most important theories of the short-term behavior of faults was developed by Harry Reid after the great California earthquake of 1906. A segment of the San Andreas fault 435 km in length suffered a sudden displacement, with the western block of the fault moving north relative to the eastern by as much as 6 m in places. Fortunately, there had been two geodetic surveys made of the area before the earthquake, one between the years 1851 and 1866 and a second between 1874 and 1892. A third survey was made the year following the earthquake. Points relatively far from the fault—for example, 20 to 30 km away—had moved 1.4 m parallel to the fault in a right-lateral sense relative to equivalent points on the other side of the fault between the first and second surveys. Between the second and third surveys, a similar movement of 1.8 m took place. Nevertheless, the two walls of the fault surface suffered no relative displacement until the time of the earthquake. What happened is illustrated in Figure 8-36. Because of frictional locking across the fault, there is a slow elastic distortion in the vicinity of the fault with an associated gradual increase in the stored elastic energy. When the associated stresses due to this elastic distortion finally exceeded the static friction at some point along the fault, slip began and produced the earthquake. This is Reid's *elastic rebound theory* of earthquakes, and it has not been superseded as the general explanation of most shallow-focus plate-boundary earthquakes.

Note that during the earthquake, the largest displacements are immediately adjacent to the fault, and the amount of slip during this catastrophic event decreases away from it. Points far from the fault undergo relatively little displacement because they had relatively little elastic strain and underwent their displacement slowly in the period prior to the earthquake. Thus the area near the fault moves during earthquakes, whereas the area far from the fault plane moves constantly.

We have earthquakes because faults have friction acting across their surface and because the static friction is greater than the dynamic friction. This difference provides the instability needed for catastrophic slip. Figure 8-37 presents a simple mechanical analog that illustrates the behavior of a fault over many earthquake cycles. A fault is represented by the contact between the blocks $N_1, N_2, \ldots N_n$ and the adjacent footwall. The blocks are joined together by a series of coil springs

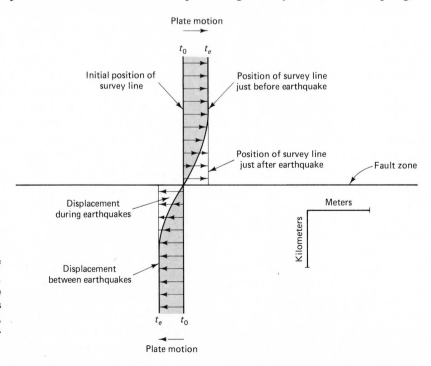

FIGURE 8–36 Schematic illustration of the elastic rebound theory of faulting, based on a right-lateral fault zone. Note that far from the fault, displacement takes place gradually between earthquakes, whereas near the fault most displacement takes place during earthquakes.

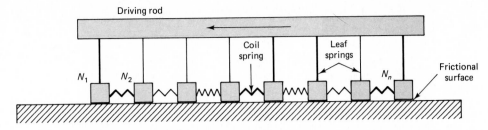

FIGURE 8–37 Analog model of fault slip. The driving rod moves at a constant rate, whereas the blocks slip from time to time, sometimes singly, sometimes in groups of two or more. Each block may have a different friction, mass, and stiffness of the attached springs. (Based on theory and model of Burridge and Knopoff, 1967.)

and are attached to a driving rod by leaf springs. The driving rod is displaced parallel to the footwall at a constant rate; however, the blocks are not able to slide freely because their contact with the footwall is made through sandpaper of varying grades and thus with varying amounts of frictional contact. Natural variations along actual faults might be due to variations in coefficient of friction, fluid pressure, or larger-scale irregularities in the shape of the fault surface. Once the driving rod has started into steady motion to the left, the leaf springs will be flexed until the friction at the base of one of the blocks is exceeded and that block slips, causing an "earthquake" in proportion to its mass and displacement. This slip of the block will cause a compression in the adjacent spring to the left and a tension in the adjacent spring to the right. As the driving rod continues its steady plate-tectonic motion, other blocks will slip, sometimes one at a time; but at other times the slip will spread to several adjacent blocks through the coil springs causing larger-magnitude earthquakes. The relationship between frequency of "earthquakes" to their magnitude for this analog device is rather similar to actual faults and gives a reasonable intuitive model of fault slip (Burridge and Knopoff, 1967).

MECHANICS OF THRUST FAULTING

The subject of the mechanics of thrust faulting has been the source of much animated discussion and controversy from the beginnings of structural geology in the last century, right up to the present. It was initially difficult for people to believe that great thrust sheets exist, and once their existence was well-established, people were hard-pressed to provide viable mechanical explanations of how relatively thin, extensive sheets of rock could possibly move.

Perhaps the first person to recognize large thrust sheets was Arnold Escher von der Linth, a professor at Zürich, who in the 1840s observed field relationships in Switzerland that required great overthrusting of Mesozoic and Paleozoic strata over Lower Tertiary sediments. The existence of great thrust sheets is particularly clear in the Glarus area because of later broad anticlinal folding that exposes the older-over-younger juxtaposition over a large region (Fig. 8-38). Nevertheless, Escher was reluctant to publish many of his ideas because he felt he would be ridiculed.

It was a number of decades before enough systematic regional mapping was done in the Alps and elsewhere to make the existence of large thrust sheets an unavoidable conclusion for most geologists. In the 1880s and 1890s great thrust sheets began to be recognized in Switzerland and the Scottish Highlands, ushering in the Golden Age of Alpine Geology, which was a time of considerable excitement lasting for several decades, during which many of the implications of thrust structure to the fabric of mountain belts were pursued.

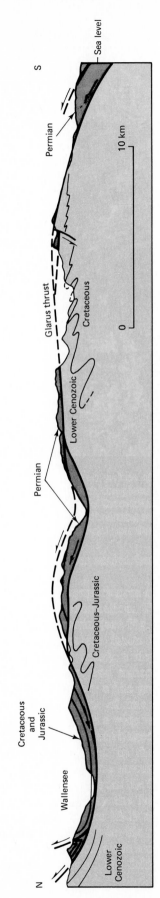

FIGURE 8–38 Folded Glarus thrust sheet of eastern Switzerland.

298

As geologic data requiring great thrust sheets accumulated, controversies blossomed over their mechanics. For example, T. Mellard Reade wrote in the *Geological Magazine* of 1908:

> In attempts to unravel some of the weightier problems of geology it has lately been assumed that certain discordances of stratification are due to the thrusting of old rocks over those of a later geological age. Without in any way suggesting that the geology has in any particular instance been misread, I should like to point out the difficulties in accepting the explanation looked at from a dynamical point of view when applied on a scale that seems to ignore mechanical probabilities. Some of the enormous overthrusts postulated are estimated at figures approaching 100 miles. If such a movement has ever taken place, would it not require an incalculable force to thrust the upper over the lower . . .? I venture to think that no force applied in any of the mechanical ways known to us in Nature would move such a mass, be it ever so adjusted in thickness to the purpose, even if supplemented with a lubricant generously applied to the thrust-plane. These are the thoughts that naturally occur to me, but as my mind is quite open to receive new ideas I shall be glad to know in what way the reasoning can be met by other thinkers.

Mr. Reade's comments did elicit further discussion, which has in effect continued throughout the century (for example, Voight, 1976). Through most of this discussion, the mechanics of thrust faulting has been reduced to the elementary physics problem of pushing or sliding a rigid block over a rigid base (Fig. 8-32). For example, if a thrust sheet has a length of 100 km, a thickness of 4 km, and a density of 2600 kg/m^3, then a 1-m-wide cross section of the thrust sheet has a weight of about 10^{13} ($kg \cdot m/s^2$). The coefficient of friction of rock is about 0.85; therefore, the frictional resistance of a 1-m-wide strip of this thrust sheet is about 0.85×10^{13} N. To push the sheet, we must apply an equivalent force to the back end of the sheet. The average force per unit area is 2.2×10^9 N/m^2, or 2.2 GPa, which is about ten times the crushing strength of the rock (Table 5-1). Therefore, if we pushed as hard as we could, the block (Fig. 8-32) would not move; it would fracture near where we applied the force. The mechanical problem is that thrust sheets are very thin relative to their apparent frictional resistance; the force with which we apparently have to push the thin sheet implies stresses that are sufficient to break it.

There have been a variety of attempts to get around this enigma. One has been to conclude that thrust sheets do not exist, which unfortunately flies in the face of geological observations. The others have been to conclude that the simple mechanical model of frictional sliding of a rigid block as presented above (Fig. 8-32) is in some way inappropriate and needs modification. The principal modifications that have been attempted are: (1) The basal resistance is not Coulomb friction, but is viscous or plastic (Smoluchowski, 1909; Kehle, 1970); (2) the effect of pore-fluid pressure greatly modifies the problem (Hubbert and Rubey, 1959); (3) thrust sheets are not pushed from behind, but slide down an inclined plane (Smoluchowski, 1909; Hubbert and Rubey, 1959); (4) the rectangular shape shown in Figure 8-32 is inappropriate; (5) mountain belts and their associated thrust sheets are plastic masses that flow outward under their own weights like ice caps (Bucher, 1956; Price, 1973; Elliott, 1976); and (6) mountain belts and their associated thrust sheets are mechanically analogous to the deformed wedge of snow or soil that develops in front of a bulldozer blade (Chapple, 1978; Davis, Suppe, and Dahlen, 1983). Each of these alternatives represents a physically valid process; at present the primary concern is which of these processes are actually quantitatively important in the mechanics of thrusting. In the following sections we outline what appear to be the most-successful and widely applicable theories of mechanics of thrusting.

Hubbert and Rubey Fluid-Pressure Theories

By far the most successful theories of thrust faulting have been those of Hubbert and Rubey (1959), who applied knowledge of the effects of fluid pressure on rock strength and frictional resistance to two problems: (1) pushing a rigid thrust block from behind, similar to the classic problem outlined above (Fig. 8-32), and (2) sliding a thrust sheet under gravity down an inclined plane. Hubbert and Rubey recognized that the frictional stress, σ_τ, is not a function of the normal stress, σ_n, acting across the fault surface, but rather of the effective normal stress $\sigma_n^* = \sigma_n - P_f$ (see Chapter 5):

$$\sigma_\tau = \tau_0 + \mu_f(\sigma_n - P_f) \tag{8-6}$$

and

$$\sigma_\tau = \tau_0 + \mu_f \sigma_n (1 - \lambda_f) \tag{8-7}$$

where μ_f is the coefficient of friction (Eq. 8-3) and $\lambda_f = P_f/\rho gz$ is the ratio of fluid pressure to solid overburden pressure, called the *Hubbert-Rubey fluid-pressure ratio*. Therefore, if the pore-fluid pressure, P_f, approaches the solid overburden pressure due to the weight of the overlying thrust sheet ($\sigma_n = \rho gz$), then λ_f approaches 1 and the frictional resistance, σ_τ, approaches the small cohesive strength of the fault surface τ_0, which is nearly negligible. Hubbert and Rubey showed that fluid pressures approaching the solid overburden pressure are widely encountered in drilling for petroleum and postulated that thrust faults lie in zones of high fluid pressure and associated low frictional resistance.

The problem is more complicated, however, because the fluid pressures also reduce the fracture strength of the overlying thrust sheet (Eq. 5-11):

$$\sigma_1 - P_f = C_0 + K(\sigma_3 - P_f) \tag{8-8}$$

and

$$\sigma_1 = C_0 + \sigma_3[K + \lambda(1 - K)] \tag{8-9}$$

Therefore, if the pore-fluid pressure, P_f, approaches the solid-overburden pressure ($\sigma_3 = \rho gz$), then the fluid-pressure ratio within the sheet ($\lambda = P_f/\rho gz$) approaches 1 and the deviatoric stress for fracture ($\sigma_1 - \sigma_3$) approaches the cohesive strength, C_0, which is 50 MPa or less.

From the above discussion, there is obviously a trade-off between the reduction of friction and the reduction of rock strength as a result of fluid pressure. Which effect dominates in the problem of pushing thrust sheets? Hubbert and Rubey illustrated the effect of fluid pressure on pushing a thrust sheet with the simplified model of a rigid thrust sheet of thickness H and length l (Fig. 8-39). In this simple model we do not consider all the forces acting on the rigid thrust sheet, but for illustrative purposes consider the frictional resisting force of the base, F_f, and the maximum horizontal driving force, F_{max}, that can be applied to the back end. The frictional resistance is the sum of the frictional stresses (Eq. 8-7) over the length of the thrust sheet, l:

$$\text{Frictional resistance} = F_f = \int_0^l [\tau_0 + \mu_f \sigma_z(1 - \lambda_f)]dx \tag{8-10}$$

The maximum possible horizontal driving force is the sum of the fracture strengths (Eq. 8-9) along the back end from the surface down to the base at depth H:

$$\text{Maximum horizontal force} = F_{max} = \int_0^H [C_0 + \sigma_z[K + \lambda(1 - K)]]\,dz \tag{8-11}$$

The maximum possible length of thrust sheet, ignoring the resisting forces of the edges and front, is then the one for which the maximum possible driving force is just equal to the frictional resistance:

$$F_{\max} = F_f \qquad (8\text{-}12)$$

If we consider the rock properties C_0, K, λ, μ_f, τ_0, and λ_f to be constant, then from Equations 8-10, 8-11, and 8-12, the maximum possible length of the thrust sheet is

$$l_{\max} = \frac{2HC_0 + \rho g H^2\,[K + \lambda\,(1 - K)]}{2\tau_0 + 2\mu_f\,(1 - \lambda_f)\,\rho g H} \qquad (8\text{-}13)$$

which is essentially a ratio of rock-strength terms to frictional-resistance terms.

Once we choose reasonable values for the parameters in equation 8-13, we may solve for the maximum fault length under various conditions and observe the effect of fluid pressure on size of intact thrust sheets. This has been done using the fracture strength and frictional properties of typical sedimentary rocks under various conditions of fluid pressure and thickness of the thrust sheet. Two cases have been considered in Figure 8-39. In one case, the fluid-pressure ratio on the fault, λ_f, and within the overlying block, λ, are equal; this case represents a conservative estimate of the effect of fluid pressure on thrust faulting because it represents a maximum weakness for the thrust sheet. In the second case, fluid pressure within the block is hydrostatic, $\lambda = 0.435$, whereas the fluid pressure

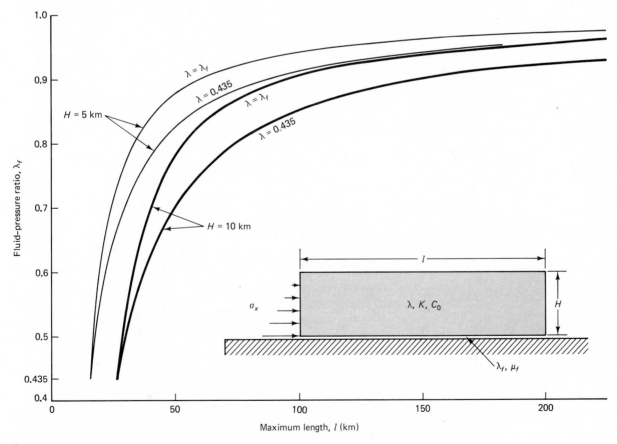

FIGURE 8–39 Maximum length l of a rectangular thrust sheet than can be pushed across a horizontal décollement without fracture, as a function of fluid-pressure ratio λ_f (Eq. 8-13). Two end-member fluid-pressure distributions are considered. The weak case has the same fluid-pressure ratio within the block as along the fault: $\lambda = \lambda_f$. The strong case has hydrostatic fluid pressures within the block: $\lambda = 0.435$. $C_0 = 50$ MPa, $K = 3$, $\mu_f = 0.85$, and $\rho = 2400$ kg/m^3.

along the fault may be elevated, or $\lambda_f \geq \lambda$. This second situation is an estimate of the maximum effect of fluid pressure on thrust faulting, because it maximizes the strength of the thrust sheet by having a low fluid-pressure ratio, λ, within the block while minimizing the frictional resistance by having a high fluid-pressure ratio, λ_f, along the fault. The geologic situation corresponding to this second case ($\lambda = 0.435$) would be a thrust sheet in which the thrust fault is localized along the base of the fluid-pressure transition zone (Fig. 4-24), thereby maximizing the strength of the thrust sheet and minimizing the frictional resistance. Figure 8-39 illustrates the maximum and minimum effect of fluid pressure (λ, λ_f) and thickness H on thrust faulting. The two cases are similar, except that substantially greater fault lengths are possible in the maximum case ($\lambda = 0.435$) at very high fluid pressures ($\lambda_f \geq 0.8$ to 0.9).

The effect of fluid pressure is substantially greater than the effect of thickness. At low fluid pressures, a tenfold increase in thickness produces only a doubling of fault length, whereas an increase of fluid pressures to $\lambda = 0.8$ to 0.9 produces about a fivefold increase in fault length over the dry (Reade's case, $\lambda = 0$) or hydrostatic ($\lambda = 0.435$) cases. For example a 5-km-thick thrust sheet could be about 15 km long in the dry case, but could be over 100 km long with $\lambda_f \cong 0.95$.

The graphs and theoretical arguments given above suggest that fluid pressures could be quantitatively important in the mechanics of faulting. It remains to be shown that fluid pressures play an important role in actual earth situations. Two lines of evidence suggest that fluid pressures are in fact important: (1) experimental control of faulting through control of fluid pressures by pumping, and (2) observations of the relationship between fluid pressures and positions of thrust faults in active fold-and-thrust belts. These are discussed below.

It might be questioned whether the effective-stress law for friction (Eq. 8-6) is valid for faulting on a geologic scale. An important test of this law has been performed using a strike-slip fault at Rangely oilfield of western Colorado. *In situ* stress measurements were made in the vicinity of the strike-slip fault: $\sigma_1 = 59$ MPa, $\sigma_2 = 43$ MPa, and $\sigma_3 = 31.5$ MPa at the depth of the Weber Sandstone reservoir with σ_2 vertical and σ_1 oriented N70°E. Using a Mohr diagram we calculate the resolved shear stress, σ_τ, on the vertical fault striking N67°E to be 8 MPa and the normal stress σ_n to be 35 MPa. Using the effective-stress law of Coulomb friction (Eq. 8-6),

$$\sigma_\tau = \tau_0 + \mu_f(\sigma_n - P_f) \qquad (8\text{-}14)$$

we calculate the fluid pressure necessary for frictional sliding to be $P_f \geq 26$ MPa given laboratory measurement of $\tau_0 \cong 1$ MPa and $\mu_f = 0.81$ for the Weber Sandstone. The fluid pressure was experimentally increased in the Weber Sandstone reservoir near the fault by pumping (Raleigh, Healy, and Bredehoeft, 1972). It was observed that frequent small earthquakes began in the region when the fluid pressure reached about 27.5 MPa, which is in good agreement with the fluid pressure of 26 MPa predicted from *in situ* stress measurements and laboratory friction measurements through the effective-stress equation for friction (Eq. 8-14). This experiment shows quantitatively that fluid pressures play an important role in mechanics of faulting, as Hubbert and Rubey suggested.

Fluid-pressure measurements have been made in the course of petroleum exploration in regions of active thrust faulting. It is observed that elevated fluid pressures do exist, and that much of the thrust faulting is confined to overpressured zones; both observations suggest that fluid pressures are important in actual thrust fault situations. For example, Figure 8-40 shows a cross section through an active thrust structure in western Taiwan showing that the major thrusts lie within the overpressured zone ($\lambda = 0.7$). The frictional resistance in the overpressured

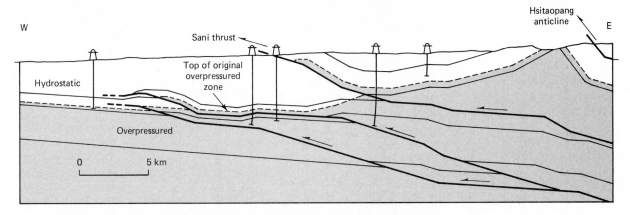

FIGURE 8–40 Cross section of part of the active fold-and-thrust belt, western Taiwan, showing the distribution of hydrostatic and overpressured sediments, prior to thrusting. Note that most of the fault surfaces lie within the overpressured zone. During thrusting the top of the over-pressured zone generally moves downward in the thrust sheets as a result of removal of the overburden load by erosion, which may cause frictional locking of more eroded sheets.

zone is almost half that in the normally pressured zone. Therefore, we conclude that fluid pressures play an important role in the mechanics of faulting.

Hubbert and Rubey (1959) also considered the role of fluid pressure in the mechanics of gravity sliding. The mechanics of moving a detachment sheet by gravity sliding is substantially different from the mechanics of pushing a thrust sheet. A detachment sheet will slide if the angle of inclination, θ, of the décollement equals a critical value that is a function of the coefficient of friction, μ_f, and fluid-pressure ratio along the fault, λ_f. The critical inclination for sliding is independent of length and thickness, in contrast with the case of pushing thrust sheets.

Consider the block of weight W and basal area A sitting on a plane of inclination θ (Fig. 8-41). The average normal stress on the plane is

$$\sigma_n = \left(\frac{W}{A}\right)\cos\theta \tag{8-15}$$

and the average shear stress is

$$\sigma_\tau = \left(\frac{W}{A}\right)\sin\theta \tag{8-16}$$

We equate the frictional effective-stress equation (8-6) with Equations 8-15 and 8-16 to obtain

$$\left(\frac{W}{A}\right)\sin\theta = \tau_0 + \mu_f\left[\left(\frac{W}{A}\right)\cos\theta - P_f\right] \tag{8-17}$$

and neglecting τ_0 we obtain

$$\tan\theta = \mu_f(1 - (\lambda_f/\cos\theta)) \approx \mu_f(1 - \lambda_f) \tag{8-18}$$

which gives the critical angle θ for initiation of sliding as a function of the coefficient of friction and the fluid-pressure ratio $\lambda_f = P_f A/W$ along the fault. This equation is graphed in Figure 8-41 for a typical coefficient of friction of $\mu_f = 0.85$. The critical angle for dry sliding ($\lambda_f = 0$) is 40°. For hydrostatic fluid pressure ($\lambda_f = 0.435$), the critical angle is 24°. At very high fluid-pressure ratios, the critical angle for sliding reaches a very low value; for example, with $\lambda_f = 0.95$, the critical angle is 2.4°. Very large gravity slides, such as the Heart Mountain detachment and the Bearpaw Mountain slides (Fig. 8-30), apparently moved down inclines of a

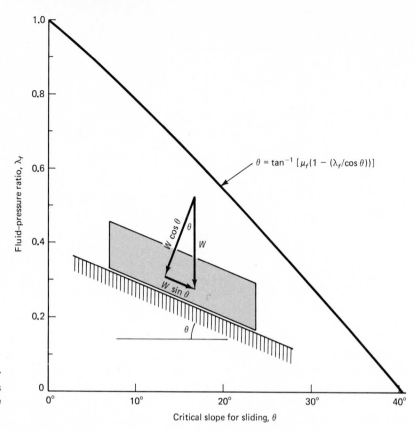

FIGURE 8-41 Critical inclination, θ, for sliding a block down an inclined plane, as a function of fluid-pressure ratio along the fault, λ_f (Eq. 8-18); $\mu_f = 0.85$.

few degrees. The commonly accepted explanation of sliding at these low inclinations is very high fluid-pressure ratios.

Mechanics of Fold-and-Thrust Belts

Our preceding discussion has dealt primarily with the mechanics of pushing rectangular thrust sheets that are not so long that they break up, as well as with the mechanics of sliding a block down an inclined plane. Neither of these theories is particularly useful in considering thrust faulting in fold-and-thrust belts and accretionary wedges along convergent plate boundaries. The structure in these regions consists of a wedge-shaped mass in cross section composed of strongly imbricated thrust slices lying above a basal décollement (Figs. 8-25, 12-4, 13-20) that eventually roots into the subduction zone and the aesthenosphere (Figs. 1-15 and 1-27). Therefore, the thrust mass is being pushed by the converging plates and the thrust mass has failed in compression breaking into many imbrications, in contrast with the no-failure criterion of the single thrust-sheet theory given above (Fig. 8-39).

Chapple (1978) proposed that fold-and-thrust belts and accretionary wedges have an overall mechanics analogous to the accretionary wedge of soil or snow that develops in front of a bulldozer blade (Fig. 8-42). The soil deforms, increasing its surface slope, until the wedge attains its critical taper and then slides stably, continuing to grow at constant taper as additional material is encountered. The *critical taper* is the shape for which the compressive force in every segment of the wedge is balanced by the frictional resistance of the part of the wedge lying in front of that segment. For example, in Figure 8-42 the compressive force, F_x, acting across the wedge at any point x is equal to the resisting force along the base between point x and the tip of the wedge. Because the wedge has deformed in compression from a thinner shape to its present wedge shape, the horizontal

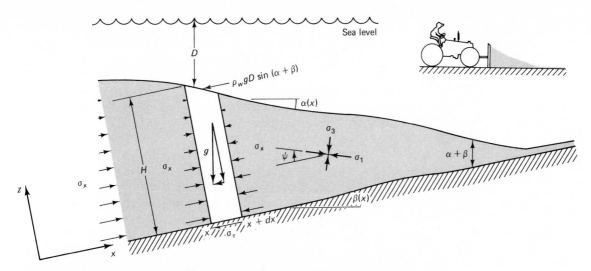

FIGURE 8–42 Schematic cross section of an active fold-and-thrust belt or accretionary wedge, showing the stresses and component of gravity acting on a small element of the wedge in the x-direction. (From Davis, Suppe, and Dahlen, J. Geophys. Research, v. 88, p. 1153–1172, 1983, copyrighted by the American Geophysical Union.)

compressive force, F_x, at any point x in the wedge is just the compressive strength integrated over the thickness H, which is just Equation 8-11 of the Hubbert-Rubey problem (Fig. 8-39)

$$F_x \cong \int_0^H [C_0 + \sigma_z (K + \lambda (1 - K))]dz \qquad (8\text{-}19)$$

Here, however, we are trying to determine the wedge shape to which the rocks deform for thrust sheets longer than the Hubbert-Rubey maximum. The deformed mass must be tapered—that is, H decreases toward the front—because the total resisting force along the base decreases toward the front of the mass. The tapered wedge is therefore a result of the geometry of the problem and the fact that the material is at failure. A general wedge shape develops for a variety of constitutive relations, including viscous, plastic, and Coulomb. Chapple (1978) developed a mathematical model of a wedge in compression for perfectly plastic material and applied it to fold-and-thrust belts and accretionary wedges. In the following paragraphs we outline a theory of the critical taper of an accretionary wedge of Coulomb material (Davis, Suppe, and Dahlen, 1983). This is the appropriate theory to apply to most fold-and-thrust belts and accretionary wedges in the upper 10 to 15 kms of the crust where silicate and carbonate rocks are brittle (for example, Fig. 6-14).

The geometry of the problem of the Coulomb wedge is shown in Figure 8-42. The topographic slope is designated by α and the dip of the décollement by β. The stress everywhere within the wedge is approximated as that given by the Coulomb fracture strength (Eqs. 5-11 and 8-8), because the wedge is assumed to be at failure everywhere:

$$(\sigma_1 - P_f) = C_0 + K(\sigma_3 - P_f) \qquad (8\text{-}20)$$

where $\sigma_3 \cong \rho gz$. In the present simplified theory, we ignore the effect of cohesive strength, C_0, which affects the shape of the wedge only in secondary ways (Dahlen, Suppe, and Davis, 1984). If the wedge is sliding along its base at equilibrium, then the sum of the forces acting in the x-direction on any small element of the wedge between x and $x + dx$ in Figure 8-42 is zero:

$$F_g + F_w + F_f + F_x + F_{x+dx} = 0 \qquad (8\text{-}21)$$

This is the fundamental equation governing the mechanics of the wedge. The forces are as follows: F_g is the gravitational body force in the x-direction, $-\rho g H \, dx \sin \beta$; F_w is the weight of the overlying water acting in the x-direction on the top of a submarine accretionary wedge (Fig. 8-42), $-\rho_w g D \, dx \sin(\alpha + \beta)$, where ρ_w is the density of the water. For simplicity, we ignore the effects of this water in the following equations; it makes a slight difference in wedge taper, as shown in Figure 8-43 (compare subaerial and submarine cases). Also, F_f is the frictional resistance of the basal décollement between x and $x + dx$, which is $-\mu_f(1 - \lambda_f)\rho g H \, dx$ (compare with Eq. 8-10 of the Hubbert-Rubey problem). All three forces just enumerated have a negative sign, indicating that they resist forward motion of the wedge. The sum of the last two forces $(F_x + F_{x+dx})$ acting on the front and back faces of the small element shown in Figure 8-42 is the net driving force and balances the resisting forces (Eqs. 8-19 and 8-21). In the limit as $dx \rightarrow 0$, Equation 8-21 becomes

$$\rho g H \sin \beta + \mu_f (1 - \lambda_f)\rho g H \, dx + \frac{d}{dx} \int_0^H \sigma_x \, dz = 0 \qquad (8\text{-}22)$$

Once the orientation of the principal stresses defined by the angle ψ (Fig. 8-42) is determined within the wedge (see Davis, Suppe, and Dahlen, 1983, for details), σ_x may be determined from the rock strength (Eq. 8-20). The final result is an equation for the critical taper of the wedge $(\alpha + \beta)$ as a function of the strength of the wedge and the décollement

$$(\alpha + \beta) = \frac{(1 - \lambda_f)\, \mu_f + \beta}{(1 - \lambda)k + 1} \qquad (8\text{-}23)$$

where k is largely a measure of rock strength in the wedge, closely related to the earth-pressure coefficient, K, in Equations 5-11 and 8-20 but also a function of the décollement coefficient of friction (see Davis, Suppe, and Dahlen, 1983, for details).

Let us look at Equation 8-23 for the critical taper and see what it tells us about the behavior of fold-and-thrust belts. The taper $(\alpha + \beta)$ can be seen to be approximately a ratio of frictional terms over wedge-strength terms. Thus we see

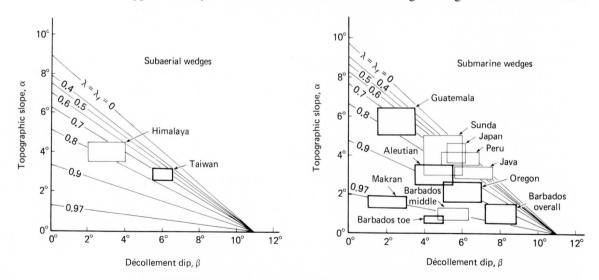

FIGURE 8–43 Surface slopes predicted for subaerial and submarine accretionary wedges, assuming $\lambda = \lambda_f$ compared with observed wedge shapes. The wedges for which fluid-pressure data are available are shown with heavy boxes, which are in good agreement with predictions. (From Davis, Suppe, and Dahlen, J. Geophys. Research, v. 88, p. 1153–1172, 1983, copyrighted by the American Geophysical Union.)

that if the basal friction increases, either by drop in fluid-pressure ratio, λ_f, or increase in friction, μ_f, then the wedge taper increases. This increase in surface slope is what we would expect intuitively to be the effect of increased friction. Alternatively, an increase in wedge strength, either by a drop in fluid-pressure ratio, λ, or an increase in k, produces a decrease in taper, as expected. Finally, it is important to note that the taper of an actively moving fold-and-thrust belt is independent of size. Equation 8-23 holds on the scale of mountain belts, as well as on the scale of bulldozers or laboratory experiments with sand or soil.

If the taper of a wedge is less than the critical value given by Equation 8-23, the material will deform as it is pushed until the critical taper is reached; then it will slide along the basal décollement without internal deformation. This property indicates an important difference in structural behavior between subaerial fold-and-thrust belts and submarine accretionary wedges. Erosion is continually reducing the taper of subaerial wedges below the critical value; therefore, they are continually deforming internally to maintain the critical surface slope, α. In contrast, submarine wedges undergo little internal deformation once their critical taper is established. Continued deformation of submarine wedges results only from new material accreted at the toe and from fluctuations in fluid pressure, friction, or rock strength.

When we consider actual rock strengths and friction coefficients, we find that surface slopes, α, of fold-and-thrust belts and accretionary wedges should lie between about 1° and 8°, which is in good agreement with observed slopes of active wedges (Fig. 8-43). We also see from Figure 8-43 that the surface slope is insensitive to fluid pressure ($\lambda = \lambda_f$) for $\lambda < 0.7$, but is quite sensitive for $\lambda > 0.8$. Wedges with a surface slope of less than 2°—such as Barbados and Makran—have extremely high fluid pressures according to Figure 8-43, which agrees with extreme pressures observed in wells. Other wedges, such as Taiwan, have only moderately high fluid pressures.

EXERCISES

8–1 What criteria may be legitimately used to distinguish between active and inactive faults for the purposes of land-use planning and engineering geology? Consider mountainous regions, alluvial plains, coastal areas, and continental margins.

8–2 Anderson's theory of faulting equates Coulomb fractures with near-surface faults. If we instead use the entire fracture envelope (Fig. 5-5), what can we predict for the change in orientation of normal, strike slip, and thrust faults with depth, still presuming a homogeneous, unlayered rock? What aspects of real structures might this explain?

8–3 In the course of making a road cut, the lateral support for several massive sandstone beds was unexpectedly removed, thereby exposing their underlying contacts, which dip 12°–15° toward the road. Will the beds slide if they become saturated with water in an episode of heavy rain, assuming a coefficient of friction of 0.6?

8–4 Prepare a structural interpretation of the cliff outcrop in Figure 13-19 in the form of a sketch showing the important faults and the structure of the beds.

8–5 Draw a map of stress trajectories in the vicinity of a schematic strike-slip pull-apart basin (compare Fig. 8-23), labeling the horizontal principal-stress directions (σ_1, σ_2, and σ_3, as appropriate).

8–6 As an accretionary wedge or fold-and-thrust belt increases in thickness toward the back, the décollement will eventually reach the depth of the brittle-plastic transition. What will be the effect of this change in material behavior on the wedge taper and on the surface slope of the mountain belt? Do presently active mountain belts show this effect?

SELECTED LITERATURE

ALLEN, C. R., 1975, Geological criteria evaluating seismicity: Geol. Soc. Amer. Bull., v. 86, p. 1041–1957.

ANDERSON, E. M., 1942, *The dynamics of faulting and dyke formation, with applications to Britain,* Oliver and Boyd, Edinburgh, 191 p.

BALLY, A. W., BERNOULLI, D., DAVIS, G. A., AND MONTADERT, L., 1981, Listric normal faults: Oceanologica Acta, Proc. 26th International Geol. Congress, Geology of Continental Margins Symposium, No. SP, p. 87–101.

BOYER, S. E., AND ELLIOTT, D., 1982, Thrust systems: American Association of Petroleum Geologists Bull., v. 66, p. 1196–1230.

BYERLEE, J., 1978, Friction of rocks: Pure and Applied Geophysics, vol. 116, p. 615–626.

DAVIS, D., SUPPE, J., AND DAHLEN, F. A., 1983, Mechanics of fold-and-thrust belts and accretionary wedges: J. Geophysical Research, v. 88, p. 1153–1172.

FERTL, W. H., 1976, *Abnormal Formation Pressures,* Elsevier, New York, 382 p.

HUBBERT, M. K., AND RUBEY, W. W., 1959, Mechanics of fluid-filled porous solids and its application to overthrust faulting: Geol. Soc. Amer. Bull., v. 70, p. 115–166.

MACDONALD, K. C., 1982, Mid-ocean ridges: fine scale tectonic, volcanic and hydrothermal processes within the plate boundary zone: Ann. Rev. Earth Planet. Sci., v. 10, p. 155–190.

MANDL, G., AND CRANS, W., 1981, Gravitational gliding in deltas: 41–54, in McClay, K. R., and Price, N. J., ed., *Thrust and Nappe Tectonics,* Geol. Soc. London, Spec. Pub. 9, 539 p.

McCLAY, K. R., AND PRICE, N. J., eds., 1981, *Thrust and Nappe Tectonics,* Geol. Soc. London, Spec. Pub. 9, 539 p.

RICE, J. R., 1980, The mechanics of earthquake rupture: Proc. Int. School Physics "Enrico Fermi," Course LXXVIII, Italian Physical Society, North-Holland Pub. Co., Amsterdam, p. 555–649.

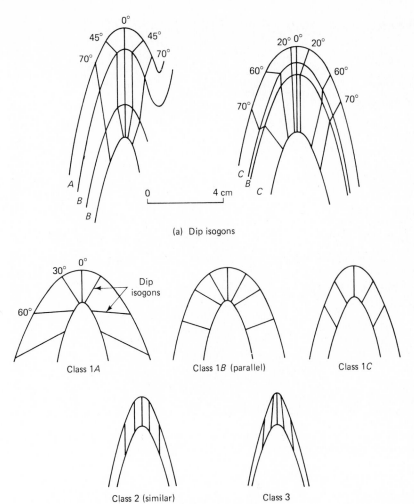

(a) Dip isogons

Class 1*A*   Class 1*B* (parallel)   Class 1*C*

Class 2 (similar)   Class 3

(b) Classification of folds in profile

FIGURE 9–17 a) Dip isogons (lines connecting points of equal dip) in two small folds in gneiss. *A* is quartz-feldspar gneiss, *B* is biotite-quartz-feldspar gneiss, and *C* is hornblende plagioclase-epidote gneiss. (From Tobisch and Glover, Geol Soc. Amer. Bull., 1971, v. 82, p. 2209–2230.) b) Classification of single-layer folds in profile based on the patterns of dip isogons. (From Ramsay, 1967.)

stratigraphic interval that is large relative to the amplitude. For example, the folded layers in Figure 9-23 exhibit a strongly harmonic stacking. In contrast, *disharmonic stacking* is characterized by a substantial change in fold shape between adjacent layers and a general uncoupling of folded layers through the stratigraphic sequence. For example, the folded layers in Figure 9-13 exhibit moderately disharmonic stacking, with a substantial change in fold shape over a short stratigraphic interval. The regular fold train in Figure 9-18 exhibits strongly disharmonic stacking; the folds are missing in nearby gneiss layers.

The degree of stacking harmony is largely a function of layer thicknesses and competence. The folding within a layered sequence is dominated by the stiffer, more-competent layers or stratigraphic intervals, whereas adjacent softer, less-competent layers conform to the adjacent stiffer folds (for example, Fig. 9-8). Disharmonic folding requires strong mechanical uncoupling between nearby stiff layers; therefore, disharmonic folding is commonly associated with stiff layers bounded by intervening weak layers that are thick relative to the competent layers.

Much folding on a map scale and some folding on an outcrop scale terminates rather abruptly in surfaces of décollement, which is an extreme form of disharmonic behavior (for example, Fig. 9-19).

The stacking of folded layers is ultimately neither harmonic nor disharmonic, but rather only relatively more or less harmonic because folds do not continue

FIGURE 9–18 Fold train in quartzo-feldspathic layer in Monson Gneiss, Quabben Reservoir, Massachusetts, northern Appalachians. (Photograph by Dave Meyer.)

forever. The most harmonic folding that is geometrically possible is that of stacked Ramsay Class 2 (similar) folds because each bed surface has exactly the same shape (for example, Fig. 9-21). Stacking of curved parallel folds (Class 1*B*) may be locally quite harmonic, as for example, in Figure 9-4, but disharmonic

FIGURE 9–19 Fold train of sedimentary slump folds. Pleistocene lake sediments, Lisan Formation, west side of Dead Sea, Israel. (Photograph by R. Hargraves.)

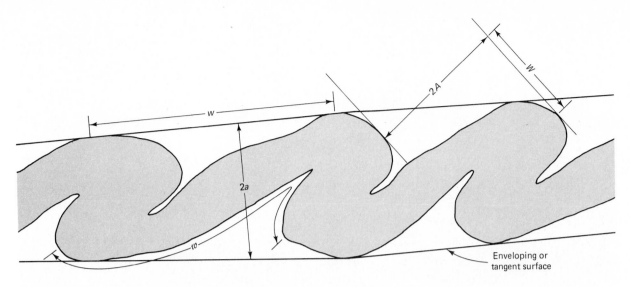

FIGURE 9–20 Several ways in which wavelength (ω, w, W) and amplitude (a and A) are measured in fold trains in structural geology. Note that, whatever method is used, the values vary along the fold. Drawing based on fold train in Figure 9-18.

aspects must exist regionally because the curvature becomes progressively higher with each successive layer toward the core of the fold.

One widespread, regular stacking geometry is an alternation of folds of Ramsay Class 3 and Class 1B or 1C geometry (Figure 9-17(a)). The more competent layers dominate the folding and form Ramsay Class 1B or 1C folds with a higher curvature on the inner bed surface. Intervening less-competent layers

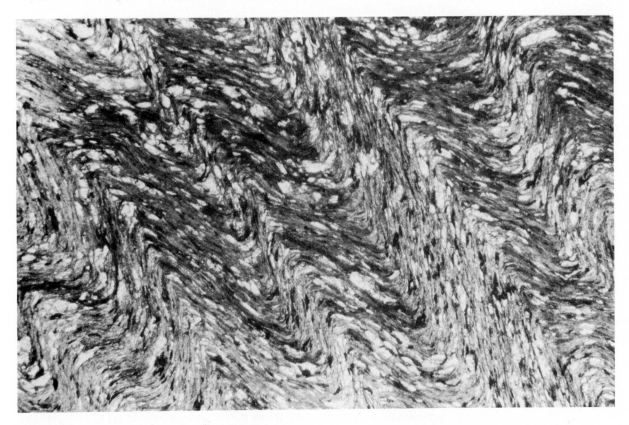

FIGURE 9–21 Regular fold trains in phyllite (photomicrograph).

FIGURE 9–22 Hierarchy of fold wavelengths displayed in gneiss, Precambrian Grenville belt, northwest Adirondacks, New York State.

that conform to the adjacent competent layers must have a higher curvature on the outer bed surface; therefore, they form Ramsay Class 3 folds. For example, the stacking of layers in the recumbent fold in Figure 9-24 is composed of an alternation of both angular and curved Class 1B and 1C folds in more-competent dolomite and Class 3 folds in less-competent limestone.

The most disharmonic folding develops in volumes of rock in which the layers do not exist originally in parallel stacks. These include rocks that contain competent dikes and veins of a variety of orientations in addition to competent and incompetent beds. Each competent layer may attempt to buckle in a different direction, producing strongly disharmonic local structure even though the overall deformation of the larger volume of rock may be simple.

Complex disharmonic structure also may develop during the redeformation of a folded and metamorphosed rock in which both bedding and cleavage are buckling, but only buckle where they are in the proper orientation with respect to the compression. These complexities lead us quite naturally to a consideration of noncylindrical and refolded folds.

Noncylindrical and Refolded Folds

Most of the preceding discussion on fold geometry has focused appropriately on cylindrical folds, because they represent the most-important and most-straightforward class of folds, those for which the least principal curvature is everywhere approximately zero and the associated least principal directions are parallel (Fig. 9-2). Here we discuss the important classes of noncylindrical folds which include: (1) noncylindrical aspects of otherwise nearly cylindrical folds, (2) basins and domes, and (3) most classes of refolded folds.

FIGURE 9–23 Angular isoclinal folds in Upper Triassic limestones and shales in western Himalayas. The ridge crest is about 6000 m elevation. Chharap Valley, Lahaul and Spiti District, Hunachal Pradesh, India. (Photography courtesy of Deba Bhattacharyya.)

1. *Noncylindrical Aspects of Nearly Cylindrical Folds.* Although most folds are approximately cylindrical, with a single well-defined fold axis parallel to the direction of least principal curvature, several important deviations from cylindrical geometry commonly exist. Folds do not extend forever with all local fold axes parallel; folds eventually terminate by a loss of amplitude with an associated change in orientation of the fold axis. The termination of a fold generally involves a plunging or change in plunge of the axis, a branching of hinge lines, and non-zero least principal curvatures.

It is common for fold axes and hinge lines in approximately cylindrical folds to have some slight curvature, but it is substantially less than the associated maximum principal curvature. The point of maximum upward curvature along the hinge line is called a *culmination,* whereas the point of maximum downward curvature of the hinge line is called a *depression.* Note that synclinal culminations and anticlinal depressions are hyperbolic points (saddle points), whereas anticlinal culminations and synclinal depressions are elliptic points. The words *culmination* and *depression* are usually applied in areas of roughly horizontal fold axes; therefore, a culmination is usually locally higher and a depression lower. Figure 9-25 shows a culmination of a map-scale anticline. Figure 9-26 shows a regional zone of culmination that involves otherwise approximately cylindrical anticlines and synclines in the Appalachian Valley-and-Ridge province of Pennsylvania.

2. *Basins and Domes.* *Basins* and *domes* are more-or-less equant folds in which the principal curvatures are of similar magnitude and neither set of principal directions is parallel over most of the structure. Basins or domes may contain spherical points of equal principal curvature.

FIGURE 9–24 Recumbent fold in Ordovician Beekmantown Group, exposed in quarry wall about 10 m high, Rheems, Pennsylvania. The more competent beds are dolomitic, whereas the less competent beds are limestone.

Basins and domes are a common fold shape in structures that do not form in response to a directed tectonic layer-parallel compression. The most spectacular include the 500-km-wide basins and domes of the continental craton—for example, the Michigan and Williston basins (Fig. 1-8). On a less-spectacular scale of 10 km or less, the continental craton exhibits low amplitude basins and domes of considerable economic importance in petroleum exploration, apparently caused by differential compaction of the locally inhomogeneous stratigraphy. The folded country rock associated with intrusion of salt, mud, and magma commonly displays a noncylindrical basin-and-dome geometry, as we discussed in Chapter 7.

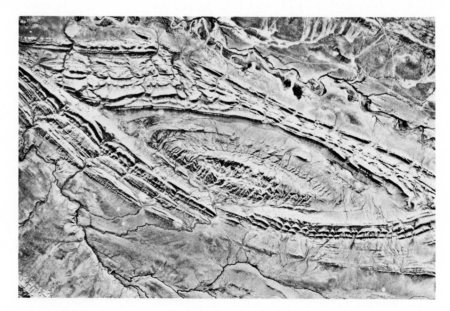

FIGURE 9–25 Culmination of doubly plunging Little Dome anticline, Wind River Basin, Wyoming.

FIGURE 9–26 Regional culmination in central Appalachian fold belt, Pennsylvania.

3. *Refolded Folds.* The most-complex regular fold shapes in nature are those that result from successive folding and refolding of the rock layers about nonparallel axes or axial surfaces; these are called *refolded* or *superposed* folds. Superposed folds may have several distinct tectonic and kinematic origins, including the following: (1) Successive unrelated deformations widely separated in time and associated with superposed orogenic belts or discrete orogenies within a single belt. Superposed folding in this case exists only in deeper levels of the crust below angular unconformities that intervene between orogenic episodes. (2) Successive related folding associated with the ongoing progressive deformation of a single orogeny. For example, several successively superposed foldings already have been produced in the ongoing Pleistocene orogeny of Taiwan. (3) Simultaneous folding in several directions during one constrictive deformation involving major compression in more than one principal direction. In this case coeval superposed folding may display different local sequences of superposition from outcrop to outcrop.

The folded surfaces that are produced through superposed folding are noncylindrical in overall geometry if successive fold axes are nonparallel; only local segments of superposed folds may be cylindrical. Therefore, cross sections of refolded folds in general do not give a complete understanding of the shape of the folded surface. Three-dimensional block diagrams and models are particularly helpful in visualizing refolded folds and considerable use is made of them in the literature (for example, Ramsay, 1967; O'Driscoll, 1962; Julivert and Marcos, 1973; Weiss, 1959).

Our visualization of the refolded shapes of originally cylindrical folds can be helped considerably if we systematically consider the possible geometries of refolding. In doing this we use a classification of refolding geometries based on the angle between the two sets of fold axes, α, and the angle between the directions of layer deflection, β, where the *direction of layer deflection* is defined to be the line within the axial surface that is perpendicular to the fold axis (Fig. 9-27). This classification of the interference patterns of regular superposed folding is that of Ramsay (1967), but with a slight change in definition of β. The angles α and β can vary between 0° and 90°, giving rise to four end-member interference types, as well as intermediate geometries, as follows:

(a) *Type 0.* Type 0 is a degenerate case in which the folds of both generations have parallel layer deflection ($\beta = 0°$) and parallel fold axes ($\alpha = 0°$). The old folds are simply amplified in the second compression. No superposed folds form. Type 0 refolding (amplification) is sometimes identified in map-scale structures by the presence of an intervening unconformity that demonstrates the amplification of a preexisting fold.

(b) *Type 1.* Type 1 corresponds to parallel layer deflection ($\beta = 0°$) and perpendicular fold axes ($\alpha = 90°$), giving rise to a basin-and-dome interference geometry such as that shown in Figure 9-28.

(c) *Type 2.* Type 2 corresponds to perpendicular layer deflection ($\beta = 90°$) and perpendicular fold axes ($\alpha = 90°$), as shown diagrammatically in Figure 9-29. An example from Massachusetts in the Appalachian mountain belt is the refolded Oak Hill syncline shown in Figure 12-16(c).

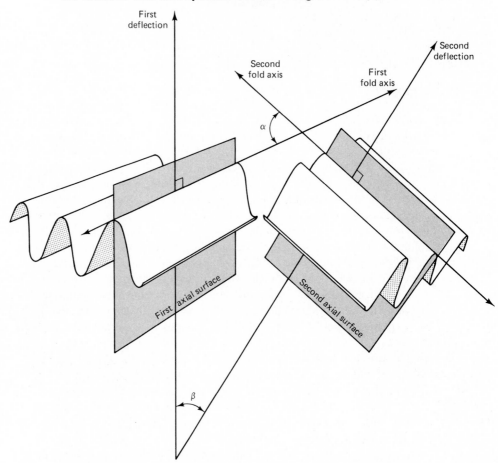

FIGURE 9–27 Definitions of angles α and β used to describe geometric classes of regularly refolded folds, based on classification of Ramsay (1967). The angle between the two sets of fold axes is α. The angle between the two directions of "layer deflection" is β, where the "layer deflection" is defined to be the line within the axial surface that is perpendicular to the fold axis.

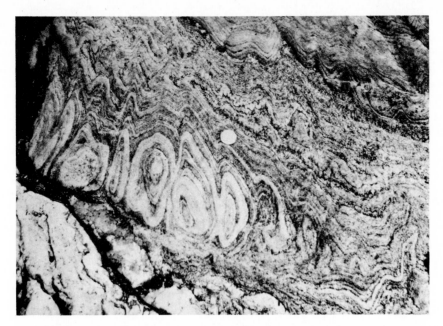

FIGURE 9–28 Type 1 basin-and-dome interference folds Loch Monar, Scotland. (Photograph by Jane Selverstone.)

(d) *Type 3.*   Type 3 corresponds to perpendicular layer deflection ($\beta = 90°$) and parallel fold axes ($\alpha = 0°$), as shown diagrammatically in Figure 9-29. Both Types 0 and 3 are called *coaxial refolding*.

Most natural examples of regular superposed folds represent intermediate types and not the precise end members listed above; furthermore, their appearance in map view, cross section, or outcrop depends greatly on the orientation of the cut through the structure. These facts may suggest a forbidding complexity to the study of refolded folds; but, in fact, we may quickly identify the broad type of refolding situation in most cases based on the traces of layering and axial surfaces in arbitrary cuts, as discussed in the following paragraphs. With further knowledge of the orientations of fold axes and axial surfaces, the complete refolding situation may be identified.

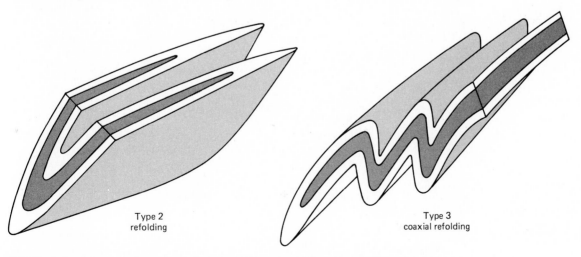

Type 2
refolding

Type 3
coaxial refolding

FIGURE 9–29 Schematic diagrams of Type 2 and Type 3 refolding classes. These two fold classes are called anticurvate because second-generation axial surfaces cut the same bed twice such that one hinge is anticlinal and one is synclinal on the same bedding surface.

The most-fundamental distinction in refolding geometries is between Types 0 and 1 and their intermediates, which we shall call concurvate types, and Types 2 and 3 and their intermediates, which we shall call anticurvate types. *Concurvate refolding* is characterized by second-generation hinge lines that are everywhere of the same sign of maximum curvature; that is, a second-generation hinge line is everywhere anticlinal or everywhere synclinal (for example, Fig. 9-28). In contrast, *anticurvate refolding* is characterized by second-generation hinge lines that change sign of maximum curvature every time they cross a first-generation hinge line; that is, a single second-generation hinge line is in some places anticlinal and in other places synclinal (for example, Fig. 9-29, Type 2).

Concurvate (0 to 1) and anticurvate (2 to 3) refolding can be distinguished in most arbitrary cuts because the traces of second-generation axial surfaces cut across the same layer more than once with a change in sign of curvature in anticurvate refolding, but only once in concurvate refolding. For example, in Figure 9-30 we see the axial surfaces of second-generation folds cutting across both limbs of an earlier isoclinal fold; therefore, the refolding is anticurvate. If we can measure the angle between the two fold axes in the vicinity of a second-generation hinge line, we can discriminate within the range of Types 2 to 3. A more-complex example involving several generations of refolding is shown in Figure 9-31, the analysis of which is left as a study problem.

The type of refolding pattern that develops depends largely on the orientation of the new compression relative to the earlier structure. If the first and second compressions are largely parallel, Type 0 refolding (amplification) results. If the second compression is nearly parallel to the first fold axis, Type 1 refolding develops if the first folds are relatively open, whereas Type 2 refolding develops if

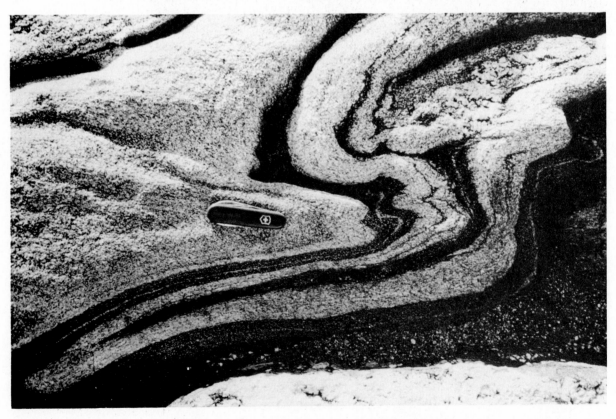

FIGURE 9–30 Anticurvate refolding in the Monson Gneiss of the Bronson Hill anticlinorium, northern Appalachians. Quabben Reservoir, Massachusetts. (Photograph by Dave Meyer.)

FIGURE 9–31 Anticurvate refolded folds in marble, Central Mountains, Taiwan.

the first folds are tight to isoclinal; the boundary between Type 1 and 2 refolding is not gradational. If the second compression is nearly parallel to the first layer deflection, Type 3 refolding develops if the first folds are tight to isoclinal; however, if the first folds are open, they may deamplify. In summary, rather open folds can be refolded into Type 1 concurvate structures, whereas tight to isoclinal folds can be refolded into Type 2 to 3 anticurvate structures.

Superposed folds are relatively uncommon in little-metamorphosed rocks, especially at an outcrop scale, but they are important in some regions on a map scale. For example, in the Cantabrian zone of northwest Spain, north-south trending folds related to thrust faults have been refolded in a major east-west trending synclinorium, producing Type 1 and Type 2 interference structures (Fig. 9-32).

Superposed folds in metamorphic terrains are common, indeed the rule, at both an outcrop and a map scale. Examples of map-scale refolding in metamorphic terrains are given in Figures 2-6, 10-26, 12-13 and 12-16(c). Superposed folds and related metamorphic fabrics provide an important chronologic key to deciphering the broader geologic history of metamorphic terrains, as discussed in Chapter 10, "Fabrics." Three, four, or more generations of folding can be recognized in many metamorphic terrains.

SOME SPECIAL CLASSES OF FOLDS

Some kinds of folds are so distinctive, yet widespread, that they should properly be described as separate special classes of folds. For example, the domal blisters

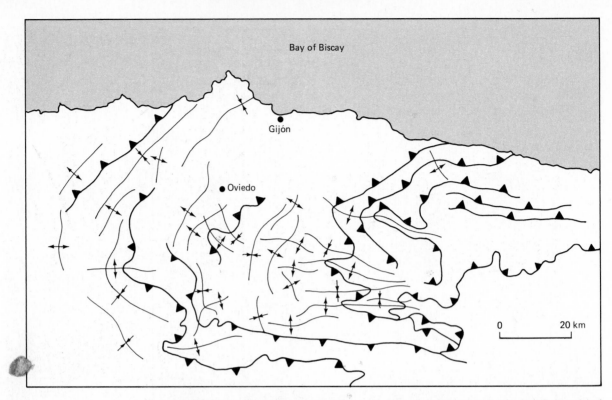

FIGURE 9–32 Map-scale fold-interference structures in the Cantabrian zone (late Paleozoic) of northwest Spain. (Simplified after Julivert and Marcos, 1973.)

above laccoliths are so distinctive in shape and separate in origin from other folds that they were treated in Chapter 7. In this section we describe three distinctive and well-known special classes of folds: kink bands, chevron folds, and fault-related folds.

Kink Bands

A *kink band* is a discrete band passing through the rock, within which the layers have been rotated with respect to their orientation outside the band; for example, Figure 9-33 shows a kink band passing through a laminated dolomite. Many kink bands are lense-shaped, such as the one shown in Figure 9-34. The kinked material is bounded by two nearly parallel axial surfaces, which are called the *kink-band boundaries* (Figure 9-35). The boundaries are typically very sharp relative to their thickness, hence the name kink; however, some have distinctly rounded hinges and diffuse boundaries.

Kink bands are most commonly formed at relatively low temperatures in relatively regularly laminated materials. They have been created in the laboratory at low temperatures in deformation of slates and phyllites by compression roughly parallel to their layering. They have also been produced experimentally with laminated paper, metals, and plastics. Natural kink bands develop in some tree bark, such as cedar where the bark is placed under compression where limbs branch. Kink bands also develop in plastic deformation of crystals, as discussed in Chapter 4 (Fig. 4-21).

Natural and man-made kink bands can develop in two orientations relative to the stress and layer orientation, with the two orientations displaying opposite senses of rotation of the layers within the band—clockwise and counterclockwise (Fig. 9-35). A pair of kink bands is called a *conjugate kink band*. Conjugate kink bands allow us to determine the approximate orientation of the principal stresses

FIGURE 9–33 Kink band in laminated dolomite, Ordovician, Beekmantown Group central Appalachians, Annville, Pennsylvania.

in a way analogous to determination of stress orientation from conjugate fault or Coulomb fracture orientations (Fig. 8-34). The line of intersection of the two kink orientations is approximately parallel to σ_2. The bisector of the conjugate pair that is approximately parallel to the layering (direction of shortening) is σ_1, whereas the bisector that is approximately perpendicular to the layering is σ_3.

The geometry of kink bands is simple relative to the complex shapes of many other folds. For this reason their shapes can be described relatively easily and in a way that gives some insight into their kinematic origin. Let us make a simplified geometric description of kink bands, ignoring their size and any curvature they may exhibit, describing their shape solely in terms of two angles (Fig. 9-35), γ, the *external kink angle* between the undeformed layering and the band boundary ($0° < \gamma < 90°$), and γ_k, the *internal kink angle* between the rotated layering and the band boundary ($0° < \gamma_k < 180°$). One might wish to distinguish between right- and left-handed kink bands (clockwise and counterclockwise), but in the following discussion we shall not.

There exist more data on the geometry of kink bands than for any other kind of fold, principally because of their simple shapes and because of Anderson (1968), who studied kink bands of the Ards Peninsula, Ireland. Figure 9-36 is a graph showing the pairs (γ, γ_k) for 526 kink bands measured by Anderson. We notice that not all geometrically possible kink shapes exist. The most common shape has an external kink angle, γ, of 50° to 65° and an internal kink angle, γ_k, of 75° to 85°.

Let us now consider some of the possible kinematic origins of kink bands. Figure 9-36 has the property that undeformed rock is represented by line for which $\gamma + \gamma_k = 180°$; that is, the layering in and outside of the kink band is everywhere

FIGURE 9–34 Lense-shaped kink band in Ordovician slate, New World Island, northern Newfoundland.

parallel. Therefore, every kink band had to start initially somewhere along the line $\gamma + \gamma_k = 180°$ and travel away from it along some path, finally stopping at the point representing its present shape. There are, of course, an infinite variety of possible paths that a given fold might take, but there is a particular path that in fact was taken, which is governed by the mechanics of kink-band formation. Let us consider two specific kinematic paths of kink formation.

The kinematics of kink formation during plastic deformation in crystals is best studied (Chapter 4). During plastic deformation, the kinking crystal under-

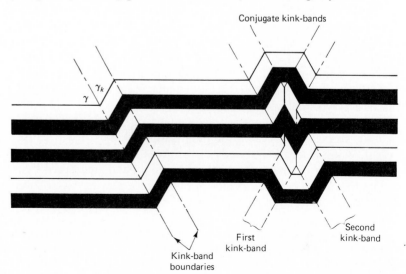

FIGURE 9–35 Geometry of kink bands showing the kink-band boundaries and the external and internal kink angles γ and γ_k. If conjugate kink bands intersect, the younger band offsets the older bands, forming a pair of chevron folds where they intersect.

goes no change in volume, which requires that the internal and external kink angles always must be equal: $\gamma = \gamma_k$ (Eq. 2-1). Therefore, crystallographic kinks involving plastic deformation begin with $\gamma = \gamma_k = 90°$ and travel along the line $\gamma = \gamma_k$ in Figure 9-36. The crystal kink-band boundaries begin at right angles to the slip plane and slip direction and rotate through the crystal as deformation proceeds, as shown schematically in Figure 9-37(a).

We immediately see from the shapes of kink bands from the Ards Peninsula (Fig. 9-36) that outcrop-scale kink bands in rocks can be very different from kink bands in crystals; they do not lie along the constant-volume line $\gamma = \gamma_k$, but rather generally lie in the region between the undeformed line and the constant-volume line. This is the region of net increase in volume of the kink band during deformation; therefore, natural kinks in rocks generally involve an increase in volume. If we recall that one of the hallmarks of brittle behavior is an increase in volume (Chapter 5), we may consider kink bands in rocks a kind of brittle behavior.

Of the various possible dilational deformation paths that might be plotted on Figure 9-36, let us consider as an example kink formation in which the kink-band boundary is fixed in the material early in the deformation such that the external angle γ is constant. The position of the kink band might be fixed by yielding of the hinges—for example, by fracture—at an early stage in the folding. The formation of a kink band with a fixed band boundary (γ = constant) is shown schematically in Figure 9-37(b). As the kink band starts to form, it begins to increase in volume until it reaches a maximum dilation, when the layering within the band is perpendicular to the band boundary ($\gamma_k = 90°$). As deformation continues, the kink band begins to collapse, with the net dilation decreasing as the internal angle, γ_k, decreases below $90°$. Finally, at $\gamma_k = \gamma$, there is no net dilation. We note that the natural kink bands of the Ards Peninsula (Fig. 9-36) generally lie to the right of the line $\gamma_k = 90°$, indicating that whatever their deformation path, they have

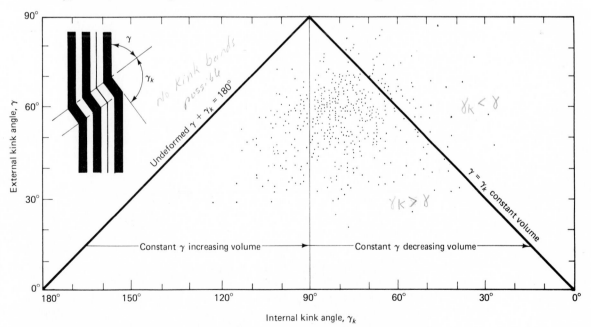

FIGURE 9–36 Measured shapes of a number of kink bands in Silurian slate, Ards Peninsula, Ireland. The triangular area below the lines $\gamma = \gamma_k$ and $\gamma + \gamma_k = 180°$ encloses all shapes that require an increase in volume within the kink band, indicating that most kinks have increased in volume. All deformation must start from the line $\gamma + \gamma_k = 180°$, which is the undeformed state. The constant volume and constant γ deformation paths are illustrated in Figure 9-37. (Data from Anderson, 1968.)

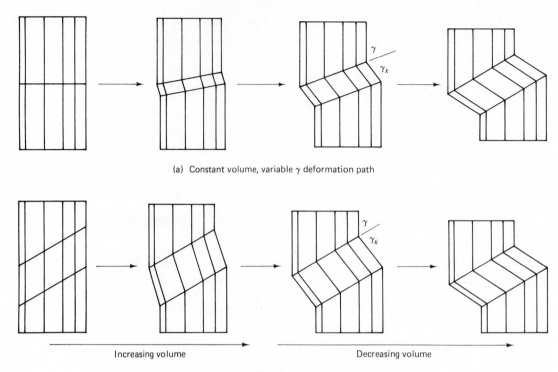

(a) Constant volume, variable γ deformation path

Increasing volume   Decreasing volume

(b) Constant γ, variable volume deformation path

FIGURE 9–37 Illustration of the constant volume and constant γ (fixed kink boundaries) paths of kink-band formation (see also Fig. 9-36).

generally exceeded their maximum dilation and started to collapse. The net increase in volume of many natural kink bands in rocks is confirmed by the common occurrence of vein material filling cracks within the kink bands.

The actual dynamics of kink-band formation in rocks is beyond our scope (see Johnson, 1977). We should note, nevertheless, that kinks are easily produced in laboratory experiments with the maximum compression, σ_1, roughly parallel to the layering (Paterson and Weiss, 1966; Donath, 1968; Weiss, 1968; Anderson, 1974). The kink bands typically form with fracturelike rapidity, involving a sudden stress drop. The deviatoric stress required for kink formation is pressure-dependent in these experiments, indicating a brittle behavior. The yield stress for kink formation can be plotted on a Mohr diagram in the same way as fracture strength. Kinks have a lower pressure dependence ($\phi \cong 15°–20°$) than Coulomb fractures ($\phi \cong 25°–40°$) and a higher cohesive strength.

In strongly deformed metamorphic terrains with polyphase folding, kink bands very commonly represent the last phase of folding and postdate the last metamorphic recrystallization. They typically are rather rare structures accounting for very little shortening. For example, in Anderson's study of kink bands of the Ards Peninsula, the 526 kink bands account for only 13 m of shortening over a distance of about 30 km. This amounts to less than a tenth of 1 percent. Even in the region of most-concentrated kink band development, the shortening amounted to only 3 percent. These very small strains are similar to elastic strains and therefore suggest that the stress was dissipated in kink formation and was not renewed. Apparently many kink bands form by the dissipation of residual stresses (Chapter 6), perhaps caused by anisotropic thermal contraction or compressibility. In contrast, tectonic loading would be expected to produce arbitrarily large strains.

Not all kink bands involve such small regional strains. For example, the outcrop in Figure 9-38 displays a high density of kink bands, rather than just a few.

FIGURE 9-38 Kink bands viewed perpendicular to layering, in Eocene slate, Central Mountains, Taiwan.

A question naturally arises: Into what fold geometry do kink bands evolve as the compression becomes very large, rather than stopping at just a few percent? This question leads us to the next special class of folds.

Chevron Folds

Kink bands and chevron folds are closely related. Both are characteristically angular parallel folds, but *chevron folds* are roughly symmetric with approximately equal limb lengths, whereas kink bands are strongly asymmetric. Some chevron folds are shown in Figure 9-21.

Chevron folds may form by several mechanisms (Johnson, 1977); one of the important mechanisms observed in rocks and laboratory experiments is the interference of conjugate kink bands (Paterson and Weiss, 1966). If one kink band develops across a preexisting kink band, the previously kinked layers in the older band will be sheared in the second kink band (Fig. 9-35). Note that the layering in the old kink band maintains a constant orientation, but the band itself jogs as a result of shear. The important feature to note in the present context is that a small pair of chevron folds has developed as a result of the interference of the kink bands. As deformation continues by either widening of the existing kink bands or by propagation of new kink bands, the proportion of the rock containing the chevron interference folds increases. When the shortening reaches about 50 percent, most of the rock has undergone chevron folding, as is shown in Figure 9-39.

Fault-Related Folds

It is a widespread observation that some folds are closely related spatially and geometrically to faults. In some cases, the faults are simply secondary structures that form in response to folding. For example, normal faults commonly form in competent layers that are stretched at the outer arcs of folds in brittle rocks (Fig. 8-19). In other cases, however, the folds are a direct result of displacement of fault

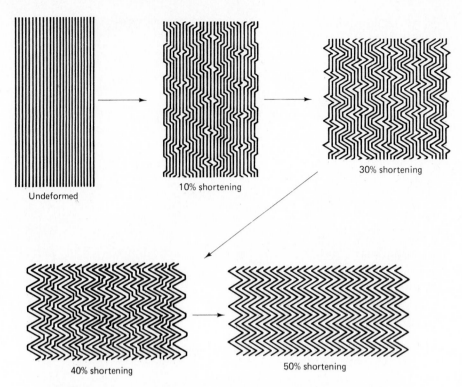

FIGURE 9–39 Schematic development of chevron folds by interference of conjugate kink bands. (From Paterson and Weiss, 1966.)

blocks. Four types are particularly important: drag folds, fault-bend folds, fault-propagation folds, and décollement folds.

1. *Drag Folds.*   The frictional force acting across a fault may exert sufficient drag to cause folds to form; these are traditionally called *drag folds*. The definition of drag folding focuses on the mechanism of formation, rather than the shape of the folds. For this reason, drag folds may be difficult to identify unambiguously in some cases. Some fault-related folds, especially fault-bend folds and fault-propagation folds, are commonly misinterpreted as drag folds. In any case, the prime necessary property for a fold to be a drag fold is that it be closely localized near the fault surface. This is required because the fault-related frictional forces decay rapidly away from the fault, as shown by the elastic-rebound theory of earthquakes (Fig. 8-36). A second necessary property of drag folds is that they show orientations and asymmetries appropriate to the orientation and sense of slip on the fault.

An example of probable drag folds is a set of folds developed in a 5-m-thick zone of highly deformed slate that lies immediately below the Champlain thrust in the northern Appalachians (Fig. 8-11). These folds have two distinctive properties that are especially worth noting: They are noncylindrical and asymmetric. The folds show a variety of orientations, but display a systematic asymmetry similar to the folds that form in a large cloth as it is pushed over a floor by hand (Fig. 9-40(a)). Such noncylindrical folds whose axes lie in a plane and whose asymmetry shows a systematic vergence are called *discon folds* (Hansen, 1971). Individual fold axes exhibit significant curvature; as a group they are strongly nonparallel, but lie roughly in a plane. If their orientations are plotted on a spherical projection, we see that they lie roughly in a plane that is parallel to the fault surface (Fig. 9-40(b)). The folds exist as asymmetric syncline-anticline pairs that have a clockwise or anticlockwise sense of rotation (S- or Z-shape) when viewed in cross section. These senses of rotation or vergence, when viewed down the

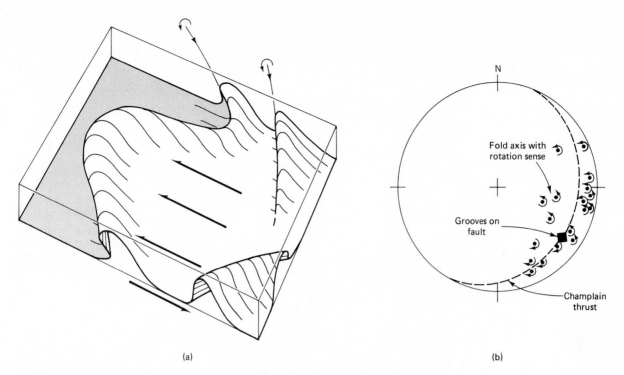

(a)   (b)

FIGURE 9–40 a) Schematic development of drag folds in a fault zone. b) Spherical projection of drag folds from just below the Champlain thrust, near Burlington, Vermont, northern Appalachians. Note that the fold axes lie fairly close to the plane of the fault. The small arrows show the sense of rotation of the folds, viewed down plunge. This sense of rotation reverses across the direction of fault slip, as shown by the direction of grooves on the fault surface (see Fig. 8-11). (After Stanley and Sarkisian, 1972.)

plunge, are indicated on the spherical projection as small arrows drawn around the fold axes (Figure 9-40(b)). The folds below the Champlain thrust that plunge to the south or southeast show a clockwise rotation, whereas those that plunge to the east or northeast shown an anticlockwise rotation. The zone of changeover between the two sets of rotations, the *separation arc,* represents the orientation of overall slip of the entire set of folds—in this case, northwest-southeast with a northwest vergence. This orientation is nearly parallel to the orientation of the large grooves on the surface of the Champlain thrust (Fig. 8-11) that record the orientation of slip on the fault, which also has a northwest vergence. These discon folds below the Champlain thrust appear to be drag folds because of their local development near the fault and their separation arc and vergence parallel to the fault slip and vergence.

2. *Fault-Bend Folds.*   Faults are not perfectly planar surfaces of slip; they generally have gentle undulations and may display substantial curvature or sharp bends (for example, Fig. 8-1). As two fault blocks slip past one another, there must be deformation in at least one fault block, because rocks are not strong enough to support large voids. For this reason many major folds in layered rocks exist within hanging-wall fault blocks, formed by bending of the fault blocks as they slip over nonplanar fault surfaces. This mechanism of folding is called *fault-bend folding.* The two most-widely recognized structural settings of fault-bend folding are (1) flattening of normal faults with depth, and (2) steps in décollement along thrust faults. It should be noted that this folding is not a result of frictional drag, but is a result of bending.

Many normal faults are listric, flattening at depth and often merging with bedding, particularly in continental margin settings such as the northern Gulf of

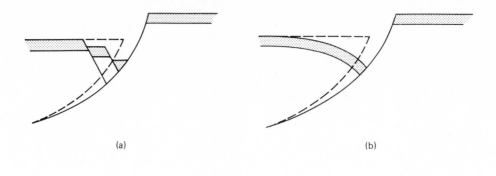

(a)   (b)

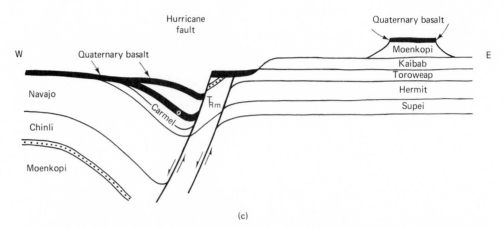

(c)

FIGURE 9–41 (a) and (b) Schematic diagram showing the development of antithetic normal faults and rollover anticlines as a result of slip on normal faults that flatten with depth. (c) Cross section of Hurricane fault, southern Utah, showing rollover anticlines. The folded Quaternary basalt shows that the folding took place during fault slip. (After Hamblin, Geol. Soc. Amer. Bull., 1965, v. 76, p. 1145–1164.)

Mexico and Niger Delta (Chapter 8). As the hanging wall moves down and away from the footwall, the steeper part of the fault tends to open up (Figure 9-41(a)), much like a bergschrund at the head of a glacier or the pull away at the head of a landslide. The hanging wall tends to collapse against the footwall where it has pulled away, creating an anticline. This kind of fault-bend fold is commonly called a *rollover,* or *reverse drag.* Rollovers are well known and documented in the Gulf Coast because they are an important site of petroleum accumulation. An example of a rollover is shown in Figure 9-41(b).

By far the best-known fault-bend folds are those associated with the stepping up of thrust faults from lower to higher décollement horizons (Fig. 9-42), which is common in fold-and-thrust belts and accretionary wedges. The basic concept of fault-bend folding associated with thrust faulting was introduced in 1934 by J. L. Rich with reference to the Pine Mountain thrust sheet (Cumberland thrust sheet) in the foreland of the Appalachian mountain belt (Fig. 9-42). The thrust sheet is approximately 175 km long parallel to strike (Fig. 8-24), with a 14-km-wide flat-bottomed syncline in the western half and a 17-km-wide flat-topped anticline in the east. Note in Figure 9-42 that the strata in the floor of the syncline are at their normal elevation, the same as their undeformed equivalents on the edge of the craton west of the surface trace of the Pine Mountain thrust. J. L. Rich recognized that the major folds of the Pine Mountain thrust sheet are the result of its translation and bending over a nonplanar fault surface. The thrust fault rides along a décollement in the Cambrian in the east and steps up to a décollement in the Devonian Chattanooga Shale. Rich recognized that the Powell Valley anticline is a fault-bend fold resulting from this 2500-m step in décollement. The west flank

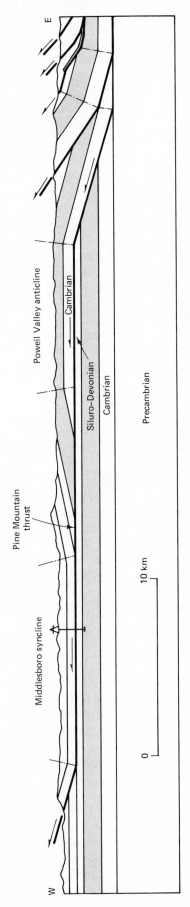

FIGURE 9–42 Cross section of Pine Mountain thrust sheet, southern Appalachians, showing the development of the Middlesboro syncline and Powell Valley anticline by fault-bend folding (compare with schematic diagram in Fig. 9-43). The location of the cross section is shown in Figure 8-24.

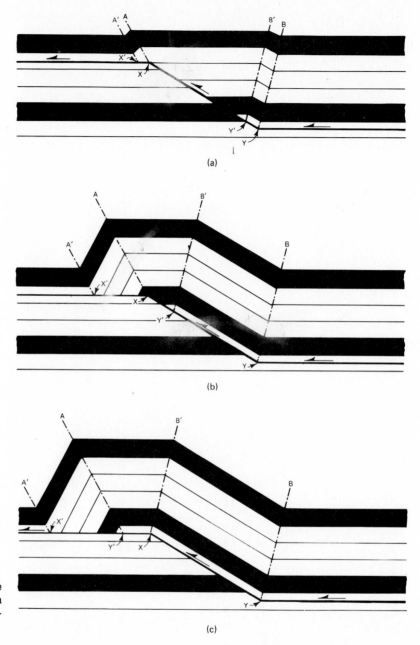

FIGURE 9–43 Schematic progressive development of fault-bend folds as a thrust sheet rides over a step in décollement. (From Suppe, 1983.)

of the syncline is formed by the bending of the thrust sheet as it steps from the Devonian décollement up to the surface (Fig. 9-42). Since this initial discovery, fault-bend folds have been widely recognized in fold-and-thrust belts throughout the world, particularly as a result of reflection seismic profiling and deep drilling in petroleum exploration.

The kinematics of fault-bend folding of a thrust sheet exhibit several properties that might not be immediately suspected, as is illustrated in Figure 9-43. At the initiation of slip, once the fault surface has propagated, two kink bands suddenly form (A-A' and B-B') in association with the two bends in the fault (Fig. 9-43(a)). Note that the folds are confined to the upper fault block and terminate at the fault surface. The axial surfaces of the kink bands terminate at the bends in the fault because it is the bends that cause the folding. Axial surfaces A and B terminate at bends in the footwall at points X and Y where the fault changes from bedding plane to cross cutting. In contrast, axial surfaces A' and B' terminate at the transitions between crosscutting and bedding-plane fault segments in the hanging wall at points X' and Y', which are the matching points of X and Y in the footwall. The behavior of folds that terminate at the footwall cutoffs X and Y is very different from the folds that terminate at the hanging-wall cutoffs X' and Y', as discussed below.

As slip on the fault continues (Fig. 9-43(b)), the kink bands A-A' and B-B' grow in width and the anticline grows in amplitude. Axial surfaces A' and B' move with the thrust sheet because they are fixed to the hanging-wall cutoffs X' and Y'. In contrast, axial surfaces A and B are fixed to footwall cutoffs X and Y, so that beds must roll through the axial surfaces, first bending, then unbending.

The kinematics of a simple step in décollement is still a bit more complex because when the hanging-wall cutoff, Y', reaches the footwall cutoff, X, the fold has reached its maximum amplitude, which is the height of the step in décollement (Figure 9-43(c)). At this point in the deformation, axial surface A, which has been fixed to the footwall, is suddenly released to move with the hanging-wall cutoff, Y', whereas axial surface B', which has been moving with the hanging wall, is suddenly locked to the footwall cutoff, X.

The characteristic ideal shape of a simple-step fault-bend fold has a steeper front dip and a more-gentle back dip, which is equal to the angle of fault dip. Most fault-bend folds in fold-and-thrust belts are more complex, particularly because of multiple imbricated thrust sheets. For example, even the relatively simple Pine Mountain thrust sheet involves two imbrications in some areas (Fig. 9-44(a)). The axial surfaces associated with lower and younger thrust faults fold the overlying thrust sheets (Fig. 9-44(b)). The final shape and complexity of the resulting structure depends greatly on the exact position and amount of slip on each fault.

Parallel fault-bend folds, such as the Pine Mountain thrust sheet (Fig. 9-42) or the schematic fault-bend fold (Fig. 9-43), have some well-defined geometric properties. In particular, there is a trigonometric relationship between the fold shape (γ in Fig. 9-45) and the fault shape (ϕ and θ). Parallel folding requires that the two axial angles are equal ($\gamma_1 = \gamma_2$ in Fig. 9-3) as discussed earlier (Eq. 2-1). If bed length and layer thickness are conserved during fault-bend folding, the relationship between fold shape and fault shape is

$$\phi = \tan^{-1}\left\{\frac{-\sin(\gamma-\theta)[\sin(2\gamma-\theta)-\sin\theta]}{\cos(\gamma-\theta)[\sin(2\gamma-\theta)-\sin\theta]-\sin\gamma}\right\} \qquad (9\text{-}7)$$

where ϕ is the angle of fault bend and θ is the initial cutoff angle. This relationship holds for both anticlinal and synclinal fault-bend folds. The final cutoff angle, β, is labeled in Figure 9-45, where

$$\beta = \theta - \phi + (180 - 2\gamma) \qquad (9\text{-}8)$$

If the fold is produced by a simple step in décollement, such as Figure 9-43, then the fault bend and initial cutoff angle are equal ($\phi = \theta$) and Equation 9-7 simplifies to

$$\phi = \theta = \tan^{-1}\left\{\frac{\sin 2\gamma}{2\cos^2\gamma + 1}\right\} \qquad (9\text{-}9)$$

Equations 9-7, 9-8, and 9-9 are useful in studying subsurface geology in areas of fault-bend folding. It should be noted that Equations 9-7 and 9-9 are double-valued for anticlinal folds; there are two possible fold shapes, γ, for a given fault shape (ϕ, θ). For example, if $\phi = \theta = 20°$, two fold shapes are possible geometrically: $\gamma = 78°$, which is relatively open, and $\gamma = 32°$, which is tight. The more open shape appears to be the only shape to form in nature. If the shape of a fold γ is known, for example, from surface mapping or dipmeter surveys, it is possible to

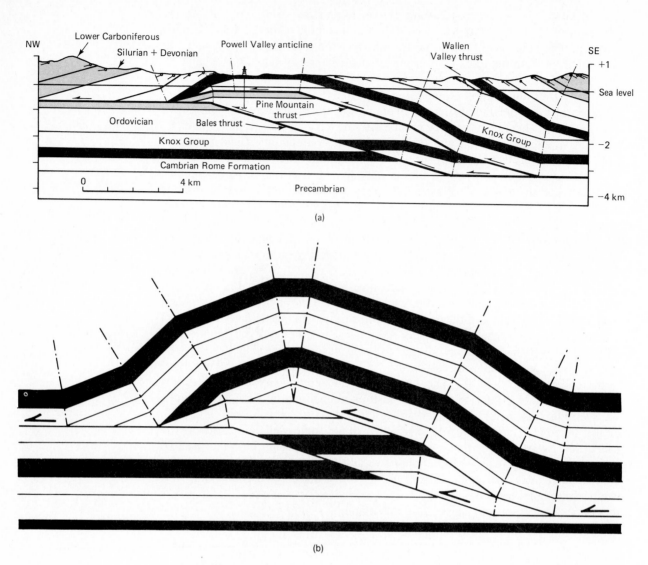

(a)

(b)

FIGURE 9–44 (a) Cross section of the Pine Mountain thrust sheet where it is deformed by an underlying imbrication on the Bales thrust. (b) Schematic cross section showing multiple imbrication fault-bend folding. (From Suppe, 1983.)

estimate the fault shape using Figure 9-45. The cross sections of the Pine Mountain thrust sheet in Figures 9-42 and 9-44(a) were solved using surface dip and stratigraphic data, the two wells, and the graphs. Examples of more complex fault-bend folding are given in Suppe (1983).

Two other special types of parallel folding are closely related geometrically to parallel fault-bend folding: folding of faults (Fig. 9-44) and folding of angular unconformities. In a purely geometric sense, there is no difference between folding of a fault block through slip on a nonplanar fault and folding of a fault block in place. Similarly, there is no geometric difference between the angular discordance of the cutoff angles that is produced by slip on a crosscutting fault and that produced by folding of an angular unconformity. Therefore Equations 9-7, 9-8, and 9-9 apply equally to fault-bend folding, folding of faults, and folding of angular unconformities.

3. *Fault-Propagation Folds.* Some folds form as part of the process of fault propagation; they represent deformation that takes place just in front of the

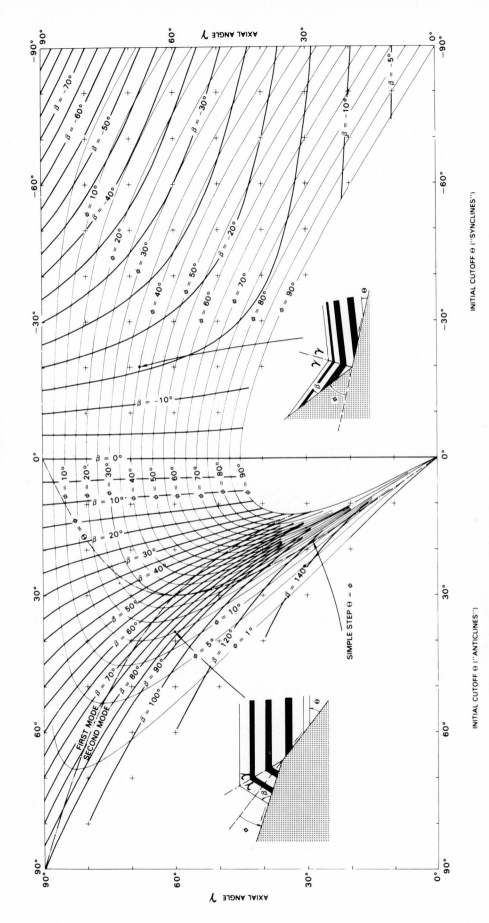

FIGURE 9–45 Graph giving relationship between fold shape γ and fault shape θ, φ for parallel angular fault-bend folds (Eqs. 9-7, 9-8, 9-9). (From Suppe, 1983.)

349

propagating fault surface. Important field evidence for fault-propagation folds is the observation that some faults, particularly thrust faults, die out in the cores of folds. Some of the major thrust faults of the Canadian Rockies die out along strike in the cores of folds, which in turn die out a short distance beyond the end of the fault. Fault-propagation folds are occasionally observed in outcrop; Figure 9-46 is a photograph of a fault-propagation fold in the southern Appalachians. Note that the thrust fault terminates at the syncline and that all the fault slip is consumed in the folding.

The growth of a fault-propagation fold associated with the stepping up of a thrust fault is shown schematically in Figure 9-47. As the propagating fault tip begins to diverge from the bedding-plane décollement and step up through the section, two kink bands, A-A' and B-B', immediately form (Fig. 9-47). The fold shape is superficially similar to a simple-step fault-bend fold (Fig. 9-43). Axial surface B is pinned to the footwall cutoff and beds roll through it. Axial surface A' terminates at the fault tip and beds roll through it as the fault propagates. Axial surface AB', formed by the merging of A and B', moves with the velocity of the thrust sheet and is the only axial surface fixed in the material. A fault propagation fold similar to the schematic fold is shown in Figure 9-48. A more-complex fault-propagation fold is shown in Figure 2-30.

Commonly, fault-propagation folds will become locked—for example, because the bending resistance of some formation may be too great. In this case the fault may propagate along the anticlinal or synclinal axial surfaces (A and A' in Fig. 9-47) or somewhere in between. This process is apparently the cause of tight synclines below many thrust faults, as shown in Figures 9-49 and 13-25.

FIGURE 9–46 Photograph of outcrop-scale fault-propagation fold. The thrust fault dips gently to the right and terminates at the base of the syncline. The anticline-syncline pair consumes the slip on the fault. Upper Carboniferous Gizzard Group near Dunlap, Tennessee.

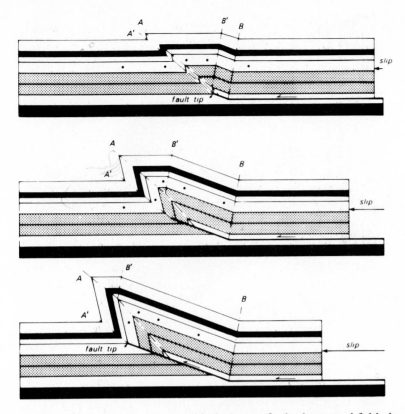

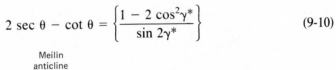

FIGURE 9–47 Schematic progressive development of a fault-propagation fold at the tip of a thrust fault.

A trigonometric relationship between fault shape and fold shape for angular fault-propagation folds can be derived, which is analogous to the relationship for fault-bend folding (Eqs. 9-7, 9-8, and 9-9). In the case of a simple step-up of a thrust sheet (Fig. 9-47) with conservation of bed length and layer thickness, we have

$$2 \sec \theta - \cot \theta = \left\{ \frac{1 - 2 \cos^2 \gamma^*}{\sin 2\gamma^*} \right\} \qquad (9\text{-}10)$$

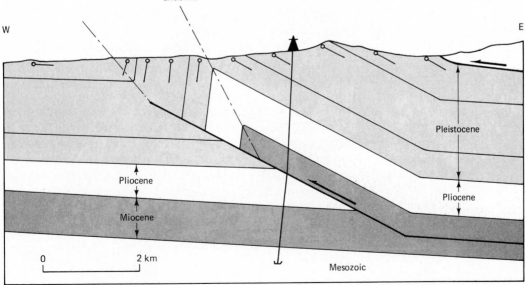

FIGURE 9–48 Cross section of a fault-propagation fold similar to the schematic diagram in Figure 9-47. Meilin anticline, western Taiwan.

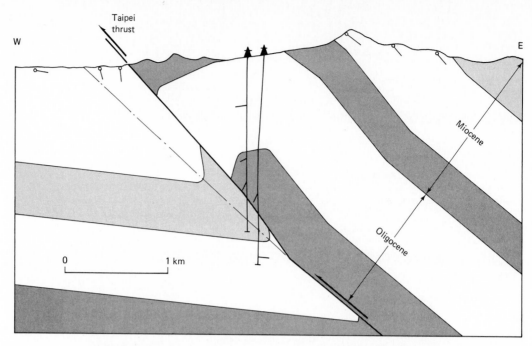

FIGURE 9–49 Cross section of broken fault-propagation fold along Taipei thrust, western Taiwan.

where $\theta = \phi$ is the angle of fault step-up and γ^* is the axial angle of the fold, as shown in Figure 9-50. Equation 9-10 is graphed in Figure 9-50, together with the simple-step fault-bend fold equation, 9-9. Let us compare the geometries of these two types of fault-related folding.

Two important properties should be noted. First, parallel folding is possible to much-higher angles of fault step-up for fault-propagation folds ($\theta = \phi = 60°$) than fault-bend folds ($\theta = \phi = 30°$). Second, only one shape, γ, of fault-propagation fold is possible for a given step-up angle $\theta = \phi$, in contrast with the two geometrically possible shapes of fault-bend folds. Natural folds associated with thrust step-up appear to be generally either open fault-bend folds or tight fault-propagation folds.

4. *Other Fault-Related Folds.* There are several other types of fault-related folds that should be noted here. It is common for sedimentary strata that unconformably overlie old faults in the basement to deform by flexure during episodes of reactivation of the basement faults. Folds that are formed by flexure over a buried fault are called *drape folds*.

Folding of sedimentary layers over reactivated basement faults is widespread in the Colorado Plateau of the western United States. Here many of the folds are *monoclinal flexures,* which are anticline-syncline pairs in which the beds are horizontal—but at different elevations—on opposite sides of the fold. At the Grand Canyon these monoclines can be traced downward through the Paleozoic section and are seen to correspond to reactivated faults in the Precambrian basement, exposed at the bottom of the canyon. A photograph of a typical monocline is shown in Figure 9-51.

BOUDINAGE

Boudinage is a structure in which a competent layer that is embedded in an incompetent material is segmented as a result of stretching parallel to its length. For example, a boudinaged layer of granitic gneiss is shown in Fig. 9-52, in which

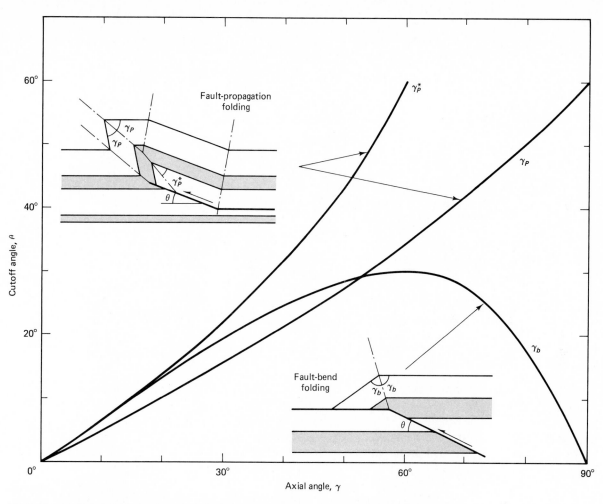

FIGURE 9–50 Relationship between fold shape γ and γ^* and fault shape $\theta = \phi$ for parallel-angular fault-propagation folds (Eq. 9-10). The equivalent relationship for fault-bend folds (Eq. 9-9) is shown for comparison.

FIGURE 9–51 Monocline in Colorado Plateau, northeastern Arizona. The beds are flat lying in the foreground and at the top of the mountain but at different elevations. They are connected through the monoclinal flexure.

FIGURE 9–52 Boudinaged layer of granitic gneiss in quartzo-feldspathic biotite gneiss. Precambrian Grenville belt, northwest Adirondak Mountains, New York State.

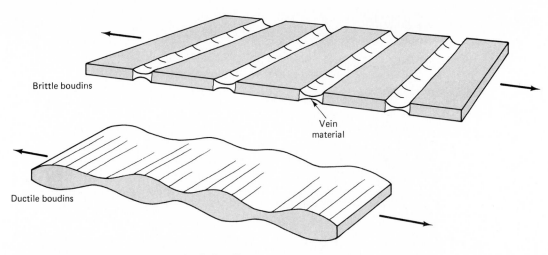

Brittle boudins

Vein
material

Ductile boudins

FIGURE 9–53 Schematic brittle and ductile boudins.

the competent layer has necked, but has not segmented to the point of separation. The name for this necking structure derives from the French *boudin,* meaning *blood sausage*. However, boudins generally do not have the shape of a sausage in three dimensions; they are usually long narrow strips of rock that have a sausage shape only in cross section (Fig. 9-53). If the rock layer has been stretched in two directions, the boudins may be tablet-shaped instead of forming strips.

At high metamorphic grade, when both the competent and incompetent layers are ductile, the boudins form with wavelike regularity, as in Figure 9-52. In fact, boudins of this type are sometimes called *inverse folds* and have a mechanics closely related to the mechanics of folding in ductile rocks (Smith, 1975, 1977). The competent layer is extended parallel to its length in boudinage, in contrast with folding, in which the layer is compressed parallel to its length.

At lower temperature, when the competent layer is brittle but the incompetent layer is ductile, tensile fractures (joints) form in the competent layer. As extension continues, vein material commonly fills the fissures if the incompetent material is unable to flow into the fissure. An example of this more brittle boudinage is given in Figure 9-54, in which dolomite forms the competent layers, limestone forms the incompetent layers, and calcite fills the veins.

In the preceding sections of this chapter, we have been concerned with the shapes and, to some extent, the kinematics of natural folds in rocks. We now move on from the geometric description of folds to explore the important physical mechanisms by which folds nucleate and grow.

MECHANISMS OF FOLDING

Introduction

Three general physical mechanisms are responsible, singly or in combination, for the majority of folds we observe in rocks. They are (1) buckling, (2) bending, and (3) passive amplification, as illustrated in Figure 9-55. If a compressive force is applied parallel to rock layers, they normally will not shorten indefinitely parallel to the length of the layers but will be deflected perpendicular to the layering. This transverse deflection in response to layer-parallel compression is called *buckling*. In contrast, *bending* is the transverse deflection of layers in response to an applied transverse force couple (Fig. 9-55). The vertical displacement of horizontal beds above a laccolithic intrusion is an example of bending (Fig. 7-19). Finally, if a

FIGURE 9–54 Boudinaged dolomitic layers interlayered with less competent limestone, showing calcite veins filling the fissures (see also Figs. 5-1 and 9-24). Ordovician Beekmantown Group, central Appalachians, Rheems, Pennsylvania.

curved surface or fold already exists in a rock, it may grow by *passive amplification* solely as a result of the homogeneous flow of the rock even if the folded layers are only geometric markers and not layers in a mechanical sense (Fig. 9-55).

Specific folds are generally the result of at least one of the three major mechanisms of folding: buckling, bending, and passive amplification. Within a stack of folding layers, different mechanisms may dominate in different parts of the stack. For example, a relatively stiff competent layer may buckle at early stages in the deformation, whereas adjacent softer layers may deform by bending in response to the sideways deflection of the adjacent stiff layer. Furthermore, the dominant mechanism may change during the course of deformation. Once the folds reach high amplitude and steep limb dips, buckling and bending are of little importance; at this stage the folds commonly grow largely by passive amplification associated with relatively homogeneous strain of the rock. Passive amplification exerts an important control on the final fold shapes in highly strained metamorphic terrains.

Buckling, bending, and passive amplification are the *primary mechanisms of folding* rock layers. In addition, however, there are important *secondary mechanisms* that operate during folding and affect the shapes of the folds. These secondary mechanisms include yielding of fold hinges, crestal fauting, interference of intersecting folds, and flow of soft material into hinge zones. We discuss the important primary mechanisms in the following sections.

Buckling

Let us recall some of the basic observations we made concerning common fold shapes in our earlier description of folds in this chapter and in Chapter 2. The first important observation is that many—in fact, probably most—folds exist in rock volumes that have undergone shortening roughly parallel to the layers that are folded. The rock layers have deflected approximately transversely to the overall shortening. This observation suggests that buckling may be an important process in the formation of many folds.

Mechanisms of folding

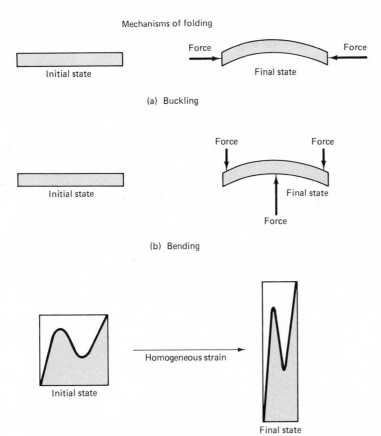

(a) Buckling

(b) Bending

(c) Passive amplification

FIGURE 9–55 Three mechanisms of folding. (a) Buckling, which is transverse deflection in response to a layer-parallel compression. (b) Bending, which is transverse deflection in response to a transverse applied force couple. (c) Passive amplification, which is the distortion of existing folds through a general flow of the rock.

A second important observation was that folds in many cases exist in reasonably regular trains for which a wavelength might be defined (Fig. 9-20). For example, the folds shown in Figures 9-1, 9-13, 9-18, 9-19, 9-21, and 9-22 all display reasonably regular fold trains. Not all these folds are characterized by a single wavelength (for example, Figs. 9-22 and 9-13); instead, several wavelengths are superposed, reflecting the control of layer thickness on fold wavelength. If the rock layering exhibits a hierarchy of layer thicknesses, a hierarchy of fold wavelengths is often observed. Microscopic layers display microscopic wavelengths (Fig. 9-21), outcrop-scale layers display outcrop-scale wavelengths (Fig. 9-1), and formation-scale layers display map-scale wavelengths. This important relationship between layer thickness and fold wavelength is graphed in Figure 9-56 for a number of sedimentary folds ranging in wavelength from 10 cm to 10 km. The ratio of wavelength to layer thickness in these multilayer folds is reasonably constant at about 27. In a later section of this chapter, on mechanical theories of folding, it will be shown that theories of folding based on buckling predict that folded rock layers should exist in trains of reasonably regular wavelength. The predicted wavelength is proportional to layer thickness for a variety of buckling theories. Therefore, the observation of regular fold trains in rocks that have undergone overall shortening parallel to bedding is evidence suggesting that buckling is an important primary mechanism of folding.

Single-layer folds, such as Figures 9-1 and 9-18, are examples of single, relatively stiff layers imbedded in a softer medium; the competent layer has buckled with the adjacent matrix merely responding to the deflecting stiff layer and offering an overall resistance to its deflection. The wavelength of the stiff fold is largely a function of its thickness; single-layer folds, such as Figure 9-1,

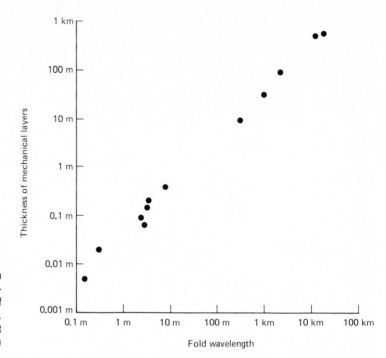

FIGURE 9–56 Relationship between fold wavelength and thickness of dominant mechanical layers for a number of multilayer folds in sedimentary strata. The wavelength-thickness ratio is about 27. (Data from Currie and others, 1962.)

commonly have a wavelength-thickness ratio of between 4 and 6, in contrast with the higher ratio for multilayer folds (Fig. 9-56).

Many naturally folded layers in rock are not imbedded in a thick, softer medium, but are close to other similarly stiff layers. How does the buckling of nearby stiff, competent layers affect each other? The effect of nearby layers is illustrated in Figure 9-57, which is a set of drawings of fold shapes produced in laboratory experiments. In Figure 9-57(a), three stiff layers are far-enough separated that they buckle independently, the middle layer being twice as thick as the outer two. As the three stiff layers are brought closer together in a series of experiments, they begin to interfere with each other because of their different characteristic wavelengths (Fig. 9-57(b)). As the three layers are brought very close together, they begin to act as a single, thicker layer, buckling with a single, longer wavelength (Fig. 9-57(c)). Figures 9-57(d) and 9-57(e) also illustrate the effect of bringing a stack of more-competent layers close together; when the layers are very close together, they act as a composite, thicker layer. These effects of nearby layers are illustrated for natural rock folds in Figure 9-58, which shows

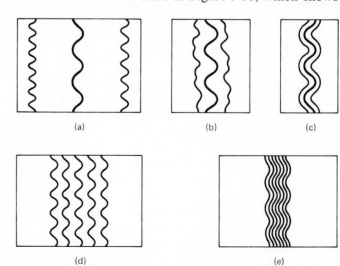

FIGURE 9–57 Buckle folds of rubber strips imbedded in gelatin, vertically compressed, illustrating the effects of layer thickness and proximity of other competent layers on fold shape. (Drawn from photographs of Currie and others, 1962.)

FIGURE 9–58 Folded limestone layers imbedded in shale showing the effect of layer proximity on fold shape (compare Fig. 9-57 (b) and (c)). Ordovician, Chaleurs Bay, Canadian Appalachians.

three stiff layers that are close enough together to fold as a unit, whereas a slightly more-separated layer has folded somewhat more independently.

We now see, particularly from the experiments in Figure 9-57, that the distinction between harmonic and disharmonic stacking of multilayer folds, which was discussed earlier in this chapter, is largely an effect of distance between stiff layers. If the stiff layers are separated by softer layers that are thick relative to the wavelength of the folds, then the stiff layers will fold independently of one another. If the soft layer is very thin or nonexistent, there will be harmonic stacking. If the soft layers are of intermediate thickness and the stiff layers have a variety of thicknesses, the folds interfere in a highly disharmonic fashion. In summary, layer thickness, relative stiffness of adjacent layers, and the details of the stacking control the gross development of multilayer buckles.

The observation of multilayer buckle folds give an important insight into the relative strength of rocks. By examining the folding of interlayered rock types, it is possible to discover which types are relatively stiffer and which are softer. Some of the common stiff/soft pairs are sandstone/shale, limestone/shale, chert/shale, sandstone/limestone, dolomite/limestone, gneisss/schist, and calcsilicate/marble. Furthermore, an overall ranking of rock competence can be made (Ramsay, 1983). Under the lowest grades of metamorphism, a typical ranking, starting with the most competent, is:

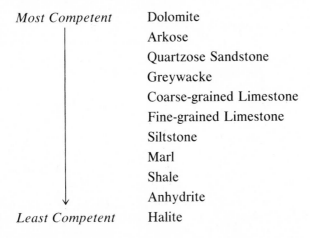

Most Competent   Dolomite

Arkose

Quartzose Sandstone

Greywacke

Coarse-grained Limestone

Fine-grained Limestone

Siltstone

Marl

Shale

Anhydrite

Least Competent   Halite

Under greenschist or lower amphibolite facies, a typical ranking is:

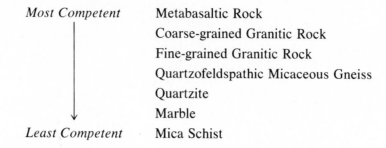

Most Competent

Metabasaltic Rock

Coarse-grained Granitic Rock

Fine-grained Granitic Rock

Quartzofeldspathic Micaceous Gneiss

Quartzite

Marble

Least Competent   Mica Schist

Bending

Bending has a mechanics closely related to buckling. They could just as well be lumped together. The main distinction is the direction and distribution of the applied compression that produces the deflection of the layers; the compression is transverse to the deflection in buckling, but a force couple is applied parallel to the deflection in bending (Fig. 9-55). Horizontal orogenic compressive shortening is commonly the cause of buckling, whereas bending deflections are commonly the effects of vertically applied forces and have no single dominant cause.

Some of the important causes of bending are as follows: (1) Flexure of lithospheric plates at the edges of major gravitational loads such as mountain belts, seamounts, continental margin sedimentary deposits, and ice sheets are all examples of bending on a major scale (Chapter 1). (2) Epeirogenic warping of the continents and ocean basins is a similar example of major bending, apparently in response to differential thermal contraction of the lithosphere (Chapter 1). (3) Differential sedimentary compaction is a common cause of bending; for example, a reef, a channel sand, or a calcareous concretion undergoes much less compaction than adjacent mudstones or coal. Differential compaction causes bending of the overlying beds. (4) Intrusion of magma and salt into a sedimentary sequence commonly involves bending, as in the formation of laccoliths (Chapter 7). (5) Fault-bend folding and drape folding are major mechanisms of bending.

Passive Amplification

We have seen in earlier sections of this chapter that unmetamorphosed sediments that have deformed under largely brittle conditions display parallel folding. The rock layers have deformed by buckling, bending, and frictional sliding with little internal distributed flow.

Rock layers begin to display nonparallel folding only as they begin to show microscopic evidence of important plastic or viscous deformation and metamorphism. What is the primary cause of the nonparallel metamorphic fold shapes? Are the shapes solely the result of buckling of viscous or plastic layers or are the nonparallel shapes largely a result of overall flow of the rock, that is, passive amplification (Fig. 9-55)? The available evidence suggests that passive amplification is an important process in producing metamorphic fold shapes; three observations are given.

1. It has been shown that the commonly encountered nonparallel metamorphic fold shapes (Fig. 9-17) and their associated patterns of strain can be produced geometrically by the superposition of a homogeneous strain on a parallel buckle-fold shape. For example, Figure 9-59 shows an initial parallel concentric buckle fold. After superposition of a 50 percent

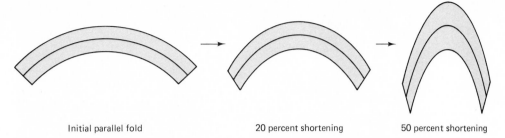

Initial parallel fold   20 percent shortening   50 percent shortening

FIGURE 9–59 Modification of initial parallel concentric buckle fold by superposition of homogeneous horizontal shortening. The final result is a Class 1*C* fold. (Modified after Hobbs, 1971.)

horizontal shortening, the fold has the common nonparallel shape and strain pattern of Ramsay Class 1*C* folds (see Hobbs, 1971).

2. We expect that continued growth of multilayer buckle folds involving layer-parallel slip will become more and more difficult as amplitude increases because the fold limbs become rotated to a high angle to the principal compressive stress. The shear stress acting parallel to layering reaches a maximum at a dip of 45° to the principal compression (Eq. 3-23) and thereafter becomes progressively lower, inhibiting layer-parallel slip. We can expect fold growth by layer-parallel flow or slip to become strongly inhibited as limb dips steepen relative to the direction of maximum compression. Continued shortening of the rock mass is then accomplished by bulk strain of the rock mass, with passive amplification of the existing folds.

3. Final evidence for the importance of passive amplification in the growth of folds is the observation in polydeformed metamorphic terrains that each successively older fold generation is more and more flattened, more nearly isoclinal. This observation suggests that general flow of already folded rock is a prime cause of flattened nonparallel fold shapes.

Passive amplification does not necessarily involve homogeneous flow of all the layers in the rock mass. Contrasts in layer stiffness during passive amplification of nearly isoclined folds causes the softer layers to flow from the fold limbs into the hinges and causes the stiffer layers to boudinage. For example, in Figure 9-24 we see that the stiffer dolomite layers on the limbs of a recumbent isoclined fold have boudinaged, whereas the interbedded softer limestones have flowed.

MECHANICS OF FOLDING

Introduction

The formation and growth of many geologic structures involves some kind of instability—for example, the sudden growth of a joint when the tensile strength of the rock is exceeded—as described by the Griffith theory of fracture (Eq. 5-6). Instabilities are particularly important in theories of initiation of folding, and two distinct types are involved. For this reason it will be useful first to introduce briefly the topic of instability before proceeding to mechanical theories of folding.

The familiar illustration of a ball resting on an irregular surface is given in Figure 9-60(a). The ball resting at point *A* is at a position of *stable equilibrium*. It is at equilibrium because it is not accelerating and it is stable because if you push it an infinitessimal distance to either side, it will roll back to the stable position. In

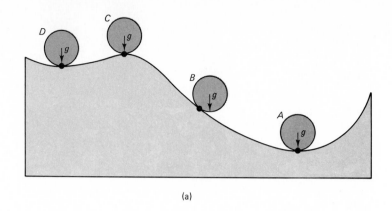

(a)

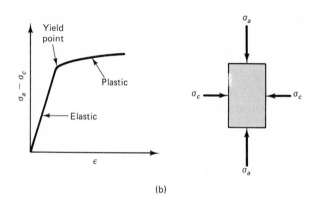

(b)

(c)

FIGURE 9–60 Illustrations of geometric and material instabilities. (a) Geometric instability: *A* is point of stable equilibrium, *B* is unstable disequilibrium, *C* is unstable equilibrium, and *D* is metastable equilibrium. (b) Material instability illustrated with yield-point transition between elastic and plastic behavior in triaxial test. (c) Instability in folding. The change from *F* to *F'* is a geometric instability whereas the change from *F'* to *F"* is a material instability.

contrast, the ball at point *B* is in a position of *unstable disequilibrium;* it will roll downhill and eventually come to rest at the stable position of point *A.* Point *C* is interesting as a point of *unstable equilibrium;* as long as the ball is unperturbed, it will stay put, but with a little shove it will roll to the stable positions *A* or *D.* The ball at point *D* is in stable equilibrium, but it is *metastable* with respect to point *A* because of its higher elevation.

This illustration of balls sitting on an irregular surface is an example of *geometric instability.* The stability or instability involves only the geometric arrangement of material. Some theories of folding, as we shall see, make use of geometric instability and predict the most stable geometric shape for a layer or set of layers in response to certain applied forces. These theories assume that sufficient initial perturbations exist, and, of a wide range of wavelengths, that only the most stable wavelength will form. Some folded shapes are more stable under layer-parallel compression than is the unfolded shape.

A second class of instabilities important in folding is *material instability,* involving a permanent change in the material behavior. The sudden yield point (Fig. 9-60(b)) marking the change in rock behavior from elastic to plastic is an example of material instability.

The roles of geometric and material instability in folding can be illustrated by compressing a strip of stiff card stock with your hands (Fig. 9-60(c)). As the compression is increased, the strip suddenly buckles, which is an example of geometric instability analogous to pushing the ball in Figure 9-60(a) from point D to point C, where it becomes unstable and rolls to a new stable position A. The strip of card stock has remained elastic during the geometric buckling instability; if the compression is released, the layer unfolds. In contrast, if we continue to compress the card stock, it will suddenly yield in a material instability, producing a permanent sharp hinge in the layer. Thus the final fold shape in this illustration is controlled by both a geometric and a material instability. Most of the theories of folding outlined below predict the shape of the layer that results from the geometric instability; it is assumed that continued deformation will lead to yielding of the rock, which will permanently imprint the folds on the rock.

Buckling of Single Layers

We introduce the mechanics of buckling initially with the theory of a single free layer unconfined by adjacent layers—for example, the card stock in Figure 9-60(c). The theory is closely related to the engineering theories of bending beams, plates, and thin shells. Consider some elastic plate of rock of thickness T and width normal to the page b under a compression F (Fig. 9-61); what are the possible equilibrium shapes that this layer may assume?

If the layer is in a state of equilibrium, then there are no unbalanced forces or torques acting on any subelement of the layer (Fig. 9-61). For example, the sum of the forces in the 1-direction is zero:

$$\Sigma F_1 = 0 \qquad (9\text{-}11)$$

Also, the sum of the torques or moments about point A in Figure 9-61 must be zero,

$$\Sigma M = \Sigma M_D + \Sigma M_R = 0 \qquad (9\text{-}12)$$

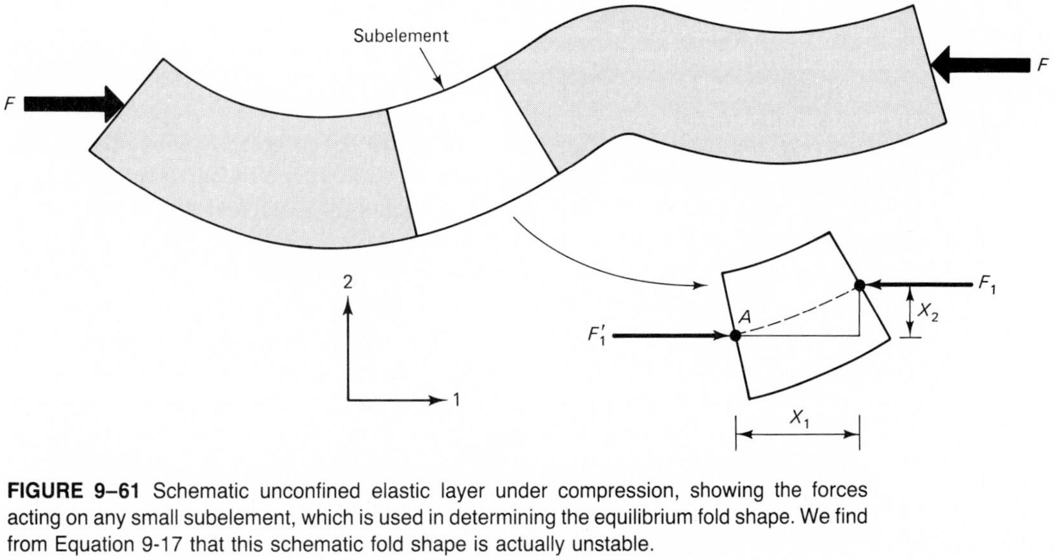

FIGURE 9–61 Schematic unconfined elastic layer under compression, showing the forces acting on any small subelement, which is used in determining the equilibrium fold shape. We find from Equation 9-17 that this schematic fold shape is actually unstable.

where there are two contributions to the moments acting on this small element: M_D, the driving moment acting on the exterior of the element, and M_R, the elastic bending moment acting within the element and resisting the bending. The driving moment in this case is the length of the moment arm (Fig. 9-61) times the force:

$$M_D = F_1 X_2 \qquad (9\text{-}13)$$

The resisting moment is the product of curvature, C, and the intrinsic elastic resistance of the material to bending, BI,

$$M_R = (BI)C \qquad (9\text{-}14)$$

where $B = [E/(1 - v^2)]$ is an elastic modulus and I is the elastic moment of inertia, which describes the distribution of elastic resistance across the layer. For example, I *beams* are used in construction because their shape has a very high elastic moment of inertia, maximizing their resistance to bending (Eq. 9-14). For the beam in Figure 9-61 $I = (bT^3/12)$.

At the beginning of the chapter we introduced curvature C (Eq. 9-1) to help us in the description of fold shapes. We now see that curvature enters explicitly into the mechanics of buckling; the resisting moment, M_R, of an elastic layer is proportional to its imposed curvature (Eq. 9-14). This curvature may also be expressed approximately as the second derivative of the deflection, X_2, as a function of position X_1 along the layer:

$$C \cong \frac{d^2 X_2}{dX_1^2} \qquad (9\text{-}15)$$

Inserting the various definitions given above into the equations of equilibrium, we obtain a single differential equation that is satisfied by any equilibrium fold shape of an unconfined elastic layer subject to a compression along its length:

$$\frac{d^2 X_2}{dX_1^2} + \frac{F_1 X_2}{BI} = 0 \qquad (9\text{-}16)$$

A solution to this equation (Johnson, 1970) is

$$X_2 = X_0 \sin \left[\left(\frac{F_1}{BI} \right)^{1/2} X_1 \right] \qquad (9\text{-}17)$$

and

$$X_2 = X_0 \sin \left[\frac{2n\pi}{L} X_1 \right] \qquad (9\text{-}18)$$

which gives the vertical deflection, X_2, as a function of position X_1 along the layer. The periodic nature of folding by buckling begins to emerge in Equations 9-17 and 9-18 through the sine function: L is the wavelength, and n is twice the wave number and may take on any integer value ($n = 1, 2, 3, \ldots$). Each value of n corresponds to a different folded state of geometric equilibrium. The most-stable shape is the one requiring the lowest force, F_1, to sustain the deflection, X_2; equating 9-17 and 9-18, we obtain

$$F_1 = BI \left[\frac{2n\pi}{L} \right]^2 \qquad (9\text{-}19)$$

Therefore, the required force, F_1, increases with wave number, and the most-stable shape is $n = 1$, which is a single arc, shown by the folded card stock in the middle of Figure 9-60(c). This mode of buckling, with $n = 1$, is very unusual in rocks because most rock layers are not free from confinement. The principal natural example in structural geology of free-layer folding is the buckling of exfoliation sheets at the surface of the earth (Fig. 6-23).

How do the mechanics of buckling change when the layer is embedded in rock? First, let us consider why the free layer (Fig. 9-60(c)) buckled at all. If the layer had not buckled, it would have shortened by an amount δl and the applied force would be

$$F_1 = ETb \left(\frac{\delta l}{l}\right) \tag{9-20}$$

As the shortening, δl, gets large, the force required for homogeneous shortening (Eq. 9-20) becomes much larger than the force required for buckling (Eq. 9-19) because the length of a buckled beam is longer than the length of equivalent homogeneously shortened beam. Buckling reduces the shortening and total elastic strain within the beam for a given overall shortening δl. This is why the free layer buckles to the side in a single arc ($n = 1$). The upper part of the layer stretches and lower part compresses, but the total elastic strain energy is less than in a homogeneously shortened layer.

Now consider the effect of a confining medium, shown schematically in Figure 9-62 as a set of springs. We now have a force resisting the vertical deflection of the layer so that the most-stable shape will be one that minimizes elastic strain energy in both the layer and the surrounding medium. Under these conditions a sinusoidal fold train is more stable than the single one-sided deflection. Solving equations analogous to 9-11, 9-12, and 9-16, we obtain an expression for the wavelength, L, of the sinusoidal fold train as a function of layer thickness, T (Johnson, 1970):

$$L = 2\pi T \left[\frac{B}{6B_0}\right]^{1/3} \tag{9-21}$$

where B is the elastic modulus of the buckling layer and B_0 is the modulus of the surrounding medium. This result provides a relationship between layer thickness T and fold wavelength L, which qualitatively agrees with field observations.

According to the elastic buckling theory, the ratio of wavelength to layer thickness is proportional to the cube root of the ratio of elastic stiffness

$$\frac{L}{T} = 2\pi \left[\frac{B}{6B_0}\right]^{1/3} \tag{9-22}$$

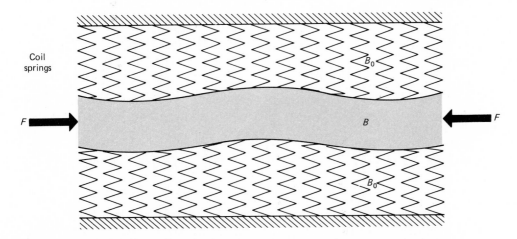

FIGURE 9–62 Schematic diagram of stiff elastic layer buckling within a less stiff elastic medium, shown diagrammatically as a set of springs. The elastic moduli of the layer and medium are B and B_0.

Such relationships are, however, not unique to elastic layers. For example, theories of viscous buckling produce an analogous result:

$$\frac{L}{T} = 2\pi \left[\frac{\eta}{6\eta_0}\right]^{1/3}$$

(9-23)

where η is the viscosity of the stiffer buckling layer and η_0 is the viscosity of the softer medium. Theories of buckling of power-law layers also yield a proportionality between layer thickness and wavelength, but in some cases an independence of competence ratio (Smith, 1979; Fletcher, 1974).

In summary, mathematical theories of geometric instability at the onset of

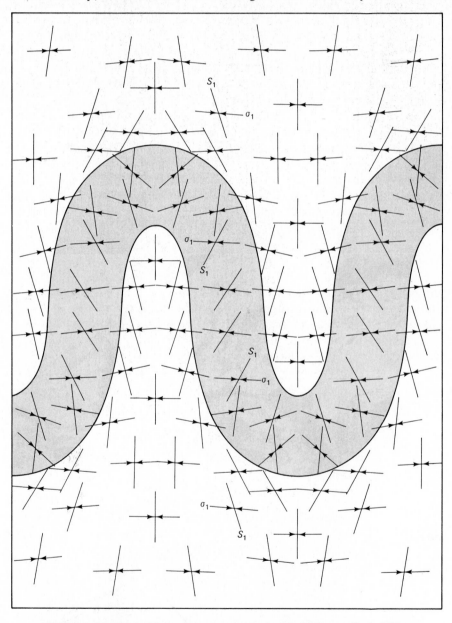

FIGURE 9–63 Result of finite-element computer model of finite-amplitude folding of a single stiff viscous layer imbedded in a less stiff viscous medium. The line segments, which are steeply dipping, are the directions of maximum finite elongation S_1; the line segments with arrows, which are generally horizontal, give the current direction of the maximum compressive stress σ_1. (Based on Dietrich, 1970.)

buckling predict some of the important properties of natural folds, including regular fold trains and a proportionality of wavelength to layer thickness.

Large-Amplitude Folding

Theories of the initiation of folding based on geometric instability include mathematical approximations that make them invalid for anything beyond shortening of a few percent. Therefore, study of the mechanics of large-amplitude folds has generally involved other methods, such as laboratory and numerical modeling. These models are particularly useful because they show the growth of analog folds through all stages in deformation and illustrate the important processes involved (for example, Johnson, 1977). Computer models are useful because they can compute the stresses as well as the finite strains within the folds at any stage in the deformation. For example, Figure 9-63 shows the results of a finite-element numerical model of viscous folding of a stiff layer in a soft medium. We see that compressive stress is concentrated at the concave inner hinges, whereas the compression is reduced in the outer hinges. The patterns of finite strain predicted by numerical models such as this may be compared with patterns of strain recorded in natural folds; they appear quite similar, as is discussed in the next chapter.

EXERCISES

9–1 Classify the photographs of folds given in this chapter according to Ramsay's classification of single-layer folds in profile.

9–2 The map in Figure 9-64 shows a region of map-scale refolded folding. Interpret the folding showing the traces of the axial surfaces of the various folds, classifying them according to fold generation. Is the refolding concurvate or anticurvate?

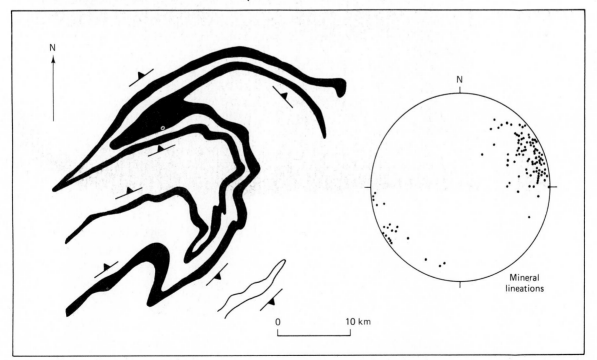

FIGURE 9–64 Map of refolded gneiss layers in Charlotte belt of southern Appalachian Piedmont, Virginia and North Carolina. Strikes and dips of foliation or schistosity are shown. Spherical projection shows the orientations of mineral lineations measured in this area. (From Tobisch and Glover, Geol. Soc. Amer. Bull., 1971, v. 82, p. 2209–2230.)

FIGURE 9–65 Photograph of folded quartzofeldspathic sandstones showing changes in layer thickness through the fold. Kii Peninsula, southwestern Japan. (Photography by W. R. Dickinson.)

9–3 Suppose the state of stress in a region is such that the effective stresses are $\sigma_1^* \leq 5$ σ_3^*, where σ_3 is vertical. Suppose also that a set of upright chevron folds are growing with their axial surfaces perpendicular to σ_1. Illustrate the state of stress on bedding as a function of dip during the growth of the folds. Suppose that growth of the folds is limited by frictional sliding according to Byerlee's law; at what limb dip will the folds become locked?

9–4 Using the theory of buckling of a single competent layer in a less-competent medium, what elastic stiffness or viscosity ratio would be predicted for the folds shown in Figures 9-1 and 9-18?

9–5 Derive Equation 9-3 for the relationship between simple shear and change in dip across an angular fold hinge.

9–6 Experimentally test Equations 9-3 and 9-5 (Fig. 9-9) using a telephone book.

9–7 Interpret the folds in Figure 9-31 showing the traces of the axial surfaces, classifying them according to fold generation. Is this refolding concurvate or anticurvate?

9–8 The sandstone layers in Figure 9-65 show changes in layer thickness. Should the folds therefore be considered nonparallel folds? Why?

SELECTED LITERATURE

JOHNSON, ARVID M., 1970, *Physical Processes in Geology,* Freeman, Cooper and Co., San Francisco, 577 p.

JOHNSON, ARVID M., 1977, *Styles of Folding,* Elsevier, New York, 406 p.

RAMSAY, J., 1967, *Folding and Fracturing of Rocks,* McGraw-Hill, New York, 568 p.

RAMSAY, J., 1983, Rock ductility and its influence on the development of tectonic structures in mountain belts: p. 111–127, in Hsü, K. J., ed., *Mountain Building Processes,* Academic Press, London, 263 p.

TURCOTTE, D. L., AND SCHUBERT, G., 1982, *Geodynamics: Applications of Continuum Physics to Geological Problems,* Wiley, New York, 450 p.

WEISS, L. E., 1972, *The Minor Structures of Deformed Rocks: A Photographic Atlas,* Springer, New York, 431 p.

10

FABRICS

INTRODUCTION

Deformation, especially under metamorphic conditions, irrevocably changes the orientation, homogeneity, and textural relationships of the mineral constituents of rock. For example, deformation may change a homogeneous and isotropic granite into a compositionally layered gneiss with strongly anisotropic mineral orientation, such as shown in Figure 10-2. As another example, the original bedding, texture, and composition of the sandstone and mudstone in Figure 10-1 has been considerably overprinted during deformation by a strong planar anisotropy and new compositional layering. Both of these examples illustrate the changes in rock fabric that accompany and record important aspects of the deformational history of rock; these changes in fabric are the subject of this chapter.

We apply the word *fabric* as a general term for the internal arrangement of the constituent particles of the rock, including texture, packing, preferred orientation, and homogeneity. The same term for cloth and tapestry implies a structure in which the constituent threads are interwoven, attached, repeated, and oriented on various scales to build up the overall design. The term for rock carries similar connotations of building up the rock through the arrangement of the constituent elements.

The fabrics of rocks reflect the primary formation of the rock and secondary processes such as deformation and metamorphism. The primary processes of fluid flow, grain settling, and crystallization build up the initial fabric at the time of formation of igneous and sedimentary rocks. Deformation and recrystallization superpose a new fabric on the rock at both a macroscopic and a microscopic scale. In this book we are naturally most concerned with fabrics produced through deformation.

FIGURE 10–1 Slaty cleavage cutting across beds of mudstone and fine-grained sandstone in the Ordovician Martinsburg Formation, central Appalachians. Figure 10-5 shows photomicrographs of this rock. Delaware River, northern New Jersey.

Deformation and metamorphism cause the development of fabric at a variety of scales. We traditionally divide this fabric into its macroscopic and microscopic parts for purposes of study simply because of the characteristics of the human eye. The *macroscopic fabric,* called *mesoscopic* by some geologists, is that part of the fabric visible on a hand-specimen or outcrop scale—for example, as seen in Figures 10-1 and 10-2. These two examples display the most-widespread and most-important macroscopic fabric phenomena, namely, cleavage, schistosity, and gneissosity, which are *planar fabrics,* commonly giving rise to repeated parallel planes of weakness within the rock. These planar weaknesses are a result of two important aspects of the deformational fabric, which will be discussed later in this chapter: (1) preferred orientation of anisotropic minerals, such as mica and amphibole, and (2) development of new compositional layering, commonly with an alternation of mica-rich layers with quartz-rich layers.

Another important aspect of many macroscopic fabrics, not shown in Figures 10-1 and 10-2, is lineations, or *linear fabrics,* which are the linear analogs of planar fabrics. Many lineations lie within the plane of an associated planar fabric (Fig. 10-3). Strong linear fabrics are not as widely developed as strong planar fabrics, but nevertheless they are important. Lineations as seen at an outcrop scale have a number of underlying causes, including oriented mineral grains, microscopic folds, intersection of cleavage and bedding or older cleavages, and stretching of particles in the rock.

Some fabrics are everywhere developed throughout a rock mass and are therefore called *penetrative fabrics.* Other fabrics are developed only on discrete surfaces or in discrete zones or areas; these are called *nonpenetrative fabrics.* These definitions depend on the scale; for example, a schistosity may be penetrative on an outcrop scale but show a discrete spacing of bands of schistosity on a microscopic scale (Fig. 10-5(a)). An example of a nonpenetrative fabric is the strong lineation developed only on the surfaces shown in Figure 10-4; it is not present throughout the rock.

Planar and linear fabrics on a macroscopic scale are very important in structural geology because they are useful in deciphering outcrop-scale to regional-scale structure. Planar and linear fabrics display close geometric relationships with coeval larger-scale folds, shear zones, and intrusions; for example, outcrop-scale lineations are commonly parallel to map-scale fold axes (Figs. 2-18 and 9-64). Therefore, the mapping of macroscopic fabrics is useful in determining

FIGURE 10–2 Foliated granitic gneiss with *augen* (German, *eyes*) of potassium feldspar. Precambrian Baltimore Gneiss, central Appalachians, Baltimore, Maryland.

FIGURE 10–3 Lineated cleavage surface in phyllite, which has been refolded. Eocene phyllite, Central Mountains, Taiwan.

the geometry of the associated map-scale structures and plays a central role in structural mapping of metamorphic terrains.

Planar and linear fabrics on a macroscopic scale also are important because of the mechanical role of these anisotropies in polyphase deformation. For example, an early planar fabric is often the layering that buckles in a later deformation rather than the primary stratigraphic layering.

The *microscopic fabric,* or *microfabric,* is that part of the fabric that is visible under the optical or electron microscope. The same features of fabric that are visible on a macroscopic scale are visible in many rocks under the optical microscope, including preferred orientation of anisotropic minerals and compositional layering of a deformational origin. However, these features are visible with such great clarity that the microscopic study of fabric is substantially different from macroscopic study. Furthermore, there are additional aspects of fabric that are visible only at a microscopic scale—for example, the shapes of grain boundaries and their angles of intersection, which record important information about the annealing history of a rock.

One of the useful properties of microscopic fabrics is that they commonly record interrelationships between the history of metamorphic recrystallization of a rock and its history of deformation. Metamorphic fabrics therefore allow the deformation of the rock, including its temperature-pressure history as determined from metamorphic mineral assemblages, to be placed into the broader history of the rock mass. For example, we can often tell from the fabric whether or not a

FIGURE 10-4 Lineated surfaces in metamorphosed dolomite. Note that this lineation is nonpenetrative. PreCenozoic, Central Mountains, Taiwan.

particular mineral-producing chemical reaction took place after a fabric-producing deformation.

MACROSCOPIC CLASSIFICATION OF PLANAR FABRICS

By far the most obvious result of plastic deformation of rocks is the widespread development of new planar fabrics. These new fabrics display a variety of structural phenomena that are used in their classification, including planar preferred orientation of platy and rod-shaped minerals, such as mica and amphibole; compositional layering, such as alternating quartz-rich and mica-rich layers; flattening of preexisting mineral grains and rock fragments; and fracture anisotropy. A general, nonspecific field term for all these planar fabrics is *foliation,* derived from *folium,* the Latin word meaning *leaf.*

In the nineteenth and early twentieth centuries, the most-important classifications of rock fabrics were largely economic in origin, based on the usefulness of different fabrics in the quarrying and shaping of rock for use in construction. The homogeneity and strength of the fracture anisotropy, or *rock cleavage,* was the main concern in classification, because in the quarrying of slate and dimension stone it was most important that the rocks be able to split consistently and uniformly in known directions to any desired size (see also Chapter 5). It was economically important that the fracture anisotropy be penetrative. The fine-grained metamorphosed mudstones that could be split smoothly along their foliation to any desired thickness to make uniform roofing shingles were said to have a slaty cleavage. The foliations of any rocks that would not produce a commercial slate were given other names such as false cleavage, fracture cleavage, close-joints cleavage, or spaced cleavage, whether or not fundamental fabric phenomena were different from those of slaty cleavage. Much of our vocabulary dealing with fabrics, especially slates, dates from these early days, so we need to look a bit more critically to see what slates actually are.

A photomicrograph of a commercial slate cut perpendicular to the cleavage shows that the foliation is associated with three phenomena (Fig. 10-5): (1) *preferred orientation* of mica and chlorite crystals parallel to the cleavage, based on their platy crystallographic shape, (2) *flattening* of quartz grains in the plane of the cleavage by pressure solution, producing a noncrystallographically controlled grain shape, and (3) *differentiation* of a new compositional layering composed of thin mica- and oxide-rich layers and thicker quartz-feldspar-carbonate-rich layers. The mica-oxide-rich layers are called *cleavage bands,* whereas the intervening generally thicker quartz-feldspar-carbonate-rich layers are called *microlithons.* These three properties are the most obvious microscopic properties of typical slaty cleavage.

As we examine other rock fabrics that do not fracture in the proper way to be called slaty cleavage, we often find the same three basic phenomena of preferred orientation, grain flattening, and metamorphic differentiation. For example, as the slaty foliation passes from muddier beds into coarser-grained and less-micaceous sandy beds, we find that the spacing of the cleavage bands increases to the point that foliation would traditionally be called a fracture cleavage rather than a slaty cleavage because the rock would no longer make a commercial slate. Nevertheless, the basic phenomena remain. In Figure 10-5(b) we see that the spacing of cleavage bands decreases and increases as the cleavage passes through slightly muddier and sandier beds.

Cleavages in fine-grained rocks that display a discrete macroscopic spacing are traditionally given a variety of names, including fracture cleavage, strain-slip cleavage, and crenulation cleavage. We use the name *spaced cleavage* to subsume all these, in contrast with *slaty cleavage,* which displays its spaced aspect only on a microscopic scale. The names for many of the varieties of spaced cleavage are part of the active vocabulary in structural geology. For this reason we discuss some of the more-important terms and see the associated structural phenomena in the following paragraphs.

Fracture cleavage, also called close-joints cleavage, was traditionally defined to be a cleavage without a fabric, simply composed of closely spaced fractures. However, under the microscope we see that fracture cleavages in fact have an associated fabric, including a compositional differentiation into cleavage bands and microlithons. For example, Figure 10-6 is a photomicrograph of what was mapped as a fracture cleavage based on field examination. The rock displays a fabric composed of (1) an older, slaty cleavage preserved in the microlithons and (2) a new planar fabric developed in the macroscopically spaced cleavage bands that define the fracture cleavage. The cleavage bands are compositionally different from the rest of the rock, displaying a strong loss of quartz, feldspar, and carbonate by pressure solution. What is called fracture cleavage is a spaced cleavage in which the cleavage bands are very narrow.

Strain-slip cleavage and crenulation cleavage are two nearly synonymous terms for spaced cleavages that display an associated folding of the older cleavages preserved within the microlithons. For example, the older slaty cleavage in Figure 10-6 bends as it passes across the cleavage band. Cleavages that display such asymmetric bending of an earlier foliation are traditionally called *strain-slip cleavages.* Strain-slip cleavage was traditionally thought to form by a shear or slip parallel to the cleavage band. Other spaced cleavages display a more-pronounced buckling of the older fabric within the microlithons; these cleavages are more commonly called *crenulation cleavages.*

As we examine fabrics in rocks of higher metamorphic grade, we generally observe the well-known coarsening of grain size from slate to phyllite to schist.

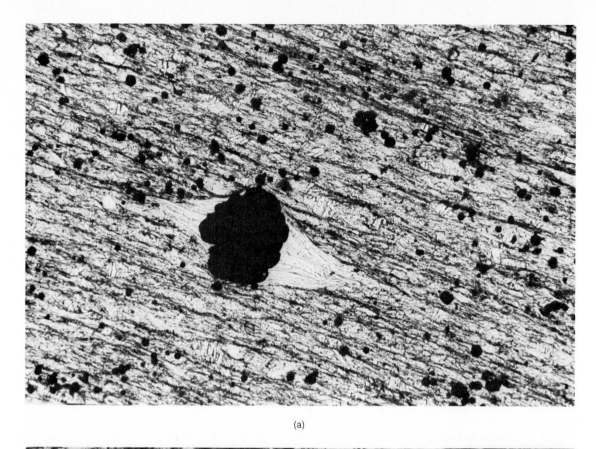

(a)

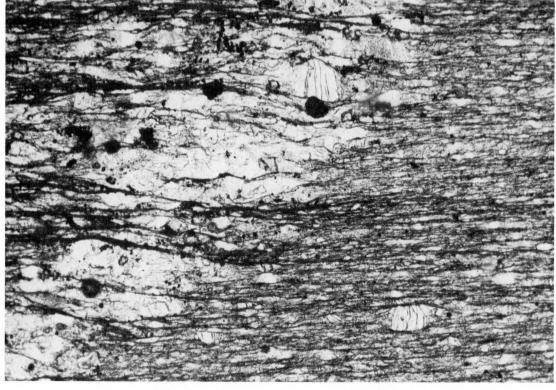

(b)

FIGURE 10–5 Photomicrographs of Ordovician slate from Martinsburg Formation, Central Appalachians. (a) Homogeneous slate showing discrete microscopic cleavage bands and microlithons. Black particles are pyrite. (High magnification.) (b) Cleavage passing from muddier to sandier layers. (Low magnification.) Samples from same outcrop as Figure 10-1.

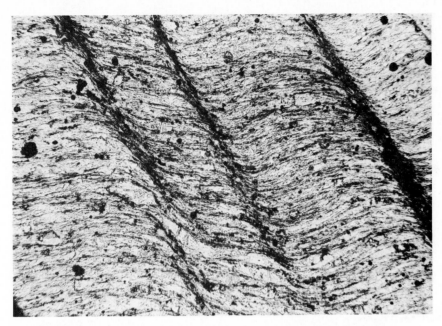

FIGURE 10–6 Photomicrograph of a spaced cleavage in Ordovician Martinsburg Formation which deforms an earlier slaty cleavage (compare Fig. 10-5).

Nevertheless, the same general assemblage of fabric phenomena remain, including the differentiation of rock into microlithons and cleavage bands, the development of preferred orientation of mica and amphibole within the cleavage bands, and the flattening of grains in the plane of the cleavage. For example, Figure 10-7 shows a photomicrograph of a quartz-mica schist. The rock displays a metamorphic differentiation into more quartz-rich microlithons displaying an older schistosity and into mica- and oxide-rich cleavage bands displaying a strong preferred orientation of mica.

AXIAL PLANAR FOLIATIONS

Many foliations are oriented roughly parallel to the axial surfaces of associated folds (Chapter 2); these are called *axial planar foliations*. Many of the metamorphic folds display these axial planar foliations (for example, Figs. 2-8 and 9-16). Even though these foliations are called axial planar, they are typically parallel to the axial surface of a fold only where they actually coincide with the axial surface. Axial planar foliations typically *fan* through a fold in the manner shown in Figure 10-8.

The details of the cleavage pattern correlate closely with the fold geometry. The cleavage fan is most pronounced in more competent layers displaying Ramsay Class 1*C* folds (Chapter 9). Less-competent layers that exhibit close to similar fold shapes (Class 2) normally have foliations that are close to parallel throughout. Foliations in less-competent layers that are next to competent layers—Class 3 folds adjacent to Class 1*B* or 1*C* folds—typically display a fanning in the opposite sense to the normal cleavage fan. This opposite pattern displayed by less-competent rock is called *antifanning* (Fig. 10-8). A natural example is shown in Figure 9-1; the spaced cleavage in the competent limestone bed (Class 1*C*) displays a typical cleavage fan, whereas the cleavage in the immediately adjacent slate (Class 3) displays antifanning. Cleavage in nearly similar folds shows essentially no fanning.

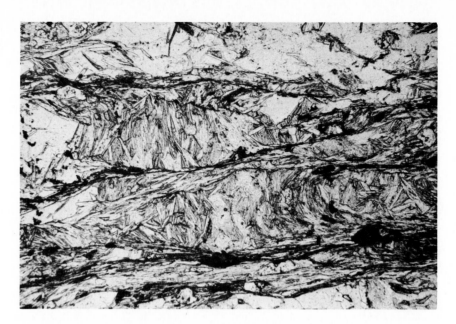

FIGURE 10–7 Photomicrograph of cleavage bands and microlithons in quartzose schist. Note the preservation of an earlier foliation within the microlithons.

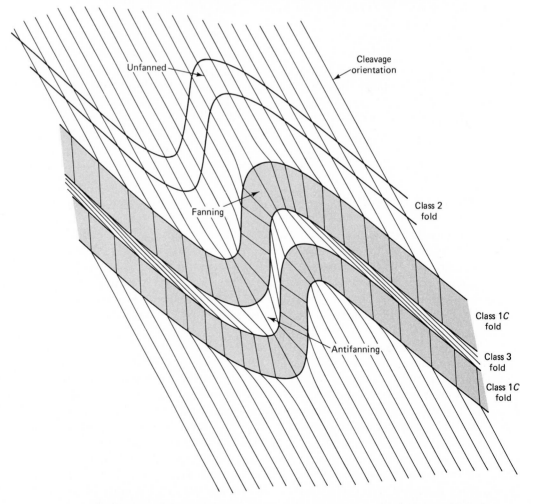

FIGURE 10–8 Schematic arrangement cleavage as it passes through a set of folds in more-and-less competent rocks.

We see from this discussion of fanning of axial planar foliations that fold shape and foliation geometry are closely tied together. This conclusion is reinforced when we recall from Chapter 2 that axial planar foliations fan about the same axis as the fold axis and the local intersections between foliations and folded layers are parallel to the fold axis. Therefore, we can conclude from all these geometric observations that folded layers and axial planar foliations are two interrelated manifestations of the same deformation.

What is the origin of these close geometric relationships between fold shape and axial planar foliations? Considerable insight into this question comes from studies of patterns of finite strain within folds. Two types of studies have been important: (1) observation of strain indicators in natural structures and (2) deformation of laboratory and numerical-model structures.

Patterns of finite strain around a number of folds have been determined using the normal finite-strain markers such as ooids, reduction spots, and fossils discussed in Chapter 3. The plane of principal flattening (S_1-S_2 plane) shown by these strain markers is typically approximately parallel to any associated cleavage and displays a pattern of fanning close to the typical patterns of cleavage fanning shown in Figure 10-8. For example, Figure 10-9 shows that the pattern of finite strain determined from ooid deformation is parallel to the cleavage orientation in two folds from the Alps.

A similar approximate parallelism of foliation and the plane of finite flattening is observed when we compare the patterns of finite strain in numerical- or laboratory-model folds with natural patterns of axial planar foliation. For example, Figure 9-63 shows the orientation of the finite flattening plane S_1-S_2 in a numerical model of a stiffer layer imbedded in a softer medium. The pattern of strain is very similar to the pattern of cleavage in the natural fold of Figure 9-1; the stiffer model layer displays typical fanning, the immediately adjacent softer material displays antifanning, and the softer medium far from the fold shows no fanning. Therefore, we can conclude from these studies that axial planar foliations in many cases are approximately parallel to the plane of finite flattening or at least to the finite flattening associated with the same generation of folding.

The approximate parallelism of axial planar foliations and the plane of finite flattening is very important because it allows us to infer the approximate orientations of the principal plane of flattening in the field from the cleavage orientation in rocks that do not contain actual strain markers, such as ooids or fossils. However, this result should be used with caution because some foliations are known not to be parallel to the plane of finite flattening; for example, many mylonitic foliations are parallel to a plane of simple shearing. Nevertheless the hypothesis that foliation or cleavage is parallel to the plane of finite flattening (S_1-S_2) is a useful rule of thumb, which we shall use in the following sections.

REFRACTION OF FOLIATIONS

One of the important properties of foliations is that they change orientation as they pass from stiffer to softer layers. This change in orientation of foliation with rock type is called *refraction* of foliation or cleavage. For example, the slaty cleavage in Figure 10-1 systematically refracts as it passes from muddy to sandy layers. Similarly, the change in orientation of cleavage from stiffer to softer beds around a fold, which is associated with cleavage antifanning, is an example of refraction of cleavage. This phenomenon of refraction is easily understood as an effect of strain compatibility between layers of different stiffness (for example, Treagus, 1983).

Refraction of cleavage exists in interlayered materials because, for a given overall compressive strain, the softer layers take up more of the compression than

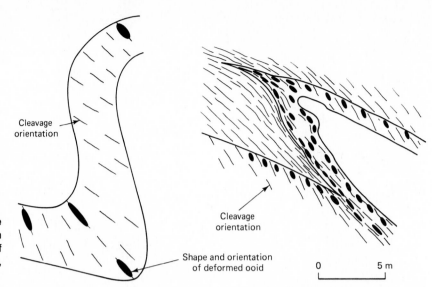

FIGURE 10-9 Parallelism of cleavage and principal elongation axis of strain in two outcrop-scale folds in the Alps of eastern Switzerland. (After Pfiffner, 1978.)

Cleavage orientation

Cleavage orientation

Shape and orientation of deformed ooid

0   5 m

the stiffer layers. Nevertheless, the layers still must fit together after they have changed shape differently. For example, on the left side of Figure 10-10 an interlayered medium of stiff and soft materials is deformed with the orientations of the principal strains being parallel throughout; this hypothetical mode of deformation gives rise to mismatches in bed length, which makes this an impossible mode of deformation in nature. It would require voids to open up in the softer beds. In a second example, on the right side of Figure 10-10, the interlayered material deforms with the requirement that bed lengths match between the stiffer and softer layers, which requires that the components of the strain in the plane of the bedding must be the same in both beds. Therefore, the principal strains must have different orientations, as shown on the right side of Figure 10-10.

The pattern of strain in this theoretical example is very similar to the pattern of refracting cleavage, which is to be expected if cleavage approximates the plane of finite flattening. The cleavage in the stiffer layer is always at a higher angle to bedding than the cleavage in the softer layer because the stiffer layer is less strained than the softer layer. Refraction of cleavage is therefore important

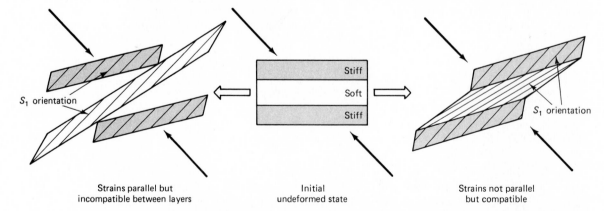

S_1 orientation

Stiff
Soft
Stiff

S_1 orientation

Strains parallel but incompatible between layers

Initial undeformed state

Strains not parallel but compatible

FIGURE 10–10 Strain compatibility and refraction of cleavage. The initially undeformed interlayered stiff and soft materials shown in the center are subjected to an overall compression. If the flattening S_1 is everywhere perpendicular to the overall compression, the layers will no longer fit together because of the different magnitudes of S_1 in the layers of different stiffness, as shown to the left. In contrast, if the strains are compatible, the flattening S_1 in general will not be perpendicular to the overall shortening and S_1 makes a higher angle to bedding in the stiffer layer, as shown to the right. This is the principal cause of refraction of cleavage. $S_1 = 2$ and $S_2 = 0.5$ in the soft layer; $S_1 = 1.3$ and $S_2 = 0.77$ in the stiff layer.

because it shows us which layers were stiffer and which were softer during deformation.

FOLIATIONS IN HETEROGENEOUS LUMPY MATERIALS

Most foliated rocks display an earlier layering that is now deformed and crosscut by the foliation. Any differences in finite strain between adjacent layers is manifest in a refraction of foliation. Foliation is parallel in volumes that have strained homogeneously. In this section we briefly consider the patterns of foliation in more heterogeneous rocks, such as deformed conglomerates, pebbly mudstones, olistostromes, and melanges. These rocks may be characterized mechanically as a mixture of lumps and fragments of different stiffness in a generally softer matrix. For example, they may contain sandstone pebbles or fragments in a mudstone matrix. During metamorphic deformation the mudstone is softer and therefore takes up much more of the strain than the sandstone. This heterogeneous deformation gives rise to complex foliation patterns in metamorphosed conglomerates, pebbly mudstones, olistostromes, and melanges. The patterns of foliation are characterized by anastomosing cleavage bands that are concentrated in the softer matrix and wrap around the stiffer, less-foliated fragments—for example, around the stiffer cobbles in the deformed conglomerate of Figure 3-12(a).

Anastomosing foliation patterns reflecting strain heterogeneity are not confined to a macroscopic scale. They are also very commonly observed under the microscope in rocks that show lumpy heterogeneity on the microscale—for example, metasandstones and porphyroblastic schists and phyllites. These lumpy rocks typically show wrapping of foliations in the softer matrix around the stiffer fragments. For example, in Figure 10-11 we see a photomicrograph of anastomosing cleavage in the matrix of a pebbly metasandstone; the foliation refracts around the stiffer quartz pebbles. This fabric pattern appears to be an effect of strain compatibility in a heterogeneous material.

CLEAVAGE-BEDDING RELATIONS IN STRUCTURAL MAPPING

The relationships between fold shape and cleavage orientation are so well established that we can use this knowledge to solve some practical problems of structural mapping. In particular, if the folds are larger than outcrop scale, then we can observe the angle between cleavage and bedding in an outcrop and immediately know where that outcrop lies with respect to the larger fold.

FIGURE 10–11 Photomicrograph of pebbly sandstone showing dark anastomosing cleavage bands, which is an effect of the heterogeneous lumpyness of the material. Crossed polarizers, Eocene slates, northern Taiwan.

Let us consider the schematic map-scale fold and associated cleavage shown in Figure 10-12. Three outcrops exposing parts of the fold are labeled A, B, and C. Outcrops A and B both show the cleavage to be inclined to bedding. If we compare these outcrops with the normal cleavage pattern (Figure 10-8), we see that both outcrops must lie on the limbs of larger folds. Only outcrop C lies in a hinge zone because its cleavage is perpendicular to bedding. Therefore, cleavage-bedding relations help us locate the hinge zones and traces of axial surfaces of major folds.

The next aspect of cleavage-bedding geometry to notice is that cleavage on the limbs of folds is inclined, relative to bedding, toward the convex side of the hinges. For example, in outcrop A the acute angle between bedding and cleavage opens to the left on the upper side of the bed; therefore, a convex or anticlinal hinge would be the first fold hinge to the left. Similarly, in outcrop B the acute angle on the right-hand side of the bed opens downward toward the right; therefore, the next convex or synclinal hinge on the right-hand side of the bed would be down the bed and to the right. Let us now consider the actual outcrop in Figure 10-1. Which direction along the bed is the first anticlinal hinge? Try to solve this problem before continuing to the next paragraph.

The geometry of cleavage and bedding is also very powerful in determining whether the beds are overturned or are right side up in an outcrop. For example, in the outcrop of Figure 10-1 the fact that the next anticlinal hinge is up the bed, which dips to the left, indicated that bedding is right side up. If the bedding were upside down, the cleavage would have dipped to the left. Similarly, we see that beds are overturned in outcrop B (Fig. 10-12) because the convex synclinal hinge

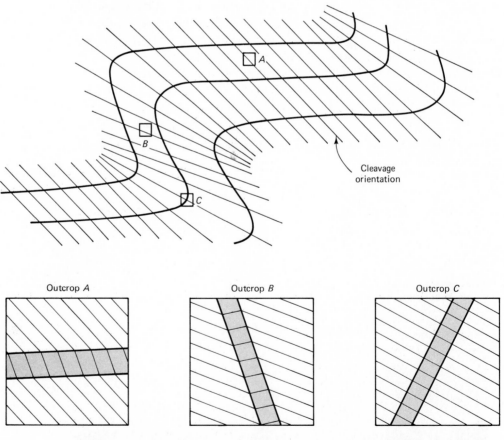

FIGURE 10–12 Deduction of map-scale structure from observation of outcrop cleavage-bedding intersections. The angles between bedding and cleavage in each outcrop indicate the direction along the bed to the next anticlinal or synclinal hinge, as discussed in the text.

orientation of the slip systems. This state of affairs in general leads to the sort of inhomogeneous plastic deformation commonly seen in rocks at moderate temperatures. In 1938 G. T. Taylor developed a theory that the crystallographic preferred orientation of polycrystalline aggregates could be predicted by specifying the independent slip systems and their stress dependence, together with the requirement of minimum work. Numerical models of polycrystalline quartz deformation based on an elaboration of Taylor's theory estimate quartz fabrics that are similar to natural fabrics (Lister, Paterson, and Hobbs, 1978; Lister and Paterson, 1979; Lister and Hobbs, 1980). These numerical models confirm the importance of slip as a major process in the development of fabrics displaying crystallographic preferred orientation.

Recrystallization

Recrystallization produces profound changes in the fabrics of deformed rocks. The recrystallization takes place because the old crystals have a higher energy than the new crystals; this energy difference drives the recrystallization. Three types of recrystallization are considered, those governed by (1) *strain energy,* (2) *surface energy,* and (3) *metamorphic phase equilibria.*

Strained crystals have more internal energy than unstrained crystals; therefore, there is a tendency for strained crystals to recrystallize. Two important types of strain energy exist: (1) line and point defects and (2) elastic. Plastic deformation greatly increases the density of line and point defects in crystals, which increases the internal energy. Recrystallization driven by the defect energy is called annealing and involves the growth of new strain-free grains and polygonization (Chapter 4). For example, the plastically deformed quartz grain in Figure 10-19 displays polygonization internally and growth of new much smaller grains around its margin. Differences in elastic strain energy may also drive recrystallization. An elastically anisotropic mineral has an elastic strain energy that depends on its orientation in an anisotropic stress field. This difference in elastic strain energy has been proposed for many years as a mechanism of development of crystallographic-preferred orientations, although the actual energy differences are small (Paterson, 1973).

Recrystallization is also driven by the contribution of the surface energies to the total energy of a rock. We have already seen the importance of surface energy to rock deformation in the Griffith theory of fracture (Chapter 5). A common generalization is that the smaller the grain size, the larger the surface area and surface energy. Therefore, there is a tendency to recrystallize to larger grain size, yet there are various energy and kinetic barriers to this recrystallization. These barriers to recrystallization generally decrease with increasing temperature; therefore, grain size generally increases with metamorphic grade. This process of minimization of surface energy—for example, by minimization of surface to volume ratio—leads to grain shapes and grain-boundary angles that are controlled by surface energies.

Highly annealed rocks display grain-scale textures that are largely controlled by surface energies. Well-annealed monomineralic rocks, particularly of minerals such as quartz, calcite, or sulfides that do not display crystallographic grain shapes, commonly display a structure of nearly equant grains with a high proportion of three-grain junctions at near 120°, which is called *comb structure.* Comb structure is similar to the structure of foams of soap bubbles. Comb structure is an arrangement of grain boundaries that minimizes surface energy. For example, Figure 10-22 displays this characteristic highly annealed texture in a calcite polycrystalline aggregate.

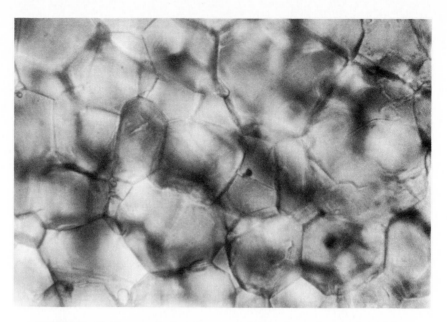

FIGURE 10–22 Photograph of comb structure in experimentally annealed calcite marble. (Sample courtesy B. Evans)

The interfacial angles in well-annealed rocks differ from 120° if different minerals are in contact because each mineral will have its own surface energy. The topic is complex in detail because most rock-forming minerals have anisotropic surface energies. For simplicity we can treat the surface energies as isotropic and displaying liquidlike surface tension, particularly for minerals that do not display crystal faces in polycrystalline aggregates. If three crystals of two minerals A and B are in contact, as shown in Figure 10-23, then the relation between their surface energies and interfacial angles is

$$\frac{\gamma_{AA}}{\sin\theta_{AA}} = \frac{\gamma_{AB}}{\sin\theta_{AB}} \tag{10-3}$$

where γ_{AB} is the interfacial surface energy of mineral A against mineral B and θ_{AB} is an interfacial angle emanating from an A-B contact (see Fig. 10-23). This

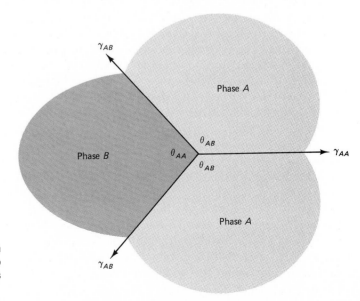

FIGURE 10–23 Relationship between well-annealed interfacial angles θ for two phases A and B and surface energies γ_{AA} and γ_{AB}. See equation 10-3.

TABLE 10–1   **Grain-Boundary Angles and Energies***

Mineral A	Mineral B	θ_{AA}	γ_{AA}/γ_{AB}
Quartz	Plagioclase	110°	1/0.87
Quartz	Garnet	135°	1/1.31
Quartz	Calcite	130°	1/1.18
Quartz	Apatite	145°	1/1.66

*Data from Vernon, 1968

relationship allows us to measure the ratio of interfacial surface energies between grains by measuring interfacial angles. Rearranging, we get

$$\frac{\gamma_{AA}}{\gamma_{AB}} = 2 \cos \left[\frac{\theta_{AA}}{2} \right] \qquad (10\text{-}4)$$

Natural interfacial angles and ratios of interfacial surface energies for some important minerals in contact with two quartz grains are given in Table 10-1 as an example of these observations.

The final driving force for recrystallization that we consider briefly is metamorphic phase equilibria. As temperature, solid pressure, and fluid pressure and composition change due to tectonic or other processes, the assemblage of minerals that makes up a rock often becomes unstable relative to another assemblage of minerals of the same composition. For example, a sandy limestone (quartz-calcite rock) will react to form a wollastonite-bearing marble with the release of carbon dioxide (CO_2):

Quartz + calcite (in excess) → wollastonite + calcite + carbon dioxide

$$SiO_2 + n CaCO_3 \rightarrow CaSiO_3 + (n - 1)CaCO_3 + CO_2 \qquad (10\text{-}5)$$

Metamorphic reactions such as these consume some minerals and produce others, thereby modifying the fabric of the rock.

Metamorphic reactions in some cases profoundly alter the mechanical properties of the rock, for at least two reasons: (1) Differences in mineral strength between reactants and products can produce major changes in style of deformation. For example, muddy carbonates may be relatively soft at low metamorphic grades, but with increasing temperature they react to form skarn rocks composed of calcium-aluminum silicates, such as garnet, diopside, and epidote, which are much less ductile. (2) Metamorphic reactions that produce or consume fluids—such as the reaction of 10-5—can produce major changes in fluid pressure, which affects all deformation involving brittle behavior (Chapters 5 and 8). For example, one of the mechanisms proposed for producing overpressured shales is clay-mineral dehydration reactions. Furthermore, these reactions can bring about changes in the depth of the brittle-plastic transition because that depth is a function of fluid pressure (Chapter 6). Dehydration and decarbonation reactions therefore indirectly influence the deformational fabric.

The final point to be mentioned under the topic of recrystallization is its effects on crystallographic-preferred orientation and grain size. Data from experiments and natural rocks suggests that recrystallization is important in developing preferred-orientation fabrics, although the topic is not well understood. Annealing experiments show that new grains have a systematic orientation relative to the old, deformed grains. For example, adjacent old and new grains of calcite commonly have a crystallographic axis in common. Experiments on recrystallization during deformation also suggest that recrystallization affects both crystallographic preferred orientation and grain size. Increasing stress decreases the grain size. An experimental relationship between grain size and stress for steady-state plastic deformation of quartz is shown in Figure 10-24.

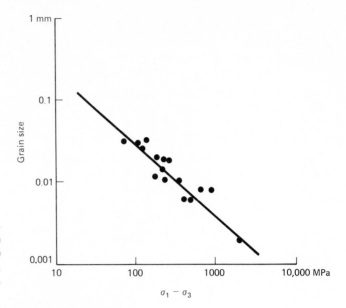

FIGURE 10–24 Experimentally determined relationships between grain size and differential stress for steady-state deformation in quartz. (Data from J. M. Christie, P. S. Koch, A. Ord and R. P. George.)

FABRICS IN ANALYSIS OF THE HISTORIES OF POLYDEFORMED TERRAINS

Superposed deformations are common, indeed the rule, in metamorphic terrains at the map, outcrop, and microscopic scale. The fabrics produced in these deformations in many cases provide an important key to the broader geologic history of the metamorphic terrain because the superposition of fabrics is easily observed.

Deformational structures that are observed on an outcrop or microscopic scale, especially fabrics and minor folds, can be classified into a relative time sequence for each outcrop or small area, based largely on their superposed relationships. For example, fold and fabric of generation 2 is older than generation 3 if the axial-surface schistosities of its minor folds are consistently folded or crosscut by those of generation 3 wherever the two are observed together. For

FIGURE 10–25 Dark chlorite porphyroblast that has grown across an earlier microscopic fold in quartz-mica schist.

example, in Figure 10-3 an earlier axial planar foliation and lineation in phyllite is seen in outcrop to be refolded with the weak development of a new lineation parallel to the younger fold axis. Similar observations can be made in thin sections of known orientation and can be related to larger-scale structures. For example, mica and chlorite oriented horizontally in Figure 10-7 define a younger spaced cleavage or schistosity that crosscuts an older vertical schistosity. As another example, a large chlorite grain has grown across an earlier microscopic fold in Figure 10-25. Observations such as these at single outcrops or in single thin sections are easily made and provide the primary data used in establishing a regional chronology of a metamorphic terrain.

The next step in establishing a regional chronology is to identify the same fold-and-fabric generations in nearby outcrops and gradually build up a network of outcrops containing correlated superposed structures. The correlation of fold-and-fabric generations between outcrops requires considerable care because this correlation offers opportunities for mistakes in reasoning. Correlations are most-commonly based on such properties as axial-surface and hinge-line orientation, sense of rotation of asymmetric folds, associated distinctive metamorphic miner-alogy, and style of folds and fabrics. This last property, style of folding and fabric, holds perhaps the most pitfalls for mistakes in reasoning because different rock types commonly deform with quite different styles.

Eventually a regional network of superposed fold-and-fabric generations can be established, based on observations of superposition at many locations. This structural network of outcrop and thin-section observations is in effect a relative geologic time scale to which other events can be correlated—for example, unconformities, igneous intrusions, metamorphic mineral assemblages, and major faulting and folding. An example of a regional network from the New England Appalachians is shown in Figure 10-26.

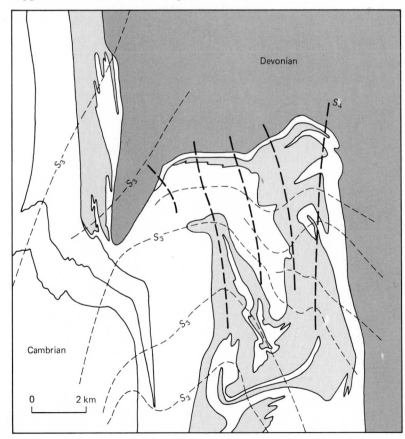

FIGURE 10–26 Network of generations of deformation, mapped in eastern flank of the Berkshire Highlands, New England Appalachians, showing the orientations of F_3 and F_4 axial surfaces, labeled S_3 and S_4. (Simplified from Stanley, 1975.)

Within a regional network such as this, the structures of each generation are commonly referred to using the following subscript notation. The fold axes or generations of folding are labeled F_1, F_2, . . ., F_n, where F_1 is the oldest recognized folding. Foliations, cleavages, and layering are labeled S_1, S_2, . . ., S_n, where S_1 is the oldest recognized generation. If primary sedimentary layering is recognized, it is commonly labeled S_0. Similarly, generations of lineations are labeled L_1, L_2, . . ., L_n.

EXERCISES

10–1 It is often said as a rule that if bedding and the associated cleavage dip in the same direction and cleavage dips more steeply, then the beds are right side up. Conversely if the cleavage dips more gently, then the bedding is overturned. Show that this rule is correct by means of sketches of cleavage-bedding geometry assuming it is true and assuming it is false. What is a special situation in which this rule could give results in conflict with facing determined from sedimentary structure?

10–2 What is the direction of the next syncline along the beds in Figure 2-8? Are the beds right side up or upside down? Show your reasoning with a sketch of the entire fold.

10–3 The association of foliations and folding is so widespread that some geologists have assumed that for every pervasive foliation, there is an associated folding, even without direct evidence of folding. If you observed pervasive foliation everywhere nearly parallel to bedding in a region, would it be reasonable to assume that the beds are isoclinally folded? Outline your reasoning, making use of your knowledge of origin and significance of foliation and the mechanics of folding.

SELECTED LITERATURE

BORRADAILE, G. J., BAYLY, M. B., AND POWELL, C. McA., 1982, *Atlas of Deformational and Metamorphic Rock Fabrics*, Springer-Verlag, Berlin, 551 p.

HOBBS, B. E., MEANS, W. D., AND WILLIAMS, P. F., 1976, *An Outline of Structural Geology*, John Wiley, New York, 571 p. (Chapters 2, 5, and 6).

NICOLAS, A., AND POIRIER, J. P., 1976, *Crystalline Plasticity and Solid State Flow in Metamorphic Rocks*, John Wiley, New York, 444 p.

ROSENFELD, J. L., 1970, Rotated garnets in metamorphic rocks: Geol. Soc. Amer. Spec. Paper 129, 105 p.

SCHMID, S. M., 1983, Microfabric studies as indicators of deformation mechanisms and flow laws operative in mountain building: p. 96–110 in Hsü, K. J., ed., *Mountain Building Processes*, Academic Press, London, 263 p.

SPRY, A., 1969, *Metamorphic Textures*, Pergamon Press, Oxford, 350 p.

TURNER, F. J., AND WEISS, L. E., 1963, *Structural Analysis of Metamorphic Tectonites*, McGraw-Hill, New York, 545 p.

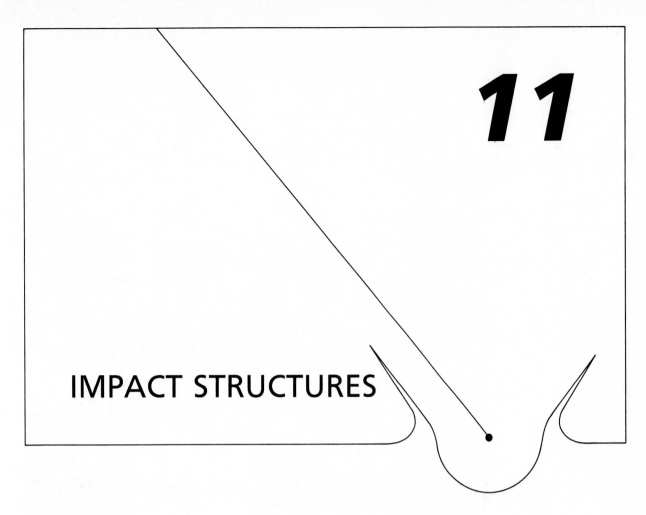

IMPACT STRUCTURES

INTRODUCTION

Exploration of outer space has demonstrated the importance of high-velocity impact of meteoritic and cometary material in shaping the surface of the terrestrial planets and moons. Impact craters are ubiquitous on the Moon, Mercury, and all known satellites of Mars, Jupiter, and Saturn possessing geologically old surfaces. Craters are somewhat less visible on Mars because of wind erosion and deposition. By analogy, we expect high-velocity impacts to have occurred widely on Earth, although the effects should generally be obscure because of erosion, sedimentation, and tectonics. Furthermore, the number of very small craters formed should be reduced on Earth by the decelerating effects of the atmosphere and oceans. The atmosphere effectively shields Earth from impacting bodies of initial mass smaller than about 1000 kg, which are broken up and ablated before they reach the surface. Somewhat-larger meteorites are also decelerated; for example, the 3×10^5 kg Sikhote-Alin meteorite (1947) was decelerated by the earth's atmosphere from its preentry velocity of about 15 km/s to 0.5 km/s when it struck the solid earth. Large meteorites are little affected by the atmosphere because of their large ratio of mass to surface area.

The rate of crater formation on the earth may be estimated from the observed flux of small meteors and their size-frequency distribution. Data on the size-frequency distribution of smaller bodies in this part of the solar system suggest that the present population of bodies is sufficient to account for the observed impact structures on Earth. One estimation gives approximately 100,000 craters larger than 1 km for the past 2 billion years, 6000 larger than 5 km, and 20 larger than 100 km.

The number actually observed at the surface is greatly reduced by the effects of erosion and burial. The average rate of erosion of the continents is about 0.1 mm/year. Thus a crater 1 km wide with shocked rocks extending to a depth of 1 km, for instance, would be erased in 10 m.y. Impact structures buried by continued sedimentation are discovered occasionally in the subsurface in regions of extensive petroleum exploration. For example, three structures ranging in diameter from 6 to 12 km have been recognized in the subsurface of the Williston basin (Figure 1-8), two of which provide commercial petroleum reservoirs (Sawatzky, 1978).

Most impact structures exposed at the earth's surface exist in a deeply eroded state, so we should be concerned with the criteria by which they are recognized and their origin assigned. The number of unequivocal impact craters, those with associated meteoritic material, is small; only 12 sites, some with multiple craters, are listed in Table 11-1. All the craters are young, little eroded, and quite small; the largest is only 1.2 km in diameter. The study of these young craters of clear impact origin has provided much insight into the phenomena of high-velocity impact cratering and has provided criteria for the recognition of more-deeply eroded and larger structures. Much of this insight has come from the study of Meteor Crater, Arizona (Fig. 11-1).

Meteor Crater, Arizona

Meteor Crater (also called Cañon Diablo or Barringer Crater) is a circular to subrectangular depression about 1.2 km in diameter and 180 m deep in the Colorado Plateau of the western United States, about 150 km southeast of the Grand Canyon. The crater is a closed depression with steep inner walls and a rim that rises about 50 m above the surrounding plain (Fig. 11-1). Meteor Crater lies within a swathe of iron meteorite discoveries that spreads across Arizona for 500 km in a northeast-southwest direction. Within a radius of about 5 km of the crater, thousands of small iron meteoritic fragments ranging from a few grams to several kilograms have been found, together with a large quantity of finely dispersed meteoritic material.

The first recorded hypothesis for the origin of the crater is that of a company of shepherds who camped on the slopes of the crater in 1866. They thought that the crater was produced by an explosion and that the material forming the rim had been thrown out of the cavity. They also found pieces of iron around the crater, which were later shown to be meteoritic.

TABLE 11-1   Impact Craters with Associated Meteoritic Fragments*

Name	Latitude	Longitude	Number of Craters	Diameter of Largest
Aouelloul, Mauritania	20°15′N	012°41′W	1	250 m
Boxhole, N.T., Australia	22°37′S	135°12′E	1	175 m
Campo del Cielo, Argentina	27°28′S	061°30′W	9	70 m
Delgaranga, Western Australia	27°45′S	117°05′E	1	21 m
Haviland, Kansas, U.S.A.	37°37′N	099°05′W	1	11 m
Henbury, N.T., Australia	24°34′S	133°10′E	14	150 m
Kaalijarvi, Estonian SSR	58°24′N	022°40′E	7	110 m
Meteor Crater, Arizona, U.S.A.	35°02′N	111°01′W	1	1200 m
Odessa, Texas, U.S.A.	31°48′N	102°30′W	3	168 m
Sikhote Alin, Siberia, USSR	46°07′N	134°40′W	22	27 m
Wabar, Saudi Arabia	21°30′N	050°28′E	2	90 m
Wolf Creek, Western Australia	19°18′S	127°47′E	1	850 m

*After Dence (1972)

(a)

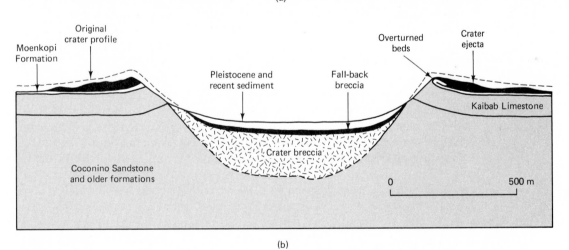

(b)

FIGURE 11–1 Oblique aerial photograph and geologic cross section, Meteor Crater, Arizona. (Photograph by N. H. Darton; section adapted from Shoemaker, 1960.)

In 1896 G. K. Gilbert considered the possibility that the crater was of impact origin; indeed, 2 years earlier he had persuasively argued that lunar craters are products of meteoritic bombardment. Nevertheless, Gilbert eventually abandoned the impact theory for Meteor Crater because he was unable to discover an excess iron mass buried in the crater. Gilbert showed quantitatively that the volumes of the rim and of the crater were equal; therefore, no extraneous mass existed below the crater. Furthermore, magnetic surveys failed to detect a buried iron meteorite. For these reasons, Gilbert was driven to the less-likely hypothesis of a steam explosion of volcanic origin and had to consider the meteoritic material to be fortuitous.

The choice between extraterrestrial and internal origin has been controversial for many strongly deformed circular structures, especially those deeply eroded. Each structure must be evaluated on its own merits. In the case of Meteor Crater confidence in the extraterrestrial origin came from a fuller exploration for meteoritic material, together with a more-detailed examination of the deformed rocks and a greater understanding of the mechanics of meteorite impact. For example, Gilbert had considered that a meteorite capable of producing the crater would have to be 250 to 500 m in diameter; this extraneous mass was just not observed. It was later realized that meteors encounter the earth at extremely high velocities (approximately 10 to 15 km/s). The energy of the impact is derived from the kinetic energy of the meteor, which depends on the square of the velocity. Gilbert underestimated the velocity and, therefore, greatly overestimated the mass. If the density of the meteorite were 7850 kg/m^3, the diameter would be no larger than about 25 m, much smaller than Gilbert's estimate.

The rim of Meteor Crater is composed of slightly upturned strata, as well as a completely overturned flap of the uppermost stratum, and is covered by debris blown out of the crater (Figure 11-1(b)). Outwardly overturned flaps of near-surface layers are present a number of places around the crater rim and have been observed in several other impact craters, as well as in some underground nuclear explosion craters. These flaps are evidence for the outward explosive character of the craters and rule out a collapse origin, such as a caldera or sinkhole.

The floor of Meteor Crater is filled with several hundred meters of breccia that includes melted rock and dispersed meteoritic material. The uppermost breccia consists of a layer of thoroughly mixed breccia, glass, and meteoritic material that is almost perfectly massive but exhibits a distinct upward gradation from coarse to fine. This graded bed of breccia probably formed by fallback of fragments that were ejected to great height. Pleistocene and Recent alluvium, lake beds, and talus overlie the fallback breccia.

The lowest part of the debris thrown out onto the flanks of the crater is almost entirely derived from the Moenkopi Formation, which is the uppermost stratigraphic unit in the region. Above Moenkopi debris lies debris or breccia of Kaibab Limestone. Finally, the oldest formations, the Coconino and Toroweap Sandstones, compose the highest thrown-out breccia. Thus the bedrock stratigraphy is preserved in the thrown-out breccia in inverted order. Mappable layers of breccia extend as far as 1 km beyond the crater rim. Blocks as large as 30 m are present. The iron meteoritic fragments found in the breccia close to the rim are generally highly deformed and some contain diamond, a very high pressure mineral, whereas the more widely dispersed meteoritic material does not. The ejected rock fragments also exhibit a broad range of distinctive structural features and high-pressure and high-temperature mineral transformations. Before we discuss these effects of high-velocity impact, it is appropriate to discuss the mechanics of impact.

MECHANICS OF HYPERVELOCITY IMPACT

Massive meteorites entering the earth's atmosphere experience little deceleration from their outer-space velocity of 10 to 20 km/s; they strike the solid earth at a velocity greater than the speed of sound in rock (about 5 km/s). Such impacts are called *hypervelocity impacts* and produce shock waves.

The response of both the meteoritic projectile and the target in hypervelocity impact is essentially fluid. At an early stage in the impact, within microseconds, stresses are set up that locally far exceed the strength of either the meteorite or the earth. For this reason the highly stressed parts of both materials behave as though they had no strength, that is, as though they were fluid. The very high pressures generated on impact are released explosively, producing the crater. This is the explanation of one of the simplest observations that may be made of impact craters: Essentially all craters are circular, without any sign of the direction of incidence of the meteorite. The crater is produced by explosion following impact, the energy being derived from the mass and velocity of the projectile:

$$E = \tfrac{1}{2} Mv^2 \tag{11-1}$$

Because of the very high stresses, the behavior of the impacting system may be approximately described by hydrodynamic equations plus equations that describe the relationship between pressure and density for the material (equations of state). Figure 11-2 shows the result of an approximate numerical calculation using such equations for a 1.2×10^4 kg iron meteorite impacting tuff. Pressures that locally exceed 10^3 GPa (Figure 11-2(a)) are initially generated and large volumes experience pressures above 1 GPa for 10 or more milliseconds (Figs. 11-2(b) and 11-2(c)). Note that as the shock wave spreads downward and to the sides, the material above decompresses and is thrown upward. The volume of material affected by the impact is many times larger than the original projectile; little of the meteoritic material is commonly evident. The transient high pressures, followed by decompression, give rise to the structural and mineralogic effects discussed in the next section.

Calculations of the sort shown in Figure 11-2, as well as analogies drawn from nuclear explosion craters, have led to approximate empirical relationships between the diameter of a crater and the kinetic energy of the projectile (energy release). We might expect the crater volume and energy release to be roughly proportional, so it is not surprising that the diameter is proportional to about the one-third power of the energy release:

$$D = 0.55(\tfrac{1}{2}Mv^2)^{0.294} \tag{11-2}$$

where D is the diameter in meters, M is the mass in kilograms, and v is the velocity in kilometers per second. The effect of material properties is contained in the proportionality constants. Thus the size and shape of the initial explosive crater formed within the first second of impact depends mainly on the energy of the explosion and on the material properties, which is to say that gravity and other properties are unimportant in determining the ejection velocities and flow of material during initial impact. Craters will initially be the same on Earth, the Moon, and the various terrestrial planets given the same impact velocity and rock materials.

In contrast with the lack of gravitational effects during initial impact and excavation, gravity plays an important role in the evolution of the crater during the seconds, minutes, hours, and days following the impact.

The distribution of ejecta about the crater depends strongly on gravitational acceleration, as illustrated by the ballistic equation, which describes the distance fragments are thrown, for velocities well below the planetary escape velocity:

$$R = \frac{v^2 \sin 2\alpha}{g} \tag{11-3}$$

where α is the takeoff angle of the ejected fragment relative to the horizontal, v is the initial velocity, and R is the horizontal distance, or range, the fragment travels.

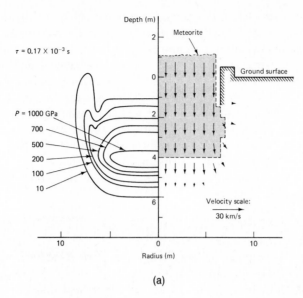

(a)

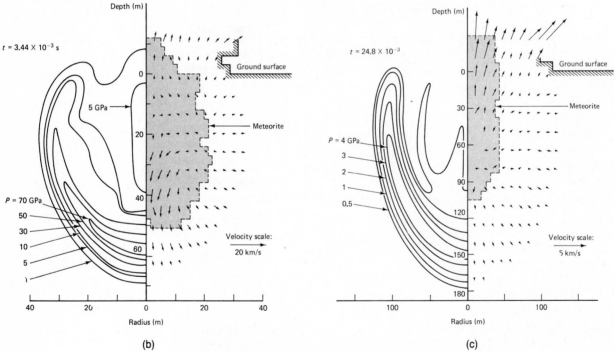

(b)   (c)

FIGURE 11–2 Results of numerical model of hypervelocity impact involving 1.2×10^4 kg iron projectile, a tuff medium, and 30 km/s initial velocity. Velocities are shown to right and pressures to left. (a) 0.17×10^{-3}s after impact. (b) 3.44×10^{-3}s. (c) 24.8×10^{-3}s. (From Bjork, 1961.)

Therefore, an equivalent meteoritic projectile on different planets produces approximately the same initial excavation, but on less massive planets where gravity is less, the ejecta are thrown more widely and fewer fragments fall back into the crater.

On Mercury, extensive secondary craters, which are formed by secondary impact of the ejecta, lie closer to the primary crater than on the Moon; this difference is apparently an effect of gravitation, which is 2.3 times greater on Mercury (Appendix B).

Large-scale gravitational collapse of the walls of a crater, involving surficial slumping and large-scale normal faulting, probably continues for days and weeks following impact, although stratigraphic relationships of some of the larger lunar basins suggest that much faulting occurred as ejecta were falling back. Large-scale slumping or faulting produce two important morphologic features: *concentric slump terrraces* around the rim and a *central uplift,* both of which are displayed by the lunar crater Aristarchus (Fig. 11-3). The central uplift is apparently produced by the converging toes of the concentric slumps. Small craters do not have high-enough rims relative to the rock strength and gravity to allow large-scale slumping. Craters without large-scale slumping and central uplift are called *simple craters.* Examples of simple craters include Meteor Crater (Fig. 11-1) and New Quebec (Table 11-3). Larger craters exhibiting the slump terraces, a central uplift, and a marginal valley are called *complex craters.* In addition, within complex craters, there are steps in complexity corresponding to increasing diameter:

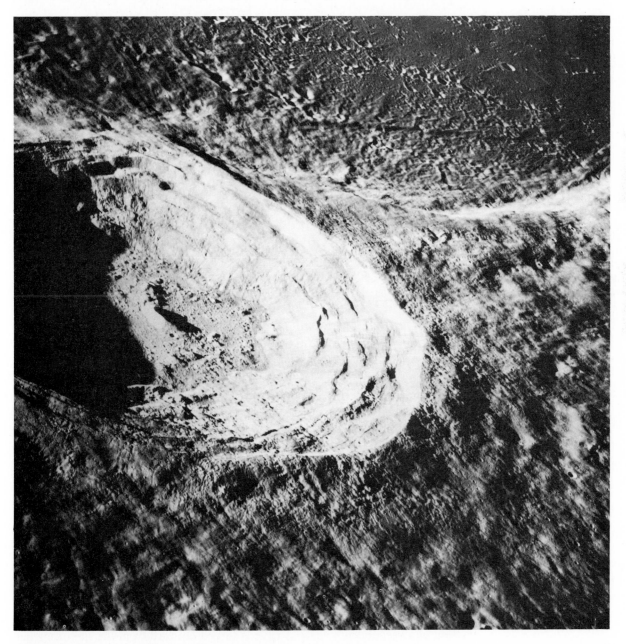

FIGURE 11–3 Oblique photograph of lunar crater Aristarchus (40 km) showing slump terraces and central peaks. (Apollo 15, National Space Science Data Center.)

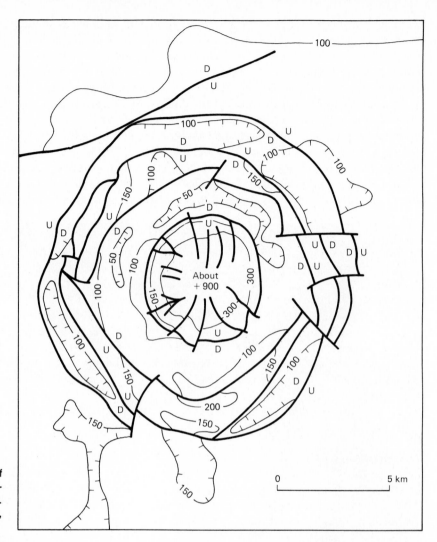

FIGURE 11–4 Structure-contour map of Wells Creek structure, Tennessee showing a central uplift. Contours in meters. (Modified after Wilson and Stearns, 1968.)

central peak → peak-ring → peak and rings → multi-ring basins. The largest basins undergo long-term isostatic adjustment.

A critical diameter exists for the transition between simple and complex craters, which varies with gravity and rock type. The critical diameter is about 2 km for sedimentary rocks and 4 km for crystalline rocks on Earth, 20 km on the Moon, and 10 km on Mercury and Mars, which have about the same gravity (Appendix B). The critical diameter on Ganymede and Callisto is small relative to their gravity because of their weak, icy crusts. Detailed structural mapping of deeply eroded craters allows the recognition of concentric normal faults and central uplifts; thus the distinction between simple and complex craters is not limited to morphologically intact craters. For example, Manicouagan, Sierra Madera, and Wells Creek are eroded complex craters (Table 11-3 and Fig. 11-4).

MICROSCOPIC EFFECTS OF IMPACT

A distinctive suite of microscopic structural and mineralogic changes takes place in the target rocks as a result of the transient high pressures that exist for a fraction of a second as the shock wave is forced outward from the site of impact (Fig. 11-2). The effects shown by a given sample in thin section depend strongly on its history and hence on its initial position within the target area. Three principal classes of effects are observed: high-pressure, high-strain-rate, and high-temperature effects.

High-Pressure Effects

High-pressure effects consist of the formation of high-density minerals that are stable only at high pressure and are preserved kinetically upon decompression. Quartz recrystallizes to coesite and stishovite. Coesite is stable between about 3 and 7.5 GPa at 25°C and has a density of 2920 kg/m^3 at room temperature and pressure. Stishovite is stable above about 8 GPa and has a density of about 4300 kg/m^3 at room temperature and pressure. In contrast, the density of quartz is 2650 kg/m^3. These high-density minerals were first produced experimentally and then searched for and found in some impact structures. Coesite and stishovite could be expected to form statically only within the upper mantle, but they could not be brought to the surface geologically without inverting to quartz because of the high inversion rates at high temperatures. Thus, coesite and stishovite are considered excellent criteria for hypervelocity impact. Diamond is occasionally found in impacts—for example, in some of the meteoritic fragments on the rim of Meteor Crater. In contrast with coesite and stishovite, diamond is more resistant to inversion and is brought to the surface in some diatremes without destruction (Chapter 7). Not all minerals recrystallize to high-pressure phases during impact. For example, the breakdown of the sodic plagioclase albite ($NaAlSi_3O_8$) to jadeitic pyroxene ($NaAlSi_2O_6$) and quartz (SiO_2), which occurs in high-pressure static metamorphism, is not found in impact metamorphism because the reaction requires diffusion to separate the albite into jadeite and quartz. There is insufficient time during impact for reactions requiring diffusive transport.

High-Strain-Rate Effects

High-strain-rate effects involve the progressive dislocation and destruction of the crystal structure. Some minerals, especially quartz, exhibit closely spaced microscopic lamellae, which are layers that have been partially or completely transformed to glass or are arrays of closely spaced dislocations (see Figure 11-5). Quartz also exhibits deformation lamellae in tectonically deformed rocks, but the tectonic lamellae have different crystallographic orientations from impact lamellae. Strain of the crystal structure during impact can also produce anomalous

FIGURE 11–5 Photomicrograph of shocked quartz sandstone showing abundant deformation lamelli and fractures, Wanapitei structure, Ontario, Canada.

optical properties. In its most extreme form, the minerals maintain their original morphologies but exhibit isotropic optical properties. For example, isotropic plagioclase, called *maskelynite,* has been observed in many impact structures and was first recognized a century ago in the Shergotty meteorite.

High-Temperature Effects

High-temperature effects result from pressures so high that upon sudden decompression the minerals have temperatures hundreds of degrees above their melting points. For example, quartz fuses to silica glass, called *lechatelierite,* and zircon fuses to zircon glass, called *baddeleyite;* these reactions require temperatures above 1500°C, which are considerably higher than normal earth-surface igneous temperatures. Highly melted impact glass that has been above 1500 to 1700°C may exhibit droplets of glasses of refractory minerals such as ilmenite, rutile, and zircon; these minerals are little affected by lower shock pressures. With extensive melting, magma is produced that is close to or identical to the composition of the country rock. These impact melts may have compositions distinctly different from typical terrestrial igneous rocks, whose compositions are determined by partial melting and crystal fractionation mechanisms rather than complete *in situ* melting. The melt formed may exhibit extensive flowage (Fig. 11-6). In large impacts extensive bodies of magma may be produced; for example, the Onaping Tuff in the Sudbury structure, Canada (100 km diameter), is over 1 km thick.

Ejected fragments of the Coconino Sandstone on the flanks of Meteor Crater exhibit a variety of microscopic effects of shock in quartz, including fractured grains, deformation lamallae, coesite, stishovite, and quartz glass (lechatelierite). No single sample exhibits all these features because they require different positions within the impact. The assemblage of microscopic effects in any sample may be correlated with the theoretical behavior of the rock during an event of shock compression followed by decompression. Table 11-2 summarizes the behavior of the quartz-rich Coconino Sandstone. The compression is characterized by progressive reduction of pore space, followed by behavior similar to that of pure quartz, followed by the formation of high-density polymorphs of quartz. Shocked Coconino Sandstone at Meteor Crater contains as much as 30 percent by weight of coesite and trace amounts of stishovite. In most cases material that was stishovite has melted to quartz glass or recrystallized to coesite upon decompression (Kieffer, 1971).

TABLE 11–2   **Regimes of Shock Compression for the Coconino Sandstone, Meteor Crater***

Pressure Range	Phenomena
0 to 0.2–0.9 GPa	Elastic deformation; very little grain damage.
0.2–0.9 to 3 GPa	Density of compressed sandstone less than quartz at zero pressure. Pores are still present. Grain fracturing.
3 to 5.5 GPa	Density of compressed sandstone greater than quartz at zero pressure and approaches the density of shocked single-crystal quartz at the higher end of the pressure range. Porosity is reduced to zero.
5.5 to 13 GPa	Pressure-density behavior of sandstone is essentially the same as single-crystal quartz. Possible formation of small amounts of coesite.
13 to 30 GPa	Density of shocked sandstone is greater than the density of shocked quartz, high-pressure phases are present.
Greater than 30 GPa	Density of shocked sandstones is that of stishovite. Quartz grains shocked above 35 GPa melt to silica glass upon release of pressure.

*Modified from Kieffer (1971)

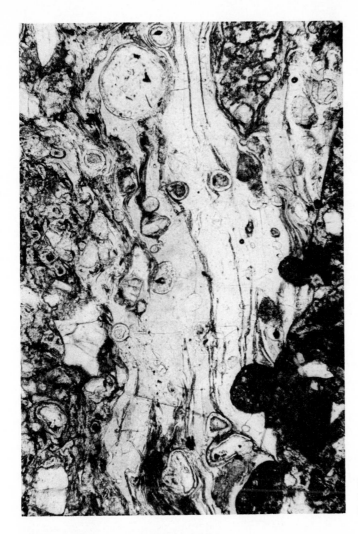

FIGURE 11–6 Photomicrograph of vessiculated impact glass, Wanapitei structure, Ontario, Canada.

OUTCROP-SCALE STRUCTURES

The microscopic effects of hypervelocity impact discussed earlier have been particularly useful in recognizing and documenting many of the older and more-deeply eroded structures, most of which are much larger than Meteor Crater and range up to 100 km (Sudbury, Canada, and Vredefort, South Africa). These more-deeply eroded structures also display a variety of impact-related structures on an outcrop scale, including fractures, faults, veins of impact melt (pseudotachylites), and veins, dikes, or lenses of mixed breccia and melted rock (Fig. 11-7). Among the most interesting outcrop-scale structures, however, are shatter cones.

Shatter cones are conical fractures that are decorated with diverging and branching longitudinal striae composed of rounded ridges and sharp troughs (Figs. 11-8 and 11-9). Complete conical fractures are rare; only conical segments, which may be nested in groups, are normally observed (Fig. 11-9). The conical segments range in size from a few centimeters to as much as 7 m. Naturally formed shatter cones are apparently confined to impact structures and are widespread in the more-deeply eroded structures. Shatter cones have also been formed in shock experiments and in underground nuclear explosions. The conical fractures are thought to form by an interaction between the shock wave and heterogeneities of the rocks, with the conical axis possibly oriented normal to the wave front.

FIGURE 11–7 Outcrop of dike of impact breccia, Sudbury structure, Ontario, Canada.

The shatter cones of a single outcrop are observed as intersecting conical segments, only a few of which may be observed at any moment during the progressive excavation of an outcrop with a hammer. This fragmentary view may be expanded if the orientations of the striae on each conical fracture segment are measured. If all the orientations from a single outcrop are plotted on an equal-angle (stereographic) projection, it is generally found that they define a small circle (Fig. 11-10). A small circle is the projection of a cone on an equal-angle spherical projection (Chapter 2); thus the striae and the conical surfaces containing them have a uniformly oriented conical symmetry throughout an entire outcrop. The diameter of the small circle is the apical angle of the cone. The cone axis should record the normal to the shock wave as it passed through the rock.

Shatter cones are widespread in the Witwatersrand Sandstone on the Vredefort ring structure in the middle of the Witwatersrand basin, South Africa (Fig. 11-11), which has a diameter of 100 km. The sandstones have been overturned outward and now dip inward toward the center of the structure at angles commonly ranging from 45° to 80°. The axes of the shatter cones in these

FIGURE 11–8 Shatter cone, Charlevoix structure, Quebec, Canada. (Photograph D. W. Roy.)

FIGURE 11–9 Shatter cones from Wells Creek structure, Tennessee. See also Figure 11-4. (Photograph courtesy of R. Dietz.)

overturned strata are somewhat dispersed but are roughly radial and plunge toward the center, with the apex of the cone pointing in the opposite direction, upward and outward (Fig. 11-11). If the shatter cones are considered to have formed during the initial impact, before the beds were folded over in the explosion, then the shatter cones may be rotated back to their initial orientation by rotating the beds back to horizontal about axes tangential to the ring structure. When this is done, the cone axes around the structure point inward and upward to a focus roughly 10 km above the base of the Witwatersrand System. These observations suggest that the shock occurred before any major structural deformation took place and that the shock was directed downward and outward from a point that must have been close to the original land surface. Similar reconstructions have been made at Gosses Bluff, Australia; Sierra Madera, Texas; and Charlevoix, Canada.

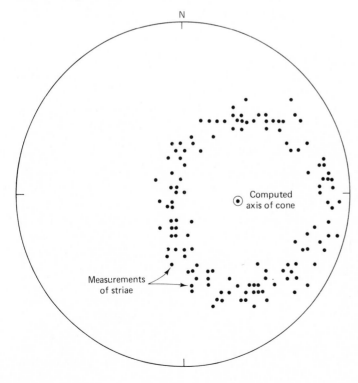

FIGURE 11–10 Equal-angle spherical projection showing measurements of shatter-cone striae from outcrop in Charlevoix structure, Quebec, Canada. (Courtesy D. W. Roy.)

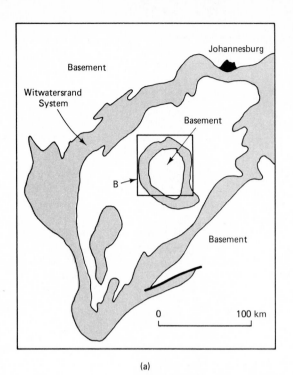
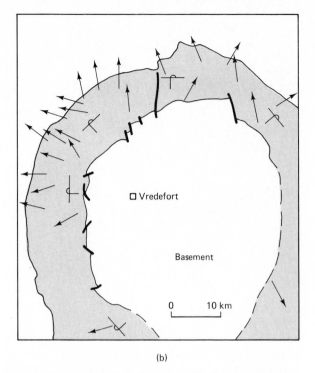

(a)   (b)

FIGURE 11–11 Vredefort structure in the center of the Witwatersrand basin near Johannesburg, South Africa. (a) Simplified map. (b) Map showing *in situ* orientations of shatter-cone apices. The axes plunge in the opposite direction, that is, inward. Note the outward overturning of the entire Witwatersrand sedimentary sequence. (Data from Manton, 1965.)

RECOGNITION OF IMPACT STRUCTURES

In the absence of meteoritic material, how can we be sure that a deeply eroded circular region of highly disturbed rocks is of hypervelocity impact origin? Strongly deformed areas with a circular plan and perhaps a circular depression are also produced by internal processes. These are primarily intrusive or extrusive structures and include salt domes, diatremes, ring dikes, and mud diapirs, as well as sink holes. The presence of igneous rocks may not indicate an internal origin, particularly at an early stage of exploration, because melting is widespread in larger impact structures. Indeed, many structures now known to be of impact origin—even Meteor Crater, Arizona—were originally thought to be of igneous origin.

On close examination, impact structures are quite different from the alternatives of internal origin. This is to be expected. The impact of a body smaller than an asteroid excavates a crater tens of kilometers in diameter in a few seconds with peak pressures exceeding 100 GPa and many cubic kilometers subjected to pressures of 1 to 10 GPa. In contrast, crustal pressures of internal origin rarely exceed 1 GPa, and even catastrophic volcanic eruptions take tens of minutes to hours. Temperatures during impact locally reach 10,000°C, and reaction products are locally preserved that record temperatures in the range 1500 to 2000°C, whereas volcanic temperatures are generally below 1300°C. The following list of features commonly displayed by impact structures is based on previous discussion: (1) strong structural disturbance in a roughly circular region with outward overturning, (2) high-pressure and high-temperature shock metamorphism, (3) breccia with shock metamorphosed fragments of country rock, (4) veins of

TABLE 11–3   **Probable Impact Structures***

Name	Diameter (km)	Longitude	Latitude
Canada			
Brent (Ont.)	4	078°29′W	46°05′N
Carswell Lake (Sask.)	32	109°30′W	58°27′N
Charlevoix (Que.)	35	070°18′W	47°32′N
Clearwater Lakes (Que.)	30, 15	074°30′W	56°13′N
Deep Bay (Sask.)	10	102°59′W	56°24′N
Holleford (Ont.)	2	076°38′W	44°28′N
Lac Couture (Que.)	10	075°18′W	60°08′N
Lake Mistastin (Lab.)	20	063°18′W	55°53′N
La Malbaie (Que.)	37	075°15′W	47°35′N
Manicouagan (Que.)	65	068°42′W	51°23′N
New Quebec (Que.)	3	073°40′W	61°17′N
Nicholson Lake (N.W. Terr.)	13	102°41′W	62°40′N
Pilot Lake (N.W. Terr.)	5	111°01′W	60°17′N
St. Martin (Man.)	24	098°33′W	51°47′N
Steen River (Alb.)	25	117°38′W	59°31′N
Sudbury (Ont.)	100	081°11′W	46°36′N
Wanapitei (Ont.)	9	080°44′W	46°44′N
West Hawk Lake (Man.)	3	095°11′W	49°46′N
United States			
Crooked Creek (Mo.)	5	091°23′W	37°50′N
Decaturville (Mo.)	6	092°43′W	37°54′N
Flynn Creek (Tenn.)	4	085°37′W	36°16′N
Glassford (Tenn.)	4	089°49′W	40°35′N
Howell (Tenn.)	2	086°35′W	35°15′N
Jeptha Knob (Ky.)	4	085°06′W	38°06′N
Kentland (Ind.)	6	087°24′W	40°45′N
Manson (Iowa)	31	094°31′W	45°35′N
Middlesboro (Ky.)	7	083°44′W	36°37′N
Serpent Mound (Ohio)	6	083°25′W	39°02′N
Sierra Madera (Tx.)	13	102°55′W	30°36′N
Wells Creek (Tenn.)	14	087°40′W	36°23′N

*After Dence (1972)

pseudotachylite, (5) shatter cones, (6) fine-grained igneous rocks with compositions similar to average country rock and often different from standard igneous rocks, and (7) undeformed rocks at depth, found by drilling.

Many circular structures have been discovered that display enough features indicative of impact that an extraterrestrial origin seems highly probable. A list of probable impact structures, most of which display shock metamorphism, is given in Table 11-3.

Canada leads the list in numbers of probable impact structures, which is not due to chance. Following the discovery of New Quebec structure (3 km) in 1950, Beals and his colleagues at the Dominion Observatory began a systematic search for ancient impact structures in Canada using topographic maps and aerial photographs, with geophysical study and drilling of a number of promising sites.

Some impacts appear to form in swarms, as is seen with the young meteoritic craters (Table 11-1). Ries (24 km) and Steinheim (3.5 km) craters in Germany apparently formed at the same time, 14.8 m.y. ago. Steinheim crater is about 40 km west of Ries and several other depressions 25 to 100 km northeast of Ries have been suggested to be impact structures. The craters are the same age as the Moldavite tectites. Most structures are insufficiently well dated to determine which may have formed in swarms.

Most probable and possible impact structures were first recognized from their circular topographic forms. Perhaps in the future many less-obvious structures will be discovered from outcrop-scale and microscopic effects of impact. This was the case with the Charlevoix structure (35 km) in Quebec, Canada, whose circular form is partially obscured by the St. Lawrence River. In 1966 Rondot discovered a peculiar type of widespread conical striated fractures during his mapping of the Charlevoix area. These fractures were later identified as shatter cones, which—together with an abundance of breccia zones and pseudotachylites—suggested an impact origin. This origin was later confirmed with more-detailed study.

In addition to the proven and probable impact structures of Tables 11-1 and 11-3, many features of possible impact origin have been identified that have been insufficiently explored to assign an unambiguous origin. Almost all the probable impact structures have been found undeformed at the surface in cratonic areas. The observed present flux of meteorites on the earth and past fluxes determined from crater densities on surfaces of varying ages on the Moon suggest that many more craters are to be found on Earth. A number of them will be buried, and others should be found in a deformed state within mountain belts.

Lunar crater densities suggest much higher rates of cratering prior to about 3.9 b.y. ago at the time of "final bombardment" of planetary accretion. This is roughly the age of the oldest rocks yet identified on Earth. We can expect that buried within the palimpsest histories of the Precambrian shield there are much-deformed and modified scars of asteroidal impacts that lie in wait of recognition.

EXERCISES

11-1 Formulate an optimal and inexpensive program for searching for undiscovered impact structures, concentrating on regions most easily searched where greatest densities of structures would be expected, at the scales appropriate to the methods of search.

11-2 Meteoritic fragments are found around Meteor Crater, Arizona, to a radius of about 5 km. If we were to assume that they are fragments of the original meteorite ejected from the crater, estimate their initial velocities of ejection. In contrast, the blanket of continuously ejected country rock extends to a radius of 2 km; estimate its velocity. Compare these estimates with the calculated velocities in Figure 11-2.

SELECTED LITERATURE

FRENCH, B. M., AND SHORT, N. M., ed., 1968, *Shock Metamorphism of Natural Materials,* Mono Book Co., Baltimore, Md., 644 p.

GRIEVE, R. A. F., AND HEAD, J. W., III, 1983, The Manicouagan impact structure: analysis of its origins, dimensions and form: J. of Geophys. Research, v. 88, p. A807–A818.

MUTCH, T. A., ET AL, 1976, *The Geology of Mars,* Princeton University Press, Princeton, N.J., 400 p.

MELOSH, H. J., 1982, A schematic model of crater modification by gravity: J. of Geophys. Research, v. 87, p. 371–380.

RODDY, D. J., PEPIN, R. O., AND MERRILL, R. E., ed., 1978, *Impact and Explosion Cratering,* Pergamon, Oxford, 1300 p.

SHORT, N. M., 1975, *Planetary Geology,* Prentice-Hall, Englewood Cliffs, N.J., 369 p.

WILSON, C. W., JR., AND STEARNS, R. G., 1968, Geology of the Wells Creek structure, Tennessee: Tenn. Div. of Geology, Bull. 68, 236 p.

WILSHIRE H. T., OFFIELD, W., HOWARD, K. A., AND CUMMINGS, D., 1972, Geology of the Sierra Madera cryptoexplosion structure, Pecos County, Texas: U.S. Geol. Survey Prof. Paper 599-H, 42 p.

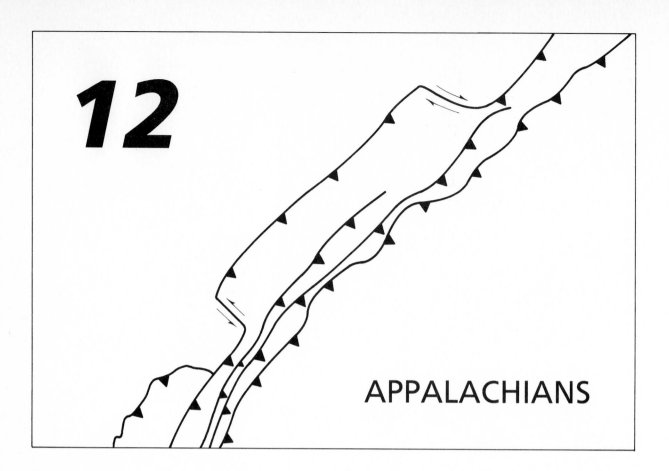

12

APPALACHIANS

INTRODUCTION

In these last two chapters on the Appalachian and Cordilleran mountain belts, we introduce the field of *regional structural geology,* which deals with the structure of the large deformed zones of the earth, especially compressive mountain belts. Regional structural geology is a subject quite different in flavor from the rest of structural geology because it is first of all concerned with unique historical events and only secondarily with general classes of deformational processes. The goal of workers in regional structural geology is primarily to decipher the history and the associated present structural geometry and prior paleogeographies of some area of our planet. Regional structural geology also leads to important insights into the underlying physical processes and recurring historical themes that are repeated in space and time among the various mountain belts of the world. Nevertheless, the field of regional structural geology is especially concerned with the deformational history of the earth.

Accordingly, we outline the regional structural geology of parts of two specific mountain belts, the Appalachian and Cordilleran belts, rather than treat the subject from the point of view of generalities. Even within these mountain belts, we choose the most throughgoing and least-controversial features. This overview gives some perspective on the fabric and history of parts of two of the many orogenic belts of the earth. In the process, some of the important methods used in regional structural geology will become apparent.

The Paleozoic Appalachian orogenic belt is one of the great linear mountain chains of the world. Of course, it does not rival the currently active Alpine-Himalayan or Circum-Pacific chains, but it is nevertheless substantial, extending some 3200 km from Newfoundland to Alabama (Fig. 12-1), with an exposed width

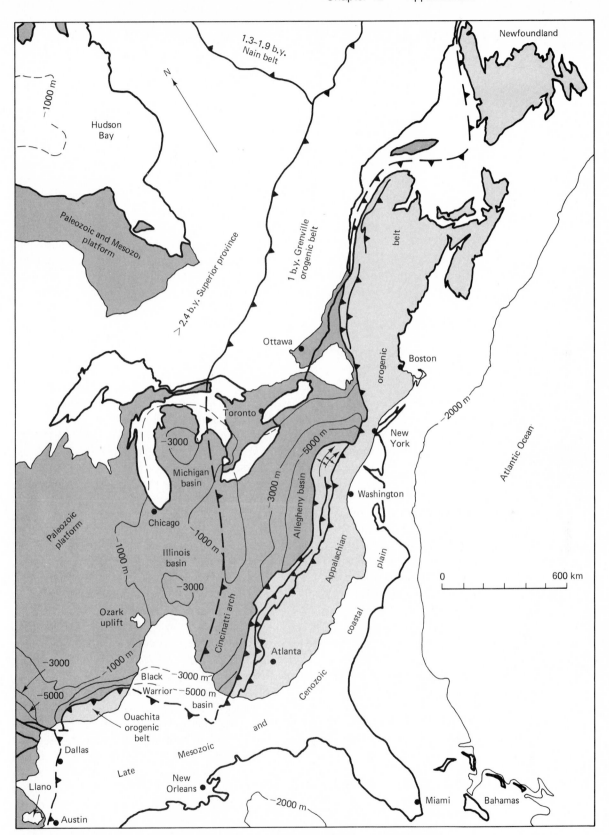

FIGURE 12–1 The regional setting of the Paleozoic Appalachian orogenic belt and its southwestward continuation, the Ouachita orogenic belt. The adjacent craton to the northwest is underlain by rocks of the billion-year-old Grenville orogenic belt, which strikes subparallel to the Appalachians.

of 150 to 650 km. The narrowest part is at New York City. Furthermore, the Appalachians today are just a fragment of a much larger orogenic system of nearly global scale that has been dispersed through Mesozoic and Cenozoic continental drift. The Appalachians have a direct northern continuation in the Caledonian mountain belt of Great Britain (Johnson and Stewart, 1963), east Greenland (Haller, 1971), and northwestern Scandinavia (Gee, 1975). Other branches extend, perhaps less clearly, into western continental Europe, northwest Africa, the Canadian Arctic, and the Urals. The southern Appalachians have an incompletely understood southern extension under the coastal plain of the Gulf of Mexico to the Ouachita Mountains of Arkansas and Oklahoma (Fig. 12-1) and the Marathon area of west Texas. Further continuations of these zones of Paleozoic deformation may exist in Mexico, western United States, and northern South America. Even without all these extensions and possible extensions, the contiguous Appalachians are an orogenic belt that is more than most geologists can encompass in a full and rewarding career. In this chapter we emphasize only some of the most throughgoing and important features. Furthermore, the Appalachians are a region of extremely active research; therefore, we emphasize mainly the better-established results.

WESTERN MARGIN OF THE APPALACHIANS

The western edge of the Appalachian mountain belt is a foreland fold-and-thrust belt impinging on the stable North American craton. This craton in Canada is eroded to the level of the Precambrian basement, whereas thin, undeformed platformal sediments overlie the basement in the United States (Fig. 12-1). The Precambrian basement immediately adjacent to the Appalachians that is exposed in the shield of Canada and the Adirondack Mountains of New York State is part of the billion-year-old *Grenville orogenic belt,* which extends to the northeast in surface exposures as far as southern Greenland and extends under the platformal sediments of the eastern and southern United States to reappear in the Llano uplift of Texas. These Precambrian basement rocks also extend into the western Appalachians under the deformed Paleozoic sedimentary rocks of the marginal fold-and-thrust belt and reappear at the surface as redeformed and remetamorphosed allochthonous basement massifs on the inner edge of the Appalachian fold-and-thrust belt (Fig. 12-2). By *allochthonous* we mean the rocks are displaced significantly from their original site; the word comes from the Greek for *other ground* or *earth* and is most commonly applied to far-travelled thrust sheets. In contrast, by *autochthonous* we mean the rocks are *in situ;* the word comes from the Greek for *same ground.* These allochthonous massifs of redeformed Grenville basement within the Appalachians form a discontinuous belt including the Great Smoky Mountains, Blue Ridge, New Jersey Highlands, Berkshire Highlands, Green Mountains, and the Long Range of Newfoundland (Fig. 12-2). A few strongly deformed and remetamorphosed fragments of basement rocks of possible Grenvillian affinity lie a bit farther east of this belt, but as we continue farther into the Appalachians, we quickly run into rock assemblages suggestive of closed ocean basins, island arcs, and fragments of continental lithosphere exotic to North America. Therefore, the deformation at the western edge of the Appalachian mountain belt involves older basement rocks of the North American continent, but as we move deeper into the mountain belt to the east we encounter deformed and metamorphosed rocks that have been added onto the continent as a result of plate motions and orogenic processes.

The western edge of Appalachian deformation, when seen in map view, describes a series of great arcs arranged like the salients and recesses of late

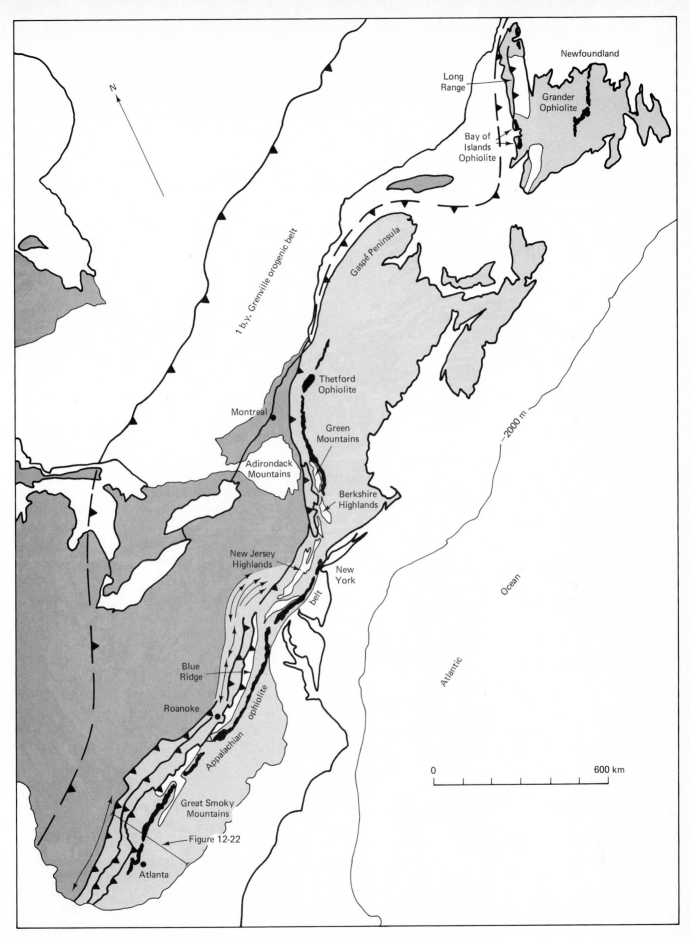

FIGURE 12–2 Map showing the principal areas of exposure of deformed billion-year-old Grenville basement within the Appalachians and the Appalachian ophiolite belt, representing the site of a closed ocean basin.

419

medieval fortifications. The *salients* form broad arcs that are convex to the craton, whereas the *recesses* are tight, sharp, and concave (Fig. 12-2). The Appalachian mountain belt is commonly divided on the basis of these salients and recesses. The *southern Appalachians* form the salient from Alabama to about Roanoke, Virginia; the *central Appalachians* form the salient from Roanoke to New York City. The sharp boundary between the southern and central Appalachians is seen in the satellite image of Figure 12-3 as an abrupt change in strike. The *northern Appalachians* form a salient extending from New York City to a recess in the Gulf of St. Lawrence between the Gaspé Peninsula and Newfoundland. *Newfoundland* may be part of an additional salient that may have extended into Great Britain prior to the opening of the North Atlantic in Cenozoic time. These four salients provide a natural framework for discussing the North American Appalachian foreland.

Each foreland salient exhibits a different structural style, stratigraphic sequence, or time of major deformation. Nevertheless, each salient exhibits major horizontal compressive deformation dominated by west- or northwest-vergent thrusting toward the craton. The deformation is thin-skinned or detached from the older sediments and Grenville basement along a basal décollement. Figure 12-4 shows the thin-skinned imbricate thrust structure of the southern Appalachian foreland. The deep horizon of décollement is in the Middle Cambrian Rome Formation, which contains abundant shale. The horizontal compression of the foreland structure is enormous, with the Grenville basement massif of the Blue Ridge having been displaced from at least 200 km to the southeast based on seismic-reflection profiling (to be discussed in a later section). This strong deformation involves rocks as young as Carboniferous and is part of the Appalachian orogeny of latest Carboniferous and Permian age, which is responsible for the Appalachian foreland deformation in both the southern and central Appalachians.

North of New York City, in the northern Appalachian and Newfoundland salients, the most-important period of foreland deformation is substantially older, during the middle and late Ordovician Taconic orogeny. The Taconic foreland also displays important thin-skinned deformation. Important middle and late Paleozoic deformation dominates the interior of the northern Appalachians and makes much of the Ordovician Taconic deformation rather obscure in New England.

Considering the entire length of the Appalachian mountain system in North America, four major periods of orogenic activity are well documented, although the exact timing and manifestation varies in detail from place to place. They are the latest Precambrian–earliest Cambrian *Avalonian orogeny,* the middle and late Ordovician *Taconic orogeny,* the Devonian *Acadian orogeny,* and the latest Carboniferous and Permian *Alleghenian* or *Appalachian orogeny.* The first three are best displayed in the northern Appalachians and Newfoundland. The last one is best known in the central and southern Appalachians.

FORELAND SEDIMENTARY RECORD

The sedimentary rocks along the western edge of the Appalachian mountain belt span virtually the entire Paleozoic and provide an extensive indirect record of nearby mountain building. For example, the times of orogenic activity are recorded in the foreland by more-rapid rates of sedimentation with extensive clastic influx from the east, whereas times of no orogenic activity are recorded by lower rates of sedimentation, with much of the sediment being carbonates and minor clastic influx from the continental interior. These times of orogenic and

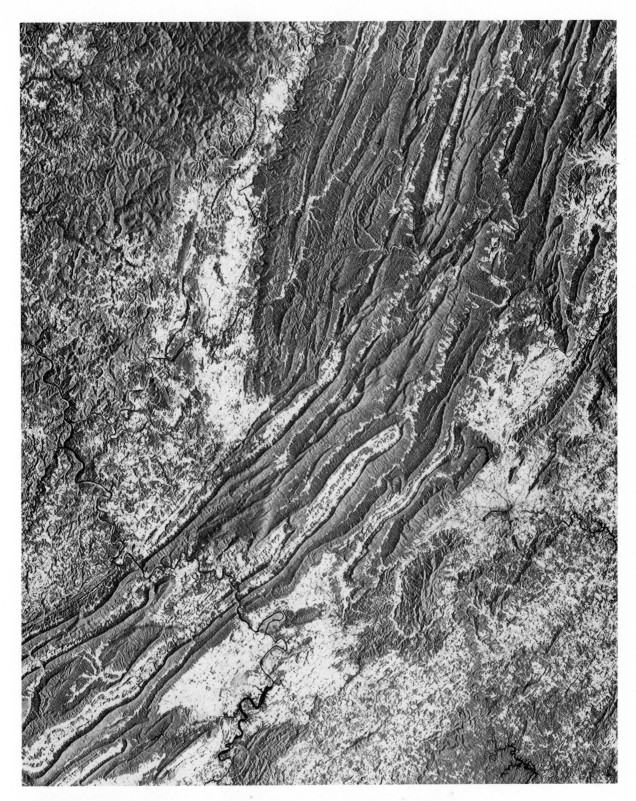

FIGURE 12–3 Satellite image illustrating the abrupt change in strike between the southern and the central parts of the Appalachian fold-and-thrust belt near Roanoke, Virginia. The smooth topography to the northwest (upper left) is underlain by flat-lying undeformed Carboniferous sediments at the periphery of the Appalachians. The smooth topography to the southeast (lower right) is underlain by strongly deformed Paleozoic metamorphic rocks of the Appalachian Piedmont. The meandering river in the lower left is the New River.

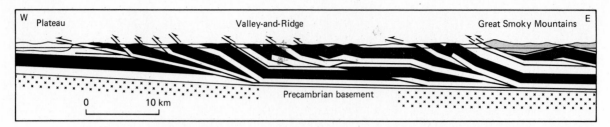

FIGURE 12–4 Cross section of the southern Appalachian fold-and-thrust belt in Tennessee. (Modified after Roedder, Gilbert, and Witherspoon, 1978.)

nonorogenic sedimentation may be seen in Figure 12-5, which is a stratigraphic diagram of the Appalachian foreland with the deformation removed. The base of the Silurian is arbitrarily chosen to be horizontal. This stratigraphic cross section extends from the edge of the craton near the West Virginia–Ohio border to the western part of the Blue Ridge. It displays several important features of the Appalachian foreland and, incidentally, of forelands in general.

The most-obvious property of the restored stratigraphic diagram is the enormous increase in thickness of the entire Paleozoic sequence from the edge of the craton to the Blue Ridge basement massif. The Cambrian through Devonian section is 1300 m thick in the west and 8000 m thick in the east. Most individual time periods also display an eastward-thickening sedimentary sequence. This fact of eastward-thickening of the foreland stratigraphic sequence was recognized by James Hall in the 1830s and 1840s, early in the study of the Appalachians. A correlation between orogenic activity and increased stratigraphic thickness has been recognized in a majority of mountain belts, an observation that has led to the hypothesis that thick sedimentation is genetically related to orogenic activity. By the 1860s the thick accumulations of sediments in deformed mountain belts were called *geosynclines* by Dana. A variety of geosynclinal theories of mountain building were developed in the next 100 years. With the rise of plate tectonics, it has become apparent that important orogenic activity commonly develops when compressive plate boundaries encounter thick sedimentary sequences and deform them into mountainous topography.

When we examine the stratigraphic section of the Appalachian foreland (Figure 12-5) in more detail, we see that there are four major eastward-thickening wedges of clastic sediment with intervening carbonate-rich sequences, also generally eastward-thickening. The upper three clastic sequences represent detritus shed from three important orogenic events of the Appalachian mountain belt. The Ordovician clastic wedge ·records the Taconic orogeny. The Devonian clastic wedge records the Acadian orogeny. The Upper Carboniferous and Permian clastic wedge records part of the Alleghenian or Appalachian orogeny, which eventually deformed this foreland sequence into a fold-and-thrust belt.

One of the best-known clastic wedges is the Acadian wedge centered in New York State; it is commonly called the *Catskill delta,* although some of the sediments were deposited beyond the deltaic environment. Figure 12-6 is an east-west stratigraphic cross section of the Devonian sediments of New York State, extending from Lake Erie on the west to the eastern edge of the Catskill Mountains, overlooking the Hudson River on the east. The preorogenic Lower Devonian section is very thin and dominated by carbonate and orthoquartzite. The Middle and Upper Devonian strata display a substantial increase in rate of sedimentation with the detritus derived from the east. We see that the detritus builds out into the foreland with time. Consequently, traveling along any time line we pass eastward into shallower-water and nonmarine facies; we also pass into shallower-water facies by climbing up any stratigraphic section. Most of the rock

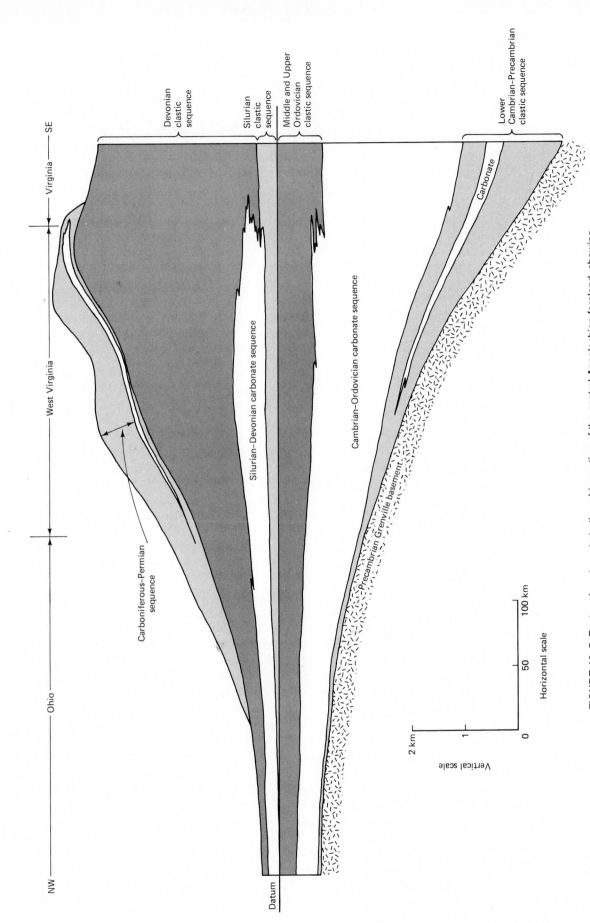

FIGURE 12–5 Restored east-west stratigraphic section of the central Appalachian foreland, showing the main periods of influx of clastic detritus. The base of the Silurian is arbitrarily chosen as the horizontal datum. (Simplified after Colton, in Fisher and others, ed., Studies in Appalachian Geology: Central and Southern, © 1970, Wiley-Interscience.)

423

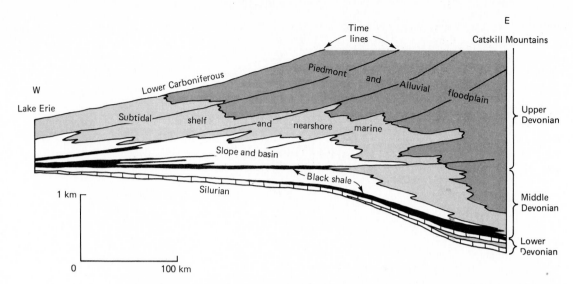

FIGURE 12–6 East-west stratigraphic section of the Devonian of New York State, showing the westward progradation of Middle and Upper Devonian clastic sediments derived from the active Acadian mountain belt to the east. (Simplified after Moore, 1941.)

stratigraphic units are time-transgressive because they represent the outbuilding of the sediment wedge and the filling of the foreland basin as the Acadian orogeny proceeded. The Ordovician clastic wedge also displays a filling of the foredeep basin as the Taconic orogeny proceeded.

Studies of sediment-transport directions in the Ordovician through Carboniferous clastic wedges show that the sediments were indeed derived from the east or southeast, from the interior of the mountain belt (McIver, 1970). In contrast, similar studies of the Upper Precambrian and Lower Cambrian clastic sequence (Fig. 12-5) show that it is derived largely from the west or northwest, from the interior of North America (Schwab, 1970). These oldest sediments of the western margin of the Appalachians appear to represent an eastward-thickening continental-margin sequence derived from the west and not an orogenic clastic wedge derived from the east. The transition into late Cambrian and early Ordovician carbonate deposition apparently reflects the Cambrian worldwide rise in sea level, which flooded the craton by late Cambrian time. The same change from Upper Precambrian and Lower Cambrian clastics to Upper Cambrian carbonates is seen on the opposite side of North America, in the Cordilleran mountain belt (Fig. 13-4). Sedimentological studies of the carbonate sequence show that the rocks were deposited very close to sea level, with many of the beds showing intertidal characteristics. During the late Cambrian and early Ordovician, eastern North America was fringed by a great carbonate bank. Modern-day analogies might be the Bahamas Bank, Florida Shelf, or Yucatan Platform.

The change from late Precambrian clastic sedimentation to Cambrian and early Orodovician carbonate-bank sedimentation also apparently reflects a change in latitude of North America due to continental drift. The Upper Precambrian sequence in both the southern Appalachians and Cordillera contains glacial sediments, suggesting a high latitude, whereas by early Ordovician, the Appalachian margin of North America was close to the equator and oriented roughly east-west according to paleomagnetic data.

The Upper Precambrian through Ordovician clastic and carbonate sequence has many similarities with sedimentary sequences of present-day stable continental margins, such as the rifted margins of the Atlantic Ocean. The Appalachian sequence thins continentward and passes gradationally into the cratonic sequences of the continental interior. Furthermore the oldest sediments are apparently rift-related, containing basalts and rhyolites, and the Precambrian basement is intruded by basaltic dikes in many places. These various igneous rocks near the base of the Upper Precambrian section generally give isotopic dates of about 800 m.y., suggesting this as the age of initial rifting of the Proto-Atlantic (IAPETUS) ocean. The Paleozoic deformation has been so extensive that important late Precambrian grabens have been largely obscured, but they are known to exist in some areas and to have remained moderately active through the Cambrian. The best-known normal-fault system of the Appalachian foreland is the Rome trough fault system (Fig. 12-7), which is known through deep drilling and seismic-reflection profiling for petroleum exploration. Late Precambrian grabens that

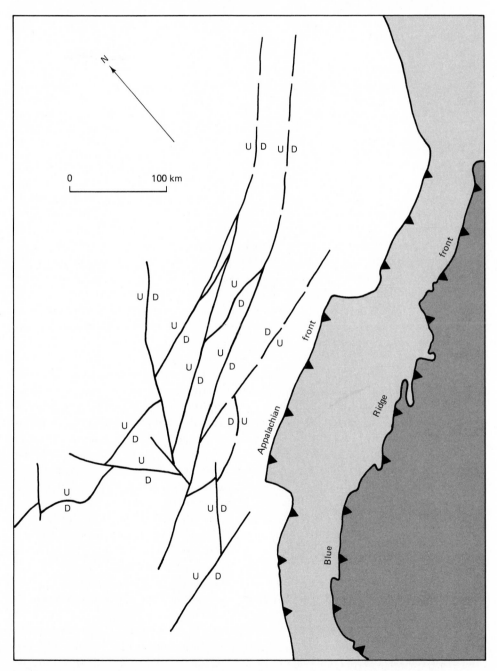

FIGURE 12–7 Map of the Late Precambrian and Cambrian Rome trough normal-fault system of the central Appalachian foreland. (After Harris, 1975.)

record the opening of the Proto-Atlantic ocean are also seen in seismic reflection profiles below the great thrust sheets of the inner Piedmont of the southern Appalachians (Figs. 12-21 and 12-22).

There is no important tectonic disturbance of the Appalachian foreland between the initial opening of the Proto-Atlantic ocean about 800 m.y. ago and the middle to late Ordovician Taconic orogeny. This is significant because important orogenic activity is recorded during this period in more interior parts of the Appalachians. Near the Cambrian-Precambrian boundary, orogenic activity is recorded in the eastern Appalachians in the Avalon Peninsula of eastern Newfoundland, in the Boston area of eastern New England, and in the eastern Piedmont of the southern Appalachians. These areas display important igneous activity and metamorphism about 550 to 600 m.y. ago in a tectonic event generally called the Avalonian orogeny, named for the Avalon Peninsula of eastern Newfoundland. It is likely that the rocks involved in the Avalonian orogeny were far from North America in earliest Cambrian time and that they were transported into North America during one of the three Paleozoic orogenies recorded in the Appalachian foreland.

TACONIC OROGENY IN THE FORELAND

The first signs of the destruction of the early Paleozoic stable continental margin of the Appalachian foreland are an increase in water depth of the carbonate bank and an influx of a thick wedge of mud and sand from the east and southeast (Fig. 12-5). During Cambrian through early Ordovician time, the average rates of deposition were typically 10 to 20 m/m.y. with much of the material in the stable continental margin being carbonate. The thicker parts of the overlying Middle Ordovician Taconic clastic wedge display rates of deposition more than an order of magnitude faster, not less than 200 to 400 m/m.y., with much of the material being eroded detritus. Studies of sediment transport show that this detritus was derived from the interior of the Appalachians (McIver, 1970). Therefore, what was previously open ocean to the east had become a mountainous sediment source.

This sediment source did not arise simultaneously throughout the length of the Appalachians. The Ordovician clastic wedge can be accurately dated by graptolites, although the exact ages of the graptolite zones in millions of years is poorly known. The transition from carbonate-bank to clastic-wedge sedimentation began very low in middle Ordovician during graptolite Zone 9 in both western Newfoundland and in eastern Tennessee, whereas it began well up in middle Ordovician, graptolite Zone 12, in eastern New York State. The Taconic clastic wedge reached its maximum extent in late Ordovician, spreading to Lake Ontario and beyond (see Fig. 12-8). Therefore, the Taconic orogeny was not wholly synchronous throughout the length of mountain belt (Rodgers, 1971).

As we continue to examine the Taconic orogeny from the perspective of the Appalachian foreland, we find deformation encroaching from the east. Moving upward stratigraphically or structurally through the clastic wedge in many parts of the Appalachian foreland, we encounter thrust sheets composed of older rocks—Cambrian and Lower Ordovician red, green, and black muddy rocks, now slates (Fig. 9-8)—lying structurally above the Taconic clastic wedge, which in turn overlies the carbonate shelf sequence (Fig. 12-8). These allochthonous muddy rocks, called the *Taconic sequence,* are the same age as the underlying Cambrian and Lower Ordovician carbonate bank but are a distinctly different type of sediment, implying very substantial displacement on the bounding thrust faults.

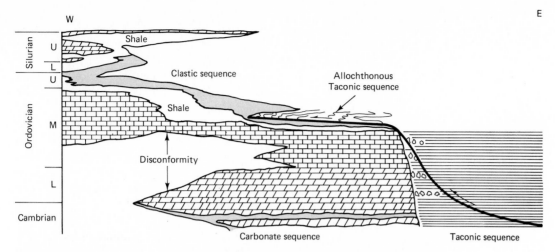

FIGURE 12–8 East-west schematic stratigraphic section across central New York State from Lake Ontario to western New England, showing the stratigraphic relationships surrounding the Taconic orogeny. The vertical dimension is time; therefore, nondeposition and eroded rock is shown as blank space. (After Rodgers, Geol. Soc. Amer. Bull., 1971, v. 82, p. 1148.)

The evidence bearing on the original paleogeographic position of the allochthonous Taconic sequence is discussed below.

The allochthonous Cambrian and Lower Ordovician slates of the Taconic sequence are best known from a stack of erosionally isolated klippen underlying the Taconic Mountains east of the Hudson River in eastern New York State and southwestern Vermont (Fig. 12-9). The Taconic sequence is clearly allochthonous because it is nearly everywhere seen structurally to overlie Middle Ordovician flysch of the clastic wedge (Figs. 12-8 and 12-9), which passes downward stratigraphically into the Lower Ordovician and Cambrian carbonate-bank sequence. This may be seen in the Vermont Marble Valley between the Taconic klippen and the Green Mountains Precambrian massif in southwestern Vermont (Fig. 12-15). Furthermore, the autochthonous carbonate sequence can be traced around the northern end of the klippen. The allochthonous nature of the Taconic sequence is thus well documented.

Similar allochthonous Upper Precambrian through Middle Ordovician slate masses of the Taconic sequence, analogous to the Taconic klippen of New York State and Vermont, are widespread in the Appalachian foreland, being well documented in Pennsylvania, New Jersey, the Gaspé Peninsula, and western Newfoundland. Allochthonous sheets of the Taconic sequence are also possibly present in the southern Appalachians.

Most of these allochthonous masses are known to have been emplaced during the Taconic orogeny, based on the presence of submarine slide and mudflow breccias, called *wildflysch breccias,* containing blocks of the Taconic sequence but interbedded as olistostromes within the flysch of the Middle Ordovician clastic wedge. These breccias apparently slumped off the front of the advancing Taconic thrust sheets in the submarine realm. In contrast, most of the clastic wedge was derived from erosion of subaerially exposed parts of the advancing mountain belt.

The Middle Ordovician clastic wedge and underlying Lower Paleozoic carbonates and Precambrian basement were also enveloped in the encroaching deformation. The most straightforward evidence of this fact is the Taconic angular unconformity (Fig. 12-10). Indeed, this unconformity, first discovered in the 1830s, was long considered to bracket the Taconic orogeny. The deformed beds below the unconformity were considered to predate the orogeny and the overlap-

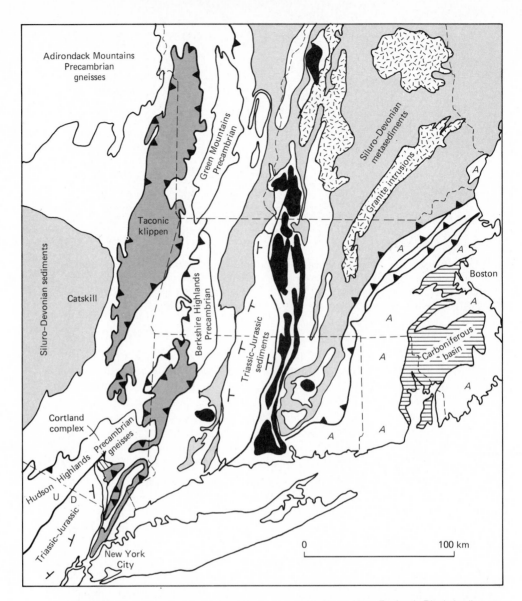

FIGURE 12–9 The Appalachian orogenic belt in southern New England. Black bodies are Ordovician and older gneisses lying along the Bronson Hill anticlinorium. Unlabeled white areas are generally Cambrian and Ordovician sediments and metasediments, except possibly in the east where ages are less certain. Areas labeled *A* are underlain by metamorphic rocks of known or probable Avalonian affinities, that is, Upper Precambrian and Cambrian metasedimentary volcanic and plutonic rocks exotic to North America; the metasediments contain rare fossils of European affinities. Light screened areas represent Siluro-Devonian strata above the Taconic unconformity. Dark screened areas are underlain by Ordovician, Cambrian, and Upper Precambrian strata of the allochthonous Taconic sequence. (Compiled from maps of Williams, 1978, and Robinson and Hall, 1980.)

ping beds to postdate the orogeny. Gradually, our perspective on the various manifestations of orogeny have broadened significantly. For example, the deformed beds below the angular unconformity at Catskill, New York (Fig. 12-10), are Middle Ordovician, Zone 13, flysch of the clastic wedge, which is just as much a manifestation of the orogeny as the deformation and the angular unconformity.

The Taconic unconformity sits on a considerable variety of deformed rocks. A few kilometers east of the outcrop at Catskill (Fig. 12-10), across the Hudson River, the unconformity sits on deformed rocks of the Taconic klippen (Fig. 12-9), whereas 150 km to the south, in the area west of New York City, it sits on

FIGURE 12–10 The Taconic angular unconformity, Catskill, New York, later deformed so that the overlying Upper Silurian dolomites now dip to the west (left) and the underlying Middle Ordovician Normanskill Graywacke, part of the Taconic clastic wedge, stands vertically.

Precambrian basement, deformed Cambrian and Lower Ordovician carbonate-bank sequence, and the deformed Middle Ordovician slates of the clastic wedge. Therefore, this unconformity shows that the Taconic orogeny, which was first manifest in the foreland as a peripheral clastic wedge, eventually expanded to deform and to metamorphose mildly the rocks of the foreland. Later in this chapter we shall see that the Taconic angular unconformity also exists in a strongly metamorphosed and deformed state in central New England and that it is a key to understanding the structural history of the metamorphic core of the Appalachians.

If we trace the Taconic unconformity to the west, the time gap and angularity decrease until it becomes a disconformity, finally dying out completely in the Upper Ordovician clastic wedge. The western limit of the clastic wedge extends beyond the western limit of deformation, as we would expect. At this western limit of the unconformity, it appears that the overlying Upper Ordovician and Silurian clastic sediments were also derived from the Taconic landmass to the east (Fig. 12-8); however, the effects of the orogeny were clearly on the wane. The uppermost Silurian strata that overlie the Middle Ordovician flysch and Taconic klippen near Catskill on the Hudson River (Figs. 12-9 and 12-10) are nonorogenic carbonate rocks. By the end of the Silurian, the Taconic mountain range had completely decayed away.

STRUCTURE AND SOURCE OF THE TACONIC SEQUENCE

We have noted that allochthonous sheets of Upper Precambrian through Lower Ordovician red, green, and black slates (called the *Taconic sequence*) structurally overlie the coeval carbonate-bank sequence in many parts of the Appalachian foreland between Newfoundland and Pennsylvania (Figs. 12-8 and 12-9). In this section we examine the internal structure and total displacement of these allochthonous masses. This total displacement is sufficiently large that it can be best estimated from studies of the depositional paleogeography. First, we consider the structure.

The great mass of shale and slate that underlies the Taconic Mountains east of the Hudson River is largely unfossiliferous and some of the rocks are at least superficially similar to Middle Ordovician autochthonous shales and slates of the clastic wedge. Only gradually, through the last hundred years, has enough fossil evidence accumulated to demonstrate unequivocally that the shale and slate of the Taconic Mountains spans the same time range of Cambrian through early Ordovician as the underlying autochthonous carbonate-rich sequence. For many years it was possible to consider unfossiliferous parts of the Taconic sequence to be the same age as the surrounding and underlying autochthonous Middle Ordovician shales and slates, especially in local quadrangle-scale studies. Eventually, a regionally consistent stratigraphy tied to the few fossil localities was developed within the slate mass. This regional stratigraphy has led to more-consistent solutions to the large-scale regional structure.

It has long been known from outcrop observations that the Taconic sequence was strongly deformed on a small scale, but regional mapping of the slate stratigraphy now has shown that substantial structural complexity exists in the Taconic sequence on a large scale. The Taconic sequence exhibits large-scale recumbent folding and internal thrusting. It is also imbricated with the structurally underlying carbonate and clastic-wedge sequences, giving rise to thrusts that locally place carbonates over the Taconic sequence. The Taconic sequence also displays important refolding. An example of the large-scale structure of the Taconic sequence is shown in Figure 12-11.

We have seen that the fossils, stratigraphic sequence, and regional detailed mapping of the shale and slate mass underlying the Taconic Mountains east of the Hudson River show it to be a strongly deformed allochthonous mass. Similar allochthonous masses of Cambrian and Lower Ordovician shale and slate structurally overlie the coeval carbonate-bank sequence and associated autochthonous Middle Ordovician clastic wedge in many parts of the foreland between Newfoundland and Pennsylvania. From where did these allochthonous masses come? They clearly came from the east or southeast because the coeval carbonate platform extends far into the continental interior merging with the cratonic sequence to the west (Fig. 12-5). But how far to the east might the slates have been derived and what sort of tectonic environment do they record? Like so many problems in regional structural geology, the most-significant insight is provided by

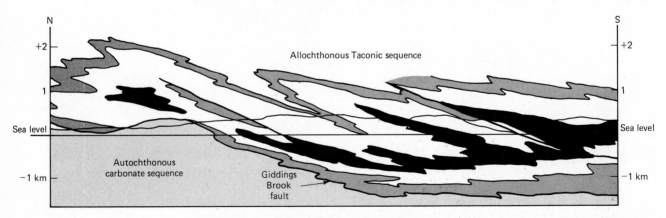

FIGURE 12–11 Cross section of the deformed Cambrian Taconic sequence displaying large-scale recumbent folds. The black layer is the oldest within the allochthonous sequence. Northern end of the Taconic klippen in southern Vermont. (Simplified after Zen, Geol. Soc. Amer. Bull., 1961, v. 72, pp. 293-338.)

stratigraphy and sedimentology because it is the sedimentary record that provides the most complete record of ancient environments and paleogeography.

Sedimentological studies of the Taconic sequence show that it records a deep-water environment far below wave base, in contrast with the autochthonous carbonate bank. It is difficult to estimate water depths precisely if they are below wave base, but parts of the Taconic sequence appear to be pelagic and deposited near or below the calcite compensation depth, which imply oceanic or near-oceanic water depths. Other parts of the Taconic sequence contain moderate amounts of turbiditic sandstone, particularly the Lower Cambrian and Upper Precambrian parts that are coeval with the autochthonous clastic sequence below the Cambrian to Lower Ordovician carbonate-bank sequence (Fig. 12-5). This observation suggests some affinities between the shallow-water autochthonous carbonate sequence and the deep-water allochthonous Taconic sequence. Perhaps they are facies equivalents (Fig. 12-8).

The most-important evidence for original paleogeographic setting of the Taconic sequence relative to the autochthonous carbonate bank is provided by deposits of carbonate breccia, which are very widely distributed within the Taconic sequence. They extend from Newfoundland to Pennsylvania and are one of the distinctive characteristics of many parts of the Taconic sequence. The deposits typically are composed of poorly sorted beds of angular carbonate clasts, interbedded with pelagic calcium-carbonate-free muds. The outcrop in Figure 12-12 near Quebec City is typical, with a variety of limestone fragments. Most are angular to subangular, including many tabular bed fragments. The largest fragments in breccias of this sort are as big as box cars. Other deposits are less angular and are properly called limestone-pebble conglomerates. Some of the breccias have a muddy matrix, whereas others have a clean quartz-sandstone matrix. Occasional angular fragments of Precambrian gneisses are present. Calcarenite

FIGURE 12–12 Bedded contact between subangular carbonate breccia and fine-grained slaty mudstone in the Taconic sequence near Quebec City.

turbidites are also associated with the limestone breccias and conglomerates in some areas.

The widespread angular breccias interbedded in deep-water pelagic muds must place severe constraints on the paleogeographic setting of the Taconic sequence. The source of such coarse breccias could be no more than a few kilometers away. Apparently the Taconic sequence represents the deep-water continental slope-and-rise deposits immediately to the east of the autochthonous carbonate bank that formed the early Paleozoic continental shelf of eastern North America (Fig. 12-8). This interpretation is supported by successful attempts to match limestone fragments with lithologic equivalents and in some cases formational equivalents in the carbonate sequence. The Taconic sequence was apparently deposited very close to the edge of the early Paleozoic carbonate bank, but in deep water. By analogy, the Florida and Bahamas banks have extremely steep edges dropping within a few kilometers to oceanic depths; they apparently give a modern-day image of the paleogeographic situation prior to the Taconic orogeny.

The breccias of the Taconic sequence apparently represent at least two depositional settings at the foot of the carbonate bank. The very angular, unsorted breccias with a muddy matrix may represent submarine rockslide deposits directly off the bank edge. The subangular breccias and limestone conglomerates, particularly those with a quartz-sand matrix or fragments of basement rocks, were probably transported to deep water down submarine canyons that cut deeply into the continental shelf. Some of these conglomerates and associated calcarenitic turbidites may have been transported moderate distances along submarine-fan channels out onto the continental rise.

Summary of Taconic Orogeny in the Foreland

Let us summarize our insight into the Taconic orogeny, so far. Prior to the Middle Ordovician, a stable continental margin existed facing an open ocean to the east. The orogeny was first manifest in the moderate down-bending of the shelf and the trapping of muddy sediments in a clastic wedge derived from the east, which was previously open ocean. Next, the deep-water Lower Ordovician through upper Precambrian continental slope-and-rise deposits, forming the Taconic sequence, were deformed and thrust up onto the coeval carbonate bank and its overlying clastic wedge. As deformation spread westward, rocks of the continental shelf, including the clastic wedge and Precambrian crystalline basement, were involved in west-vergent folding and thrusting. The clastic wedge reached its maximum extent in the late Ordovician. During the Silurian the mountains produced by deformation in the Taconic orogeny were gradually eroded until the residual topography lay at sea level and was covered by the transgressive Siluro-Devonian sequence, thereby establishing the Taconic unconformity (Fig. 12-10).

The orogenic history outlined above from the perspective of the Appalachian foreland very likely reflects the collision of a west-vergent compressive plate boundary with the early Paleozoic stable continental margin of eastern North America. This supposition raises many questions. For example, what was the other plate, and what rock masses, such as oceanic fragments, island arcs, or continents, might have been accreted to the Appalachians during the Taconic orogeny? To attempt to answer these questions, we must leave the Appalachian foreland and consider geology farther to the east within the metamorphic core of the Appalachians, looking for rocks involved in the Taconic orogeny that might be exotic to North America. Our task immediately becomes much more difficult because of strong post-Taconic deformation and metamorphism and an almost complete lack of fossils.

Transition into the Metamorphic Core

The billion-year-old Grenville basement of the Appalachian foreland, together with its autochthonous lower Paleozoic carbonate cover, can be traced a moderate distance into the metamorphic core in some parts of the Appalachians. It is present in the chain of basement massifs, including the Long Range of Newfoundland, the Green Mountains, and the Blue Ridge (Fig. 12-2). Some of these massifs are known to be allochthonous because the autochthonous basement and carbonate cover have been traced beneath the massifs through seismic reflection profiling. Even where the basement massifs are missing in Quebec, the autochthonous basement and carbonate sequence have been traced below the Taconic sequence far into the mountain belt by seismic profiling in petroleum exploration.

The basement massifs are generally at the edge of the metamorphic core at approximately greenschist metamorphic facies. In this structural position the basement rock displays a relatively stiff deformational behavior, in general. However, just interior to the belt of large basement massifs are a number of plastically redeformed and remetamorphosed bodies of basement with an unconformably overlying cover of marble-bearing metasedimentary rocks. These bodies include the Chester and Athens domes in Vermont east of the Green Mountains (Figs. 12-9 and 12-15), the billion-year-old Baltimore gneiss domes in Maryland and similar rocks near Philadelphia, and the gneisses in the New York City area.

Let us look at the geology near New York City as an example of these strongly deformed and metamorphosed bodies of Precambrian basement and their Lower Paleozoic cover (Fig. 12-13). The oldest rocks of the region are a complex sequence of high-grade gneisses collectively called the Fordham and Yonkers Gneisses. The mappable layering within the gneiss complex is truncated by the presumed Cambrian Lowerre Quartzite and Inwood Marble (see Fig. 12-13). This folded contact appears to be an angular unconformity, with the previously deformed Precambrian gneisses below and Cambrian and Lower Ordovician carbonate-bank sequence above, now isoclinically folded and metamorphosed to amphibolite facies. Overlying the Inwood Marble, and deformed and metamorphosed with it, is the Manhattan Schist, which underlies much of Manhattan Island and is the bedrock on which many New York City skyscrapers are built. The Manhattan Schist, by analogy with the foreland stratigraphy, appears to be the metamorphosed equivalent of the Middle Ordovician clastic wedge (Fig. 12-8). Part of the deformation and metamorphism in this area is cut by the Upper Ordovician (435 m.y.) Cortland Intrusive Complex along the Hudson River, just downstream from West Point (Fig. 12-9). The deformation, metamorphism, and magmatism therefore all appear to be manifestations of the Taconic orogeny.

As we move up stratigraphically through the Manhattan Schist, we encounter rock types that are unfamiliar in the foreland. These include metamorphosed basaltic igneous rocks (now amphibolites) and serpentinites. These mafic and ultramafic rock types cannot be ignored as simply unusual, because they are part of a belt of discontinuous bodies of mafic and ultramafic rocks that generally lie close to the western edge of the metamorphic core of the Appalachians and extend from Newfoundland to Georgia (Figs. 12-2 and 12-15). This serpentinite or ophiolite belt has been suspected to be of fundamental importance in some way to Appalachian tectonics ever since it was emphasized by Harry Hess in the late 1930s. Some of the larger and less-metamorphosed bodies of the ophiolite belt have now been studied in detail. One of the best known and largest is the Bay of Islands Complex in western Newfoundland, which actually structurally overlies the Taconic sequence along a major thrust fault (Fig. 12-14). Mapping, petrologic,

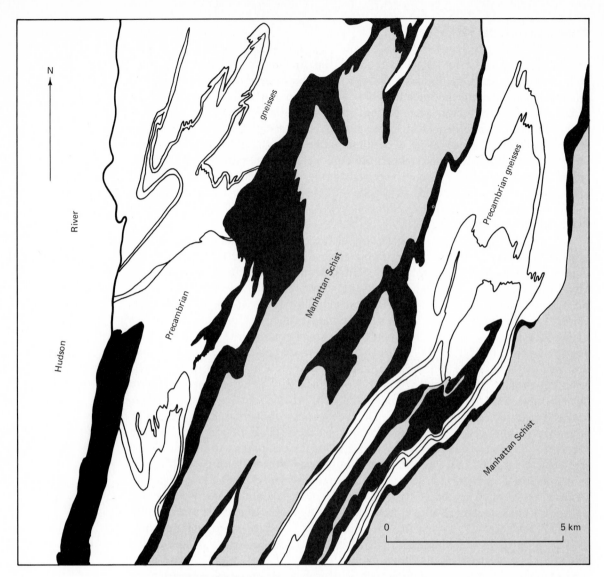

FIGURE 12–13 Geologic map of the White Plains area of northern New York City showing the angular unconformity between Precambrian gneisses (Fordham and Yonkers Gneisses) and overlying Cambrian and Ordovician sediments, now multiply-deformed quartzite, marble, and schist. Cambrian quartzite and marble above the unconformity is shown in black and include the Lowerre Quartzite and Inwood Marble. (After Hall, 1968.)

and geochemical studies show the klippen to be complex in origin but fundamentally to be fragments of Lower Paleozoic oceanic crust and mantle. The complex includes the remains of an oceanic transform-fault system.

Most of the fragments of the western Appalachian ophiolite belt have not been studied in detail and in any case do not preserve such a complete history as the Bay of Islands Complex. A variety of oceanic environments, including ocean floor, fracture zone, backarc basin, seamount, and oceanic island arc, could be potentially represented among the fragments of this belt. Incomplete information on their igneous age suggest that they range between late Precambrian and early Ordovician.

What is the probable origin and paleogeography of these rocks of the ophiolite belt? The most-certain observation is that many—if not all—are of oceanic origin, yet they appear high in a "stratigraphic" sequence that starts with Precambrian basement, passes upward through the Cambrian unconformity into

metamorphosed carbonate-bank deposits, and then moves into the pelitic schists
that contain the ophiolite. The base of the sequence is North American continent,
whereas the top is oceanic; therefore, the top of the sequence must be allochthonous relative to the base. A likely general interpretation is that the base of the
pelitic sequence near New York City, the Manhattan Schist, may indeed be part
of the Middle Ordovician clastic wedge, but as we pass upward we go into
allochthonous oceanic rocks, including the Taconic sequence and the ophiolites.
This is, in fact, what is observed in the weakly metamorphosed sequence of
western Newfoundland (Fig. 12-14). Detailed mapping within part of the ophiolite
belt in eastern Vermont shows numerous thrust faults within the schists.

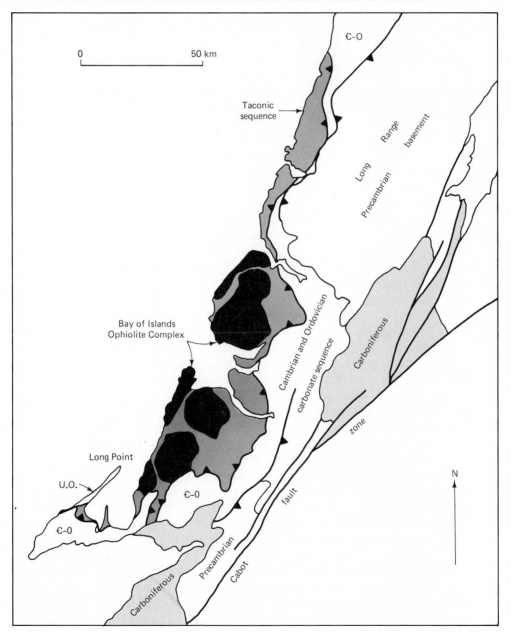

FIGURE 12–14 Simplified geologic map of western Newfoundland showing the Bay of Islands
Ophiolite Complex structurally overlying the Taconic sequence, which in turn structurally overlies
autochthonous Cambrian and Lower Ordovician carbonates and Middle Ordovician clastic
wedge. The Taconic unconformity lies below the Upper Ordovician sediments exposed at Long
Point.

The complexly deformed ophiolite belt apparently provides confirmation for our notion that an early Paleozoic ocean basin east of the carbonate bank was compressed and accreted onto the eastern margin of North America during the Taconic orogeny, although much of the accreted material, except for the ophiolitic fragments, is a largely undifferentiated terrain of schist.

Taconic Orogeny in the Metamorphic Core

As we pass farther east, particularly in the United States, we encounter rocks at sufficiently high metamorphic grade and sufficiently high state of deformation that the original stratigraphic relationships and the associated regional structure are exceedingly difficult to decipher. The significance of these metamorphic rocks to Appalachian tectonics was for many years unclear and controversial. On the 1932 edition of the Geologic Map of the United States, almost all the schists and gneisses of New England and the southern Appalachian Piedmont were simply called "Precambrian basement." By this interpretation the Appalachian mountain belt was considered to comprise merely the Appalachian foreland; the metamorphic rocks to the east were considered the eroded roots of a highland of Precambrian basement rocks called "Appalachia." This interpretation is now known to be incorrect, and it is instructive to see how this metamorphic terrain has been deciphered.

Already in the nineteenth century, several localities of metamorphosed Paleozoic fossils were known in central New England, suggesting that New England is, in fact, a Paleozoic metamorphic core to the Appalachians. This Paleozoic view of New England was expressed by B. K. Emerson, a professor at Amherst College, in his 1917 Geologic Map of Massachusetts and Rhode Island. In contrast, the 1932 Geologic Map of the United States chose to equate age with metamorphic grade, which has been a common mistake in many orogenic belts. A modern understanding of the metamorphic rocks of New England began in the 1930s, when Marland Billings, a young professor at Harvard, and his students began mapping out from these metamorphosed fossil localities and gradually deciphered the stratigraphic sequence and the regional structure. In fact, the understanding of stratigraphy and structure developed hand-in-hand in stages because the stratigraphy and structure of metamorphic terrains are intensely intertwined. We illustrate this fact as we explore the geology of central New England in the vicinity of the fossil localities, along the Connecticut River in Massachusetts, New Hampshire, and Vermont (Fig. 12-15).

The region just east of the Connecticut River is characterized by a north-south belt of domes of gneiss and granitic plutons; some of these are elongated in the same north-south direction, but others are more or less equant. This belt of domal structure is now thought to expose older rocks, particularly relative to the area just west of the Connecticut River; therefore, it is called an anticlinorium, the *Bronson Hill anticlinorium. Anticlinoria* and *synclinoria* are words that refer to large-scale linear highs and lows within the regional structure of mountain belts. They need not be specifically anticlines or synclines; indeed, they may be superposed on smaller-scale recumbent folds, basins and domes, and other often-unrelated structures. The belt of large Precambrian basement massifs at the western edge of New England is called the *Green Mountains anticlinorium.* The structural low between the Green Mountains anticlinorium and Bronson Hill anticlinorium is called the *Connecticut Valley–Gaspé synclinorium.*

The early mapping of central New England clearly recognized the domal structure along the Bronson Hill anticlinorium. For example, Emerson's 1917 map shows the domal bodies of gneiss, called the *Oliverian Gneiss,* and the intervening

FIGURE 12–15 Geologic map of part of western New England between the Taconic foreland and the Bronson Hill anticlinorium. The belt of Ordovician and older gneiss domes shown in black marks the axis of the Bronson Hill anticlinorium. White areas are underlain by Ordovician and older rocks lying below the Taconic unconformity. Dashed pattern represents Devonian granitic intrusive rocks. Cross sections *AA'* and *BB'* are shown in Figure 12-17. (Compiled from Doll and others, 1960, and Thompson and others, 1968.)

synformal keels of schist, as well as the granitic plutons, in north central Massachusetts (Fig. 12-16(a)). The Paleozoic fossil localities are associated with quartzites interlayered with the schist in the keels between the domes. Most of the fossils are barely recognizable as brachiopods, corals, or crinoids because of

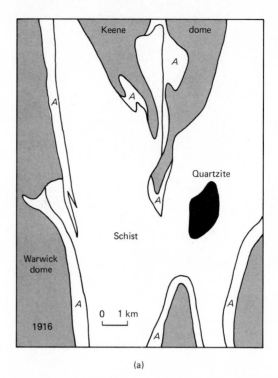

(a)

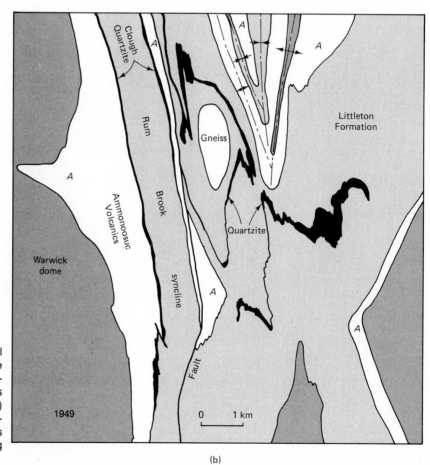

FIGURE 12–16 Evolution in structural and stratigraphic understanding of the Bronson Hill anticlinorium in part of north-central Massachusetts: (a) Emerson's 1917 map, (b) Hadley's 1949 map, (c) Robinson's 1967 map. *A* represents amphibolites of the Ammonoosuc Volcanics and *P* represents rusty-weathering schists of the Partridge Formation.

(b)

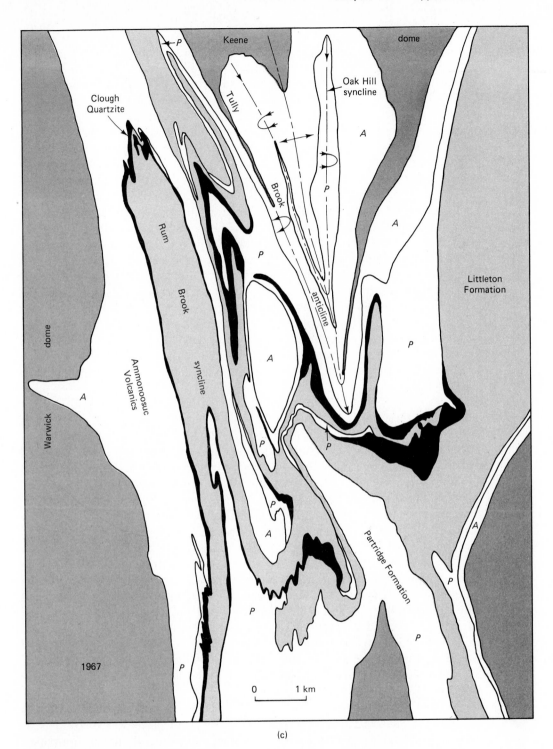

(c)

FIGURE 12–16 Continued

metamorphism and deformation. Nevertheless, some are well-enough preserved for generic identification—even some brachiopods that are now composed of calcium garnet rather than calcite because of metamorphism. Their identification establishes the fossiliferous rocks as Silurian and Lower Devonian, about the same age as the strata overlying the Taconic unconformity in the foreland west of the Green Mountains–Berkshire Highlands anticlinorium (Figs. 12-8 and 12-10). These fossils turn out to be a key to the regional structure of New England.

In the late 1930s, the mapping of Marland Billings and his students established that there is a consistent stratigraphy within the schists overlying the Oliverian gneiss domes of the Bronson Hill anticlinorium. This stratigraphic understanding is illustrated in a 1949 map by Jarvis Hadley (Fig. 12-16(b)), which mapped the same area as Emerson's 1917 map. Hadley was a student of Billings, and his map shows the same consistent stratigraphy first developed by Billings 10 years earlier at Littleton, New Hampshire, 175 km to the north along the Bronson Hill anticlinorium. The success of the stratigraphy was that the same sequence of schists could be found above the gneiss over such a great distance. The first stratigraphic unit directly overlying the Oliverian Gneiss was amphibolites, which were considered to be metamorphosed volcanic rocks because of their stratigraphic properties (Fig. 12-16(b)). The amphibolites were named the Ammonoosuc Volcanics, a unit that eventually takes on considerable importance in our understanding of Appalachian tectonics. A thin discontinuous quartzite and conglomerate overlies the Ammonoosuc Volcanics in some places; it is called the *Clough Quartzite* and in nearby areas contains metamorphosed Lower Silurian fossils. The next overlying unit is a heterogeneous assemblage of pelitic schists, which was called the *Littleton Formation,* first defined in the Littleton area 175 km to the north. The Littleton Formation in Hadley's map contained scattered lenses of schistose quartzite, some of them quite similar to the Clough Quartzite (Fig. 12-16(b)).

The Bronson Hill anticlinorium in north central Massachusetts was mapped yet another time, in the 1950s and 1960s by Peter Robinson, also a student of Billings. This third-generation mapping is shown in Figure 12-16(c) and is instructive to examine in detail in comparison with Emerson's and Hadley's earlier maps of the same area. As Robinson examined the stratigraphy and structure in more detail, he found that the scattered quartzites within what Hadley mapped as the Littleton Formation (Fig. 12-16(b)) were all the same layer—in fact, the same layer as the Silurian Clough Quartzite, far more deformed than Hadley realized. Furthermore, the pelitic schists above and below the Clough Quartzite were in fact different. Below was a rusty weathering sulphidic schist, now called the Partridge Formation, and above was nonrusty weathering graphitic schist, the true Littleton Formation. Furthermore, by now it was possible to divide the Ammonoosuc Volcanics into a sequence of distinctive layers that represent different volcanic layers and allow some of the structure involving the volcanics to be better understood. The revised stratigraphic and structural interpretation of the 1950s and 1960s, of which Robinson's map (Fig. 12-16(c)) is an example, provided two fundamental insights. The first is the tectonostratigraphic significance of the Silurian Clough Quartzite, and the second is the large-scale structure of the Bronson Hill anticlinorium.

If we look at Figure 12-16(c) in more detail, we see that different rocks underlie the Silurian Clough Quartzite from place to place. In some areas, it directly overlies the Ammonoosuc Volcanics, with the Partridge Formation missing. Apparently the base of the Clough is an angular unconformity, a conclusion that is consistent with the fact that it is an orthoquartzite and, in places, a boulder conglomerate. Furthermore, isotopic dating of the Ammonoosuc Volcanics and the underlying Oliverian Gneiss has shown them to be essentially the same age, Ordovician, although the Pelham dome (Fig. 12-17) is Cambrian. Therefore, the angular unconformity has Silurian and younger rocks above and Ordovician and older rocks below. Apparently the Clough Quartzite marks the same Taconic unconformity seen in the Appalachian foreland, but now strongly

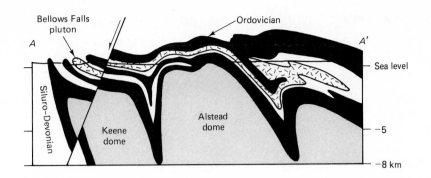

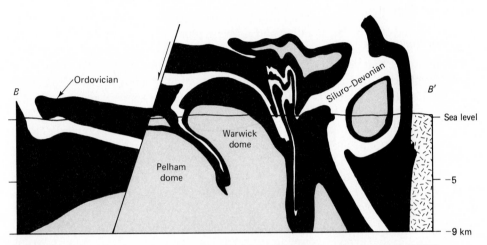

FIGURE 12–17 Cross sections of the Bronson Hill anticlinorium along section lines *AA'* and *BB'* of Figure 12-15. Note that the graphical patterns differ from Figure 12-15 for clarity of displaying the structure. (Simplified from sections of Thompson and others, in Zen and others, ed., Studies in Appalachian Geology: Northern and Maritime, © 1968, Wiley-Interscience.)

deformed and metamorphosed in later Appalachian orogenic activity. This Taconic unconformity has by now been traced through most of central and western New England, as is shown in Figure 12-9. We now see that the Connecticut Valley–Gaspé synclinorium largely exposes Silurian and Devonian schists above the unconformity, particularly in Vermont, whereas the Bronson Hill anticlinorium exposes a higher proportion of Ordovician and older metamorphic rocks.

What does the stratigraphy below the Taconic unconformity in the metamorphic core of New England tell us about the Taconic orogeny? First we note that the Ordovician Partridge Formation, Ammonoosuc Volcanics, and Oliverian gneisses are roughly contemporaneous or slightly older than orogenic events of the Appalachian foreland. The most significant observation is that the Ammonoosuc and similar volcanic rocks of the northern Appalachians show petrologic affinities to modern island arcs and the roughly coeval Oliverian granitic gneisses appear to be deformed subvolcanic plutons. Therefore, the Ordovician rocks of the Bronson Hill anticlinorium in New England and similar rocks in the Maritime Provinces of Canada and Newfoundland appear to be the remains of an island arc or island arcs. It is not yet clear whether the Bronson Hill volcanic arc was an oceanic island arc such as the present-day western Aleutians or Marianas or whether it was an arc associated with significant continental lithosphere, such as Japan, Luzon, or Sumatra. In any case it is now widely accepted that the Bronson Hill island arc was the leading edge of the plate that collided with the eastern margin of North America in Ordovician time to produce the Taconic orogeny, now so clearly displayed in the Appalachian foreland. The volcanic arc is perhaps the source of Middle Ordovician volcanic ash beds that are widespread in the Appalachian foreland to the west. By Silurian time the mountains that developed during arc-continent collision had decayed sufficiently through erosion that the remains of the Taconic orogenic belt, including the Bronson Hill island-arc, were covered by the Taconic unconformity.

POST-TACONIC DEFORMATION IN CENTRAL NEW ENGLAND—THE ACADIAN OROGENY

One of the satisfying results of stratigraphic and structural mapping of the metamorphic core of the Appalachians in New England has been the tracing of the Taconic unconformity below the Clough Quartzite through a remarkable maze of refolded folds that record the sum of post-Taconic deformation. The dome-and-keel structure of the Bronson Hill anticlinorium, already recognized on Emerson's 1917 map (Fig. 12-16(a)), is only the youngest of these plastic structures. Similar post-Taconic domal structures are also recognized in the Connecticut Valley–Gaspé synclinorium (Fig. 12-15). Both regions display Devonian granitic plutons. Some plutons are great semiconcordant plutonic sheets involved in much of the structure, whereas others are discordant plutons that were intruded relatively late in the structural history, indicating that most of the deformation and coeval metamorphism is Devonian and therefore subsumed under the name *Acadian orogeny*.

We have already encountered this Devonian orogeny west of the Hudson River in the flat-lying sedimentary rocks of the Catskill Mountains in the Appalachian foreland. There we encountered the Middle and Upper Devonian clastic wedge derived from the Acadian orogenic belt (Fig. 12-6). Now we encounter strong Devonian plastic deformation, metamorphism, and plutonism 100 km to the east in New England. Apparently, the metamorphic rocks of New England are the eroded roots of the Acadian mountain belt that was only seen indirectly in the detritus of the Catskill delta.

Stages of deformation in the Acadian orogeny in central New England that are earlier than the gneiss domes are preserved in the structure within the synformal keels. Detailed mapping of the cover sequence of the domes, such as illustrated in Robinson's map (Fig. 12-16(c)), shows good evidence of map-scale refolded folds, with the axial surfaces of the youngest dome-and-keel synclines and anticlines cross cutting the same beds more than once. For example, in Figure 12-16(c) the axial surface of the Rum Brook syncline between the Warwick and Keene domes is south-plunging with the trace of the axial surface cutting the Clough Quartzite twice; therefore, we have anticurvate refolding of Ramsay Type 2 to 3 (Chapter 9). The same style of refolding is also displayed at an outcrop scale within the region; for example, Figure 9-30 is a photograph of a Type 3 refolded structure in the Monson dome indicating an early phase of recumbent isoclinal folding.

Another map pattern indicating refolding is seen in the south-plunging southern end of the Keene dome. The first-generation mapping (Fig. 12-16(a)) clearly recognized the southward anticlinal plunge of the dome. The second-generation mapping (Fig. 12-16(b)) showed that the anticline displayed a branching toward the north because of the narrow tongues of dome rock that extend southward into the Ammonoosuc Volcanics of the cover sequence. The third-generation mapping (Fig. 12-16(c)) successfully subdivided the Ammonoosuc

Volcanics into a sequence of distinctive mappable members and documented large-scale Type 2 refolding. The two tongues of gneiss are, in fact, the same isoclinal anticlinal fold, the Tully Brook anticline, with a parallel isoclinal syncline to the north involving the Partridge Formation, the Oak Hill syncline (Fig. 12-16(c)). The Tully Brook anticline and Oak Hill syncline are refolded by the south-plunging antiform of the Keene dome to produce a Type 2 refolding geometry.

Similar refolded map-scale geometries have been documented up and down the Bronson Hill anticlinorium, indicating an early large-scale recumbent isoclinal folding with hinge-to-hinge amplitudes of 5 to 25 km for the larger folds. Several regional cross sections of the Bronson Hill anticlinorium in Figure 12-17 show these refolded recumbent folds; however, the two fold axes are generally not coaxial. Therefore, a cross section cannot give a complete picture of the structure. Nevertheless, they show several important features. First, the early folds include very large-scale recumbent isoclines. In addition, however, some of the granitic plutons—such as the Bellows Falls pluton in Figure 12-17—are, in fact, great semiconcordant sheets that are involved in much of the structure. Some of the domes are also apparently culminations in deeper refolded recumbent nappes and not gneissic diapirs analogous to salt domes (Chapter 8). It is not known at present if all the gneiss domes are culminations in the refolding of deeper recumbent gneiss-cored nappes or if some are gneissic diapirs analogous to salt domes, as they were originally interpreted.

The recumbent nappe formation and the dome-and-keel structure appear to be Devonian in age, largely based on isotopic dating of crosscutting granitic plutons that were emplaced at later stages in the deformation. For example, the Prescott Plutonic Complex, a few kilometers to the south of the areas we discussed in Figures 12-15 and 12-16, crosscuts the recumbent nappes but is deformed in the same deformation that produced the domes. The Prescott Complex has an age of about 380 m.y. (Middle Devonian), indicating that the nappe structure is part of the Acadian orogeny. Structural, metamorphic, and isotopic studies of a more-regional nature show that much of the high-grade metamorphism and plastic deformation of western and central New England is a manifestation of the Acadian orogeny. We have arbitrarily picked out a few detailed studies of central New England to illustrate the ways in which the structures of the cores of mountain belts are first deciphered locally and then built up into an image of the large-scale structure and regional orogenic events. This deciphering is an ongoing and exceedingly challenging task in every complex mountain belt, including New England. For example, as we move east of the Bronson Hill anticlinorium, we come to much less well-understood terrain that is the subject of current investigation and controversy. This region has been especially difficult to decipher because of very high grade granulite-facies metamorphism with associated partial melting of many of the metamorphic rocks.

Still farther to the east in the Boston area, we rather suddenly encounter rocks of a new and unfamiliar kind and of a substantially lower greenschist metamorphic grade. It is as if we have stumbled into a new country with unfamiliar house types, crops, and signs. To begin with, there are fault-bounded nonmarine Carboniferous basins, particularly the Narragansett basin in Rhode Island and the Boston basin (Fig. 12-9). These basins are part of an array of nonmarine Carboniferous basins extending north through the Maritime Provinces and into Newfoundland (Fig. 12-14). Similar Carboniferous basins exist in the northern continuation of the Appalachians in Ireland and Scotland.

The geologic differences of easternmost New England go deeper; unconformably below the Carboniferous, we find slates with Cambrian fauna that are unfamiliar to North America. Unconformably below the Cambrian is a complex of rhyolitic volcanic rocks and related gabbro, diabase, and granitic plutons of very latest Precambrian age. They yield isotopic ages of about 570 to 580 m.y.; the Cambrian-Precambrian boundary is generally placed at about 570 m.y. (Appendix C).

Similar uppermost Precambrian and Cambrian stratigraphy is known elsewhere in the eastern Appalachians and generally is little metamorphosed or deformed in comparison with the much more strongly deformed and metamorphosed Cambrian through Devonian rocks to the west. The Avalon Peninsula of eastern Newfoundland is the best known of these eastern terrains, but others are known in the Maritime Provinces and in the Carolina Slate belt of the eastern Piedmont in the southern Appalachians. All these terrains are characterized by latest Precambrian magmatic activity, possibly in an island arc.

The sequence in the Avalon Peninsula is marked by Upper Precambrian volcanic rocks and volcanogenic sedimentary rocks noted for their soft-bodied, coelenterate-like fossils. These are intruded by the 550 m.y. Holyrood Granite and overlain by a Cambrian sequence that is characteristically clastic with very little carbonate, in contrast with the Cambrian sequence of the western foreland. The sequence continues through the Upper Cambrian and into the Lower Ordovician and is characterized by faunas that are strikingly different from the western foreland of the Appalachians and North America in general. In contrast, the Avalonian faunas are very similar to those of central England, as well as of Scandinavia, Spain, France, and Morocco. For example, North America is characterized by the familiar *Olenellus* and *Zocanthoides* types of trilobites (Fig. 3-1), whereas Avalon Peninsula is characterized by the *Callavia* and *Paradoxides* types familiar to northwest Europe. The same distinctive European trilobites are found in the Boston sequence and the Carolina Slate belt.

Although there are important stratigraphic and structural differences among these various eastern terrains of the Appalachians, as a group they are referred to as the *Avalonian terrains,* in contrast with the coeval strata of the Appalachian foreland, which—together with the Taconic sequence—represents the late Precambrian and early Paleozoic continental margin of North America. Both terrains are known over much of the length of the Appalachians. The North American margin is characterized by a late Precambrian rifting of billion-year-old Grenville orogenic belt approximately 800 m.y. ago to establish a stable continental margin first marked by rift sediments and volcanics and followed by an Upper Precambrian and Lowest Cambrian clastic sequence grading upward into the Cambrian and Lower Ordovician carbonate bank. In contrast, the latest Precambrian–earliest Cambrian magmatic activity that is so characteristic of Avalonia is missing in the Appalachian foreland. Furthermore, the lack of carbonate within the Avalonian sequence is in marked contrast with all of North America. Apparently Avalonia was far away, perhaps at higher latitudes during the early Paleozoic.

Therefore, Avalonia represents a terrain or complex of terrains that was brought into North America from some unknown distance, but far away. The Avalonian fragments did not undergo the strong Taconic or Acadian orogenic events that are so pronounced to the west, although the western limit of Avalonia is not well known at present. For example it could conceivably extend as far west as the Bronson Hill anticlinorium and correspond to the pre-Ordovician gneisses of the Pelham dome (Fig. 12-17) and similar rocks elsewhere in central and eastern New England. These questions remain for the future. Nevertheless Avalonia appears to have been separated from North America by the early Paleozoic ocean represented by the belt of ophiolitic rocks and associated oceanic sedimentary rocks of western New England (Figs. 12-9 and 12-15).

LATEST PALEOZOIC TECTONIC ACTIVITY—
THE ALLEGHENIAN OROGENY

Historically, the orogenic episode that was considered all-important in the Appalachian mountain belt was the *Appalachian revolution* marking the close of the Paleozoic. The striking evidence for this orogeny is the strong folding and thrusting involving rocks as young as late Carboniferous in the central and southern Appalachian foreland (Fig. 12-18). However, it was gradually discovered

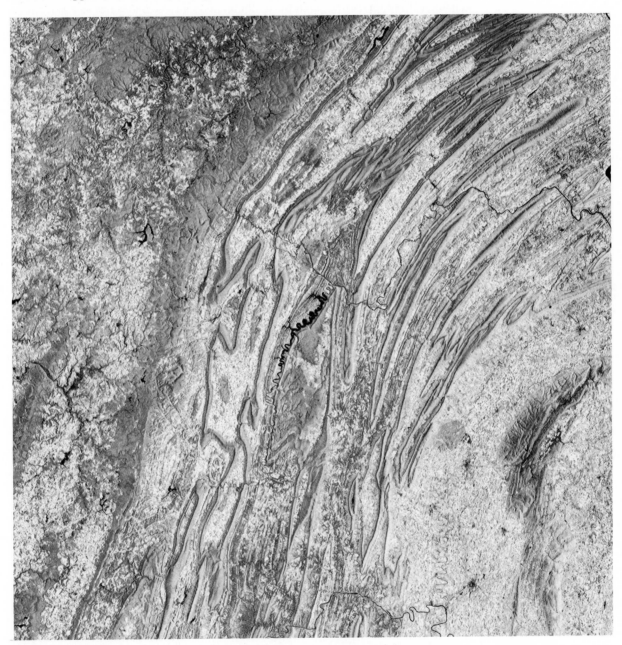

FIGURE 12–18 Satellite image of the Pennsylvania and Maryland fold-and-thrust belt. The area to the northwest (left) of the strong folding is underlain by nearly flat-lying Carboniferous sediments of the Appalachian foredeep. The isolated hills to the southeast are underlain by Precambrian basement of the Blue Ridge province.

that the most-profound deformation, metamorphism, and magmatism within the core of the Appalachians were manifestations of the Taconic and Acadian orogenies. Tangible evidence for latest Paleozoic orogenic activity has been less obvious in the Appalachian core. For these reasons the late Carboniferous and Permian orogeny, best known from the foreland fold-and-thrust belt, is now commonly given a more local name, the *Alleghenian orogeny,* rather than the name of the entire mountain system.

Nevertheless, the Alleghenian orogeny is very substantial. Let us first look at some indirect evidence given by paleomagnetism. When we compare paleomagnetic poles from the Carboniferous basins of the Avalon belt in the Maritime Provinces with Carboniferous paleomagnetic poles from North America, we find they differ systematically by about 10° to 15°. In contrast, Triassic or latest Permian North American and European poles coincide. This very general evidence suggests that there was roughly 1000 to 1500 km of relative motion between the Avalonian belt or Europe and North America in the latest Paleozoic. This relative motion must have taken place through deformation within the Appalachian mountain belt.

More-direct evidence of a substantial latest Paleozoic orogeny is the large deformation of the fold-and-thrust belt. For example the Upper Carboniferous coal of the Pennsylvania Appalachians (Fig. 12-19) has been metamorphosed to anthracite grade, implying a thick tectonic cover, and has been folded into large amplitude folds (Fig. 12-18). In the southern Appalachians, Carboniferous strata are fully involved in the great imbricate thrust structure of the fold-and-thrust belt. Carboniferous strata are deformed in both the frontal thrusts (Fig. 9-44) and the inner thrusts that bring up Upper Precambrian sedimentary rocks and its billion-year-old Grenville basement of the Blue Ridge and Great Smoky Mountains (Fig. 12-4).

If we attempt to undeform the fold-and-thrust belt, we begin to have some concept of the magnitude of Alleghenian deformation in the foreland. When we undeform the 60-km-wide cross section of the valley and ridge in Figure 12-4, we find that it must have a minimum original width of 140 km. This shortening is confined to the thrust sheets above the basal décollement in the Middle Cambrian Rome Formation.

As we pass toward the interior of the fold-and-thrust belt, at the west edge of the Blue Ridge, we find older, pre-Rome rocks riding up to the surface on thrust faults. These older rocks include the Lower Cambrian and Upper Precambrian pre-Rome sedimentary sequence and its billion-year-old Grenville basement. For

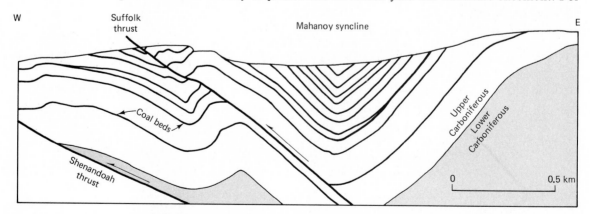

FIGURE 12–19 Detailed cross section of folded and thrust Carboniferous sediments in the anthracite basins of central Pennsylvania. The black lines are coal beds. (After Danilchik, Rothrock, and Wagner, 1955.)

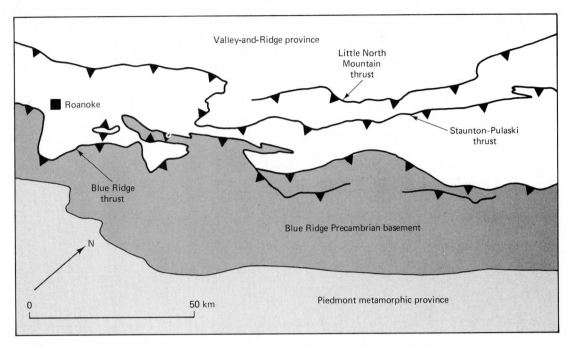

FIGURE 12–20 Map of the Blue Ridge basement thrust sheet near Roanoke, Virginia. (Simplified after Spencer, 1972.)

example, the geologic map of the Blue Ridge near Roanoke, Virginia (Fig. 12-20), shows the trace of the Blue Ridge thrust to be quite sinuous, implying a nearly flat thrust fault with the Precambrian basement and its Upper Precambrian cover in the upper plate outcropping on the ridges and the younger Cambrian sedimentary rocks below the thrust exposed in the valleys. These observations show that the Blue Ridge is at least somewhat allochthonous relative to the Valley-and-Ridge province to the northwest.

About 200 km to the southwest of the area shown in Figure 12-20, we find evidence for even more extreme thrusting of basement rocks out over the sedimentary cover (Fig. 12-21). Precambrian basement and its Upper Precambrian and Lower Cambrian sedimentary cover is present in a complex imbricated and folded thrust sheet with a great 20-km-wide window through it, the Mountain City window. The basal thrust of the Blue Ridge complex eventually comes to the surface again 15 kilometers farther to the northwest, where Precambrian Grenville basement rocks are no longer present in the upper plate. Twenty kilometers to the southeast of the Mountain City window is another window exposing rocks below the uppermost Blue Ridge thrust sheet, the famous 30-km-wide Grandfather Mountain window, which has Precambrian basement and Lower Paleozoic metamorphic rocks in the upper plate and Precambrian basement and its now-metamorphosed Upper Precambrian cover in the window. These observations show us some important aspects of the transition between the unmetamorphosed foreland thrust belt and the metamorphic core of the Appalachians in the Piedmont.

First, the important thrust sheets of the foreland, which involve only Rome and younger rocks, disappear below the Blue Ridge thrust sheets that contain pre-Rome rocks. Therefore, as we go farther into the mountain belt, we see deeper and older rocks coming up on higher thrust sheets. Everything appears to be allochthonous, with each successive thrust sheet more allochthonous relative to the craton than the last. Second, the Blue Ridge thrust brings up rocks metamorphosed in the Paleozoic, with the thrusts being postmetamorphic, cutting the

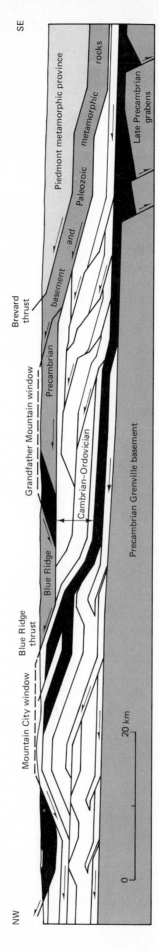

FIGURE 12–21 Simplified cross section of Mountain City and Grandfather Mountain windows, Tennessee and North Carolina. White areas are Cambrian and Ordovician sedimentary rocks lying above the décollement at the base of the Rome Formation. (Simplified interpretation based on seismic data of Harris and others, 1981.)

metamorphic isograds. Some of the thrusts cut Carboniferous strata, suggesting that much of the important deformation is Alleghenian.

How extensive is this thrusting involving basement and metamorphic rocks? Our evidence from the Blue Ridge, outlined above, suggests it is substantial. A more-complete idea is provided by seismic-reflection profiles that extend from the foreland thrust belt across the Blue Ridge and into the metamorphic core of the Piedmont. They show that the entire metamorphic core is allochthonous, with the flat-lying footwall of the Rome décollement below at a depth of about 8 to 10 km (Fig. 12-22). This observation is to be expected because the imbrications of thrust sheets riding on the Rome décollement accounted for a minimum of 140 km of shortening of the Rome and younger section (Fig. 12-4). Therefore, a matching pre-Rome section of a minimum width 140 km must be present at depth to the east. The predicted pre-Rome section is found undeformed below the Blue Ridge and Piedmont thrust sheets over a width of about 200 km, as shown by the seismic evidence (Fig. 12-22).

Eventually the basal thrust fault steps down into the pre-Rome section below the Piedmont. This, too, is predicted from the foreland structure because the thrust sheets of the Blue Ridge contain pre-Rome sedimentary cover and basement rocks; therefore, they must have come from southeast of the still-in-place pre-Rome rocks in the autochthonous footwall. From this reasoning we see that the Blue Ridge thrust sheets came from just southeast of the last autochthonous pre-Rome in the footwall—that is, below the Avalonian belt in Figure 12-22. This region is the matching footwall to the allochthonous Blue Ridge rocks, 200 km to the northwest. Therefore, approximately 200 km of compression must be present in the foreland between the Blue Ridge thrust sheet and the craton.

Our next question might be: Is there any Alleghenian metamorphic core exposed in the interior of the Appalachians to correspond with the Alleghenian foreland thrusting, perhaps by analogy with the Acadian orogeny? The answer appears to be no. Dating of major metamorphism in the Piedmont province suggests it is old, Acadian or Taconic. Nevertheless, K-Ar dating of the rocks, which records cooling through the 300° to 400° isotherm, yields widespread late Carboniferous and Permian dates. The main way in which large masses of deep-seated rocks of the crust undergo such cooling is by uplift and erosion of the cover. Apparently the mechanism of uplift and erosion was the thrusting of Piedmont rocks up over the Rome décollement, as shown in the seismic-based cross sections (Figs. 12-21 and 12-22). Therefore, the main effect of the Alleghenі-

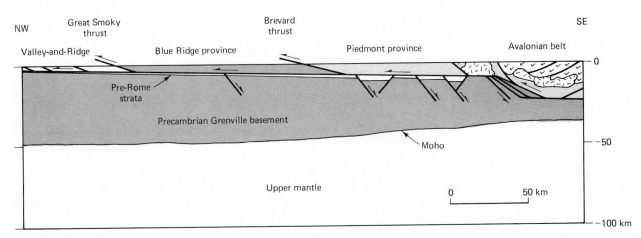

FIGURE 12–22 Regional cross section of the southern Appalachians based on surface geology and seismic reflection profiling. Location of section is shown in Figure 12-2. (Adapted from sections of Cook and others, 1979, 1981; Iverson and Smithson, 1982.)

an orogeny was mountain formation, with the foreland as the foothills and the Piedmont as the eroding interior.

Some Alleghenian metamorphic deformation and magmatism is exposed in the Appalachian mountain belt, but it is relatively minor. The system of Carboniferous fault basins in the northeastern Appalachians (Fig. 12-9) is locally strongly deformed and metamorphosed, although the rocks of the basins are generally little metamorphosed. Some Permian and late Carboniferous plutons are present. A zone of late Carboniferous magmatism and deformation also is present in the southeasternmost part of the southern Appalachians, partly covered by the Mesozoic sediments of the coastal plain.

Width of the Appalachians

We should note that there is a large hidden part of the Appalachians below the coastal plain and continental shelf (Fig. 12-1). The width of this hidden region is generally about 300 km, about the same width as much of the exposed Appalachian orogenic belt. Therefore, we have only a partial picture of the orogenic belt. We never see the southeastern limit of the Paleozoic deformation exposed at the surface on North America. Even though the Avalonian terrain has European affinities, other deformed terrains exist to the southeast—for example, the Meguma terrain of Cambrian and Ordovician deep-water clastic sediments southeast of the Avalonian terrain in Nova Scotia.

If we were to examine in detail the possible across-strike continuations of the Appalachian mountain belt in western Europe and northwest Africa, we would find that they too have Paleozoic deformation, some of which is east-vergent toward the African craton. It is in northwest Africa that we finally reach the opposite side of the great Appalachian zone of Paleozoic deformation, at least half of which lay covered by the Atlantic continental shelves.

Late Triassic and Jurassic Rifting

Virtually no Permian or Lower to Middle Triassic rocks exist in the Appalachian mountain belt, a gap of 50 m.y.; therefore, we have almost no knowledge of the end of the Alleghenian deformation. By late Triassic, however, the tectonics of

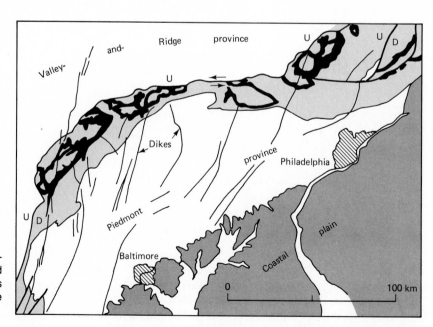

FIGURE 12–23 Triassic-Jurassic graben system, diabase sills (in black), and dike swarms in the central Appalachians near Philadelphia. (Simplified after Stose and Stose, 1944.)

the Appalachians had undergone a radical change; they had changed from the colossal thrusting of the Alleghenian orogeny to normal faulting and horizontal extension in the late Triassic, which by middle or late Jurassic had proceeded to the extent of an open Atlantic ocean running down the middle of what was the Appalachian mountain system.

The main manifestation of this rifting in the exposed Appalachians is a chain of late Triassic and early Jurassic grabens extending from Nova Scotia through New England to Georgia (for example Fig. 12-23). Other grabens are known under the coastal plain and continental shelf (Fig. 1-14). The extension was accompanied by basaltic igneous activity in the form of lava flows, sills, and dike swarms (Figs. 12-23 and 7-3).

The Triassic and Jurassic rifting marks the end of compressive deformation and the Cretaceous and Tertiary coastal plain and continental shelf (Figure 12-23) certify the death of the Appalachian mountain belt.

EXERCISES

12–1   Which strata in Figure 12-8 can be considered orogenic sediments deposited in response to the Taconic orogeny? In what way is the Taconic unconformity displayed on this stratigraphic diagram?

12–2   What is the evidence that the Clough Quartzite in Figure 12-16(c) overlies an angular unconformity? Are there any other possible interpretations?

12–3   Many rock bodies have been displaced through deformation in the area of Figure 12-15. What were the original relative positions of each of the following rock masses in early Ordovician before the Taconic orogeny? State the evidence supporting your answer.
  (a) Taconic klippen
  (b) Precambrian gneisses of the Green Mountains
  (c) Appalachian ophiolite belt
  (d) Precambrian gneisses of Chester and Athens domes
  (e) Gneiss domes of the Bronson Hill anticliniorium.

12–4   List the main structural features in Figures 12-13, 12-14, and 12-16(c) that were produced during the Grenville, Taconic, Acadian, or other orogenic periods. State the evidence for your age assignments.

12–5   In the cross section of Figure 12-21, measure the minimum shortening in kilometers of (a) the structures involving Cambrian and Ordovician strata below the Blue Ridge thrust sheet, (b) the Blue Ridge thrust sheet relative of the easternmost underlying Cambrian strata, and (c) the total minimum shortening between the Piedmont province and the westernmost Cambrian and Ordovician rocks. What percentage of the original minimum width is the shortening?

12–6   What was the orientation of σ_3 during the late Triassic and Jurassic in the central Appalachians, near Philadelphia (Fig. 12-23)? Outline your reasoning. What were the orientations of σ_1 and σ_2?

12–7   Continuing from the last problem, what was the orientation of σ_1 during the late Permian Alleghenian deformation of the central Appalachians (Fig. 12-18)? How did the state of stress change between late Permian and late Triassic?

SELECTED LITERATURE

General

BIRD, J. M. AND DEWEY, J. F. 1970, Lithosphere plate–continental margin tectonics and the evolution of the Appalachian orogen: Geol. Soc. Amer. Bull., v. 81, p. 1031–1059.

KING, P. B., 1977, *The Evolution of North America,* rev. ed., Princeton University Press, Princeton, N.J., 197 p. (Chapter 4 on the Appalachians.)

RODGERS, JOHN, 1970, *The Tectonics of the Appalachians,* Wiley-Interscience, New York, 271 p.

WILLIAMS, HAROLD, 1978, *Tectonic Lithofacies Map of the Appalachian Orogen,* Memorial Univ. of Newfoundland, St. John's (1:1,000,000).

WONES, D. R., ed., 1980, *The Caledonides in the U.S.A.,* Va. Polytech. Inst. and State Univ. Dept. Geol. Sci. Mem. 2, 329 p.

Taconic Orogeny

RODGERS, JOHN, 1971, The Taconic orogeny: Geol. Soc. Amer. Bull., v. 82, p. 1141–1178.

ZEN, E-AN, 1972, The Taconide zone and the Taconic orogeny in the western part of the northern Appalachian orogen: Geol. Soc. Amer. Special Paper 135, 72 p.

Northern Appalachians

LYONS, P. C. AND BROWNLOW, A. H., ed., 1976, Studies in New England Geology: Geol. Soc. Amer. Mem. 146, 372 p.

PAGE, L. R., ed., 1976, Contributions to the Stratigraphy of New England: Geol. Soc. Amer. Mem. 148, 445 p.

SKEHAN, J. W., S. J. AND OSBERG, P. H., ed., 1979, *Caledonides in the U.S.A.—Geologic excursions in New England Appalachians,* Weston Observatory, Boston College, Weston, Mass., 250 p.

ZEN, E-AN, WHITE, W. S., HADLEY, J. B. AND THOMPSON, J. B., JR., ed., 1968, *Studies in Appalachian Geology, Northern and Maritime,* Wiley-Interscience, New York, 475 p.

Southern Appalachians

FISHER, G. W., PETTIJOHN, F. J., REED, J. C., JR., AND WEAVER, K. N., ed., 1970, *Studies in Appalachian Geology, Central and Southern,* Wiley-Interscience, New York, 460 p.

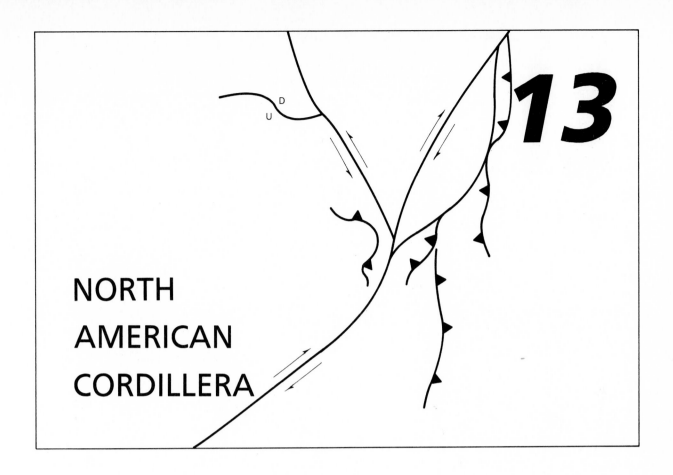

NORTH AMERICAN CORDILLERA

INTRODUCTION

The great *Cordilleran mountain belt* (Figure 13-1) composes the deformed western border of the North American continent, extending from Alaska to Central America, with further extensions of deformation into northeastern Asia, the Caribbean, and the Andes of South America. In contrast with the Paleozoic Appalachian mountain belt discussed in the preceding chapter, the Cordilleran mountain belt is undergoing present-day deformation over its entire length and over much of its width, as demonstrated by current seismicity, volcanism, and mountainous topography. The Cordilleran mountain belt has been a locus of continental-margin deformation, at least sporadically, since the late Paleozoic, about 350 m.y. Furthermore, it has never experienced the suturing of megacontinental masses in the manner of the Appalachian-Caledonide chain and parts of the Alpine-Himalayan system; it has always bordered a major "Pacific" ocean. The Cordilleran belt displays an orogenic style in marked contrast with the Appalachian style introduced in Chapter 12.

One of the most immediately instructive aspects of the Cordilleran belt is the fact that the presently active mountain belt does not represent a single plate boundary but involves deformation between the North American plate and the Pacific, Gorda, Cocos, and Caribbean plates. From place to place, the mountain belt is characterized by active compressive, strike-slip, and extensional deformation. The relative motion between the major North American and Pacific plates is right-lateral strike slip, approximately parallel to the continental margin in Canada, California, and northern Mexico (Fig. 13-1). Therefore, active strike-slip faults, such as the San Andreas in California and the Queen Charlotte offshore of western Canada, exist where the North American and Pacific plates are in

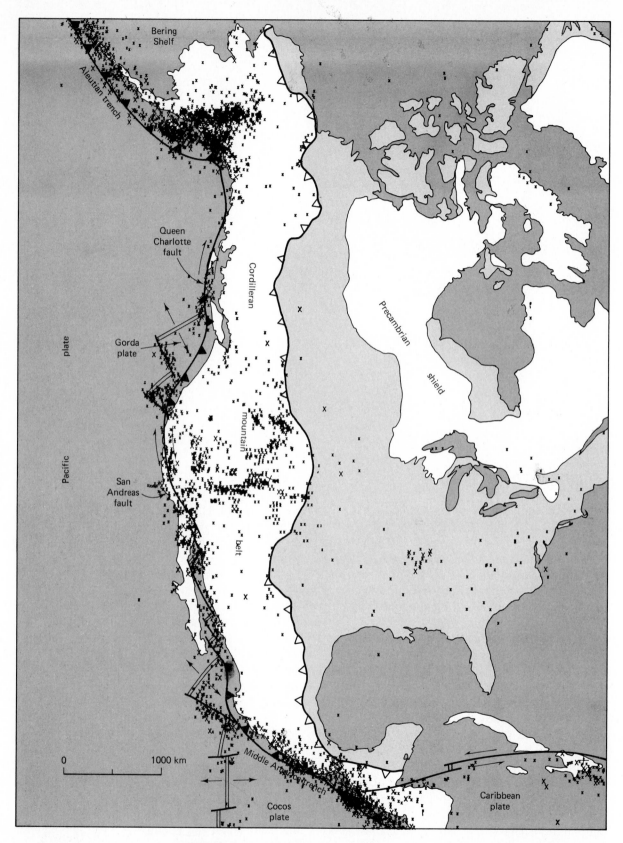

FIGURE 13–1 The Cordilleran mountain belt of western North America and its present-day seismicity and plate configuration. (Seismicity courtesy of U.S. Geological Survey.)

contact. The same North American–Pacific plate boundary becomes compressive, marked by subduction and arc volcanism, where the continental margin bends perpendicular to the plate motion in southern Alaska and the eastern Aleutians. The plate boundary leaves the North American continent with the Aleutian island arc, which gradually bends to parallelism with the plate motion as it approaches the Kamchatka Peninsula of Siberia 2000 km to the west. The Bering Shelf of western Alaska is the only part of the western continental margin of North America that is not marked by present-day plate-boundary tectonics. In summary, important variation in the style of present-day Cordilleran tectonics between right-lateral strike slip and subduction primarily reflects the angle between the continental margin and the relative-motion vector.

A second major cause of present-day tectonic variation along the length of the Cordillera is several oceanic plates that intervene between the Pacific plate and the Cordillera: the Gorda and Cocos plates (Fig. 13-1). These smaller plates give rise to subduction and volcanic-arc tectonics in the northwestern United States, southern Mexico, and Central America. South of the strike-slip fault zone in central Guatemala, the Cordillera lie on the Caribbean plate rather than the North American plate.

The history of interaction between North America and the Pacific and intervening plates has led to a similarly complex variety of tectonic behavior along the Cordillera in the Cenozoic, as will be discussed later in this chapter. For the moment, the important principle illustrated by this brief tour of present-day plate interactions is that a great mountain belt of continental or intercontinental extent should not, in general, be equated with a single plate boundary even at a single time, but rather a whole system of plate boundaries, often localized at a continent-ocean interface. This fact could equally well be illustrated with the present-day Andes or the Alpine-Himalayan chain (Fig. 1-30). The continent-ocean interface is inherently unstable, making it a common site of plate interactions.

We can expect equally complex plate interactions to have marked the tectonic history of the Cordillera in the last 350 m.y. Much of this history is yet to be deciphered over the 12,000-km length of the chain. Our purpose in this chapter is to introduce only the most-throughgoing and best-understood themes in the regional structural geology and tectonic history of the Cordillera, particularly in the United States and adjacent Canada and Mexico.

Our methods of approaching the tectonics of the Cordillera will initially be similar to those used in the Appalachians. We begin from the North American continental interior and move into the eastern marginal zones of deformation, which involve rocks of strong North American affinities. As we move west, we eventually encounter rocks that have strong oceanic affinities or show signs of being highly allochthonous. This is all very similar to the Appalachians in some ways. In contrast, other aspects of orogenic belts that have not been prominent in our discussion of Appalachian tectonics, such as strike-slip faulting, voluminous outpouring and intrusion of magma, plateau uplift, and normal faulting, are found to play a major role in the Cordillera.

CRATONIC FORELAND

The eastern edge of strong deformation in the Cordilleran mountain belt is generally close to the eastern edge of thick Paleozoic sedimentation, although there are important exceptions to be emphasized later. Paleozoic sedimentary thicknesses east of the eastern fold-and-thrust belt (Fig. 13-2) are generally 2 km or less, typical of cratonic stratigraphic sections (Fig. 1-8). For example, this

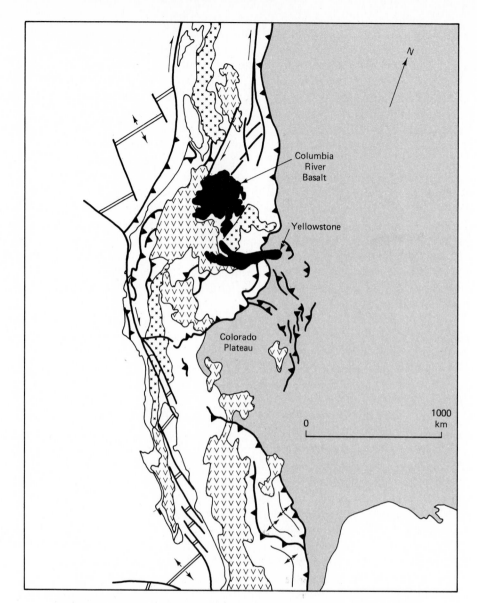

FIGURE 13–2 Major tectonic features of the Cordilleran mountain belt from southern Canada to northern Mexico. The batholith belt is shown in the x pattern and Cenozoic volcanic rocks are shown in the v pattern. The eastern edge of the mountain belt is the east-vergent fold-and-thrust belt.

cratonic stratigraphy is exposed in the Grand Canyon of the Colorado River, where the Cambrian through Permian sequence is only about 1200 m thick (Fig. 13-3). The overlying Triassic and Jurassic sediments that are exposed in nearby basins to the north increase the total foreland thickness by only another kilometer. Similar cratonic thicknesses are typical of most areas east of the fold-and-thrust belt.

In contrast, if we turn west from the Grand Canyon and travel into the core of the Cordilleran orogenic belt, we find that the Paleozoic section passes continuously into a basin of geosynclinal proportions, as shown in Figure 13-4, which is a restored stratigraphic cross section. The Permian and older section near Las Vegas is about 8 km thick, in contrast with 1200 m at the Grand Canyon (Fig. 13-3) only 250 km to the east. This thickening is quite analogous to the thickening of sediments into the Appalachian mountain belt discussed in the last chapter (Fig. 12-5). The increased thickness is accomplished in three ways: (1) greater rate of

FIGURE 13–3 The Grand Canyon of the Colorado River, Arizona, exposing the thin (1200 m) Paleozoic cratonic stratigraphy of the Colorado Plateau, east of the Cordilleran geosyncline. Precambrian basement and Proterozoic sedimentary rocks are exposed in the canyon bottom.

subsidence, (2) more-complete stratigraphic sequence with less time missing in disconformities, and (3) older rocks at the base of the section. Most of the thick Cordilleran section represents a carbonate platform at the edge of the craton, dominated by shallow-water limestone, dolomite, and quartzites. This carbonate-quartzite sequence is analogous to the Cambrian–Lower Ordovician carbonate-quartzite sequence of the Appalachian foreland, which preceded the Taconic orogeny, and might be analogous to the present-day Florida, Bahama, and Yucatan platforms on the tropical edges of North America.

Underlying the Paleozoic carbonate bank is a very thick sequence of Upper Proterozoic clastic sediments of both deep- and shallow-water deposition, including minor basaltic volcanic rocks; they are called the *Windermere Series* in Canada. This section is in many ways analogous to the Upper Precambrian clastic section of the southern Appalachians that records the initial rifting of the Proto-

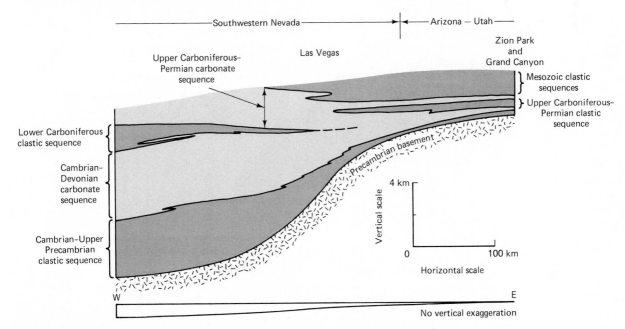

FIGURE 13–4 Restored stratigraphic section of eastern margin of Cordilleran geosyncline from the Grand Canyon to southwestern Nevada. See Figure 13-9 for location. (Simplified from Armstrong, 1968.)

Atlantic Ocean. The upper part of the Cordilleran Upper Precambrian clastic section and the overlying carbonate is fairly similar from the Yukon to California and even to the Appalachians, reflecting a stable continental margin whose sedimentation is dominated by changes in sea level and latitude. However, as we go deeper the section becomes much more heterogeneous, displaying rapid facies changes and evidence for syndepositional normal faulting and basaltic volcanism. The beginning of this episode of deposition is roughly 750 to 800 m.y. ago and is widely interpreted tectonically as a rifting event that established the late Precambrian and early Paleozoic stable continental margin of the Cordillera. It is interesting to note that the rifting of the Appalachian margin was at least roughly contemporaneous.

Only a fragmentary record of these earliest deposits of the Cordilleran stable continental margin and its Precambrian basement is exposed; however, the fragments are widely scattered from the Yukon to California. One of the places they are exposed is in the Cordilleran fold-and-thrust belt in Utah near Salt Lake City (Fig. 13-5). There in the Wasatch Mountains is a complete section extending from old Precambrian basement to Cretaceous at a paleogeographic position midway between the craton and the true Cordilleran geosyncline. The exposures

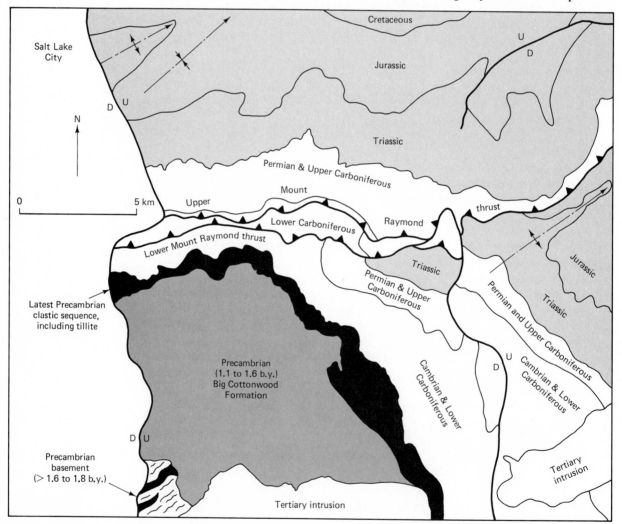

FIGURE 13–5 Simplified geologic map of Cottonwood area near Salt Lake City, Utah. (Compiled from mapping by M. Crittenden, U.S. Geological Survey.)

in map view are very unusual for the fold-and-thrust belt because the northeast-southwest-striking Cretaceous overthrust belt has been refolded about an east-west axis in the early Cenozoic Laramide orogeny (Fig. 13-24); as a result we see an oblique cross section of the overthrust belt in map view (Fig. 13-5). For example, the late Cretaceous Mount Raymond thrust has a slip vector that is toward the southeast, giving it a large component of strike slip in the present map view because of the refolding. Notice that it rides along a décollement in the Lower Carboniferous and then steps up eastward to the Jurassic, with an apparent offset of a few kilometers.

Our main concern for the moment, however, is the stratigraphy exposed in this oblique cross section because it gives a very deep look at the Cordilleran geosyncline. The deepest rocks exposed are gneissic and metamorphic Precambrian basement rocks, not less than 1.6 to 1.8 b.y. old. They are some of the few exposures of old Precambrian crust of North America within the Cordilleran mountain belt. Nevertheless, this old crust is known to extend far to the west in the Cordillera based on a remarkable isotopic property of Mesozoic and Cenozoic igneous rocks of the Cordillera. Magmas that pass through old Precambrian crust on their way up from the mantle take on a trace of the distinctive isotopic composition of old continental crust, whereas magmas that are west of the edge of Precambrian continental crust show isotopic compositions much closer to mantle compositions. Strontium isotopes are the most-useful tracer because ^{87}Sr is produced by decay of ^{87}Rb and rubidium is in the crust; therefore, old crust has a lot of ^{87}Sr. Samples of igneous rock with high ratios of ^{87}Sr to ^{86}Sr are considered to reflect Precambrian crust at depth. The line of ^{87}Sr/^{86}Sr > 0.704 shown in Figure 13-11 is often taken as the western limit of the old crust and, therefore, the edge of the continent in latest Precambrian and early Paleozoic time. This edge is far to the west of our outcrops of this crust in the Wasatch Mountains to which we now return (Fig. 13-5).

There is still more tectonic history in western North America that predates the Cordilleran rifted continental margin; this history is represented by the 4-km-thick Big Cottonwood Formation that overlies the old metamorphic basement. The Big Cottonwood Formation is just one of many thick Upper Precambrian (1.1 to 1.8 b.y.) clastic formations that are exposed along the western margin of the continent in Canada and the United States. The most widely used names are Belt and Purcell Supergroups. These formations fill grabens and more extensive rift systems and for this reason have been thought by some people to be the initial rift deposits of the Cordilleran continental margin. It is more likely, however, that they record an earlier period of extension. The main evidence is that the rifting and thermal subsidence of a stable continental margin takes about 200 m.y. (Chapter 1); therefore, the Big Cottonwood Formation and its equivalents are 0.5 to 1.2 b.y. too old to be the initial deposits of the latest Precambrian–early Paleozoic Cordilleran continental margin. This observation shows us that the continental crust that preceded the Cordilleran mountain belt was a collage of very thick, sediment-filled grabens and older metamorphic and igneous basement.

As we pass upward through the latest Precambrian clastic section, we find marine and terrestrial glacial deposits, which are widespread on the edges of North America and some other continents at this time. The deposits in the Wasatch Mountains are largely nonmarine and the unconformities on the map reflect the filling of late Precambrian glacial valleys. After the glacial periods, there was the Cambrian transgression, followed by a long period of stable deposition. In this area there is no obvious sign of tectonism between the time of the late Precambrian glaciation and the Cretaceous thrusting, a period of a half a billion years. We see a conformable stratigraphic sequence on the map (Fig. 13-5), exhibiting disconformities reflecting only sea-level and epeirogenic effects. It is deceptive, however, to try to view the history of an orogenic belt from a single geographic position. In fact, deformation was going on from time to time not far away, beginning as early as late Devonian and early Carboniferous.

THE ANTLER OROGENY (Late Devonian and Early Carboniferous)

The first sign of major Phanerozoic orogeny in the Cordillera is given by a wedge of shale, sandstone, and conglomerate derived from the west in latest Devonian and early Carboniferous (Mississippian) time. This clastic wedge represents a marginal effect of the *Antler orogeny* and can be seen west of Las Vegas in the stratigraphic diagram of Figure 13-4. The Antler clastic wedge is possibly present as far north as the Yukon territory in Canada, but it is best developed in central Nevada (Fig. 13-6), where the deformational cause of the detritus is also well exposed without an undecipherable amount of later deformation and metamorphism.

Central Nevada is part of the Basin-and-Range province marked by major late Tertiary to Recent normal faulting; nevertheless, fragments of the Paleozoic

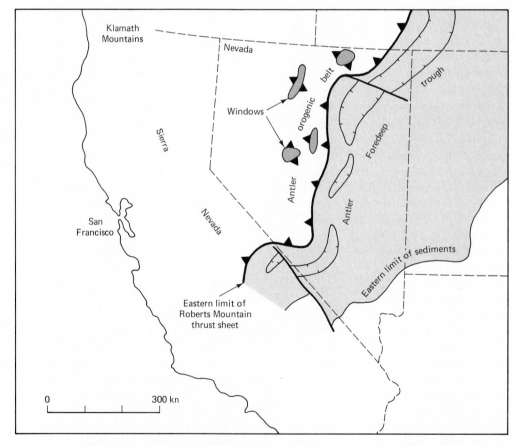

FIGURE 13–6 The Antler orogenic belt in central Nevada and foredeep trough to the east showing the extent of upper Lower Carboniferous strata. Windows through the Roberts Mountain thrust sheet expose autochthonous lower Paleozoic shallow-water carbonates, whereas the structurally overlying rocks are lower Paleozoic deep-water sediments. (Data from Stewart, 1980, and Poole and Sanberg, 1977.)

history are displayed in the mountain ranges, which are the horsts. We can trace the Cambrian through Devonian carbonate-bank sequence much of the way across Nevada, going from horst to horst. In the east the carbonate section is overlain by the westward-thickening and coarsening Antler clastic wedge, but as we move farther west we begin to find mountain ranges displaying thrust sheets of complexly deformed slate, chert, and argillite that is structurally overlying the carbonate sequence. A cross section of one of these ranges is shown in Figure 13-7. The carbonate sequence is shown at depth, exposed at the surface in a window on the west side of the range, with an overlying complex of coeval Ordovician through Devonian deep-water clastic sediments deformed in a number of east-vergent thrusts, many of which are refolded. Later igneous intrusions and normal faults of Tertiary age post-date the thrusting. The basal thrust of the allochthonous deep-water sequence is usually called the *Roberts Mountain thrust* and has been traced through many of the ranges of central Nevada. Stratigraphic relations show that it was emplaced during the time of the Lower Carboniferous (Mississipian) clastic wedge because the thrust sheet contains deformed Devonian strata and is unconformably overlain by Upper Carboniferous (Pennsylvanian) and Permian. The Roberts Mountain thrust sheet is an important structural manifestation of the Antler orogeny.

What is the larger tectonic significance of this late Devonian and early Carboniferous Antler orogeny? There is so much later tectonic disruption of the Cordillera that complete understanding may not be possible, but several conclusions can be made. First, sedimentological and stratigraphic studies show that there is an interfingering facies relationship between the autochthonous carbonate-bank sequence and the deep-water allochthonous clastic sequence (Silberling and Roberts, 1962). Apparently the Roberts Mountain allochthon is a deep-water slope-and-rise sequence, in many ways analogous to the Taconic sequence of the Appalachian mountain belt. We can conclude that some compressive plate boundary impinged on the stable continental margin of the western United States in late Devonian and early Carboniferous time and caused the thrusting of the continental rise-and-slope sediments onto the carbonate shelf.

An exact plate-tectonic scheme for the Antler orogeny is speculative. Perhaps the Antler orogeny is the result of arc-continent collision, by analogy with the Taconic orogeny of the Appalachians. This is a likely possibility and a popular interpretation, but the necessary supporting data are less complete than for the Taconic orogeny. Possible candidates for the island arc and other associated oceanic rocks are present in the Sierra Nevada and Klamath Mountains (Fig. 13-6) of California to the west (Burchfiel and Davis, 1975; Schweickert and Snyder, 1981). The principal uncertainties stem from the fact that many later orogenic events are known in this region, some of which are discussed later in this chapter; these orogenies show signs of producing such profound displacements that the oceanic Paleozoic rocks of California may have nothing to do with the Roberts Mountain thrust sheet and the Antler orogeny. We shall not pursue this uncertain path; many such paths exist in this and all mountain belts. Our concern in this chapter is the better-established aspects of the Cordilleran system. What is clear about the Antler orogeny is that it represents the destruction of the late Precambrian through Devonian stable continental margin and marks the beginning of Cordilleran orogenesis, at least in the western United States.

The Antler orogeny was not long lived, lasting less than 25 m.y. in late Devonian and early Carboniferous time. By late Carboniferous time, the orogenic belt appears to have been well eroded and began to be covered unconformably by a new cycle of stable continental margin sedimentation.

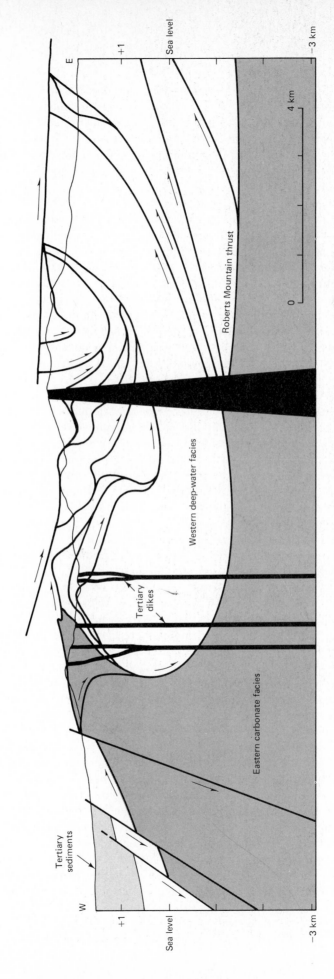

FIGURE 13-7 Cross section of early Carboniferous Roberts Mountain thrust sheet, Shoshone Range, Nevada. Thrust imbrications display folding during thrusting. Tertiary structure includes normal faulting and igneous intrusion. (Simplified after Gilluly and Gates, 1965.)

ANCESTRAL ROCKY MOUNTAINS (Late Carboniferous and Permian)

The part of the Cordilleran mountain belt exposed in the western United States is remarkable in that it displays a fairly clear record of a number of discrete orogenic events, extending as far back as the Antler orogeny. They are spread out spacially to a sufficient extent that younger deformation, metamorphism, and magmatism does not extensively obscure older events. In contrast, Tertiary magmatism and Mesozoic deformation, sedimentation, and magmatism have been so extensive in the Mexican Cordillera that Paleozoic and Precambrian rocks are exposed only in small enclaves and record only a very fragmentary pre-Mesozoic history. The late Paleozoic deformational history of the Canadian Cordillera is also fairly obscure for reasons we shall see later. Therefore, we continue our discussion in the U.S. Cordillera but, perhaps surprisingly, move nearly a thousand kilometers east, well into the Paleozoic craton near Denver, Colorado, to find the next locus of Paleozoic orogenic activity.

The continental interior east of the Cordilleran miogeosyncline had been an area of cratonic stability, exemplified by the Grand Canyon (Fig. 13-3), since the Cambrian. During the Antler orogeny in the early Carboniferous, the continental interior was a stable cratonic carbonate sea unaffected by the deformation to the west in Nevada; to the southeast in the Ouachita orogenic belt of Arkansas, Oklahoma, and the Gulf Coast; or to the north in the Innuitian orogenic belt of the Canadian Arctic.

Seemingly without warning, the locus of deformation stepped out into the continental interior in late Carboniferous (Pennsylvanian) time with the deformation of the Precambrian basement in a system of fault blocks, largely marked by major strike-slip faults with associated thrust and normal faults. We see the effects of this deformation in the Rocky Mountain area of Colorado as a major unconformity in areas of fault-block uplift and as thick orogenic sedimentation in the adjacent basins. Only in a few local areas have the actual late Paleozoic faults been clearly identified because the fault blocks partly coincide with the basement uplifts that form the present-day Rocky Mountains and with the latest Cretaceous–earliest Tertiary block uplifts of the Laramide orogeny. The late Carboniferous structures are called the *Ancestral Rocky Mountains* because of this partial coincidence with the present-day mountains.

What is known of the regional pattern of Ancestral Rocky Mountain deformation is summarized in Figure 13-8. Two main belts of block uplift are known, based on associated coarse clastic rocks fringing them in the adjacent basins and based on unconformities with overlying Permian or Triassic sediments. The eastern belt of uplifts includes the Ancestral Front Range, Apishapa, and Sierra Grande uplifts, which in part coincide with the present-day Front Range west of Denver but also are buried under the Great Plains of southeastern Colorado and northeastern New Mexico. The Apishapa uplift and the Sierra Grande uplift to the south are known largely from subsurface exploration. The coarse conglomeratic Upper Carboniferous sediments that flank the east side of the Ancestral Front Range uplift are exposed as the steeply dipping red beds of the well-known Fountain Formation in the western suburbs of Denver on the flanks of the present Front Range.

The second major uplift is the Ancestral Uncompahgre uplift, which in part coincides with the present-day Uncompahgre Plateau; most of the late Paleozoic uplift is delineated by subsurface exploration. Its age appears to be slightly younger than the Ancestral Front Range based on the sedimentary history of the flanking basins, apparently latest Carboniferous and Permian as opposed to middle and late Carboniferous. The intervening Colorado trough is a site of

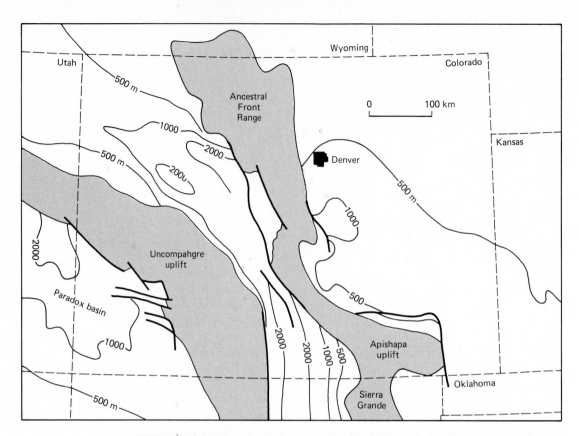

FIGURE 13-8 Uplifts, adjacent basins, and faults of the late Carboniferous and early Permian Ancestral Rocky Mountains in Colorado. Thickness of Upper Carboniferous strata in meters. (Simplified after De Voto, 1980.)

massive orogenic sedimentation amounting to more than 3000 m in the late Carboniferous and Permian. The sediments of the Colorado trough are now exposed in several of the present Rocky Mountain uplifts. By Permian time the orogenic source of sediment had diminished substantially, so that the orogenic troughs began to receive finer sediments, including substantial evaporites in the Paradox basin, southwest of the Uncompahgre uplift (Figs. 13-8 and 7-38).

The Ancestral Rocky Mountains are not a wholly Cordilleran orogenic event. This zone of basement deformation can be traced eastward through Oklahoma and Texas by drilling and seismic methods as a series of buried mountain ridges and flanking basins, which are important for their petroleum deposits. Some of them finally come to the surface in the low Arbuckle and Wichita Mountains of southeastern Oklahoma, where this zone of late Carboniferous and Permian basement deformation impinges on the Ouachita orogenic belt north of Dallas (Fig. 12-1). The Ancestral Rocky Mountains are, therefore, part of a zone of deformation that cuts across the North American craton between the Ouachita and Cordilleran geosynclinal mountain belts.

Other zones of deformation of the craton are present at the same time elsewhere in North America, including the Kentucky–Rough Creek fault zone extending west from the Appalachians into the Illinois basin, the Nemaha ridge in eastern Kansas and Nebraska, and an early Devonian zone of basement deformation extending from the Innuitian mountain belt of the Canadian Arctic Islands south across the craton to the center of Hudson's Bay. Apparently North America was caught between active mountain belts on so many sides that the cratonic interior began to fail, much like central Asia today north and east of Tibet, which is cut by zones of active basement block uplift, marked by major strike-slip, with associated thrust faults with flanking deep basins and grabens that are forming in response to the collision of India and Eurasia. The Ancestral Rocky Mountains and the Arbuckle and Wichita mountains are apparently an analogous effect of collision between South America and North America along the Ouachita orogenic belt.

SONOMA OROGENY (Late Permian and Early Triassic)

The deformation of the normally stable continental interior represented by the Ancestral Rocky Mountains is somewhat exceptional, but by no means unique; this region was again subjected to major deformation in the latest Cretaceous and early Tertiary Laramide orogeny. Nevertheless, the continental margin is the normal site of orogenic instability; therefore, it is not surprising that we must retreat from the unstable craton to the now-familiar Basin and Range of central Nevada to observe the next major orogenic event, the late Permian and early Triassic *Sonoma orogeny*.

During the Sonoma orogeny, deep-water upper Paleozoic sedimentary and volcanic rocks of the *Golconda allochthon* were thrust from the west over a coeval autochthonous shallow-water sequence. The thrust contact is exposed in many of the horsts of central Nevada (Fig. 13-9) and is commonly called the *Golconda thrust*. Stratigraphic relations exposed in some of the ranges show that the time of emplacement of the Golconda allochthon must be within the late Permian or early Triassic because deformed sediments of both the autochthon and allochthon include widespread Permian strata and the Golconda thrust is overlain unconformably by Lower and Middle Triassic sediments.

The autochthonous upper Paleozoic shallow-water sequence is very thin, generally only a few hundred meters thick, and characterized by chert-pebble conglomerates, sandstones, and limestones. This sequence is called the *Antler overlap sequence* because it unconformably overlies the deformed rocks of the earlier Antler orogeny, the chert-pebble conglomerates being derived from the oceanic rocks of the Roberts Mountain allochthon. The stratigraphy of the overlap sequence suggests that in late Carboniferous and Permian time, between the Antler and Sonoma orogenies, the continental margin ceased to be a site of active deformation, for a period of about 100 m.y. Therefore, the Sonoma orogeny marks a major new episode in the progression of the Cordilleran mountain belt. In essence, what happened?

The next important observation is that the Golconda allochthon includes continental-slope-and-rise sediments that can be correlated with the autochthonous-overlap sequence. Furthermore, the other important rock units of the Golconda allochthon are pelagic oceanic rocks and widespread volcanic rocks of island-arc affinities, including pyroxene andesite, basalt, and dacite, with associated volcanogenic sediments and shallow-water limestones. Therefore, the rocks deformed and thrust onto the continental shelf in the Sonoma orogeny include a stable continental-margin sequence, overlapping the roots of the Antler orogenic belt, together with oceanic sediments and an island arc. It is generally concluded, based on these observations, that the plate-tectonic impetus for the Sonoma orogeny was an arc-continent collision. The Sonoma, Antler, and Taconic orogenies are broadly analogous in that each involves deformation of a previously stable continental margin by the thrusting of deep-water slope-and-rise sediments over the continental shelf, apparently in response to arc-continent collision.

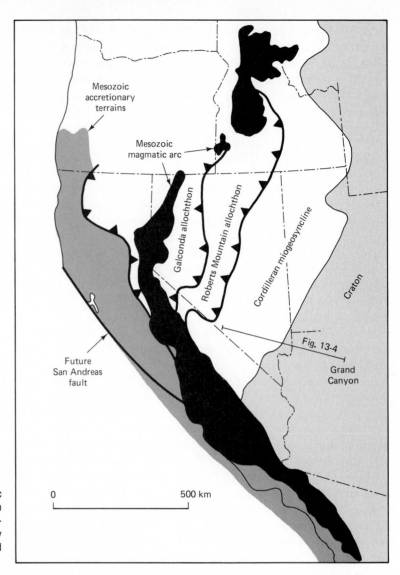

FIGURE 13–9 Truncation of Paleozoic structural trends in the southwestern United States by Mesozoic continental-margin magmatic arc and accretionary wedges. (Compiled from Burchfiel and Davis, 1975.)

EARLY MESOZOIC TRUNCATION EVENT

The Paleozoic continental margin in the western United States has a northeast-southwest orientation. For example, the boundary between the thin cratonic sequences such as the Grand Canyon and the thick cordilleran miogeosyncline, shown in cross section in Figure 13-4, has a N40°E strike (Fig. 13-9). Isopach maps showing the thickness variation of individual Paleozoic stratigraphic units show this same northeast-southwest orientation of the contours, reflecting the orientation of the Paleozoic continental margin (Stewart, 1980). The late Paleozoic continental-margin Antler and Sonoma orogenies also show this same orientation for the foredeep sedimentary trough and for the fronts of the Roberts Mountain and Golconda allochthons (Figs. 13-6 and 13-9). However, by late Jurassic, a new northwest-striking Mesozoic and Cenozoic structural grain, which cuts across and truncates the northeast-southwest Paleozoic structural grain, was well established. This structural reorganization of the western United States sometimes is called the *early Mesozoic truncation event*.

The truncation event not only represents a reorientation of continental-margin deformation, but it also marks the beginning of a new era of Cordilleran tectonics that is dominated by continental-margin volcanic and plutonic activity. Some of this magmatic activity is illustrated in Figure 13-9, which shows the belt of Mesozoic and early Cenozoic batholiths that exists along the western Cordillera. It is this belt of continental-margin magmatism and related deformation and sedimentation that cuts across the Paleozoic structural trends. In northern California the batholiths are emplaced into Paleozoic oceanic rocks related to the Antler and Sonoma orogenies. In central California the batholiths are emplaced into the Cordilleran miogeosyncline. In southern California and Arizona the batholiths are emplaced into the Paleozoic craton. The Mesozoic batholithic belt therefore cuts across the Paleozoic fabric of the western United States.

A second important observation is that the Paleozoic belts do not reappear on the southwest side of the batholith belt. Instead, we find belts of deformed Mesozoic continental-margin and oceanic rocks oriented northwest-southeast, parallel to the batholith belt (Fig. 13-9). Therefore, the continuation of the Paleozoic continental margin is apparently missing. Many workers have interpreted these observations to mean that the southwestward continuation of the Paleozoic continental margin was actually broken off and drifted away sometime in the Triassic or Jurassic. This is why the early Mesozoic tectonic change is called a *truncation event*.

This hypothesis of tectonic truncation is actually rather difficult to evaluate fully. The largest difficulty stems from the lack of extensive exposures of Paleozoic sediments and Precambrian basement rocks in the Mexican Cordillera. If the Paleozoic continental margin was not truncated in the early Mesozoic, it must have originally taken a sharp bend to the southeast in southern California and northwestern Mexico. This constitutes a second hypothesis, which is not as popular, partly in light of the available data on the few scattered exposures in Mexico but also based on considerations of the early Mesozoic plate tectonics of the North Atlantic and Gulf of Mexico (Anderson and Schmidt, 1983), which are outlined in the following paragraphs. By taking this larger view, we gain insight into the two aspects of the truncation event: (1) the actual truncation and (2) the beginning of Mesozoic continental-margin arc magmatism. Furthermore, we gain some insight into the role of plate tectonics in theories of ancient mountain belts.

By the end of the Paleozoic, North America, Europe, Africa, and South America were sutured together into one large continent along the Appalachian-Ouachita mountain belt. In late Triassic and early Jurassic time, parts of that suture began to open up again: the North Atlantic between Africa and North America and the Gulf of Mexico between North and South America. It should be noted that the North Atlantic between Europe and North America, the South Atlantic between Africa and South America, and the Caribbean are all later ocean basins. The major late Triassic and Jurassic rifting event is a splitting of North America–Europe from South America–Africa. It is generally considered significant that this time of major tectonic change between late Permian and Jurassic on the southeastern and eastern edge of North America is also a time of major tectonic change in the western Cordillera (Coney, 1978). Apparently, late Permian to early Jurassic was a time of significant absolute change in North American plate motion that affected plate interactions on both sides of the continent. This change in plate motion may account for the beginning of continental-margin arc magmatism in the Cordillera in late Triassic and Jurassic time.

The actual truncation of Paleozoic structural trends is also hypothesized by many to be a manifestation of the early Jurassic rifting. The reasoning is as follows. When we carefully attempt to fit the continents back together across the North and South Atlantic, we always find that there is an unsatisfactory overlap between South America and North America. Mexico and Central America are too far south for northern South America to fit neatly into the Gulf of Mexico, where it

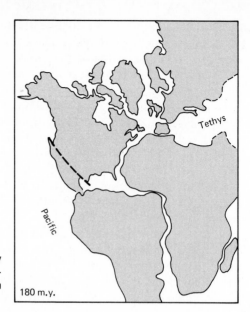

FIGURE 13–10 Reconstruction of early Jurassic positions of Eurasia and Pangea, showing problem of overlap of North and South America.

apparently belongs (Fig. 13-10). For this reason the Paleozoic and Precambrian continental crust of Mexico and Central America is generally thought to have moved southeast relative to the rest of North America as several small crustal blocks. The truncation of Paleozoic structural trends (Fig. 13-9) is hypothesized to be a result of this southeastward motion of Mexico; the bounding fault is proposed to be a hidden strike-slip fault system roughly along the U.S.–Mexican border extending between southern California and the Gulf of Mexico (Anderson and Schmidt, 1983).

In this discussion of the early Mesozoic truncation event, we have gone far beyond the basic geologic observations of the western Cordillera in search of some explanation of what happened between the Late Permian and Late Jurassic. What is clear is that some major change in tectonic orientation and style took place (Fig. 13-9). It is less clear why. Researchers in regional structural geology of mountain belts generally look to plate tectonics as the most-important cause of the major tectonic changes that are observed and documented in mountain belts. The plate-tectonic explanation of the early Mesozoic truncation event is a good example of the sorts of attempts at explanation that are made; the attempt was successful because it was able to join together otherwise unrelated observations into a comprehensive picture, but it is uncertain because all the necessary geologic data are not available. Attempts at plate-tectonic explanation generally become more satisfactory as the mountain belt becomes younger. For example, in our brief introduction to the present-day spatial changes in tectonic style along the Cordillera (Fig. 13-1), most of the important present-day tectonic changes within the mountain belt could be easily related to changes in plate boundaries along the continental margin. In contrast, map-view plate tectonic explanations of the Paleozoic Appalachian mountain belt are so uncertain that we did not present any in Chapter 12. As we proceed to the late Mesozoic and Cenozoic tectonics of the Cordillera, plate-tectonic explanations in general become more specific, testable, and enlightening.

PACIFIC-MARGIN MAGMATIC BELT

The Pacific margin of the Cordillera, particularly in the United States, was an active site of continental-margin subduction quasi-continuously from late Triassic until late Cenozoic. This fact is most-easily documented by study of the volcanic and plutonic rocks of the Pacific margin, including the batholith belt (Fig. 13-9). Three types of evidence are particularly important.

1. Petrographic and geochemical studies of the magmatic belt show that it has close petrologic affinities to present-day continental-margin volcanic arcs such as the Andes, Cascades, and Indonesia.

2. Field studies show that many of the magmatic rocks were intruded into the North American continental margin or were erupted upon it. Other magmatic rocks—for example, in the Jurassic of the western Sierra Nevada—may represent oceanic island arcs that are accreted to North America; however, there also are coeval arc rocks to the east that were clearly emplaced on the North American continent (Schweikert, 1981; Burchfiel and Davis, 1981). The emplacement of these Cordilleran granitic plutons has already been discussed in Chapter 7 (Figs. 7-20 and 7-23).

3. Isotopic and fossil dating show that the continental-margin magmatic arc was active for much of the time between late Triassic and late Cenozoic. However, the history has not been uniform and monotonous. The available data show that there are several episodes of especially abundant and widespread magmatism, whereas other times show little activity. Subduction was certainly not uniform and may not have been continuous. The data also show that the locus of magmatic activity has undergone important changes in map pattern during the last 200 m.y. Several of these changes in abundance and location of the Cordilleran magmatic arc are illustrated in the following paragraphs.

Two periods of late Mesozoic magmatism were especially important in constructing the batholith belt in the western United States and northwestern Mexico. This fact can be illustrated by histograms of potassium-argon dates on plutonic rocks of the western United States (Fig. 13-11). So many plutonic rocks have been isotopically dated in the western United States that they give a reasonably accurate impression of the temporal pattern of late Jurassic through early Cenozoic magmatic activity. The pre-late Jurassic history is largely obscured because K-Ar dates only give the time of last cooling below about 300° to 400°C, and many of the older rocks have suffered reheating after their initial crystallization. Figure 13-11 strictly gives a picture of times of major cooling. The mid-to-late Cenozoic igneous history also is not well represented on the plutonic histograms because most of this younger history is represented at the surface by volcanic rocks, which is discussed later. Two periods of especially voluminous magmatic activity are immediately apparent in Figure 13-11: late Jurassic and early Cretaceous (about 130 to 160 m.y.) and late Cretaceous (about 75 to 105 m.y.).

The late Jurassic and earliest Cretaceous magmatic activity is one manifestation of a period of widespread tectonic activity sometimes called the *Nevadan orogeny*. The late Cretaceous period of tectonic activity is sometimes called the *Sevier orogeny,* although this name is less universally applied. The period of latest Cretaceous and earliest Cenozoic magmatism that appears in southern California, Arizona, and the Pacific Northwest, but not in the Sierra Nevada, is a manifestation of the *Laramide orogeny*. Later in this chapter we shall see that each of these three periods of orogenic activity display important tectonic phenomena in other geologic provinces of the Cordilleran mountain belt in addition to the magmatic arc.

The names for each of these three orogenic periods were originally applied to locally recognized orogenic phenomena, just as *Taconic orogeny* was originally

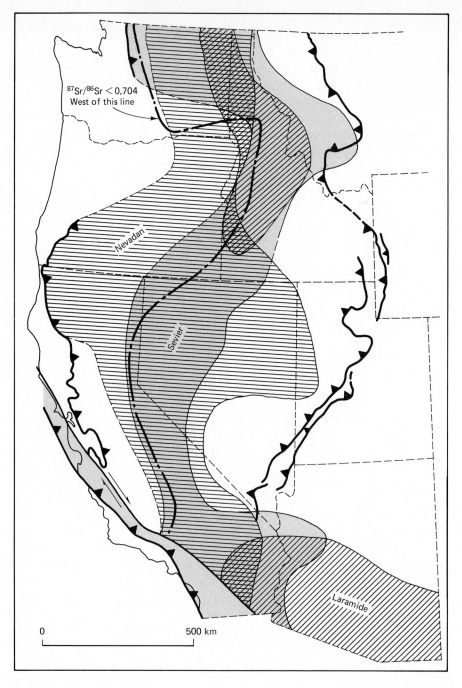

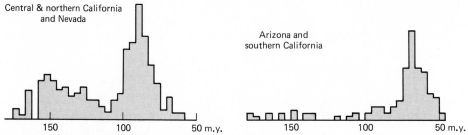

FIGURE 13–11 Loci of late Jurassic to early Cretaceous Nevadan (160 to 125 m.y.) mid-to-late Cretaceous Sevier (105 to 75 m.y.) and Laramide (50 to 75 m.y.) magmatism in the western United States. The histograms give K/Ar radiometric dates on the granitic rocks. The presumed western edge of Precambrian Continental crust is along the line defined by $^{87}Sr/^{86}Sr = 0.704$ for Cenozoic volcanic rocks and Mesozoic plutonic rocks. (Based on data of Armstrong and Suppe, 1973; Kistler and Peterman, 1973; and Armstrong, Taubeneck, and Hales, 1977.)

applied to an angular unconformity along the Hudson River in New York State (Fig. 12-10). *Nevadan orogeny* was originally applied to an angular unconformity in the Sierra Nevada and Klamath Mountains of California. *Sevier orogeny* was originally applied to the fold-and-thrust belt of western Utah. *Laramide orogeny* was originally applied to the basement-block uplift of the Laramie Range of central Wyoming and other similar nearby uplifts. In this chapter we apply these orogenic names in their broader senses to help us identify some of the major throughgoing features of Cordilleran tectonics.

The spatial distribution of magmatic activity within the arc is quite different during each of the three orogenic periods mentioned above and is illustrated in Figure 13-11. During the late Jurassic, Nevadan orogeny magmatic activity was very widely distributed across the western United States from the Pacific margin in California to the edge of the Paleozoic craton in Utah. During the late Cretaceous Sevier orogenic period, magmatism was more nearly confined to the 150-km-wide belt of closely spaced intrusions of the Sierra Nevada and Peninsular Range batholiths of California and Baja California. Much of the volume of these large composite batholiths was emplaced during a relatively short time in the late Cretaceous, creating a vast sea of granitic plutonic rock. The central zone of the batholith belt is nearly all late Cretaceous granitic rock. For example, most of the area of the central Sierra Nevada shown in Figure 13-12 is late Cretaceous granitic rock. Just before the end of the Cretaceous, there was a major change in the map distribution of magmatism (Fig. 13-11); this change marks the beginning of the Laramide orogenic period.

The early Jurassic and younger arc magmatism on the Pacific margin of North America has many petrologic similarities to the present-day continental-margin magmatism of the Andes, which points to subduction that is localized along the edge of the continent. For this reason the Cordilleran continental margin after the early Mesozoic truncation event has often been called an *Andean-type margin*. In contrast with the Andes, however, the Cordillera does not appear to have generally had the very high topography that is characteristic of much of the present-day Andes. During Jurassic time, volcanic rocks of the Cordilleran arc generally interfinger with shallow-water marine sediments. Some of the volcanic centers were actually submarine calderas. Possible analogs of this continental-margin arc magmatism at or near sea level include the present-day arcs of western Indonesia and Central America south of Guatemala. The magmatic arc also appears to have been at low elevation during most of the Cenozoic in the western United States based on evidence from fossil floras interbedded with the volcanic rocks. Local mountains, especially volcanoes, may have reached substantial elevations, but the regional topography was low. The present high elevation of the western United States is a late Cenozoic orogenic phenomenon, to be discussed near the end of this chapter. The only period of possible Andean topography to the western United States is the late Cretaceous, and even that is unlikely based on present data.

PAIRED CRETACEOUS BASEMENT BELTS

In the brief introduction to Cordilleran tectonics presented in this chapter, we skipped over essentially all the late Triassic and Jurassic tectonism of the Pacific margin, including much of the Nevadan orogeny. This period is represented in the rock record by some extremely complex and fragmented geology that is the focus of very active research (for example, see Ernst, 1981). Interpretations of the tectonics of this period are controversial and insight is rapidly evolving. We summarize this period by simply noting that a number of belts of oceanic rocks,

FIGURE 13–12 Oblique aerial view of Sierra Nevada batholith belt, Yosemite National Park, looking east. All the foreground and middle ground is underlain by Cretaceous granitic rock. The darker peaks in the background are underlain by metamorphic roof pendants of the Ritter Range, which contains extrusive equivalents of the mesozoic plutons. The peak in the foreground is Half Dome (see also Fig. 6-21). (U.S. Geological Survey.)

including ophiolites, pelagic sediments, and island-arc volcanic rocks, were accreted to parts of western North America during this period. We now move on to the Cretaceous of Oregon, California, and northwestern Mexico, which represents a time of great unity and throughgoing tectonics in the Cordilleran margin.

The title of this section, "Paired Cretaceous Basement Belts," provides a hint of the fact that one of the great unifying features of Cretaceous Cordilleran tectonics is a pair of belts of contrasting basement rocks running along the North American continental margin; they extend for more than 3000 km from southern Baja California to Oregon. In this context we mean by *basement* any metamorphic and plutonic rocks that act as basement for the overlying unmetamorphosed Cenozoic sedimentary and volcanic rocks.

The Cretaceous paired basement belts (Fig. 13-13) comprise (1) the *batholith belt,* already discussed, with its high-temperature, low-pressure metamorphic

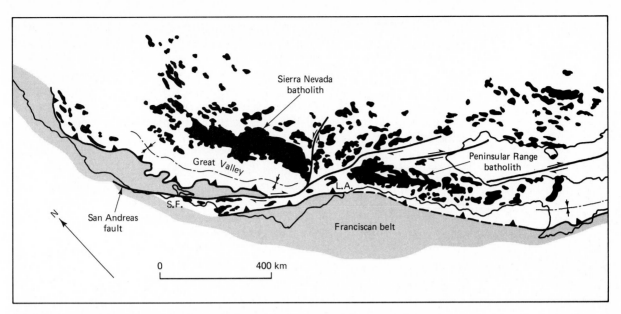

FIGURE 13–13 Cretaceous paired-basement belts, Oregon to Baja California. The high-pressure, low-temperature Franciscan belt is shown by the screened pattern along the continental margin. The granitic rocks of the high-temperature, low-pressure batholith belt are shown as black splotches.

country rock and (2) the *Franciscan belt* of low-temperature, high-pressure metamorphic rocks of oceanic origin, which exhibits little magmatic activity other than oceanic basaltic volcanism prior to metamorphism. These two contrasting basement belts are the same age and run approximately parallel to each other and to the continental margin from Oregon to southern Baja California (Fig. 13-13). The original parallel pattern has been disrupted in part by the large Cenozoic strike slip on the San Andreas fault system that causes a repetition of the basement belts in map view in central California. In going west from the Sierra Nevada in central California, we go from the batholith belt to the Franciscan belt, cross the San Andreas fault, go back into rocks of the batholith belt, and then into the Franciscan belt again near the coast. In spite of these complexities caused by later deformation, a single pair of belts existed in the late Mesozoic, with the Franciscan belt to the west.

This pair of basement belts is not unique. Similar paired-basement belts are fairly common, particularly in the young mountain systems that ring the Pacific margin (Miyashiro, 1961). For example, they exist in New Zealand, Sulawesi, Japan, Kamchatka, southern Alaska, and southern Chile. One essential feature of paired belts is that the metamorphic and plutonic activity of the two halves is of the same age, whereas the stratigraphic age of the sedimentary protoliths may differ. For example, the metamorphic rocks of the Cordilleran batholith belt range in stratigraphic age from Precambrian to Cretaceous, whereas the high-temperature metamorphism and magmatic activity is late Mesozoic. In contrast, the Franciscan belt has a stratigraphic age ranging from Jurassic through the Cretaceous and displays late Mesozoic metamorphism.

A second essential feature of the paired-basement belts is that one belt is characterized by metamorphism that records a low geothermal gradient, whereas the other records a high geothermal gradient with abundant associated arc magmatism. The low-temperature, high-pressure Franciscan belt typically displays zeolite, blueschist, high-pressure amphibolite, and eclogite facies metamorphism, whereas the high-temperature, low-pressure belt may display hornfels and other high-temperature facies.

The two coeval belts of contrasting thermal gradient running parallel to the continental margin must reflect something fundamental in the underlying tectonic processes. We can discover what this is by looking to present-day analogs of paired-basement belts. We have already noted that the batholith belt exhibits strong petrologic similarities to continental-margin magmatic arcs. Studies of present-day geothermal gradients in the environs of magmatic arcs show that they have the high thermal gradients and heat flow required for present-day high-temperature, low-pressure facies of metamorphism. Seaward of the magmatic arc in the region of the accretionary wedge (Fig. 1-15) is a region of low thermal gradient and low heat flow; this is thought to be a site of present-day low-temperature, high-pressure metamorphism. Numerical thermal models show that the high thermal gradient of magmatic arc is due largely to magmatic convection of heat and hot upper mantle, whereas the low thermal gradient of the accretionary wedge is due to the subduction process. Subduction places cooler rocks under warmer rocks, thereby causing a reversal or reduction of the geothermal gradient; little heat flows into the accretionary wedge. Therefore, a good physical basis exists for drawing an analogy between late Mesozoic paired-basement belts of the Cordillera and present-day continental-margin island arcs. The paired-basement belts record the fact that the late Mesozoic continental margin was a site of subduction.

We now turn to the Franciscan belt for a more-detailed look. What we see indeed looks like what we would expect of a large continental-margin accretionary wedge (Fig. 1-15); the Franciscan belt includes a structurally complex assemblage of deep-oceanic and continental-margin rock types. The deep-oceanic rocks are fragments of ophiolitic material and overlying pelagic sediments, including serpentinite, gabbro, basalt, radiolarian chert, submarine hot-spring manganese deposits, and pelagic limestone. At least one normal-sized submarine seamount is exposed in a deformed and metamorphosed state in the Franciscan terrain (Snow Mountain Volcanics in Fig. 13-14). The continental-margin sediments include thick sequences of arc-derived turbidites and interbedded olistostromes; they compose much of the volume of the Franciscan terrain. These sediments are widely interpreted as trench deposits, although other oceanic paleogeographic settings near the continent may also be represented. Little is known of the specific paleogeography of the Franciscan ocean basin. In fact, so much time is involved (late Jurassic through middle Tertiary, 155 to 30 m.y.) that an enormous area must be represented; for example, if the average rate of subduction were only 5 cm/y, over 6000 km of paleogeography would have been accreted into the Franciscan wedge.

Another fundamental fact contributes to the paleogeographic complexity of the Franciscan ocean basin, namely, that it never existed in its entirety at any single time. One of the most important observations about the Franciscan belt is that deposition, deformation, and high-pressure metamorphism are all broadly coeval throughout the belt. Some rocks were being deformed and metamorphosed at great depth at the same time that other, now nearby, rocks were far out to sea being deposited in an oceanic environment. For example, in the part of the northern Coast Ranges shown in Figure 13-14, some of the metamorphic rocks at Goat Mountain were undergoing high-pressure metamorphism for about 20 m.y. in the late Jurassic (155 through 135 m.y.) at the same time as the sedimentary and volcanic sequence on Snow Mountain (Fig. 13-14) was being deposited. At Snow Mountain we find a late Jurassic sequence of pelagic and abyssal-plain sediments overlain by an oceanic seamount volcano; this sequence later underwent thrusting and high-pressure metamorphism in the middle or late Cretaceous.

This juxtaposition of rock bodies displaying jarringly different histories is characteristic of the Franciscan belt. In some areas the contrasting rock bodies

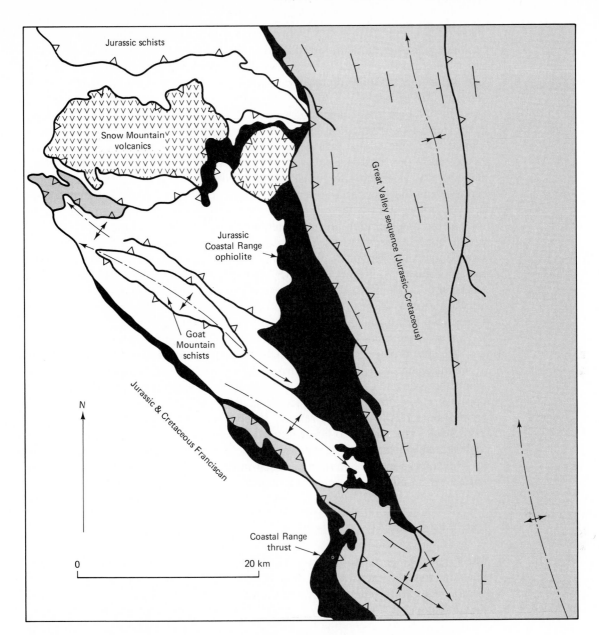

FIGURE 13–14 Map showing typical structure of Coast Range and Great Valley sequence, northern California. (Simplified after Suppe and Foland, 1978. Permission to reproduce granted by the Pacific Section, Society of Economic Paleontologists and Mineralogists.)

are thrust sheets kilometers thick, whereas in other areas they are pieces in the range 1 to 100 m, with the pieces imbedded in a chaotic matrix. For this reason many parts of the Franciscan belt are classified as mélange (Chapter 8). The two best-documented processes by which these fragments of contrasting history are juxtaposed are (1) submarine landslides (olistostromes) and (2) thrust imbrication with multiple episodes of thrusting.

The image we are building up of the Franciscan belt is one that is internally complex and in many ways chaotic, juxtaposing rocks of disparate histories. Nevertheless, it is similarly complex throughout its length and is quite different from adjacent tectonic belts in its assemblage of oceanic rocks, high-pressure metamorphism, and age. The tectonic unity of the Franciscan belt is the sort of unity we might expect of a long-lived accretionary wedge containing an assemblage of diverse oceanic and trench sediments now strongly deformed and subject to varying degrees of metamorphism, depending on their position in the accretionary wedge. Furthermore, we expect many times of metamorphism, deformation, and sedimentation to be represented, rather than a single orogenic event of short duration.

In spite of the general agreement between the structural and tectonic properties of the Franciscan belt and our image of what an accretionary wedge should look like, there are some differences. In particular, the accretionary processes in the Franciscan do not appear to have been monotonous throughout the 150 m.y. between late Jurassic and middle Tertiary; the data on history of tectonism in the Franciscan belt suggest that it contains certain discrete orogenic episodes rather than a fully continuous and uniform accretionary process. If we look at the times of high-pressure metamorphism and sedimentation over a large part of the Franciscan, we see that it is broadly continuous from late Jurassic to late Cretaceous, but on a closer look we see that much of the high-pressure metamorphism is in the range 155 to 135 m.y. (late Jurassic and earliest Cretaceous) and 115 to 95 m.y. (late early Cretaceous to late Cretaceous) with very little after 95 m.y. or between 115 and 135 m.y. Thus the high-pressure Franciscan belt displays some signs of discrete orogenic events.

If we look more carefully at the batholith belt, we find a somewhat similar but not fully in-phase set of magmatic pulses. The late Jurassic period of activity in the magmatic arc (Fig. 13-11) corresponds closely in time with the major period of late Jurassic and earliest Cretaceous metamorphism in the Franciscan belt; this episode is traditionally called the Nevadan orogeny, as discussed above. The early Cretaceous is a time of relative igneous and metamorphic inactivity in the paired belt. The middle to late Cretaceous is again a time of coeval magmatism and metamorphism, the Sevier orogeny. Tectonic activity resumed in the Cenozoic. The correspondence in time between tectonic events in the two halves of the paired belt reinforces the interpretation that they are indeed a genetic pair.

THE GREAT VALLEY SEQUENCE

The Franciscan and batholith basement belts are generally separated by 50 to 100 km along much of their length. The coeval late Mesozoic rocks that fill this space are unmetamorphosed sediments of the *Great Valley sequence* in California and its equivalents to the north and south. These sediments represent a very thick sedimentary basin, exceeding 10 km in some areas, which records an intermediate paleogeography between the magmatic arc on the east and the Franciscan ocean on the west. In terms of analogies with present-day subducting continental margins, the Great Valley sequence apparently fills a forearc basin such as the forearc basin of Guatemala shown in Figure 1-15.

For comparison, a cross section of the Great Valley sequence in northern California is given in Figure 13-15, extending from the Franciscan belt in the Coast Range on the west to the batholith belt in the Sierra Nevada on the east. Note that the Great Valley sequence onlaps the plutonic and metamorphic rocks of the batholith belt so that younger Upper Cretaceous part of the Great Valley sequence overlies Upper Jurassic and Lower Cretaceous plutonic rocks on the western edge of the batholith belt. In a few areas such as Baja California, the eastern edge of the Great Valley sequence actually includes volcaniclastic sediments; furthermore, the Great Valley sequence has a sedimentary petrography that shows a derivation from a volcanic and plutonic arc. Therefore, it is well documented that the Great Valley sequence was in its present position relative to the batholith belt at the time of its deposition.

The relationship of the Great Valley sequence to the coeval Franciscan belt is more complex. As we travel across the Great Valley of California, from the Sierra Nevada to the Coast Ranges (Fig. 13-15), the Great Valley sequence is undeformed and gradually thickens for two-thirds of the way across the valley. Then suddenly at the western edge of the valley (Figs. 13-14 and 13-15), the section turns up on end and, in a single homocline, exposes one of the thickest complete stratigraphic sections you will see anywhere, more than 10 km of Upper Jurassic and Cretaceous clastic sediments, which sits depositionally on a disturbed ophiolite sequence. This ophiolite sequence is called the *Coast Range Ophiolite* (Hopson, Mattinson, and Pessagno, 1981) and is a great sheet of Middle Jurassic (165–155 m.y.) oceanic crust, which extends from Oregon to southern California and probably southern Baja California.

So we see that the Great Valley sequence sits on several different basements. On the east it sits on the batholith belt, whereas on the west it sits on disturbed Middle Jurassic Coast Range Ophiolite; therefore, the base of the Great Valley sequence, which ranges in age from late Jurassic to middle Cretaceous, covers over the contact between continental margin basement and oceanic basement hidden under the western Great Valley. The Coast Range Ophiolite appears to have been attached to North America in the late Jurassic Nevadan orogeny, one of the great orogenic episodes of the Cordillera.

The Coast Range Ophiolite is generally not a fully intact sheet of Jurassic oceanic crust. In most areas it is disturbed with the volcanic and mafic plutonic parts of the lithosphere commonly missing; we generally pass from the basal Great Valley sequence, which contains sediments rich in ophiolitic detritus, directly into mantle rocks, serpentinized peridotites, and dunites. Apparently much of the Coast Range Ophiolite was strongly disturbed before the deposition of the basal Great Valley sequence. In other areas the ophiolitic sequence is fairly complete; for example at Point Sal, along the southern California coast, you can walk along the shore, observing a nearly complete cross section of the oceanic crust and upper mantle in the sea cliffs (Hopson, Mattinson, and Pessagno, 1981).

Immediately below the Coast Range Ophiolite in most areas are the Franciscan rocks, and the contact is generally a thrust fault, the *Coast Range thrust,* which is best exposed along the western edge of the Great Valley in the foothills of the Coast Ranges (Figs. 13-14 and 13-15). The rocks of the Franciscan have the same range in stratigraphic age as the Great Valley sequence, so the Coast Range thrust juxtaposes two coeval sequences, putting the Upper Jurassic and Cretaceous Great Valley sequence and its ophiolitic basement over the Upper Jurassic and Cretaceous Franciscan.

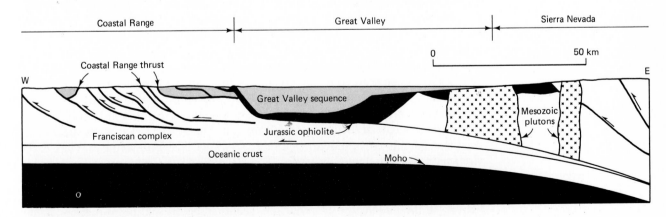

FIGURE 13–15 Cross section of Coastal Range, Great Valley, and Sierra Nevada, northern California. (Based in part on Suppe, 1979.)

The Coast Range thrust has some unusual properties for a thrust fault; in some ways it is quite different from any fault we have encountered heretofore in the Appalachians or Cordilleras. The Franciscan rocks below the thrust have been metamorphosed at various times in the late Jurassic and Cretaceous at very high pressures, equivalent to 20 to 35 km burial, whereas the Great Valley sequence has never been buried deeper than its maximum stratigraphic depth of 10 to 12 km. Therefore, from the point of view of metamorphism, the Coast Range thrust might be mistaken for a normal fault because it seems to cut out section. The cover of the Franciscan terrain at the time of metamorphism is now missing. This strong contrast in history across the Coast Range thrust serves to emphasize how very important this structure is. It has a minimum slip of 100 km, and the actual slip is probably much more because klippen of Great Valley sequence and Coast Range Ophiolite overlie the Franciscan rocks all the way across the Coastal Range for a distance of about 100 km in northern California (Fig. 13-15). Furthermore, the Great Valley sequence and Franciscan rocks are imbricated in a series of Cenozoic thrusts that account for roughly another 100 km of shortening (Figs. 13-14 and 13-15).

In summary, the Great Valley sequence represents an intermediate paleogeography in the late Mesozoic between the batholith belt on the east and the Franciscan belt on the west and apparently represents a forearc basin. Nevertheless, the late Mesozoic boundary between the forearc basin and the Franciscan accretionary wedge is now obscure because of very large displacement on the Coast Range thrust and on thrusts that deform the Coast Range thrust and imbricate the Great Valley sequence and coeval Franciscan metamorphic rocks.

SEVIER FOLD-AND-THRUST BELT

So far we have seen that the late Mesozoic tectonics of the western margin of North America from Oregon to southern Baja California are dominated by a continental-margin subduction system comprising the Franciscan accretionary wedge, Great Valley forearc basin, and batholith-belt magmatic arc. We now move east of the magmatic arc to another coeval zone of deformation, the *Sevier fold-and-thrust belt,* which is part of the great Cordilleran foreland fold-and-thrust belt, extending from Alaska to Central America (Fig. 13-1).

The Sevier fold-and-thrust belt lies at the edge of the North American craton and runs right along the boundary between the Paleozoic miogeosyncline and craton (Figs. 13-9 and 13-16). For this reason the fold-and-thrust belt is not parallel to the paired basement belts to the west. The map pattern of the paired basement belts is controlled by the edge of the continental-margin subduction, whereas the edge of the overthrust belt is controlled by the edge of thick sediments around the orogenic belt.

Directly east of the fold-and-thrust belt is a late Mesozoic foredeep, a sedimentary basin that contains a record of the nearby orogenic activity. For this reason we first look at the foredeep to get an overview of the orogenic history by noting the ages of important clastic wedges. A restored stratigraphic cross section of the western part of the foredeep is shown in Figure 13-17. Three big tongues of clastic sediments are apparent: (1) late Jurassic–earliest Cretaceous, (2) late early Cretaceous to early late Cretaceous, and (3) middle late Cretaceous. These times of orogenic sedimentation show that mountain building is broadly coeval between the paired basement belt and the overthrust belt; indeed the time of lessening activity in the middle early Cretaceous seems to be present in both.

You have probably noticed that we have generally dated orogenic activity in our discussions of the Appalachian and Cordillera tectonics using the ages of

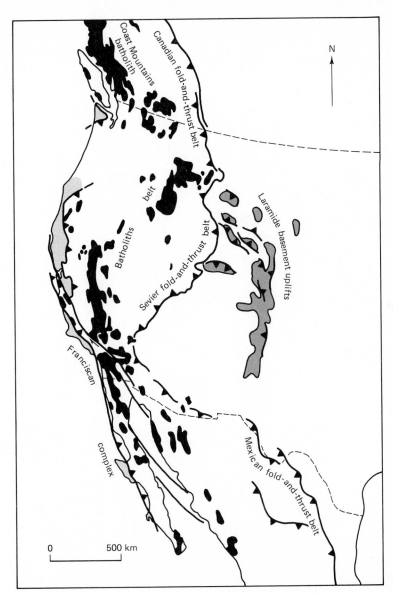

FIGURE 13–16 Map of major late Mesozoic and early Cenozoic structural belts, including the Franciscan and related accretionary complexes, the batholith belt (in black), the fold-and-thrust belt, and the Laramide basement uplifts.

elastic wedges, metamorphism, and igneous activity. Well-constrained ages of actual structures such as faults and folds are much more difficult to obtain and therefore often play a subsidiary role in discussions of tectonic history. The Sevier overthrust belt is an exception to this generality in that some of the important structures are reasonably well dated based on angular unconformities and distinctive detritus eroded from the thrust sheets.

An example of one of these well-dated structures is the great Absaroka thrust of Wyoming (Fig. 13-18). The older branch contains an overturned fault-propagation fold of Paleozoic through Lower Cretaceous strata in the upper plate, which overrides middle Santonian (Upper Cretaceous) conglomerates. These conglomerates were apparently eroded from the advancing thrust sheet because it contains detritus similar to formations in the thrust sheet, with rock types that do not resist abrasion very well. The conglomerate is very coarse, with some boulders exceeding 2 m. As you pass up through the conglomeratic sequence, clasts of older and older formations appear, indicating that uplift and denudation was going on during deposition.

A second branch of the Absaroka thrust was activated to the east of the early Absaroka thrust after the middle Santonian conglomerate was deposited and

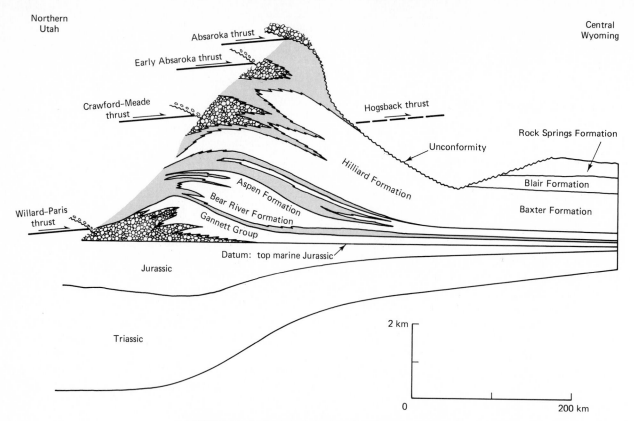

FIGURE 13–17 Restored stratigraphic diagram of mesozoic foredeep deposits in western Wyoming and northern Utah, showing relationship between clastic wedges and thrusting. (Simplified after Royse, Warner, and Reese, 1975.)

overridden. The entire region is eroded and overlain by the Lower Tertiary Evanston Formation, which places a lower limit on the youngest deformation. Similar field and subsurface relations allow fairly precise dating of many of

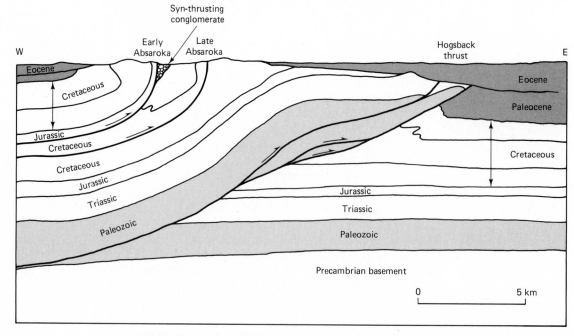

FIGURE 13–18 Cross section of early and late Absaroka thrusts and Hogsback thrust, Sevier fold-and-thrust belt, Wyoming. (After Royse, Warner, and Reese, 1975).

the major structures of the Sevier fold-and-thrust belt and show that deformation generally coincides with the times of the clastic wedges of the foredeep (Fig. 13-17).

SOUTHERN CANADIAN FOLD-AND-THRUST BELT

This brief introduction to the regional structural geology and tectonics of the Cordillera has focused almost exclusively on the segment in the western United States because it is regionally the best known and geographically the most-commonly visited. However, the best-known and best-displayed segment of the Cordilleran fold-and-thrust belt is not in the United States, but in the southern Canadian Rocky Mountains in western Alberta and eastern British Columbia. Here the structure is displayed in a grand alpine fashion on the mountain faces of Banff and Jasper National Parks (Fig. 9-6). For example, Figure 13-19 shows some complex imbricate fold-and-thrust structure in the Boule Range along the Athabaska River east of Jasper.

The structure of the foreland of the Canadian Rocky Mountains is naturally divided into a series of belts running parallel to the mountain front, each characterized by a distinctive structural style, distinctive stratigraphy, and distinctive topography. These belts may be seen in the cross section of the Canadian Rockies in Figure 13-20 and are the Alberta basin (Fig. 1-8), Foothills, Front Ranges, Main Ranges, Purcell anticlinorium, and Kootenay arc. These belts and the cross section take us about halfway across the 700-km-wide Cordilleran mountain belt in southern Canada (Fig. 13-16).

FIGURE 13–19 Frontal thrusts of the Front Ranges, Athabaska River, Jasper National Park, Canada. The foreground is underlain by Mesozoic clastic sediments of the Foothills belt, whereas the mountain exposes Upper Paleozoic carbonates of the Front Ranges.

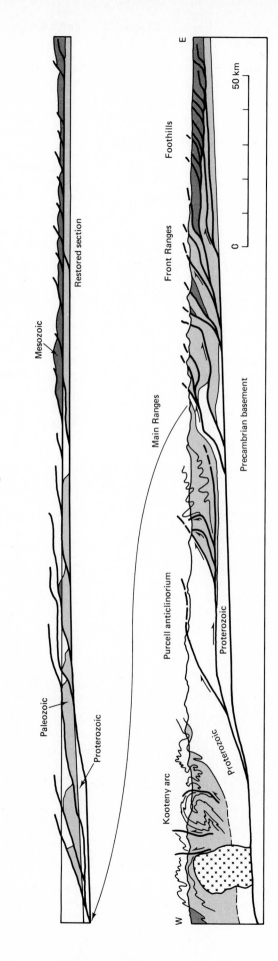

FIGURE 13–20 Cross section and restored section of the eastern Canadian Rockies. (Simplified after Price and Fermor, undated.)

The easternmost belt is the *Alberta basin,* which is the undeformed foredeep in front of the fold-and-thrust belt (Fig. 1-8). This basin and its deformed equivalents to the west contain as much as 4 km of Upper Jurassic through Lower Tertiary clastic sediments derived from the deforming Cordillera to the west. The times of orogenic activity recorded in the southern Canadian foredeep are not entirely the same as those recorded on the eastern margin of the Sevier orogenic belt in the United States (Fig. 13-17). Both are active in the late Jurassic and Cretaceous, but the fold-and-thrust belt in the western United States ceases just before the end of the Cretaceous, whereas the overthrust belt in southern Canada and Montana continued to deform through the Paleocene and into the Eocene. These differences in history will become important in the next section, when we discuss the latest Cretaceous and early Tertiary Laramide orogeny.

The *Foothills belt* of the Rocky Mountains is characterized by low topographic relief and very strong imbricate thrusting and folding of the same foredeep section that is undeformed in the Alberta basin to the east. The Foothills belt is in essence the deformed western margin of the Alberta basin. The structure is well known because of extensive exploration for petroleum, with the initial discoveries in the 1920s. One of the important contributions of this exploration to structural geology is the discovery that the strong deformation exposed at the surface, displaying steep dips, abundant fault repetitions of the stratigraphy, and tight folding, does not extend to great depth. At depths of 3 to 5 km, we pass into undeformed Paleozoic cratonic sediments, which overlie Precambrian crystalline basement (Fig. 13-21). Most of the thrust faults flatten to décollement horizons in the Mesozoic or, in some places, the upper Paleozoic section. The horizontal shortening above the basal décollement is very large; the 50-km-wide Foothills belt in Figure 13-20 was originally about 100 km wide based on retrodeformation of the cross section.

The western limit of the Foothills belt is marked by the abrupt topographic front that is the beginning of the *Front Ranges.* The Rocky Mountain front, such as that east of Jasper (Fig. 13-19), generally marks the first thrust sheet with Paleozoic rocks that rises to the surface. The thrust that serves as the frontal thrust changes along strike in *en échelon* arrangement, but each places cliff-forming Cambrian and younger carbonates over the easily erodable Cretaceous sandstones and shales. These frontal thrusts include some of the best-known thrust faults in North American geology: the Lewis thrust in Glacier-Waterton National Park (Fig. 8-26), the McConnell thrust in Banff National Park (Fig. 8-4), and the Boule thrust at Jasper National Park (Fig. 13-19).

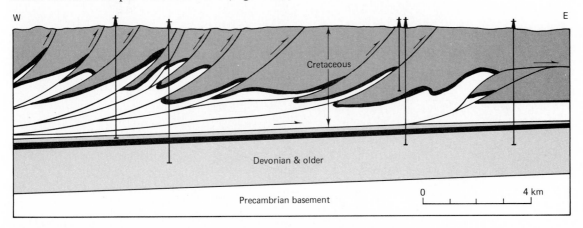

FIGURE 13–21 Detailed cross section of part of Foothills belt of the Canadian fold-and-thrust belt, showing complex imbrication of the Mesozoic strata but underformed Paleozoic at depth. (Simplified after Ollerenshaw, 1973.)

The Paleozoic and lower Mesozoic carbonate section that forms the imbricated thrust structure of the Front Ranges (Fig. 13-20) displays geosynclinal thicknesses of 4 to 10 km, thickening to the west with stratigraphy that is very similar to the Cordilleran miogeosyncline in the western United States (for example, Fig. 13-4). In contrast, the coeval section that is present in the subsurface of the Foothills Belt and Alberta basin is cratonic in character and only 1 to 1.5 km thick. Thus the stratigraphic change from Paleozoic craton to geosynclinal continental margin is now structurally compressed and displaced and represented by the Foothills–Front Range structural boundary (Fig. 13-20).

The major thrust sheets in the Front Ranges are much more widely spaced than those in the Foothills Belt—about 5 km, as opposed to 1 to 2 km (Figs. 13-21 and 8-4). This difference apparently reflects the much-larger steps in décollement that are characteristic of the geosynclinal section of the Front Ranges; the thrusts generally step from low in the Cambrian to the Upper Paleozoic or Lower Mesozoic in a single step. Each of the mountain ranges or ridges running parallel to strike in the Front Ranges is generally a single thrust sheet, with the ridge held up by Paleozoic cliff-forming carbonates. As we travel across the Front Ranges— for example, along the Trans-Canadian Highway (Fig. 8-4)—we see the same Paleozoic stratigraphic section repeated in each mountain range with the strata always dipping west. This is the first and most-obvious indication of thrust imbrication. Only after much mapping is it found that major thrust faults generally

FIGURE 13–22 Little-deformed Cambrian carbonates of the eastern Main Ranges, near Lake Louise, Banff National Park. Compare with Figure 13-23.

run along the parallel valleys, hidden by the thick forest cover. The décollement structure of the Front Ranges is substantiated by outcrops of the thrusts here and there, seismic reflection profiling, deep drilling, and retrodeformable cross sections. The shortening of the 50-km-wide Front Ranges in Figure 13-20 is about 75 km.

As we travel through the western Front Ranges and into the Main Ranges, we encounter two important changes in the Paleozoic and late Precambrian continental-margin stratigraphy that bring about major changes in the mountain structure. First, the basal décollement steps down in the section several kilometers into a thick section of late Precambrian clastic sediments, which is correlative with the late Precambrian sediments of the western United States already discussed. The main structural effect of the appearance of Upper Precambrian strata below the Cambrian is much thicker thrust sheets (Fig. 13-20).

A second, more-structurally profound, stratigraphic change is caused by a sedimentary facies change in the Cambrian. In the east the Cambrian is characterized by thick massive carbonate sequences, typical of the early Paleozoic carbonate bank. This stratigraphy has deformed stiffly, forming great thrust sheets, but displaying little deformation on the scale of a mountain face (Fig. 13-22). As we move west into the western Main Ranges, this Cambrian carbonate bank abruptly terminates and drops into a deeper-water shale basin containing carbonate submarine landslide breccias derived from the bank. This Cambrian shale sequence is much less resistant to buckling than the stiffer bank carbonates; therefore, they display very tight chevron folds that dominate the mountain faces (Fig. 13-23, see also Fig. 13-20).

FIGURE 13–23 Chevron folds of Cambrian shales of the western Main Ranges. Compare with Figure 13-22. (Geological Survey of Canada, Ottawa.)

At this point we truncate our traverse, although we are still less than halfway across the Canadian Cordillera. The rocks have reached only low greenschist facies of regional metamorphism and the shales are deformed to slates. If we were to continue westward (Fig. 13-20), we would pass into high-grade metamorphic rocks, still apparently sediments of the North American continental margin; but eventually we would pass through ophiolitic belts, recording destroyed oceanic basins, and into orogenic terrains that are far-traveled with respect to North America. These *displaced terrains* play an important role in Cordilleran tectonics, which is discussed below, but it is first appropriate to describe the latest Cretaceous and early Tertiary foreland deformation in the United States east of the Sevier fold-and-thrust belt.

LARAMIDE OROGENY (Latest Cretaceous and Early Tertiary)

Latest Cretaceous is a time of tectonic reorganization, particularly in the segment of the Cordillera within the United States. At this time the Franciscan and Batholith belts cease their active metamorphism and magmatism; furthermore, the Sevier fold-and-thrust belt becomes inactive. Orogenic deformation moved east of the Sevier overthrust belt, producing fault-bounded uplifts of Precambrian basement in the formerly stable craton. The latest Cretaceous through Eocene *Laramide orogeny* is named for one of these basement uplifts, the Laramie Range of eastern Wyoming (Fig. 13-24). In contrast, the magmatic arc and fold-and-thrust belt continued to be quite active during Laramide time to the north in the Canadian Rockies and Montana and to the south in Mexico.

The locus of latest Cretaceous through Eocene foreland deformation in the Cordillera has a remarkable map pattern (Fig. 13-16). The two active segments of the fold-and-thrust belt deform completely different stratigraphic sections. The northern segment, in the eastern Canadian Cordillera, deforms the westward-thickening Paleozoic continental margin of the Proto-Pacific ocean in great east-vergent thrust sheets (Fig. 13-20). In contrast, the Mexican fold-and-thrust belt of the Sierra Madre Oriental deforms the eastward-thickening Mesozoic continental margin of the Gulf of Mexico in great east-vergent thrust sheets. In essence the Mexican fold-and-thrust belt lies at the eastern edge of the continent and thrusts oceanward, whereas the Canadian fold-and-thrust belt lies at the western edge of the continent and thrusts cratonward. Between eastern Mexico and western Canada, the locus of Laramide foreland deformation cuts across the North American craton, giving rise to the distinctive fault-bounded uplifts of Precambrian basement.

The Laramide basement uplifts are generally 20 to 30 km across and as a group form an arcuate pattern changing from an east-west strike in the northwest to a north-south strike in the southeast. The margins of most ranges are bounded by thrust faults or sharp fault-related flexures of the sedimentary cover. The flexure on the east side of the Front Range near Denver, Colorado, is shown in Figure 13-25(a); faults of minor displacement exist within the structure, as documented by drilling, but the deformation of the sedimentary cover is accomplished largely by folding at higher levels.

The major underlying structural mechanism of the basement uplifts appears to be thrust faulting. For example, low-angle thrusting on the north side of the Uinta Mountains uplift, documented by seismic data and drilling, has displaced the Precambrian basement approximately 10 km over the adjacent sedimentary basin (Fig. 13-25(b)). Most of the structural relief of the Uinta uplift is accomplished by slip along the bounding thrust.

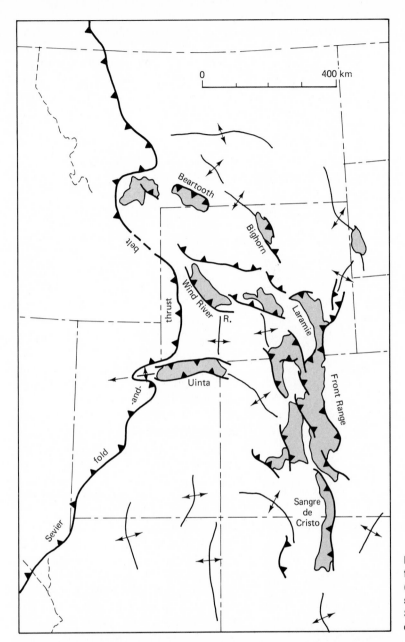

FIGURE 13–24 Map of main structural features of Laramide Rocky Mountains. (Simplified after Hamilton, 1978. Permission to reproduce granted by the Pacific Section, Society of Economic Paleontologists and Mineralogists.)

Similar bounding thrust faults have been documented for most of the major Laramide uplifts, the best known of which is the Wind River uplift in western Wyoming (Fig. 13-26). The Wind River uplift is asymmetric with a gentle homocline of the sedimentary cover on the northeast side and a major thrust fault on the southwest side, the *Wind River thrust,* which places the Precambrian basement over sediments of the flanking Green River basin. The Wind River thrust has been traced to a depth of about 20 km by seismic-reflection profiling and appears to flatten in the lower crust or along the Moho, based on considerations of retrodeformability (Fig. 13-26). The structural mechanism of uplift is slip on the Wind River thrust.

DISPLACED TERRAINS

Let us summarize the broadest features of Cordilleran tectonics as far as we have gone—up to the early Tertiary—in terms of just two contrasting regimes. (1) For

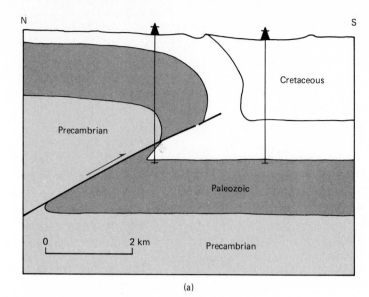

(a)

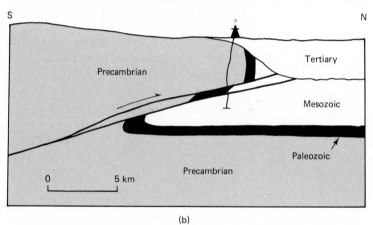

FIGURE 13–25 Structures of Laramide Rocky Mountains. (a) Willow Creek thrust, Colorado. (Modified after Berg, 1962.) (b) Uinta Mountains. (Simplified from Gries, 1983.)

(b)

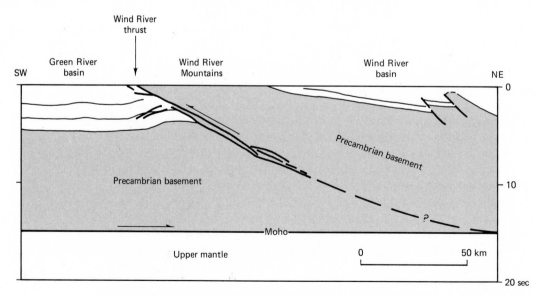

FIGURE 13–26 Cross section of Wind River thrust, based on seismic line. (*Note:* Vertical dimension is time.) (Modified after Brewer and others, 1979.)

the half-billion years beginning with late Proterozoic rifting (approximately 850 m.y. ago), western North America was the site of a stable continental shelf, slope, and rise bordering the Proto-Pacific ocean. (2) This stable tectonic regime began to change with the late Devonian and early Carboniferous Antler orogeny. By late Triassic or early Jurassic, a regime of continental-margin subduction of the Indonesian type was well established, consuming the vast late Proterozoic and Paleozoic Proto-Pacific ocean and eventually much Mesozoic and Cenozoic ocean as well.

It follows that all Paleozoic rocks in the Cordillera found west of the original Paleozoic continental margin must be far-traveled and are by some process accreted to North America. Furthermore, much Mesozoic and Cenozoic rock, such as the Franciscan belt and Coast Range Ophiolite, appears to be accreted. In all, over half of the Cordilleran mountain belt is known or suspected to be far-traveled and accreted to the edge of North America during the regime of plate convergence. The western half of the Cordillera is a vast collage of fault-bounded accreted terrains together with superjacent post-accretion sediments, such as the Great Valley sequence, and intrusive and extrusive rocks of the continental-margin magmatic arc.

More than fifty discrete *displaced terrains* have been recognized in the western Cordillera; a few of the larger ones are shown in Figure 13-27. Each displaced terrain is characterized by a distinctive internal stratigraphy, structural style, and geologic history that is foreign to North America and to the other terrains, with boundaries that cannot be interpreted as facies changes or unconformities but must be fundamentally faults of great displacement. Geologic understanding of many of these displaced terrains is limited, but enough are well known that we can provide important examples and make certain generalizations that provide us with a picture of this fundamental aspect of Cordilleran tectonics.

Displaced terrains naturally divide themselves into two fundamental types: (1) *accretionary terrains*, such as the Franciscan belt, produced by the piecewise imbrication of many small, fault-bounded fragments and (2) *stratigraphic terrains* that contain a coherent stratigraphic sequence and substantial geologic history that predates accretion of the terrain to North America as major fragments. Stratigraphic terrains appear to include island arcs, aseismic ridges, oceanic plateaus, and small fragments of probable continental crust. Two of the major stratigraphic terrains are discussed in the following paragraphs: Stikinia and Wrangellia (Fig. 13-27).

As we travel westward across the Canadian fold-and-thrust belt, we encounter Paleozoic rocks that record paleogeographies progressively farther offshore of the North American craton (Fig. 13-20). Eventually, we encounter displaced terrains exotic to North America. One of the most-obvious of these is the long, linear *Cache Creek terrain* of early Carboniferous through late Triassic age, approximately 1400 km long and up to 75 km wide (Fig. 13-27). The Cache Creek belt is a structurally complex accretionary terrain composed of radiolarian chert, argillite, submarine basaltic volcanic rocks, gabbros, and ultramafic rocks. This assemblage of oceanic rock types indicates that the Cache Creek terrain represents the remains of a late Paleozoic and early Mesozoic Proto-Pacific ocean.

Just west of the Cache Creek accretionary terrain is a major late Paleozoic and Mesozoic stratigraphic terrain, the *Stikine terrain* of northwestern and north central British Columbia (Fig. 13-27). The Stikine terrain in Carboniferous time is composed of stratigraphy indicative of an island arc: marine and subaerial basalts, andesites, and rhyolites, with associated volcaniclastic sediments and shallow-

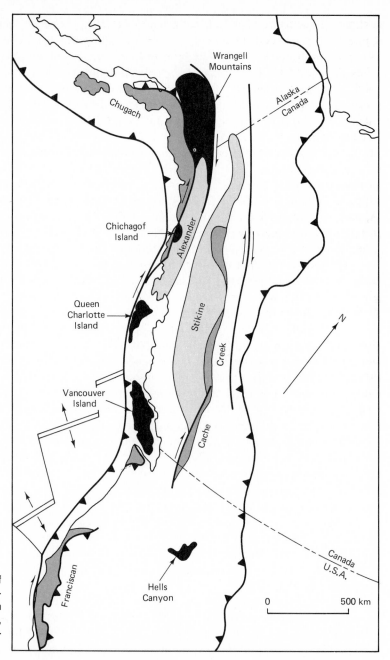

FIGURE 13–27 Map showing some of the major displaced terranes in the Cordilleran mountain belt. (Compiled from maps of Coney, Jones, and Monger, 1980, and Jones, Silberling, and Hillhouse, 1977.)

water limestones. Paleomagnetic data from overlying Lower Mesozoic volcanogenic sediments indicates a paleolatitude about 15° south of the present position of the Stikine terrain. Paleolatitude is determined from the inclination of the magnetization vector relative to the originally horizontal bedding surface (Chapter 2, Problem 2-8). The northward component of post–early Jurassic motion of the Stikine terrain relative to North America is therefore about 1300 km; the amount of longitudinal displacement is unknown. The stratigraphy of the Cache Creek and Stikine terrains seems to require that they be exotic to North America and the paleomagnetic measurements provide confirmation of substantial motion.

West of the Stikine terrain and west of the plutonic and high-grade metamorphic rocks of the Coast Mountains batholith (Figs. 13-16 and 13-27) are a set of widely dispersed fragments of the *Wrangellia terrain,* noted for a very distinctive sequence of Middle and Upper Triassic rocks, as well as underlying Upper Paleozoic rocks. Fragments of Wrangellia extend nearly 3000 km from the Wrangell Mountains of southern Alaska to the Hells Canyon of eastern Oregon (Figs. 13-27 and 13-28). The characteristic stratigraphic feature of Wrangellia is a thick 3 to 6 km sequence of Middle and Upper Triassic marine tholeiitic flood basalts, called the Karmutsen Formation on Vancouver and Queen Charlotte Islands and the Nikolai Greenstone in Alaska, overlain disconformably by Upper Triassic shallow-water limestones (Fig. 13-28). It is remarkable that such similar stratigraphy exists over a distance of 3000 km.

Wrangellia does not represent an early Mesozoic ophiolitic sequence, because where the bases of the Triassic basalts are exposed, they are underlain by Upper Paleozoic limestones, argillites, and volcanic rocks of island-arc affinities. Apparently the flood basalts poured out over a preexisting island-arc terrain and are perhaps analogous to the submerged basaltic platforms or hot-spot traces that exist today in the western Pacific and Indian Oceans.

Paleomagnetic data from the Triassic flood basalts indicate that Wrangellia is far-traveled. Data from both the Wrangell Mountains and Vancouver Island give consistently low paleolatitudes, 15° north or south of the Triassic equator, depending on the polarity epoch.

There are a number of other significant displaced terrains for which large differences in stratigraphy and paleogeography are documented. Paleomagnetic data and faunal evidence indicate significant northward motion relative to North America, similar to that shown by Stikinia and Wrangellia. It is less clear how much eastward motion accompanied the displacement of these terrains; for example, did they sail in from the central or western Proto-Pacific ocean? Some

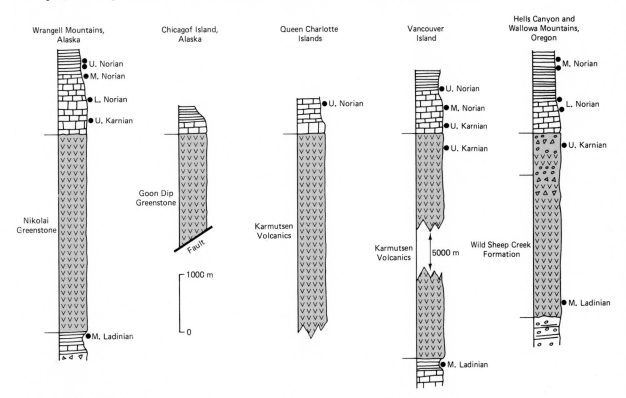

FIGURE 13–28 Comparison of stratigraphies of various parts of Wrangellia between Idaho and Alaska. See Figure 13-27 for locations. (After Jones, Silberling, and Hillhouse. Reproduced by permission of the National Research Council of Canada from the Canadian Journal of Earth Sciences, v. 14, 1977.)

eastward motion is required by the closing of the Cache Creek ocean basin. Furthermore, not all northward motion took place prior to accretion of these displaced terrains to North America. A great system of right-lateral strike-slip faults exists along the Cordilleran mountain belt from Baja California to Alaska (Figs. 13-2 and 13-29). Even today the Wrangell Mountains are moving northwestward along the Denali fault.

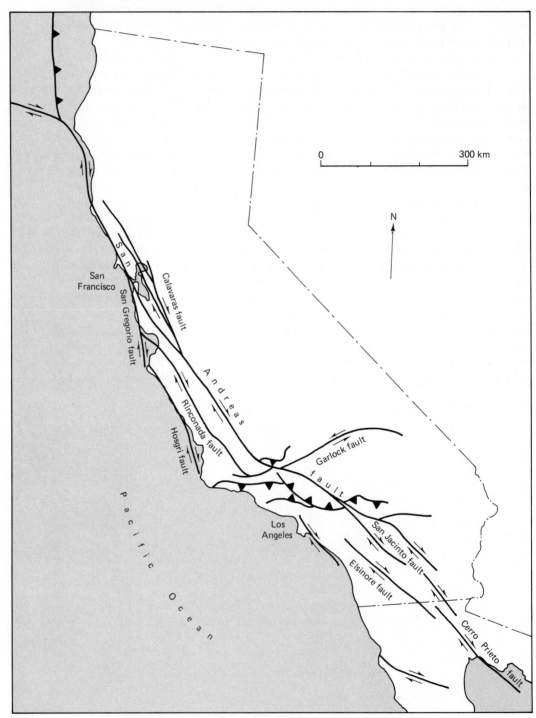

FIGURE 13–29 Major strike-slip faults of the San Andreas system, California.

STRIKE-SLIP ALONG THE PACIFIC MARGIN:
The San Andreas Fault

The large northward displacements of Wrangellia and other terrains that are suggested by paleomagnetic evidence are among the more-recent of a whole series of discoveries of Mesozoic to Recent northward displacements along the western margin of North America. The first important indication of northward displacement was the great 1906 San Francisco earthquake (Chapter 8), which showed that an active right-lateral strike-slip fault zone, the *San Andreas fault system,* runs the length of California (Figure 13-29).

At first it seemed to many people that the San Andreas fault might have only a few kilometers of slip, but as regional geologic mapping of California became more and more complete, large regional mismatches in geology across the fault were discovered (Fig. 13-30). In 1953 M. L. Hill and T. L. Dibblee showed regional mapping evidence that suggested displacements of hundreds of kilometers with the displacements becoming larger with age. The late Mesozoic Franciscan-Batholith pair of basement belts were apparently offset by about 500 km (Fig. 13-13).

The idea of such large slip was not immediately accepted by many people and became very controversial. There were two principal reasons for the controversy. First, large horizontal displacements, including both large-scale thrusting and continental drift, were very unpopular at the time among the more articulate North American geologists. Secondly, there existed some apparently conflicting evidence. Some areas displayed apparent matches in strata across the fault, whereas other areas displayed strong mismatches. Geologists were naturally impressed with the evidence most familiar to their personal experience, and the San Andreas fault is so long (1200 km) that most geologists were familiar with at most small segments. T. L. Dibblee played a very special role in the discovery of large slip on the San Andreas fault because he was intimately familiar with much of its length. He was a tall, thin, quiet man, a member of an old California pioneer

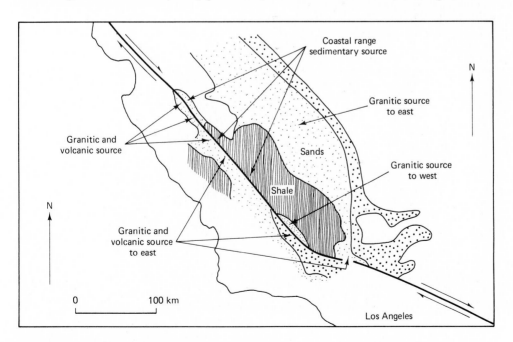

FIGURE 13–30 Discordant Upper Miocene sedimentary facies across the San Andreas fault in central California. (After Huffman, 1972).

family, who loved to spend his time in the field mapping. In the course of his career he mapped several hundred geologic quadrangles in California, many of them along or near the San Andreas fault. Dibblee's experience provided the detailed, but regional, perspective that forced people to consider seriously the uncomfortable possibility of very large slip.

By now parts of the displacement history have been very well documented and the idea of large slip is no longer controversial. For example, Figure 13-30 is a map of central California showing the mismatch in Upper Miocene sedimentary facies and potential sediment-source terrains across the San Andreas fault. There

FIGURE 13–31 Satellite image of part of Transverse Ranges. The large, light region in the upper right is the Mojave block, bounded on the south (bottom) by the San Andreas fault and on the northwest by the Garlock fault. Los Angeles is in the lower right.

are obvious mismatches along about half of the fault length. Along the other half there are sediments that might appear superficially similar, but a close examination of their sedimentary petrography and source directions show them also to be mismatched. By reversing the right slip by about 235 km, the Upper Miocene (10 m.y.) geology appears to match quite well.

Similar evidence has been used to match geology across the fault at other times. Earlier Miocene (22 m.y.), Oligocene, and Eocene strata all show a 300-km offset; therefore, in central California the San Andreas fault appears to have been inactive in early to middle Tertiary times and began slipping in late Miocene. Some evidence, such as the offset Mesozoic paired basement belts, hints at a period of strike slip during Laramide time. There is also evidence that suggests or documents important Neogene strike slip in the range 25 to 100 km on a half-dozen other faults, subparallel to the San Andreas (Crowell, 1979).

The principal exception to the pattern of the late Neogene right slip along the San Andreas fault system is the left-lateral *Garlock fault,* which strikes about 80° to the regional trend of the San Andreas (Fig. 13-29). The Garlock fault joins the San Andreas fault in the most-complex part of the San Andreas system, the Transverse Ranges of southern California (Figs. 8-23 and 13-31). A variety of aspects of the Mesozoic through Precambrian geology appear to match across the Garlock fault if it has about 65 km of left-lateral slip. The most striking is a distinctive Mesozoic diabase dike swarm (Fig. 13-32).

This apparent slip of 65 km on the Garlock fault is large relative to the total length of the fault, 260 km. Similarly, the 250 to 500 km of slip on the San Andreas fault is large relative to its length of about 1200 km. Where does the slip go?

The slip on Garlock fault at its eastern end appears to go into the system of normal faults that produce the grabens between the Sierra Nevada and Death

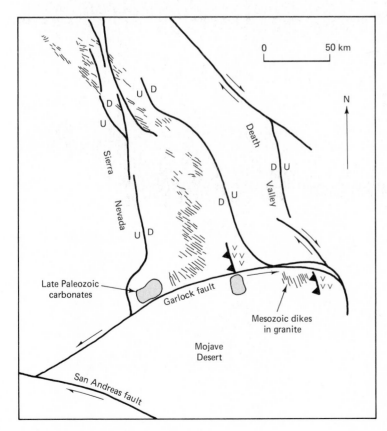

FIGURE 13–32 Offset structures along the Garlock fault, southern California. (Compiled from Smith, 1962, and Davis and Burchfiel, 1973.)

FIGURE 13–33 Oblique aerial view to the west of normal fault scarps of the eastern face of the Sierra Nevada, Owens Valley graben, California. Mount Whitney, the highest peak in the continental United States (4418 m.), is in the center skyline. Ancient shorelines of Owens Lake are in foreground. (U.S. Geological Survey.)

Valley and strike perpendicular to the Garlock fault. One of these active grabens is Owen Valley, just east of the Sierra Nevada, shown in Figure 13-33. The slip of the Garlock fault at its western end appears to deform the San Andreas system in a complex way, contributing to the great bend in the San Andreas in southern California (Fig. 13-29). The Garlock fault appears to be a transform or tear fault at the southern end of a zone of extension east of the Sierra Nevada.

Most of the 240-km slip on the San Andreas fault in the Pliocene and Pleistocene (post 4.5 m.y.) passes at its southern end into the Gulf of California rift, which shows an equivalent extension (Figs. 13-2 and 13-29). Other slip on the San Andreas system may pass into zones of extension that exist in the continental margin of southern California and northern Baja California.

The San Andreas system also displays subsidiary zones of extension and compression associated with irregularities of the fault system. Some of these subsidiary structures were described in Chapter 8 (Fig. 8-23). The most-important irregularity is the great bend in southern California (Figs. 8-23(d) and 13-29), which places this segment in compression, producing strong topographic and structural relief, thrust faulting, and folding. This irregularity is the Transverse Ranges, which juxtapose 3-km-high peaks immediately against the sea-level city of Los Angeles (Figure 13-31). The deformation within the Transverse Ranges appears to divide itself into faults dominated by thrust motion and faults dominated by strike-slip (Figure 13-29). The eastern and central Transverse Ranges are thrust-and-reverse-fault bounded sheets and blocks of Precambrian and Mesozoic basement rocks of the Batholith belt. One of these thrusts slipped in the 1971 San Fernando (Los Angeles) earthquake (Fig. 8-2). Farther west, toward the continental margin, compressive structure of the Transverse Ranges is dominated by folding and décollement thrusting of the Great Valley sequence, Franciscan belt, and their Cenozoic sedimentary cover (for example, Fig. 13-34).

PLATE MOTIONS AND CENOZOIC TECTONICS

The present-day configuration of plate boundaries along the western margin of North America, already discussed at the beginning of the chapter (Fig. 13-1), is marked by transform (strike-slip) boundaries between Pacific and North American plates in the two areas where these plates are in contact: (1) western Canada and (2) western California and northern Mexico. In contrast, the subducting boundaries of North America with their associated magmatic arcs exist where oceanic plates intervene between the Pacific and North American plates: (1) central and southern Mexico and Central America, where the Cocos plate intervenes, and (2) Oregon and Washington, where the Gorda plate intervenes. Subduction tectonics also occur along the North American–Pacific boundary in the Gulf of Alaska and the Aleutians because of the east-west orientation. Therefore, many of the first-order present-day tectonic features of the Cordilleran mountain system, such as the San Andreas and Queen Charlotte strike-slip fault

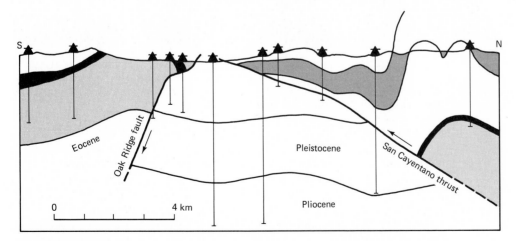

FIGURE 13-34 Cross section of compressive Pleistocene structure within the western Transverse Ranges, California. Oak Ridge fault is a folded old normal fault, now operating as a reverse fault. (After Yeats, J. Geophys. Research, v. 88, p. 569–583, 1983, copyrighted by the American Geophysical Union.)

systems and the Aleutian, Cascade, Mexico–Central American magmatic arcs, have a well-defined setting within the present-day pattern of plate interactions.

In this section we trace the patterns of plate interactions back in time, using primarily the constraints of sea-floor spreading magnetic anomalies, and see how the predicted plate interactions compare with the geologic record in the Cordillera. The most-important observation in making predictions of plate interactions is the fact that the magnetic anomalies of the Pacific plate become progressively older towards the west (Fig. 13-35). Therefore, we can reconstruct the shape of the ridge and transform boundaries of the Pacific plate for various times in the past. We see that the ridge-transform system was formerly much more extensive. A single oceanic plate, the Farallon plate, intervened between the Pacific and North American plates prior to about 30 m.y. ago. The Gorda and Cocos plates represent residual fragments of the single, much more extensive Farallon plate. The strike-slip boundary between the Pacific and North American plates did not exist in California 30 m.y. ago; the Farallon plate intervened.

These observations hold important implications for Cenozoic tectonics in the Cordillera. If sea-floor spreading was symmetric, as it usually is, every bit of the Pacific plate that was created along the Pacific-Farallon ridge system had an associated matching crust in the Farallon plate. This great expanse of Farallon crust is now missing and must have been subducted along a more extensive subduction zone along the western Cordillera; the North America–Gorda and North America–Cocos subduction zones are remnants of this more-extensive system. The area of equivalent crust on the Pacific plate (Fig. 13-35) shows us the minimum amount of Farallon crust that has been subducted.

The relative motions of the Pacific and Farallon (Gorda and Cocos) plates can be completely determined from the marine magnetic anomalies of the eastern Pacific. However, a complete reconstruction of Pacific–Farallon–North American plate interactions requires that we also know the Pacific–North American motions or Farallon–North American motions. This is difficult because of intervening subduction zones during much of the Cenozoic. The Pacific–North American motions for the last 4.5 m.y., during the time that the Gulf of California has been opening, are given by magnetic anomalies at the mouth of the Gulf of California. Prior to 4.5 m.y. (early Pliocene), the motions are only indirectly constrained;

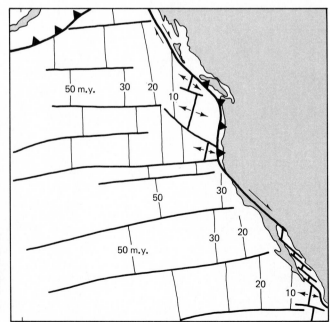

FIGURE 13–35 Map of magnetic anomalies offshore of western North America. Note that the anomalies generally become older toward the west, implying a large amount of subduction along the Cordilleran margin of North America. (Simplified after Atwater, 1970.)

nevertheless, the general pattern of plate interactions is clear and illustrated in Figure 13-36. The key point to remember is that the Pacific plate has been moving north relative to North America through much of the Cenozoic, so that strike-slip tectonics dominated the Cordilleran margin wherever and whenever the Pacific plate was in contact with North America.

According to the magnetic anomalies along the California coast (Fig. 13-35), the Pacific plate began to come in contact with North America no later than about 30 m.y. (Oligocene). This point of contact between the Pacific and North American plates divided the intervening Farallon plate into two isolated parts which are equivalent to the present-day Gorda and Cocos plates. The zone of strike slip has grown progressively (Fig. 13-36) until it is now about 2600 km long. The total Oligocene and younger strike slip predicted from plate motions is about 1000 km, whereas only a fraction of this has taken place along the San Andreas fault because the plate boundary has been jumping into the North American continent by steps. The latest and best-known step is the jumping into the Gulf of California about 4.5 m.y. After this jump, all the 5.5 to 6 cm/year slip between the Pacific and North American plates had been taken up in the Gulf of California and along the San Andreas fault system. Prior to 4.5 m.y., the slip seems to have been taken up on various faults farther to the west in the continent, such as the San Gregorio–Hosgri fault in central California (Fig. 13-29), and possibly initially

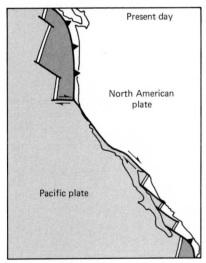
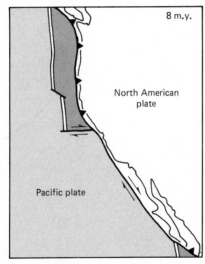
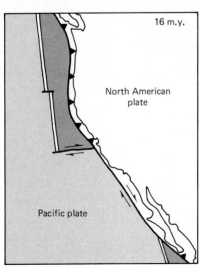
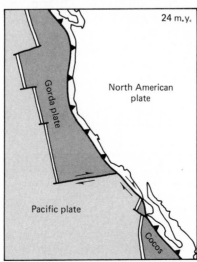

FIGURE 13–36 History of late Cenozoic plate interactions along the Pacific margin of North America. (Compiled from Atwater, 1970, and Engebretson, 1982.)

along the actual continent-ocean boundary. Only fragments of this earlier history of faulting can be assigned to specific faults.

In summary, we can say that reconstructions of plate motions provide a reasonable semiquantitative explanation of the history of late Cenozoic strike-slip that is documented on land from mismatches in geology across individual faults (for example, Fig. 13-30). A second fruitful line of inquiry is to see if the predicted changes from subduction and magmatic-arc tectonics to strike-slip tectonics are actually recorded at the proper places and times in the geologic record. The present areas of Gorda and Cocos subduction have associated magmatic arcs and accretionary wedges. Can we identify a progressive disappearance of these subduction features over the last 30 m.y. associated with the growth of the strike-slip boundary? The answer is qualitatively yes. Marine geophysical studies off the California coast show earlier Cenozoic accretionary-wedge complexes that are now inactive and covered by younger continental-slope sediments that are cut by young strike-slip faults (Fig. 13-37). It is difficult to date precisely the change in tectonics from seismic data available, but it is clear that the change in continental-margin tectonics predicted from plate reconstructions has in fact taken place.

The patterns of magmatic-arc volcanism also show a progressive disappearance in California and Baja California in general agreement with plate reconstructions. However, there is a lot of additional late Cenozoic magmatic activity in western North America, and especially the western United States, that would not have been predicted solely on the basis of plate reconstructions. This additional magmatic activity and related tectonism is the subject of the following sections.

UPLIFT, RIFTING, AND MAGMATISM EAST OF THE PLATE BOUNDARY

The region east of the subducting and strike-slip boundaries of the Cordillera has undergone a remarkable evolution in tectonics beginning about 15 m.y. (Miocene), particularly in the United States. The three most-obvious tectonic phenomena have been regional uplift, normal faulting, and a change in volcanic petrochemistry.

Regional Uplift

The western United States stands regionally high with much of the area above 1.5 km (Fig. 13-38); however, this uplift is anomalous with respect to many other mountain belts, such as the Himalayas, Alps, or Andes, which display a thickening of the crust through strong horizontal compressive deformation. The crust of the western United States is similar in thickness to the cratonic interior (about 40 km) and in some places is much thinner (20–30 km). Furthermore, the

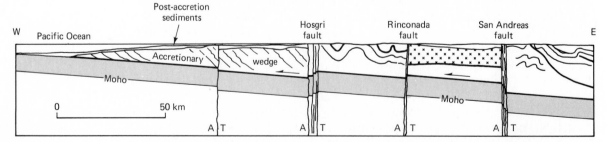

FIGURE 13-37 Cross section of continental margin in central California, showing the earlier Cenozoic accretionary wedge and the later strike-slip faulting. (Based on section of Page and others, 1979.)

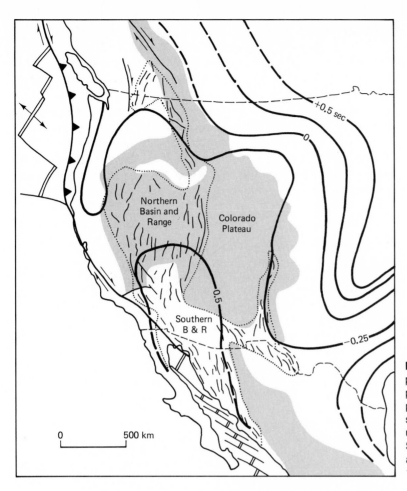

FIGURE 13-38 Map of late Cenozoic plateau uplift and normal faulting. Area of plateau uplift above 1.5 km in screened pattern. *P*-wave travel-time anomalies showing the area of anomalously hot upper mantle in the western United States. (Compiled from Suppe, Powell, and Berry, 1975, and Herrin and Taggart, 1968.)

uplift is epeirogenic in character, similar to the plateau uplifts of the Guiana or Brazilian Highlands or the East African rift system (Chapter 1). The Grand Canyon in the Colorado Plateau (Fig. 13-3) is symptomatic of much of the late Cenozoic uplift; it is a region of typical cratonic geology, but it now stands, largely undeformed, at 1.5 to 2 km above sea level.

We know from the natural experiments in isostasy, such as the unloading of glacial Lake Bonneville near Salt Lake City, that the lithosphere of the western United States is unable to support the present high elevations with its own strength. It must be isostatically supported. Furthermore, early and middle Cenozoic fossil plants from throughout the western United States show that most of the region stood within a few hundred meters of sea level prior to late Cenozoic. Therefore, some major tectonic process must have operated on the lithosphere to produce this regional uplift. There are two principal ways that the lithosphere can be uplifted (Chapter 1): (1) The crust, which is less dense than the aesthenosphere, can be thickened by horizontal compression. This has not happened in the late Cenozoic because the compressive structures and great crustal thicknesses are not observed; (2) The mantle lithosphere, which is more dense than aesthenosphere, can be thinned by heating from below or possibly removed by other processes.

Thinning of the mantle lithosphere can be looked for geophysically. For example, if the upper mantle is abnormally hot, it will transmit seismic waves at lower velocities than normal and cause anomalously late times of arrival of seismic waves from distant earthquakes at seismographic stations. The seismic waves come in nearly vertically from distant earthquakes, and most of the anomaly is caused by differences in upper-mantle velocities directly below the station, relative to the average for continental areas. It is found that the area of high elevations in the western United States shows anomalously slow arrival times (Fig. 13-38). Other types of geophysical measurements that are sensitive to the state of the upper mantle, such as magnetometer arrays, magnetotellurics, and surface-wave propagation, all show an anomalously hot upper mantle and thin mantle lithosphere under the region of high elevation. Therefore, we can conclude that the late Cenozoic uplift of the western United States is an effect of change in temperature and thickness of the mantle lithosphere. The fundamental process is controversial, but the observations show us that we should not look just to the crust to understand late Cenozoic tectonics of the Cordillera.

Normal Faulting

Crustal deformation, in the form of normal faulting, is nevertheless important over about half of the region of late Cenozoic uplift east of the plate boundaries. The bulk of this region is the *Basin and Range* physiographic province (Fig. 13-38) marked by long parallel horsts and grabens. The *northern Basin and Range* is still quite actively faulting and displays extensive internal drainage. It is the site of the 1915 Pleasant Valley earthquake, which produced the scarp shown in Figure 8-6. The *southern Basin and Range* in southwestern Arizona and New Mexico and in Sonora appears to be much less active, displaying more-subdued topography, throughgoing drainage, and fewer earthquakes. Deformation at this latitude is concentrated today in the central rift system of the Gulf of California. Another important locus of Cenozoic extension is the *Rio Grande rift* of New Mexico.

The normal faulting is accompanied by progressive tilting of the fault blocks (Fig. 8-15). For example, the seismic section of the Railroad Valley graben in eastern Nevada (Fig. 13-39) shows progressively greater tilting of deeper reflectors. The tilting of the fault blocks is thought to indicate the flattening of the normal faults (Chapter 8) into a zone of plastic flow in the lower crust.

The amount of horizontal extension is not well known but is generally considered to be substantial, on the order of 100 to 200 km in the northern Basin and Range. The thinnest crust in the western United States, 20 to 30 km thick, is found in the Basin and Range, which is in agreement with large horizontal extension.

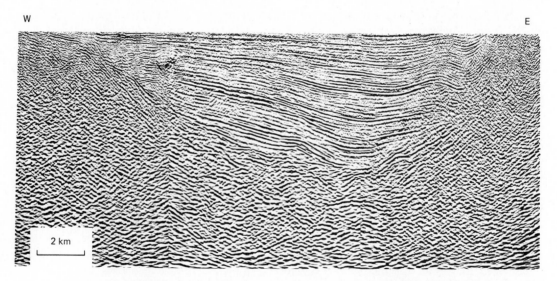

W

E

2 km

FIGURE 13–39 Seismic section of the Railroad Valley graben, eastern Nevada. (Courtesy John H. Vreeland, Northwest Exploration Company, and Rocky Mountain Association of Geologists.)

FIGURE 13–40 Basaltic lava flows of the Columbia River Basalt, Grande Ronde River, southeastern Washington. (Photograph courtesy of Washington State Department of Commerce and Economic Development.)

Magmatism

The third major aspect of late Cenozoic tectonics east of the San Andreas fault is a major change in the nature of the magmatic activity. The Cenozoic, in general, was a time of voluminous magmatism in the western Cordillera; for example, much of the Mexican Cordillera is covered by Cenozoic tuffs and lavas, which greatly obscures the Paleozoic and Mesozoic history (Fig. 13-2). Sufficient isotopic dating of Cenozoic volcanism has been done in the western United States to assure that the temporal patterns are well defined. Early to middle Cenozoic is marked by voluminous intermediate to silicic welded ash-flow tuffs erupted from great calderas overlying batholithic magma chambers—for example, the Timber Mountain caldera in Nevada (Fig. 8-21). This magmatism is broadly related to subduction along the Pacific margin, generally predating the uplift and Basin-and-Range faulting in the western United States. An important petrologic characteristic of this magmatism is that the bulk of the magma is of intermediate silica content.

About 15 million years ago, broadly coincident with the beginning of uplift and rifting, a change in magmatism began, to a basalt-rhyolite association, characterized by a bimodal distribution of silica content. This change in magmatism was not everywhere simultaneous; for example, today the intermediate magmatism is still active along the subduction-related volcanic chains of the Cascades and Mexico, while the bimodal volcanism is associated with active normal faulting in the Basin and Range and the Rio Grande graben.

A second major development in magmatic activity is the massive outpouring of flood basalts in the Columbia River (Fig. 13-40) composing about 250,000 km^3 of lava, with most of it extruded in just 3 m.y. (16 to 13 m.y.). The Columbia River flood basalts are quite comparable to the other major fields of flood basalts of the world—for example, the Deccan traps of India, the Parana basalts of South America, and the Greenland and Scottish flood basalts. Also broadly associated with the Columbia River flood basalts is the Snake River Plain–Yellowstone volcanic belt (Fig. 13-2), which exhibits some similarity to oceanic hot-spot tracks or aseismic volcanic ridges (Chapter 1). The volcanic belt has been propagating toward the northeast for the last 10 m.y. at about 2.5 cm/year and is approximately fixed with respect to other hot spots, such as Hawaii. Therefore the voluminous magmatism of the Columbia River, Snake River Plain, and Yellowstone may reflect a deeper mantle instability rather than a crustal process. This possibility is supported by observations of seismic travel-time anomalies directly under Yellowstone. A carrot-shaped zone of anomalous mantle exists under Yellowstone, with velocities about 5 percent lower than the already-anomalous surrounding upper mantle of the western United States (Iyer and others, 1981). The anomalous mantle extends to a depth of about 300 km, well below the base of the lithosphere.

EXERCISES

13–1   Compare the contrast the Sonoma and Antler orogenies of the Cordilleran mountain belt with the Taconic orogeny of the Appalachian mountain belt.

13–2   Why does the Upper Carboniferous–Permian clastic sequence thin toward the west, whereas the Cambrian–Upper Precambrian and Lower Carboniferous clastic sequences thin toward the east (Fig. 13-4)?

13–3   Estimate the percent shortening for the entire cross section of the eastern part of the southern Canadian Cordillera in Figure 13-20.

13–4 Compare the nature of the structures developed in the Laramide basement uplifts with the structures of the Cordilleran fold-and-thrust belt. What are the reasons for the similarities and differences?

SELECTED LITERATURE

BURCHFIEL, B. C. AND DAVIS, G. A., 1972, Structural framework of the Cordilleran orogen, western United States: Amer. J. of Sci., v. 272, p. 97–118.

BURCHFIEL, B. C. AND DAVIS, G. A., 1975, Nature and controls of Cordilleran orogenesis, western United States: extensions of an earlier synthesis: Amer. J. of Sci., v. 275A, p. 363–396.

ERNST, W. G., ed., 1981, *The Geotectonic Development of California,* Prentice-Hall, Englewood Cliffs, N.J., 706 p.

MONGER, J. W. H., AND PRICE, R. A., 1979, Geodynamic evolution of the Canadian Cordillera—progress and problems: Canadian J. Earth Sci., v. 16, no. 3, p. 770–791.

ROCKY MOUNTAIN ASSOCIATION OF GEOLOGISTS, 1972, *Geologic Atlas of the Rocky Mountain Region, United States of America,* Denver, Colorado, 331 p.

STEWART, J. H., 1980, *Geology of Nevada,* Nevada Bureau of Mines and Geology Special Publication 4, 136 p.

APPENDIX

Length: SI unit meter *m*

1 *m* = 3.28084 *ft*
1 *km* = 0.621371 *mile*
1 *cm* = 0.39370 *inch*
1 *angstrom* (Å) = $10^{-10}m$

Time: SI unit second *s*

1 *year* = 365.25 *days*
= $3.1556 \times 10^7 s$

Mass: SI unit kilogram *kg*

1 *kg* = $10^3 g$
= 2.2046 *pounds*

Temperature: SI unit Kelvin °K

°K = °C + 273.16°
= $\frac{5}{9}$ (°F − 32°) + 273.16°

Force: SI unit newton *N* (kg m s^{-2})

1 *N* = 10^5 *dyne* (g cm s^{-s})

Energy: SI unit joule *J* (kg m^{2}s^{-2})

1 *J* = 0.239006 *cal*
= 10^7 *erg* (dyne cm)

Power: SI unit watt *W* (kg m^2 s^{-3})

Pressure: SI unit pascal *Pa* (kg m^{-1} s^{-2})

1 *MPa* = $10^6 Pa$
= 10 *bar*
= 9.8692 *atm*
= 145.039 *lb in^{-2}*
1 *GPa* = $10^9 Pa$
= 10 *kbar*

APPENDIX B:
PHYSICAL CONSTANTS
AND PROPERTIES

Physical Constants

Boltzmann Constant
$k = 1.38066 \times 10^{-23} J \, °K^{-1}$

Gravitational Constant
$G = 6.6732 \times 10^{-11} N \, m^2 kg^{-2}$

Universal Gas Constant
$R = 8.3144 \, J \, mol^{-1} \, °K^{-1} = 1.9872 \, cal \, mol^{-1} \, °K^{-1}$

The Terrestrial Planets and the Moon

	Radius (km)	Mass (10^{22} kg)	Mean Density (kg/m³)	Surface Gravity (m/s²)
Mercury	2439	33	5430	3.62
Venus	6055	487	5250	8.61
Earth	6378	597	5515	9.78
The Moon	1738	7.3	3340	1.56
Mars	3393	64	3940	3.72

Typical Rock Properties

	Density (kg/m³)	Poisson's Ratio ν	Young's Modulus E (GPa)	Shear Modulus G (GPa)	Compressibility β (GPa⁻¹)	Linear Thermal Expansion Coefficient α (°C⁻¹)
Granite	2670	0.25	85	33	185	8×10^{-6}
Quartz diorite	2800	0.27	95	37	150	7×10^{-6}
Anorthosite	2730	0.31	93	35	120	7×10^{-6}
Diabase	2970	0.27	112	44	120	6×10^{-6}
Gabbro	2980	0.31	115	44	115	6×10^{-6}
Pyroxenite	3250		150	65		
Dunite	3300	0.27	150	70	85	
Peridotite	3250					
Limestone	2660	0.34	65	24	150	8×10^{-6}
Marble	2700	0.34	70	27	130	7×10^{-6}
Quartzite	2660	0.25	96	43	235	11×10^{-6}
Dolomite	2830		90	35	110	

Steady-State Plastic Flow Constants:

$$\dot{\varepsilon} = C_o \Delta^n \exp[-Q/RT]$$

	$\log_{10} C_o$ $(GPa^{-n}s^{-1})$	n	Q $(kJ\ mol^{-1})$
Rock Salt (Halite)	16.7	5.3	102
Anhydrite	—	1.5–2.0	114–152
Marble	25.8	7.6	418
	20.6	4.2	427
Quartzite (wet)	3.0–3.7	1.8–2.6	134–167
Aplite	2.8	3.1	163
Granite (wet)	2.0	1.9	137
Albite Rock	6.1	3.9	234
Anorthosite	6.1	3.2	238
Diabase	6.5	3.4	260
Clinopyroxenite	9.0	2.6	335
Dunite (Olivine)	4.8	3.5	533

(From Kirby, 1983)

APPENDIX C:
GEOLOGIC TIME SCALE*

Phanerozoic

Cenozoic
Neogene
Pleistocene
———————————— 1.6 m.y.
Pliocene
———————————— 5.3
Miocene
———————————— 23.7
Paleogene
Oligocene
———————————— 36.6
Eocene
———————————— 57.8
Paleocene
———————————— 66.4

Mesozoic
Cretaceous
Late
———————————— 97.5
Early
———————————— 144
Jurassic
Late
———————————— 163
Middle
———————————— 187
Early
———————————— 208
Triassic
Late
———————————— 230
Middle
———————————— 240
Early
———————————— 245

Paleozoic
Permian
Late
———————————— 258
Early
———————————— 286

Carboniferous
Pennsylvanian
———————————— 320
Mississippian
———————————— 360
Devonian
Late
———————————— 374
Middle
———————————— 387
Early
———————————— 408
Silurian
Late
———————————— 421
Early
———————————— 438
Ordovician
Late
———————————— 458
Middle
———————————— 478
Early
———————————— 505
Cambrian
Late
———————————— 523
Middle
———————————— 540
Early
———————————— 570

Proterozoic

Late
———————————— 900
Middle
————————————1600
Early
————————————2500

Archean

————————————

*Data from Palmer (1983)

REFERENCES

AHNERT, F., 1970, Functional relationships between denudation, relief, and uplift in large mid-latitude drainage basins, Amer. J. of Sci., v. 268, p. 243–263.

AIRY, G. B., 1855, On the computation of the effect of the attraction of mountain-masses, as disturbing the apparent astronomical latitude of stations in geodetic surveys: Trans. Roy. Soc. Lond., v. 145, p. 101–104.

ALBAREDE, F., 1976, Thermal models of post-tectonic decompression as exemplified by the Haut-Allier granulites (Massif Central, France): Bull. Soc. Géol. France, v. 18, p. 1023–1032.

ANDERSON, E. M., 1942, *The Dynamics of Faulting and Dyke Formation, with Applications to Britain,* Oliver and Boyd, Edinburgh, 191 p.

ANDERSON, T. B., 1968, The geometry of a natural orthorhombic system of kink bands: Geol. Surv. Canada Paper 68-52, p. 200–220.

ANDERSON, T. B., 1974, The relationship between kink-bands and shear fractures in the experimental deformation of slate: J. Geol. Soc. Lond., v. 130, pp. 367–382.

ANDERSON, T. H., AND SCHMIDT, V. A., 1983, The evolution of middle America and the Gulf of Mexico-Caribbean Sea region during Mesozoic time: Geol. Soc. Amer. Bull., v. 94, p. 941–966.

ANGELIER, J., 1979a, Determination of the mean principal directions of stresses for a given fault population: Tectonophysics, v. 56, p. T17–T26.

ANGELIER, J., 1979b, Neotectonique de L'Arc Egéen: Société Géologique du Nord, Pub. no. 3, 417 p.

ARMSTRONG, R. L., TAUBENECK, W. H., AND HALES, P. O., 1977, Rb-Sr and K-Ar geochronometry of Mesozoic granitic rocks and their Sr isotopic composition, Oregon, Washington and Idaho: Geol. Soc. Amer. Bull., v. 88, p. 397–411.

ARMSTRONG, R. L., AND SUPPE, JR., 1973, Potassium-Argon geochronometry of Mesozoic igneous rocks in Nevada, Utah, and southern California: Geol. Soc. Amer. Bull., v. 84, p. 1375–1392.

ARMSTRONG, R. L., 1968, The Cordilleran miogeosyncline in Nevada and Utah: Utah Geol. Mineral. Surv. Bull. 78, 58 p.

ARTHAUD, F., AND MATTAUER, M., 1969, Exemples de styolites d'origine tectonique dans la Languedoc, leurs relations avec le tectonique cassante: Bull. Soc. Géol. France, 7th Ser., v. 11, p. 738–744.

ATWATER, G. I. AND FORMAN, M. J., 1959, Nature and growth of southern Louisiana salt domes and its effect on petroleum accumulation: Bull. Amer. Assoc. Petrol. Geol., v. 43, p. 2592–2622.

ATWATER, T., 1970, Implications of plate tectonics for the Cenozoic tectonic evolution of western North America: Geol. Soc. Amer. Bull., v. 81, p. 3513–3536.

BABCOCK, E. A., 1973, Regional jointing in southern Alberta: Canadian J. Earth Sci., v. 10, p. 1769–1781.

BAILEY, E. B., CLOUGH, C. T., WRIGHT, W. B., RICHEY, J. E. AND WILSON, G. V., 1924, Tertiary and post-Tertiary geology of Mull, Loch Aline, and Oban: Mem. Geol. Surv. Scotland, 445 p.

BERBERIAN, M., 1981, Active faulting and tectonics of Iran: in Gupta, H. K., and Delany, F. M., ed., *Zagros, Hindu Kush, Himalaya Geodynamic Evolution,* Amer. Geophys. Union, p. 33–69.

BERG, R. R., 1962, Mountain flank thrusting in Rocky Mountain foreland, Wyoming and Colorado: Amer. Assoc. Petrol. Geol. Bull, v. 46, p. 2019–2032.

BEUTNER, E. C., 1978, Slaty cleavage and related strain in Martinsburg slate, Delaware Water Gap, New Jersey: Amer. J. of Sci., v. 278, p. 1–23.

BODINE, J. H., STECKLER, M. S., AND WATTS, A. B., 1981, Observations of flexure and the rheology of the oceanic lithosphere: J. Geophys. Research, v. 86, p. 3695–3707.

BJORK, R. L., 1961, Analysis of the formation of Meteor Crater, Arizona: a preliminary report: J. Geophys. Research, v. 66, p. 3379–3387.

BJØRN, L., 1970, Natural stress values obtained in different parts of the Fennoscandian rock masses: Proc. 3rd Congress, Int. Soc. Rock. Mech., v. 1, p. 209–212.

BOYER, S. E., AND ELLIOTT, D., 1982, Thrust systems: Amer. Assoc. Petrol. Geol. Bull., v. 66, p. 1196–1230.

BRACE, W. F., 1964, Brittle fracture of rocks: in Judd, W. R., ed., State of Stress in the Earth's Crust: American Elsevier, New York, p. 111–180.

BRACE, W. F., PAULDING, B. W., JR., AND SCHOLZ, C., 1966, Dilatancy in the fracture of crystalline rocks: J. Geophys. Research, v. 71, p. 3939–3953.

BRACE, W. F., AND JONES, A. H., 1971, Comparison of uniaxial deformation in shock and static loading of three rocks: J. Geophys. Research, v. 76, p. 4913–4921.

BRACE, W. F., AND KOHLSTEDT, D. L., 1980, Limits on lithospheric stress imposed by laboratory experiments: J. Geophys. Res., v. 85, p. 6248–6252.

BREWER, J. A., SMITHSON, S. B., OLIVER, J. E., KAUFMAN, S., AND BROWN, L. D., 1980, The Laramide Orogeny: evidence from COCORP deep crustal seismic profiles in the Wind River Mountains, Wyoming: Tectonophysics, v. 62, p. 165–189.

BUCHER, W. H., 1956, The role of gravity in orogenesis: Geol. Soc. Amer. Bull., v. 67, p. 1295–1318.

BURCHFIEL, B. C., AND DAVIS, G. A., 1975, Nature and controls of Cordilleran orogenesis, western United States: extensions of an earlier synthesis: Amer. J. Sci. vol. 275A, p. 363–396.

BURCHFIEL, B. C., AND DAVIS, G. A., 1981, Triassic and Jurassic tectonic evolution of the Klamath Mountains—Sierra Nevada geologic terrane: in Ernst, W. G., ed., The Geotectonic Development of California, Prentice-Hall, Englewood Cliffs, N.J., p. 50–70.

BURRIDGE, R., AND KNOPOFF, L., 1967, Model and theoretical seismicity: Bull. Seism. Soc. Amer., v. 57, p. 341–371.

BUSK, H. G., 1929, Earth Flexures, Cambridge University Press, Cambridge, 106 p.

BYERLEE, J. D., 1975, The fracture strength and frictional strength of Weber sandstone: International J. Rock Mechanics Mining Sci., v. 12, p. 1–4.

BYERLEE, J., 1978, Friction of rocks: Pure and Applied Geophysics, v. 116, p. 615–626.

CALDWELL, J. G., AND TURCOTTE, D. L., 1979, Dependence of the thickness of the elastic oceanic lithosphere on age: J. Geophys. Research, v. 84, p. 7572–7576.

CARTER, N. L., AND AVE'LALLEMANT, H. G., 1970, High temperature flow of dunite and peridotite: Geol. Soc. Amer. Bull., v. 81, p. 2181–2202.

CATER, F. W., 1970, Geology of the salt anticline region in southwestern Colorado: U.S. Geol. Surv. Prof. Paper 637, 80 p.

CHAPPLE, W. M., 1978, Mechanics of thin-skinned fold-and-thrust belts: Geol. Soc. Amer. Bull., v. 89, p. 1189–1198.

CHAPPLE, W. M., AND SPANG, J. H., 1974, Significance of layer-parallel slip during folding of layered sedimentary rocks: Geol. Soc. Amer. Bull., v. 85, p. 1523–1534.

CHRISTIE, J. M., 1963, The Moine thrust zone in the Assynt region northwest Scotland: Univ. Calif. Pub. Geol. Sci., v. 40, p. 345–440.

CLOOS, E., 1971, Microtectonics along the Western Edge of the Blue Ridge, Maryland and Virginia, The Johns Hopkins Press, Baltimore, 234 p.

COLEMAN, R. G., 1981, Tectonic setting for ophiolite obduction in Oman: J. Geophys. Research, v. 86, p. 2497–2508.

COLTON, G. W., 1970, The Appalachian basin—its depositional sequences and their geologic relationships: in Fisher, G. W., Pettijohn, F. J., Reed, J. R. Jr. and Weaver, K. N., ed., Studies of Appalachian Geology: Central and Southern, Wiley-Interscience, New York, p. 5–47.

COMPTON, R. R., 1955, Trondhjemite batholith near Bidwell Bar, California: Geol. Soc. Amer. Bull., v. 66, p. 9–44.

CONEY, P. J., JONES, D. L. AND MONGER, J. W. H., 1980, Cordilleran suspect terranes: Nature, v. 288, p. 329–333.

CONEY, P. J., 1978, Mesozoic-Cenozoic Cordilleran plate tectonics: Geol. Soc. Amer. Memoir 152, p. 33–50.

COOK, F. A., ALBAUGH, D. S., BROWN, L. D., KAUFMAN, S., OLIVER, J. E. AND HATCHER, R. D., JR., 1979, Thin-skinned tectonics in the crystalline southern Appalachians: COCORP seismic-reflection profiling of the Blue Ridge and Piedmont: Geology, v. 7, p. 563–567.

COOK, F. A., BROWN, L. D., KAUFMAN, S., OLIVER, J. E., AND PETERSON, T. A., 1981, COCORP seismic profiling of the Appalachian orogen beneath the Coastal Plain of Georgia: Geol. Soc. Amer. Bull., v. 92, p. 738–748.

CORNWALL, H. R., 1972, Geology and mineral resources of southern Nye county, Nevada: Nevada Bureau of Mines and Geology Bull. 77, 49 pp.

CRANS, W., MANDL, G. AND HAREMBOURE, J., 1980, On the theory of growth faulting: a geomechanical delta model based on gravity sliding: J. Petrol Geol., v. 2, p. 265–307.

CRANS, W. AND MANDL, G., 1981, On the theory of growth faulting Part II: genesis of the "unit": J. Petrol. Geol., v. 3, p. 333–355, p. 455–476.

CRITTENDEN, M. D., JR., 1976, Stratigraphic and structural setting of the Cottonwood area, Utah: Rocky Mountain Association of Geologists 1976 Symposium, p. 363–379.

CROWELL, J. C., 1979, The San Andreas fault system through time: Jour. Geol. Soc. Lond., v. 136, pp. 293–302.

CURRAY, J. R., EMMEL, F. J., MOORE, D. G., AND RAITT, R. W., 1981, Structure, tectonics and geological history of the northeastern Indian Ocean, in Nairn, A. E. M. and Stehli, F. G., ed., Ocean Basins and Margins, The Indian Ocean, v. 6, Plenum Press, New York, p. 399–450.

CURRIE, J. B., PATNODE, H. W. AND TRUMP, R. P., 1962, Development of folds in sedimentary strata: Geol. Soc. Amer. Bull., v. 73, p. 655–674.

DAHLEN, F. A., SUPPE, J., AND DAVIS, D., 1984, Mechanics of fold-and-thrust belts and accretionary wedges: Cohesive Coulomb theory: J. Geophys. Research, v. 89, in press.

DAHLSTROM, C. D. A., 1969, Balanced cross sections: Canadian J. Earth Sci., v. 6, p. 743–757.

DANILCHIK, W., ROTHROCK, H. E. AND WAGNER, H. C., 1955, Geology of anthracite in the western part of the

Shenandoah quadrangle, Pennsylvania: U.S. Geol. Surv. Coal Invest., Map C-21.

DAVIS, D., SUPPE, J., AND DAHLEN, F. A., 1983, Mechanics of fold-and-thrust belts and accretionary wedges: J. Geophys. Research, v. 88, p. 1153–1172.

DAVIS, G. A., AND BURCHFIEL, B. C., 1973, Garlock fault: an intracontinental transform structure, southern California: Geol. Soc. Amer. Bull., v. 84, p. 1407–1422.

DAVIS, W. M., 1898, The Triassic formation of Connecticut: 18th Annual Report of the U.S. Geological Survey, 1896–1897, part 2, p. 1–192.

DELANEY, P. T., AND POLLARD, D. D., 1981, Deformation of host rocks and flow of magma during growth of minette dikes and breccia-bearing intrusions near Ship Rock, New Mexico: U.S. Geological Survey Professional Paper 1202, 61 p.

DENCE, M. R., 1972, The nature and significance of terrestrial impact structures: 24th International Geological Congress, Montreal, Section 15, p. 77–89.

DETRICK, R. S. AND CROUGH, S. T., 1978, Island subsidence, hot spots, and lithospheric thinning: J. Geophys. Research, v. 83, p. 1236–1244.

DETRICK, R. S., SCLATER, J. G. AND THIEDE, J., 1977, The subsidence of aseismic ridges: Earth Planet. Sci. Lett., v. 34, no. 2, p. 185–196.

DeVOTO, R. H., 1980, Pennsylvanian stratigraphy and history of Colorado: in Kent, H. C. and Porter, K. W., ed., *Colorado Geology,* Denver, Rocky Mountain Association of Geologists, p. 71–101.

DICKINSON, W. R., 1973, Widths of modern arc-trench gaps proportional to past duration of igneous activity in associated magmatic arcs: Jour. Geophys. Research, v. 78, pp. 3376–3389.

DIETERICH, J. H., 1970, Computer experiments on mechanics of finite amplitude folds: Canadian J. Earth Sci., v. 7, p. 467–476.

DOEBL, F., AND BADER, M., 1971, Alter und Verhalten einiger Stoerungen im Oelfeld Landau/Pfalz: Oberrhein. geol. Abh., v. 20, p. 1–14.

DOLL, C. G., CADY, W. M., THOMPSON, J. B., JR., AND BILLINGS, M. P. 1961, *Geologic Map of Vermont,* Vermont Development Department, Vermont Geological Survey, (1:250,000).

DONATH, F. A., 1961, Experimental study of shear failure in anisotropic rocks: Geol. Soc. Amer. Bull., v. 72, p. 985–990.

DONATH, F. A., 1968, Experimental study of kink-band development in Martinsburg Slate: Geol. Surv. Canada Paper 68-52, p. 255–288.

ELLIOTT, D., 1976, The motion of thrust sheets: J. Geophys. Research, v. 81, p. 949–963.

ELLIOTT, D. AND JOHNSON, M. R. W., 1980, Structural evolution in the northern part of the Moine thrust belt, N.W. Scotland: Trans. R. Soc. Edinb., v. 71, p. 69–96.

EMERSON, B. K., 1917, Geology of Massachusetts and Rhode Island: U.S. Geological Survey Bull., v. 597, 289 p.

ENGEBRETSON, D. C., 1982, *Relative motions between oceanic and continental plates in the Pacific Basin,* Ph.D. dissertation, Stanford University, 211 p.

ENGELDER, T. AND GEISER, P., 1980, On the use of regional joint sets as trajectories of paleostress fields during the development of the Appalachian Plateau, New York: Jour. Geophys. Research v. 85, p. 6319–6341.

ERNST, W. G., ED., 1981, *The Geotectonic Development of California,* Prentice-Hall, Englewood Cliffs, N.J., 706 pp.

FLETCHER, R. C., 1974, Wavelength selection in the folding of a single layer with power-law rheology: Amer. J. Sci., v. 274, p. 1029–1043.

FLETCHER, R. C., 1977. Quantitative theory for metamorphic differentiation in development of crenulation cleavage: Geology, v. 5, p. 185–187.

FUNG, Y. C., 1969 *A First Course in Continuum Mechanics,* Prentice-Hall, Englewood Cliffs, N.J., 301 p.

GASTIL, R. G., PHILLIPS, R. P. AND ALLISON, E. C., 1975, Reconnaissance geology of the State of Baja California: Geol. Soc. Amer. Mem. 140, 170 p.

GEE, D. G., 1975, A tectonic model for the central part of the Scandinavian Caledonides: Amer. J. Sci., v. 275-A (Rodgers Volume), p. 468–515.

GEOLOGIC MAP OF CALIFORNIA, 1977, C. W. Jennings, ed., California Division of Mines and Geology, (1:750,000).

GEOLOGIC MAP OF PENNSYLVANIA, 1980, Topographic and Geologic Survey, State of Pennsylvania (1:250,000).

GEOLOGIC MAP OF THE UNITED STATES, 1974, P. B. King, ed., U.S. Geological Survey (1:2,500,000).

GILLULY, J., AND GATES, O., 1965, Tectonic and igneous geology of the northern Shoshone Range, Nevada: U.S. Geol. Surv. Prof. Paper 465, 153 p.

GRETENER, P. E., 1977, On the character of thrust faults with particular reference to basal tongues: Bull. Canadian Petrol. Geol., v. 25, p. 110–122.

GRIES, R., 1983, Oil and gas prospecting beneath Precambrian of foreland thrust plates in Rocky Mountains: Bull. Amer. Assoc. Petrol. Geol. v. 67, p. 1–28.

GROSS, G. A., 1967, Geology of iron deposits of Canada: Geol. Surv. Canada, Econ. Rept. 22, 179 p.

GROW, J. A., MATTICK, R. E. AND SCHLEE, J. S., 1979, Multichannel seismic depth sections and interval velocities over outer continental shelf and upper continental slope between Cape Hatteras and Cape Cod: Amer. Assoc. Petrol. Geol. Mem. 29, p. 65–83.

GUCWA, P. R. AND KEHLE, R. O., 1978, Bearpaw Mountains rockslide, Montana, U.S.A.: Voight, B., ed., *Rockslides and Avalanches,* I, *Natural Phenomena,* Elsevier, Amsterdam, p. 393–421.

HADLEY, J. B., 1949, Geologic map of Mount Grace quadrangle, Massachusetts: U.S. Geol. Surv. Quad. Map, GQ-3 (1:31,680).

HAFNER, W., 1951, Stress distributions and faulting: Geol. Soc. Amer. Bull., v. 62, p. 373–398.

HAIMSON, B. C., 1977, Crustal stress in the continental United States as derived from hydrofracturing tests: in J. G. Heacock, ed., *The Earth's Crust,* Amer. Geophys. Union Monograph 20, p. 576–592.

HALL, L. M., 1968, Bedrock geology in the vicinity of White Plains, New York: Guidebook to Field Excursions at the 40th Annual Meeting of the New York State Geological Association: p. 7–31.

HALLER, J., 1971, *Geology of the East Greenland Caledonides,* Wiley-Interscience, London, New York, 413 p.

HAMBLIN, W. K., 1965, Origin of "reverse drag" on the downthrown side of normal faults: Geol. Soc. Amer. Bull., v. 76, p. 1145–1164.

HAMILTON, W., 1978, Mesozoic tectonics of the western United States: in Howell, D. G. and McDougall, K. A., ed.,

Mesozoic Paleogeography of the Western United States, Pacific Coast Paleogeography Symposium 2; Pacific Section Society of Economic Paleontologists and Mineralogists, Los Angeles, p. 33–70.

HANDIN, J., HAGER, R. V., FRIEDMAN, M., AND FEATHER, J. N., 1963, Experimental deformation of sedimentary rocks under confining pressure: pore pressure tests: Bull. Amer. Assoc. Petrol. Geol., v. 47, p. 717–755.

HANSEN, E., 1971, *Strain Facies,* Springer-Verlag, New York, 207 p.

HARRIS, L. D., 1970, Details of thin-skinned tectonics in parts of Valley and Ridge and Cumberland Plateau provinces of the southern Appalachians: in Fischer, G. W., Pettijohn, F. J., Reed, J. C., Jr., and Weaver, K. N., ed., *Studies of Appalachian Geology: Central and Southern,* Wiley-Interscience, New York, 460 p.

HARRIS, L. D., 1975, Oil and gas data from the Lower Ordovician and Cambrian rocks of the Appalachian Basin: U.S. Geol. Surv. Misc. Invest., Map I-917D.

HARRIS, L. D. AND MIXON, R. B., 1970, Geologic map of the Howard Quarter quadrangle, northeastern Tennessee: U.S. Geol. Surv. Quadrangle Map, GQ-842 (1:24,000).

HARRIS, L. D., HARRIS, A. G., DeWITT, W., JR., AND BAYER, K. C., 1981, Evaluation of southern eastern overthrust belt beneath Blue Ridge–Piedmont thrust: Bull. Amer. Assoc. Petrol. Geol., v. 65, p. 2497–2505.

HEARD, H. C., 1960, Transition from brittle to ductile flow in Solenhofen Limestone as a function of temperature, confining pressure, and interstitial fluid pressure: Geol. Soc. Amer. Mem. 79, p. 193–226.

HEARD, H. C., 1972, Steady-state flow in polycrystalline halite at pressure of 2 kilobars: Amer. Geophys. Union Monogr. 16, p. 191–210.

HEARN, B. C., JR., 1976, Geologic and tectonic maps of the Bearpaw Mountain area, north-central Montana: U.S. Geol. Surv. Misc. Invest. Map I-919 (1:125,000).

HEARN, B. C., JR., PECORA, W. T. AND SWADLEY, W. C., 1963, Geology of the Rattlesnake Quadrangle, Bearpaw Mountains, Blaine County, Montana: U.S. Geol. Surv. Bull. 1181-B, 66 p.

HEDBERG, H. D., 1936, Gravitational compaction of clays and shales: Amer. J. of Sci., 5th Ser., v. 31, p. 241–287.

HERRIN, E., AND TAGGART, J., 1968, Regional variations in *P* travel times: Seism. Soc. of Amer. Bull., v. 58, p. 1325–1337.

HIGGINS, G. E. AND SAUNDERS, J. B., 1967, Report on 1964 Chatham mud island, Erin Bay, Trinidad, West Indies: Bull. Amer. Assoc. Petrol. Geol., v. 51, p. 55–64.

HOBBS, B. E., MEANS, W. D., AND WILLIAMS, P. E., 1976, *An Outline of Structural Geology,* John Wiley, New York, 571 p.

HOBBS, B. E., 1971, The analysis of strain in folded layers: Tectonophysics, v. 11, p. 329–375.

HOBBS, D. W., 1967, The formation of tension joints in sedimentary rocks: an explanation: Geological Magazine, v. 104, p. 550–556.

HODGSON, R. A., 1961, Regional study of jointing in Comb Ridge-Navajo Mountain area, Arizona and Utah: Bull. Amer. Assoc. Petrol. Geol., v. 45, p. 1–38.

HOLCOMBE, T. L., 1977, Ocean bottom features—terminology and nomenclature: Geojournal, v. 1.6, p. 25–48.

HOPSON, C. A., MATTINSON, J. M., AND PESSAGNO, E. A., JR., 1981, Coast Range Ophiolite, western California: in Ernst, W. G., ed., *The Geotectonic Development of California,* Prentice-Hall, Englewood Cliffs, N.J., p. 418–510.

HOSHINO, K., KOIDE, H., INAMI, K., IWAMURA, S., AND MITSUI, S., 1972, Mechanical properties of Japanese Tertiary sedimentary rocks under high confining pressures: Geol. Surv. of Japan, Rep. 244, 200 p.

HUBBERT, M. K., AND RUBEY, W. W., 1959, Role of fluid pressure in mechanics of overthrust faulting: I. Mechanics of fluid-filled porous solids and its application to overthrust faulting: Geol. Soc. Amer. Bull., v. 70, p. 115–166.

HUFFMAN, O. F., 1972, Lateral displacement of Upper Miocene rocks and the Neogene history of offset along the San Andreas fault in central California: Geol. Soc. Amer. Bull., v. 83, p. 2913–2946.

HUGHES, D. J., 1960, Faulting associated with deep-seated salt domes in the northeast portion of the Mississippi salt basin: Gulf Coast Association of Geol. Soc. Transactions, vol. 10, p. 155–173.

HUTCHINSON, W. W., 1970, Metamorphic framework and plutonic styles in the Prince Rupert region of the central Coast Mountains, British Columbia: Canadian J. Earth Sci., v. 7, p. 376–405.

IVERSON, W. P., AND SMITHSON, S. B., 1982, Master décollement root zone beneath the southern Appalachians and crustal balance: Geology, v. 10, p. 241–245.

IYER, H. M., EVANS, J. R., ZANDT, G., STEWART, R. M., COAKLEY, J. M., AND ROLOFF, J. N., 1981, A deep low-velocity body under the Yellowstone caldera, Wyoming: delineation using teleseismic P-wave residuals and tectonic interpretation: Geol. Soc. of Amer. Bull., v. 92, p. 792–798.

JAEGER, J. C., 1961, The cooling of irregularly shaped igneous bodies: Amer. J. Sci., v. 259, p. 721–734.

JAEGER, J. C., AND COOK, N. G. W., 1979, *Fundamentals of Rock Mechanics, 3rd ed.,* Methuen, London, 593 p.

JAHNS, R. H., 1943, Sheet structure in granites: its origin and use as a measure of glacial erosion in New England: J. of Geol., v. 51, p. 71–98.

JOHNSON, A. M., 1970, *Physical Processes in Geology,* Freeman, Cooper and Co., San Francisco, 577 p.

JOHNSON, A. M., 1977, *Styles of Folding,* Elsevier, New York, 406 p.

JOHNSON, M. R. W., AND STEWART, F. H., ED., 1963, *The British Caledonides,* Oliver and Boyd, London, 280 p.

JOHNSON, R. B., 1968, Geology of the igneous rocks of the Spanish Peaks region, Colorado: U.S. Geol. Surv. Prof. Paper 594-G, 47 p.

JONES, D. L., SILBERLING, N. J., AND HILLHOUSE, J., 1977, Wrangellia—a displaced terrane in northwestern North America: Canadian J. Earth Sci., v. 14, p. 2565–2577.

JULIVERT, M. AND MARCOS, A., 1973, Superimposed folding under flexural conditions in the Cantabrian zone (Hercynian Cordillera, northwest Spain): Amer. J. of Sci., v. 273, p. 353–375.

KEHLE, R. O., 1970, Analysis of gravity sliding and orogenic translation: Bull. Geol. Soc. Amer., v. 81, p. 1641–1664.

KELLY, V. C. AND CLINTON, N. J., 1960, Fracture systems and tectonic elements of the Colorado Plateau: Publication University New Mexico Geology, no. 6, 104 p.

KIEFFER, W. W., 1971, Shock metamorphism of the Coconino Sandstone at Meteor Crater, Arizona: J. of Geophys. Research, v. 76, p. 5449–5473.

KING, P. B., 1969, The tectonics of North America—A discussion to accompany the tectonic map of North America: U.S. Geol. Surv. Prof. Paper 628, 95 p.

KING, P. B. AND EDMONSTON, G. J., 1972, Generalized tectonic map of North America: U.S. Geol. Surv. Misc. Invest., Map I-688 (1:15,000,000).

KIRBY, H., 1983, Rheology of the lithosphere: Reviews of Geophysics and Space Physics, v. 21, p. 1458–1487.

KISTLER, R. W., AND PETERMAN, Z. E., 1973, Variations in Sr, Rb, K, Na, and initial Sr^{87}/Sr^{86} in Mesozoic granitic rocks and intruded wallrocks in central California: Geol. Soc. Amer. Bull., v. 84, p. 3489–3512.

KOCH, P. S., CHRISTIE, J. M., AND GEORGE, R. P., 1980, Flow law of "wet" quartzite in the α-quartz field: E⊕S, v. 61, p. 376.

KOHLSTEDT, D. L. AND WEATHERS, M. S., 1980, Deformation-induced microstructures, paleopiezometers, and differential stresses in deeply eroded fault zones: J. Geophys. Research, v. 85, p. 6269–6285.

LACHENBRUCH, A. H., 1961, Depth and spacing of tension cracks: J. Geophys. Research, v. 66, p. 4273–4292.

LACHENBRUCH, A. H., 1962, Mechanics of thermal contraction cracks and ice-wedge polygons in permafrost: Geol. Soc. Amer. Spec. Paper 70, 69 p.

LAFAYETTE GEOLOGICAL SOCIETY AND NEW ORLEANS GEOLOGICAL SOCIETY, 1973, Offshore oil and gas fields, 124 p.

LARSEN, E. S., 1948, Batholith and associated rocks of Corona, Elsinore and San Luis Rey quadrangles, southern California: Geol. Soc. Amer., Memoir 29, 182 p.

LAUBSCHER, H. P., 1961, Die Fernschubhypothese der Jurafaltung: Eclogae Geol. Helvetiae, v. 54, p. 221–282.

LAUBSCHER, H. P., 1965, Ein kinematisches Modell der Jurafaltung: Eclogae Geol. Helvetiae, v. 58, p. 231–318.

LE PICHON, X., FRANCHETEAU, J. AND BONNIN, J., 1973, *Plate Tectonics,* Elsevier, New York, 306 p.

LISTER, G. S. AND HOBBS, B. E., 1980, The simulation of fabric development during plastic deformation and its application to quartzite: the influence of deformation history: J. of Structural Geol., v. 2, p. 355–370.

LISTER, G. S. AND PATERSON, M. S., 1979, The simulation of fabric development during plastic deformation and its application to quartzite: fabric transitions: J. of Structural Geol., v. 1, p. 99–115.

LISTER, G. S., PATERSON, M. S. AND HOBBS, B. E., 1978, The simulation of fabric development in plastic deformation and its application to quartzite: the model: Tectonophysics, v. 45, p. 107–158.

MACDONALD, K. C., 1982, Mid-ocean Ridges: fine scale tectonic, volcanic and hydrothermal processes within the plate boundary zone: Ann. Rev. Earth Planet. Sci., v. 10, p. 155–190.

MACKIN, J. H., 1947, Some structural features of the intrusions in the Iron Springs district: Utah Geol. Soc., Guidebook to the Geology of Utah no. 2, 62 p.

MACKIN, J. H., 1950, The down-structure method of viewing geologic maps: J. of Geol., v. 58, p. 55–72.

MALVERN, L. E., 1969, *Introduction to the Mechanics of a Continuous Medium,* Prentice-Hall, Englewood Cliffs, N.J., 713 p.

MANTON, W. I., 1965, The orientation and origin of shatter cones in the Vredefort Ring: New York Academy of Sciences Annals, v. 123, p. 1017–1049.

MARSH, B. D., 1982, On the mechanics of igneous diapirism, stoping and zone melting: Amer. J. of Sci., v. 282, p. 808–855.

McDOWELL, S. D., 1974, Emplacement of the Little Chief stock, Panamint Range, California: Geol. Soc. Amer. Bull., v. 85, p. 1535–1546.

McDOUGALL, I. AND DUNCAN, R. A., 1980, Linear volcanic chains—recording plate motions?: Tectonophysics, v. 63, p. 275–295.

McGARR, A. AND GAY, N. C., 1978, State of stress in the earth's crust: Ann. Rev. Earth Planet Sci., v. 6, p. 405–436.

McGETCHIN, T. R. AND ULLRICH, G. W., 1973, The xenoliths in maars and diatremes with inferences for the Moon, Mars and Venus: J. Geophys. Research, v. 78, p. 1833–1853.

McIVER, N. L., 1970, appalachian turbidites: in Fisher, G. W., Pettijohn, F. J., Reed, J. C., Jr. and Weaver, K. N., ed., *Studies of Appalachian Geology: Central and Southern,* Wiley-Interscience, New York, p. 69–81.

McKENZIE, D., 1978a, Some remarks on the development of sedimentary basins: Earth Planet. Sci. Lett., v. 40, p. 25–32.

McKENZIE, D., 1978b, Active tectonics of the Alpine-Himalayan belt: the Aegean Sea and surrounding regions: Geophys. J. Roy. Astron. Soc., v. 55, p. 217–254.

McKENZIE, D. AND BRUNE, J. N., 1972, Melting on fault planes during large earthquakes: Geophys. J. Roy. Astron. Soc., v. 29, p. 65–78.

McKENZIE, D. P. AND MORGAN, W. J., 1969, Evolution of triple junctions: Nature, v. 224, p. 125–133.

MELTON, F. A., 1929, A reconnaissance of the joint systems in the Ouachita Mountains and central plains of Oklahoma: J. of Geol., v. 37, p. 729–746.

MILLER, R. L. AND FULLER, J. O., 1954, Geology and oil resources of the Rose Hill district—the Fenster area of the Cumberland overthrust block—Lee County, Virginia: Virginia Geol. Survey, Bulletin 71, 383 pp.

MINSTER, J. B. AND JORDAN, T. H., 1978, Present-day plate motions: J. Geophys. Research, v. 83, p. 5331–5354.

MIYASHIRO, A., 1961, Evolution of metamorphic belts: J. of Petrology, v. 2, p. 277–311.

MOLNAR, P. AND TAPPONNIER, P., 1975, Cenozoic tectonics of Asia: effects of a continental collision: Science, v. 189, p. 419–426.

MOORE, R. C., 1941, Stratigraphy: in *Geology 1888–1938,* 50th Anniversary Volume, Geol. Soc. of Amer., 578 p.

MORGAN, J. P., COLEMAN, J. M. AND GAGLIANO, S. M., 1968, Mud lumps: diapiric structures in Mississippi delta sediments: Amer. Assoc. Petrol. Geol. Mem. 8, p. 145–161.

MORGAN, W. J., 1968, Rises, trenches, great faults, and crustal blocks: J. Geophys. Research, v. 73, p. 1959–1982.

MORGAN, W. J., 1975, Heat flow and vertical movements of the crust: in A. G. Fischer and S. Judson, ed., *Petroleum and Global Tectonics,* Princeton University Press, Princeton, N.J., p. 23–43.

MORGAN, W. J., 1981, Hotspot tracks and the opening of the Atlantic and Indian Oceans, in The Sea, v. 7, C. Emiliani, ed., John Wiley, New York, p. 443–487.

MULLER, O. H. AND POLLARD, D. D., 1977, The stress state near Spanish Peaks, Colorado, determined from a dike pattern: Pure Applied Geophys., v. 115, p. 69–86.

MULLER, W. H., SCHMID, S. M. AND BRIEGEL, U., 1981, Deformation experiments on anhydrite rocks of different grain sizes: rheology and microfabric: Tectonophysics, v. 78, p. 527–543.

NAMSON, J., 1981, Detailed structural analysis of the western foothills belt in the Miaoli-Hsinchu area, Taiwan: I. southern part: Petrol. Geol. of Taiwan, no. 18, p. 31–51.

NELSON, C. A. AND SYLVESTER, A. G., 1971, Wall rock decarbonation and forcible emplacement of Birch Creek pluton, southern White Mountains, Calif: Geol. Soc. Amer. Bull., v. 82, p. 2891–2904.

NICKELSEN, R. P. AND HOUGH, V. D., 1967, Jointing in the Appalachian Plateau of Pennsylvania: Geol. Soc. Amer. Bull., v. 78, p. 609–630.

NYE, J. F., 1957, *Physical Properties of Crystals,* Oxford University Press, Oxford, England, 322 p.

ODÉ, H., 1957, Mechanical analysis of the dike pattern of the Spanish Peaks area, Colorado: Geol. Soc. Amer. Bull., v. 68, p. 567–576.

O'DRISCOLL, E. S., 1962, Experimental patterns in superimposed similar folding: Alberta Soc. Petrol. Geol. J., v. 10, p. 145–167.

OERTEL, G. AND CURTIS, C. D., 1972, Clay-ironstone concretion preserving fabrics due to progressive compaction: Geol. Soc. Amer. Bull., v. 83, p. 2597–2606.

OFFIELD, T. W. AND POHN, H. A., 1979, Geology of the Decaturville impact structure, Missouri: U.S. Geol. Survey Prof. Paper 1042, 48 p.

OLLERENSHAW, N. C., 1972, Geology Fallentimber Creek, Alberta: Geol. Surv. Canada, Map 1387A (1:50,000).

O'NEILL, C. A., III, 1973, Evolution of Belle Isle salt dome, Louisiana: Gulf Coast Assoc. Geol. Soc. Trans., v. 23, p. 115–135.

OWENS, W. H., 1973, Strain modification of angular density distributions: Tectonophysics, v. 16, p. 249–261.

PAGE, B. M., WAGNER, H. C., McCULLOCH, D. S., SILVER, E. A. AND SPOTTS, J. H., 1979, Geologic cross section of the continental margin off San Luis Obispo, the southern Coast Ranges, and the San Joaquin Valley, California: Geol. Soc. Amer. Map Chart Ser. MC-28G.

PALMER, A. R., 1983, The Decade of North American Geology 1983 Geologic Time Scale: Geology, v. 11, p. 503–504.

PARKER, T. J. AND McDOWELL, A. N., 1951, Scale models as guide to interpretation of salt-dome faulting: Bull. Amer. Assoc. Petrol. Geol., v. 35, 2076–2094.

PATERSON, M. S. AND WEISS, L. E., 1966, Experimental deformation and folding in phyllite: Geol. Soc. Amer. Bull., v. 77, p. 343–374.

PATERSON, M. S., 1973, Nonhydrostatic thermodynamics and its geologic applications: Rev. Geophys. Space Phys., v. 11, p. 355–389.

PATERSON, M. S., 1978, *Experimental Rock Deformation—The Brittle Field,* Springer-Verlag, New York, 254 p.

PELTZER, G., TAPPONIER, P. AND COBBOLD, P., 1982, Les grands décrochements de l'Est Asiatique: évolution dans le temps et comparaison avec un modèle expérimental: Comptes rendus de l'Académie des Sciences, Paris, v. 294, p. 1341–1348.

PFIFFNER, O. A., 1978, Der Falten- und Kleindeckenbau im Infrahelvetikum der Ostschweiz: Eclogae Geologae Helvetiae, v. 71, p. 61–84.

PFIFFNER, O. A. AND RAMSAY, J. G., 1982, Constraints on geological strain rates: arguments from finite strain states of naturally deformed rocks: J. Geophys. Research, v. 87, p. 311–321.

PIERCE, W. G., 1973, Principal features of the Heart Mountain fault and the mechanism problem: in DeJong, K. A., and Scholten, R., ed., *Gravity and Tectonics,* John Wiley and Sons, New York, p. 457–471.

PIERCE, W. G., 1979, Clastic dikes of Heart Mountain fault breccia, northwestern Wyoming, and their significance: U.S. Geol. Survey Prof. Paper 1133, 25 p.

PITCHER, W. S., 1979, The nature, ascent and emplacement of granitic magmas: J. Geol. Soc. Lond., v. 136, p. 627–662.

PLESSMANN, W., 1972, Horizontal-Stylolithen im französisch-schweizerischen Tafel und Faltenjura und ihre Einpassung in den regionalen Rahmen: Geolog. Rundschau, v. 61, p. 332–347.

POLLARD, D. D., 1973, Derivation and evaluation of a mechanical model for sheet intrusions: Tectonophysics, v. 19, p. 233–269.

POLLARD, D. D., 1976, On the form and stability of open hydraulic fractures in the earth's crust: Geophys. Res. Lett., v. 3, p. 513–516.

POLLARD, D. D. AND JOHNSON, A. M., 1973, Mechanics of growth of some laccolithic intrusions in the Henry Mountains, Utah: II. bending and failure of overburden layers and sill formation: Tectonophysics, v. 18, p. 311–354.

POLLARD, D. D. AND HOLZHAUSEN, G., 1979, On the mechanical interaction between a fluid-filled fracture and the earth's surface: Tectonophysics, v. 53, p. 27–57.

POOLE, F. G. AND SANBERG, C. A., 1977, Mississippian paleogeography and tectonics of the western U.S.: Soc. Econ. Paleon. and Mineral. Pacific Section, Pacific Coast Paleogeography Symposium 1, p. 67–85.

PRICE, N. J., 1974, The development of stress systems and fracture patterns in undeformed sediments: *Advances in Rock Mechanics,* Proceedings 3rd Congress Int. Soc. Rock Mech., National Academy of Sciences, Washington, D.C., v. 1A, p. 487–519.

PRICE, R. A., 1973, Large-scale gravitational flow of supracrustal rocks, southern Canadian Rockies: in DeJong, K. A., and Scholten, R., ed., *Gravity and Tectonics:* John Wiley, New York, p. 491–502.

PRICE, R. A. AND FERMOR, P. R., undated, Structure section of the Cordilleran foreland thrust and fold belt west of Calgary, Alberta: Geol. Sur. of Canada, Open File Report 882.

PRICE, R. A. AND OLLERENSHAW, N. C., 1971, Geology Lake Minewanka, Alberta: Geol. Surv. of Canada, Map 1272A (1:50,000).

PROFFETT, J. M., JR., 1977, Cenozoic geology of the Yerington district, Nevada, and implications for the nature and origin of Basin and Range faulting: Geol. Soc. Amer. Bull., v. 88, p. 247–266.

RALEIGH, C. B., HEALY, J. H. AND BREDEHOEFT, J. D., 1972, Faulting and crustal stress at Rangeley, Colorado: Amer. Geophys. Union Monograph 16, p. 275–284.

RAMSAY, J. G., 1967, *Folding and Fracturing of Rocks,* McGraw-Hill, New York, 568 pp.

RAMSAY, J. G., 1980, Shear zone geometry, a review: J. Struct. Geol., v. 2, p. 83–99.

RAMSAY, J., 1983, Rock ductility and its influence on the development of tectonics structures in mountain belts: p. 111–127, in Hsu, K. J., ed., *Mountain Building Processes,* Academic Press, London, 263 p.

RAMSAY, J. G. AND HUBER, M. I., 1984, *The Techniques of Modern Structural Geology,* vol. 1: Strain Analysis; Academic Press, London, 306 p.

READE, T. M., 1908, The mechanics of overthrusts: Geol. Magazine, 5th Series, v. 5, p. 518.

REED, B. L. AND LANPHERE, M. A., 1974, Offset plutons and history of movement along the McKinley segment of the Denali fault system, Alaska: Geol. Soc. Amer. Bull., v. 85, p. 1883–1892.

RICE, J. R., 1980, The mechanics of earthquake rupture: Proc. Int. School Physics "Enrico Fermi," Course LXXVIII, Italian Physical Society, North-Holland Pub. Co., Amsterdam, p. 555–649.

RICH, J. L., 1934, Mechanics of low-angle overthrust faulting as illustrated by Cumberland thrust block, Virginia, Kentucky and Tennessee: Bull. Amer. Assoc. Petrol. Geol., v. 18, p. 1584–1596.

RICHEY, J. E. AND THOMAS, H. H., 1930, *The Geology of Ardnamurchan, North-west Mull and Coll:* Memoir Geol. Surv. Scotland, 393 p.

ROBERTS, J. L., 1970, The intrusion of magma into brittle rocks: Liverpool Geol. Soc., Geol. J. Special Issue No. 2, p. 287–338.

ROBINSON, P., 1967, Gneiss domes and recumbent folds of the Orange area, west central Massachusetts: New England Intercollegiate Geological Conference, 59th Annual Meeting, p. 17–47.

ROBINSON, P., ROBINSON, R. AND GARLAND, S. J., 1963, Preparation of beta diagrams in structural geology by a digital computer: Amer. J. of Science, v. 261, p. 913–928.

ROBINSON, P. AND HALL, L. M., 1980, Tectonic synthesis of southern New England: Wones, D. R., ed., *The Caledonides in the USA,* Dept. Geol. Sci., Virginia Polytechnic and State University Memoir 2, p. 73–82.

ROCKY MOUNTAIN ASSOCIATION OF GEOLOGISTS, 1972, *Geologic Atlas of the Rocky Mountain Region, United States of America:* Denver, Colorado, 331 p.

RODGERS, J., 1970, *The Tectonics of the Appalachians,* Wiley-Interscience, New York, 271 pp.

RODGERS, J., 1971, The Taconic orogeny: Geol. Soc. of Amer. Bull., v. 82, p. 1141–1178.

ROEDER, D., GILBERT, O. E., JR. AND WITHERSPOON, W. D., 1978, *Evolution and macroscopic structure of Valley and Ridge thrust belt, Tennessee and Virginia,* Dept. Geol. Sci. Univ. Tenn., Studies in Geology 2, 25 p.

ROSENFELD, J. L., 1968, Garnet rotations due to the major Paleozoic deformations in southeastern Vermont: p. 185–202 in Zen, E-an, White, W. S., Hadley, J. B., and Thompson, J. B., Jr., ed., *Studies of Appalachian Geology: Northern and Maritime,* Wiley-Interscience, New York, 475 pp.

ROSENFELD, J. L., 1970, Rotated garnets in metamorphic rocks: Geol. Soc. Amer. Spec. Paper 129, 105 p.

ROYSE, F., JR., WARNER, M. A. AND REESE, D. L., 1975, Thrust belt structural geometry and related stratigraphic problems, Wyoming-Idaho-northern Utah: Rocky Mountain Assoc. of Geol. 1975 Symposium, p. 41–54.

RUXTON, B. P. AND MCDOUGALL, I., 1967, Denudation rates in northeast Papua from potassium-argon dating of lavas, Amer. J. of Science, v. 265, p. 545–561.

SANFORD, A. R., 1959, Analytical and experimental study of simple geologic structures: Geol. Soc. Amer. Bull., v. 70, p. 19–52.

SAWATZKY, H. B., 1978, Buried impact craters in the Williston basin and adjacent areas: 461–480 in, Roddy, D. J., Pepin, R. O., and Merrill, R. E., ed., *Impact and Explosion Cratering,* Pergamon, Oxford, 130 p.

SCHMID, S. M., PATERSON, M. S. AND BOLAND, J. N., 1980, High temperature flow and dynamic recrystallization in Carrara marble: Tectonophysics, v. 65, p. 245–280.

SCHWAB, F. L., 1971, The Chilhowee Group and the Late Precambrian-Early Paleozoic sedimentary framework in the central and southern Appalachians: 59–101 in Kanes, W. H., ed., *Appalachian Structures, Origin, Evolution, and Possible Potential for New Exploration Frontiers,* West Virginia Geol. and Econ. Survey, 322 p.

SCHWEICKERT, R. A., 1981, Tectonic evolution of the Sierra Nevada Range, p. 87–131 in Ernst, W. G., ed., *The Geotectonic Development of California,* Prentice-Hall, Englewood Cliffs, N.J., 706 p.

SCHWEICKERT, R. A. AND SNYDER, W. S., 1981, Paleozoic plate tectonics of the Sierra Nevada and adjacent regions: p. 182–201 in Ernst., W. G., ed., *The Geotectonic Development of California,* Prentice-Hall, Englewood Cliffs., N.J., 706 p.

SEELY, D. R., VAIL, P. R. AND WATSON, G. G., 1974, Trench slope model: in Burk, C. A. and Drake, C. L., ed., *The Geology of Continental Margins,* Springer-Verlag, New York, p. 249–260.

SHARP, R. V., 1975, En échelon fault patterns of the San Jacinto fault zone: California Div. Mines Geol., Special Report 118, p. 147–152.

SHAW, H. R., 1980. The fracture mechanisms of magma transport from the mantle to the surface: in Hargraves, R. B., ed., *Physics of Magmatic Processes,* Princeton University Press, Princeton, N.J., p. 201–264.

SHELTON, G. AND TULLIS, J., 1981, Experimental flow laws for crustal rocks: E⊕S, v. 62, p. 396.

SHOEMAKER, E. M., 1960, Penetration mechanics of high velocity meteorites, illustrated by Meteor Crater, Arizona: 21st Int. Geol. Congress, Norden, v. 18, p. 418–434.

SHORT, N. M., 1975, *Planetary Geology,* Prentice-Hall, Englewood Cliffs, N.J., 361 p.

SILBERLING, N. J. AND ROBERTS, R. J., 1962, Pre-Tertiary stratigraphy and structure of northwestern Nevada: Geol. Soc. Amer. Spec. Paper 72, 58 p.

SJÖSTRAND, T., 1978, Caledonian geology of the Kvarnberg-vattnet area, northern Jämtland central Sweden: Sveriges Geologiska Undersökning, C735, 107 p.

SMITH, G. I., 1962, Large lateral displacement on Garlock fault, California, as measured from offset dike swarm: Bull. Amer. Assoc. Petrol. Geol., v. 46, p. 85–104.

SMITH, R. B., 1975, Unified theory of the onset of folding, boudinage and mullion structure: Geol. Soc. of Amer. Bull., v. 86, p. 1601–1609.

SMITH, R. B., 1977, Formation of folds, boudinage, and mullions in non-Newtonian materials: Geol. Soc. of Amer. Bull., v. 88, p. 312–320.

SMITH, R. B., 1978, Geologic maps of part of the Spanish Peaks dike system, south-central Colorado: Geol. Soc. Amer. Map Chart Ser. MC-22.

SMITH, R. B., 1979, The folding of a strongly non-Newtonian layer: Amer. J. of Science, v. 279, p. 272–287.

SMITH, D. A. AND REEVE, F. A. E., 1970. Salt piercement in shallow Gulf Coast salt structures: Bull. Amer. Assoc. Petrol. Geol., v. 54, p. 1271–1289.

SMITHSON, S. B., BREWER, J. A., KAUFMAN, S., OLIVER, J. E. AND HURICH, C. A., 1979, Structure of the Laramide Wind River uplift, Wyoming, from COCORP deep reflection data and from gravity data: J. Geophys. Research, v. 84, p. 5955–5972.

SMOLUCHOWSKI, M. S., 1909, Some remarks on the mechanics of overthrusts: Geol. Magazine, 5th Series, v. 6, p. 204–205.

SPENCER, E. W., 1959, Geologic evolution of the Beartooth Mountains, Montana and Wyoming, Part 2: fracture patterns: Geol. Soc. Amer. Bull., v. 70, p. 467–508.

SPENCER, E. W., 1972, Structure of the Blue Ridge front in central Virginia: p. 39–57 in Kanes, W. H., ed., *Appalachian Structures Origin, Evolution and Possible Potential for New Exploration Frontiers,* West Va. Geol. & Econ. Surv., 322 p.

SPENCER-JONES, D., 1963, Joint patterns and their relationships to regional trends: J. Geol. Soc. Australia, v. 10, p. 279–298.

SPERA, F. J., 1980, Aspects of magma transport: in Hargraves, R. B., ed., *Physics of Magmatic Processes,* Princeton University Press., Princeton, N.J., p. 265–323.

STANLEY, R. S., 1975, Time and space relationships of structures associated with the domes of southwestern Massachusetts and western Connecticut: Chapter E in Harwood, D. S., and others, *Tectonic Studies of the Berkshire massif, Western Massachusetts, Connecticut, and Vermont,* U.S. Geol. Sur. Prof. Paper 888, p. 67–96.

STANLEY, R. S. AND SARKISIAN, A., 1972, Analysis and chronology of structures along the Champlain thrust west of the Hinesburg synclinorium: p. 117–150 in Doolan, B., and Stanley, R. S., ed., *Guidebook for Field Trips in Vermont,* 64th Annual Meeting New England Intercollegiate Geological Conference, Department of Geology, University of Vermont, Burlington, Vermont, 483 p.

STEWART, J. H., 1980, *Geology of Nevada:* Nevada Bureau of Mines and Geology, Spec. Publ. 4, 136 p.

STOCKER, R. L. AND ASHBY, M. F., 1973, On the rheology of the upper mantle: Rev. Geophys. and Space Phys., v. 11, p. 391–426.

STOSE, A. J. AND STOSE, G. W., 1944, Geology of the Hanover-York district, Pennsylvania: U.S. Geol. Sur. Prof. Paper 204, 84 p.

SUPPE, J., 1979, Structural interpretation of the southern part of the nothern Coast Ranges and Sacramento Valley, California: summary: Geol. Soc. of Amer. Bull., v. 90, p. 327–330.

SUPPE, J., 1983, Geometry and kinematics of fault-bend folding: Amer. J. of Science, v. 283, p. 648–721.

SUPPE, J., POWELL, C. AND BERRY, R., 1975, Regional topography, seismicity, Quaternary volcanism, and the present-day tectonics of the western United States: Amer. J. of Science, v. 275-A, p. 397–436.

SUPPE, J. AND FOLAND, K. A., 1978, The Goat Mountain and Pacific Ridge Complex: a redeformed but still-intact Late Mesozoic Franciscan schuppen complex: in Howell, D., and McDougall, K., ed., *Mesozoic Paleogeography of the Western United States,* Society of Economic Paleontologists and Mineralogists, Pacific Section, Pacific Coast Paleogeography Symposium 2, p. 431–451.

SVERDRUP, H. U., JOHNSON, M. W. AND FLEMING, R. H., 1942, *The Oceans, Their Physics, Chemistry and General Biology,* Prentice-Hall, New York, 1087 p.

TAN, B. K., 1973, Determination of strain ellipses from deformed ammonoids: Tectonophysics, v. 16, p. 89–101.

TAPPONNIER, P., PELTZER, G., LE DAIN, A. Y., ARMIJO, R., AND COBBOLD, P., 1982, Propagating extrusion tectonics in Asia: New insights from simple experiments with plasticine: Geology, v. 10, p. 611–616.

TECTONIC MAP OF NORTH AMERICA, 1969, P. B. King, ed., U.S. Geological Survey (1:5,000,000).

THOMPSON, J. B., JR., ROBINSON, P., CLIFFORD, T. N. AND TRASK, N. J., JR., 1968, Nappes and gneiss domes in west-central New England: in Zen, E., White, W. S., Hadley, J. B., and Thompson, J. B., Jr., ed., *Studies in Appalachian Geology, Northern and Maritime,* Wiley-Interscience, New York, p. 203–218.

TOBISCH, O. T. AND GLOVER, L., III, 1971, Nappe formation in part of the southern Appalachian Piedmont: Geol. Soc. Amer. Bull., v. 82, p. 2209–2230.

TOCHER, D. AND NASON, R., 1967, Fault creep at the Alma-den-Cienega winery, San Benito County: *Guidebook Gabilan Range and adjacent San Andreas Fault:* Pacific Section Amer. Assoc. Petrol. Geol., p. 99–101.

TREAGUS, S. H., 1983, A theory of finite strain variation through contrasting layers, and its bearing on cleavage refraction: J. of Structural Geology, v. 5, p. 351–368.

TRUSHEIM, F., 1960, Mechanism of salt migration in northern Germany: Bull. Amer. Assoc. of Petrol. Geol., v. 44, p. 1519–1540.

TUCKER, P. M. AND YORSTON, H. J., 1973, Pitfalls in seismic interpretation: Soc. Explor. Geophys. Monogr. 2, 50 p.

TURNER, F. J., AND WEISS, L. E., 1963, *Structural Analysis of Metamorphic Tectonites,* McGraw-Hill, New York, 545 p.

VERNON, R. H., 1968, Microstructures of high-grade metamorphic rocks at Broken Hill, Australia: J. of Petrology, v. 9, p. 1–22.

VOIGHT, B., 1974, Architecture and mechanics of the Heart Mountain and South Fork rockslides: p. 26–36 in Voight, B., ed., *Rock Mechanics,* The American Northwest, International Society for Rock Mechanics, 3rd Congress Expedition Guide, Special Publication, College of Earth and Mineral Sciences, Pennsylvania State University, 292 p.

VOIGHT, B., ed., 1976, *Mechanics of Thrust Faults and Décollement,* Dowden, Hutchinson and Ross, Stroudsburg, Pa., 471 p.

VOIGHT, B., ed., 1978, *Rockslides and Avalanches,* 1, *Natural Phenomena,* Elsevier, Amsterdam, 833 p.

VROMAN, A. J., 1967, On the fold pattern of Israel and the Levant: Israel Geol. Surv. Bull., v. 43, p. 23–32.

WEISS, L. E., 1959, Geometry of superposed folding: Geol. Soc. Amer. Bull., v. 70, p. 91–106.

WEISS, L. E., 1968, Flexural-slip folding of foliated model materials: Geol. Surv. Canada Paper 68-52, p. 294–357.

WELLMAN, H. W., 1962, A graphical method for analyzing fossil distortion caused by tectonic deformation: Geol. Mag., v. 99, p. 348–352.

WENDLANDT, E. A., SHELBY, T. H. JR. AND BELL, J. S., 1946, Hawkins Field, Wood County, Texas: Amer. Assoc. Petrol. Geol. Bull., v. 30, p. 1830–1856.

WILLIAMS, H., 1978, *Tectonic Lithofacies Map of the Appalachian Orogen,* Memorial University of Newfoundland (1:1,000,000).

WILSON, C. W., JR. AND STEARNS, R. G., 1958, Structure of the Cumberland Plateau, Tennessee: Geol. Soc. Amer. Bull., v. 69, p. 1283–1296.

WILSON, C. W., JR., AND STEARNS, R. G., 1968, Geology of the Wells Creek structure, Tennessee: Tenn. Div. of Geol., Bull. 68, 236 p.

WISE, D. U., 1964, Microjointing in basement, middle Rocky Mountains of Montana and Wyoming: Geol. Soc. Amer. Bull., v. 75, p. 287–306.

WOODRING, W. P., STEWART, R. AND RICHARDS R. W., 1940, Geology of the Kettleman Hills Oil Field California: U.S. Geol. Surv. Prof. Paper 195, 170 p.

WRIGHT, T. O. AND PLATT, L. B., 1982, Pressure dissolution and cleavage in the Martinsburg shale: Amer. J. of Science, v. 282, p. 122–135.

YEATS, R. S., 1983, Large-scale Quaternary detachments in Ventura basin, southern California: J. Geophys. Research, v. 88, p. 569–583.

YODER, H. S., JR., 1976, *Generation of Basaltic Magma,* National Academy of Sciences, Washington, D.C., 265 p.

ZEN, E-AN, 1961, Stratigraphy and structure at the north end of the Taconic Range in west-central Vermont: Geol. Soc. of Amer. Bull., v. 72, p. 293–338.

ZOBACK, M. L. AND ZOBACK, M. D., 1980, State of stress in the coterminous United States: J. Geophys. Research, v. 85, p. 6113–6156.

INDEX*†

How to use this index. This index is designed to be both general and specific. Therefore it has three levels of hierarchy. You can look up very specific topics, such as *stylolites*, as you would in any index, by directly looking under *S.* You can also look up very specific topics under more general categories, similar to the main subject headings of this book. For example, you can look up *deformation mechanisms* and find subcategories such as *plastic deformation* and *time-dependent mechanisms.* Alternatively, you can look up *plastic deformation* directly for more complete coverage of its subcategories. Thus, imbedded within this index is a very detailed table of contents, or outline, of this book—rearranged in alphabetical order. Reading over the index is a helpful way for a student to review before a final exam.

†Definitions are given in boldface, figures or tables in italics.